Principles of Engineering Economic Analysis

Revised Canadian Edition

Andrew J. Szonyi
Robert G. Fenton
John A. White
Marvin H. Agee
Kenneth E. Case

Wall & Emerson, Inc.
Toronto, Ontario ◆ Dayton, Ohio

Copyright © 1977 by John Wiley & Sons, Inc.
Copyright © 1982 by John Wiley & Sons Canada Limited
Copyright © 1989, 2000 by Andrew J. Szonyi

Cover design by Alexander Wall

All rights reserved. No part of this publication may be reproduced or transmitted in any form or by any means, electronic or mechanical, including photography, recording, or any information storage and retrieval system, without permission in writing from the publisher.

Orders for this book or requests for permission to make copies of any part of this work should be sent to:

Wall & Emerson, Inc.
Six O'Connor Drive
Toronto, Ontario, Canada M4K 2K1

Telephone: (416) 467-8685
Toll free (outside of Toronto) 1-877-409-4601
Fax: (416) 352-5368
E-mail: wall@wallbooks.com
Website: www.wallbooks.com

Canadian Cataloguing in Publication Data

Main entry under title:

Principles of engineering economic analysis

Rev. Canadian ed.
Includes bibliographical references and index.
ISBN 0-921332-49-1

1. Engineering economy. I. Szonyi, Andrew J., 1934-   .

TA177.4P74 2000   620'.0068'1   C99-931893-4

Printed in Canada.

2nd printing, February 2003.

Table of Contents

Preface to the Revised Canadian Edition. ix

Chapter 1. Introduction
1.1 Background 1
1.2 The Problem-Solving Process 2
1.3 Cash Flow. 5
1.4 A Fundamental Concept: The Time Value of Money 7
1.5 Economic Justification May Not Be Enough 10
1.6 Non-Monetary Considerations 11
1.7 Overview of the Text. 14

Chapter 2. Cost Concepts and Cost Estimation
2.1 Introduction 17
2.2 Cost Terminology 18
    2.2.1 Life-Cycle Cost, 19
    2.2.2 Past and Sunk Costs, 21
    2.2.3 Future and Opportunity Costs, 22
    2.2.4 Direct, Indirect, and Overhead Costs, 24
    2.2.5 Fixed and Variable Costs, 26
    2.2.6 Average and Marginal Cost, 28
2.3 Estimation 30
    2.3.1 Project Estimation, 32
    2.3.2 General Sources of Data, 34
2.4 Capital Cost Estimation 35
    2.4.1 Price Indexes , 35
    2.4.2 Cost-Capacity Relationship, 38
2.5 The Learning Curve 40
    2.6 Standard Costs, 44
Problems. 45

Chapter 3. Time Value of Money Operations
3.1 Introduction 49
3.2 Interest Calculations 49
3.3 Single Sums of Money 52
3.4 Series of Cash Flows 54
    3.4.1 Uniform Series of Cash Flows, 58
    3.4.2 Gradient Series of Cash Flows, 64
    3.4.3 Geometric Series of Cash Flows, 69
3.5 Multiple Compounding Periods in a Year 73
3.6 Continuous Compounding 76
    3.6.1 Discrete Flows, 77
    3.6.2 Continuous Flow, 80
3.7 Equivalence 83
3.8 Loan Payments 90
3.9 Special Topics 93
    3.9.1 Changing Interest Rates, 94
    3.9.2 End-of-Period Cash Flows and End-of-Period
          Compounding, 97
    3.9.3 Perpetuities and Capitalized Value, 99
    3.9.4 Bond Problems, 103
    3.9.5 Capital Recovery Cost, 110
3.10 Inflationary Effects 112
3.11 Summary 115
Problems 116

Chapter 4. Comparison of Alternatives
4.1 Introduction 137
4.2 Defining Mutually Exclusive Alternatives 138
4.3 The Planning Period 141
4.4 Developing Cash Flow Profiles 146
4.5 Specifying the Time Value of Money 148
4.6 The Measures of Merit 151
    4.6.1 Present Worth Method, 152
    4.6.2 Annual Worth Method, 153
    4.6.3 Future Worth Method, 154
    4.6.4 Payback Period Method, 155
    4.6.5 Rate of Return Method, 157

    4.6.6 Savings/Investment Ratio Method, 163
4.7 Comparing the Investment Alternatives 164
    4.7.1 Ranking Approaches, 165
    4.7.2 Incremental Approaches, 167
4.8 Supplementary Analyses 172
4.9 Selecting the Preferred Alternative 172
4.10 Analyzing Alternatives with No Positive Cash Flows 174
4.11 Classical Method of Dealing with Unequal Lives 179
4.12 Replacement Analysis............................. 183
    4.12.1 Cash Flow Approach, 184
    4.12.2 Classical Approach, 188
4.13 Computer Applications........................... 195
4.14 Summary 196
Problems.. 197

Chapter 5. Depreciation and Income Tax Considerations

5.1 Introduction...................................... 223
5.2 The Meaning of Depreciation........................ 223
5.3 Factors Used to Determine Depreciation 225
5.4 Methods of Depreciation........................... 226
    5.4.1 Straight-Line Depreciation, 226
    5.4.2 Sum of the Years' Digits Depreciation, 228
    5.4.3 Declining Balance Depreciation, 229
    5.4.4 Capital Cost Allowance, 230
    5.4.5 Double Declining Depreciation, 232
    5.4.6 Sinking Fund Depreciation, 235
5.5 Comparison of Depreciation Methods 238
5.6 Special Provisions for Accelerated Depreciation 238
5.7 Other Methods of Depreciation 240
    5.7.1 Units of Production Method, 240
    5.7.2 Operating Day Method, 240
    5.7.3 Income Forecast Method, 241
    5.7.4 Multiple-Asset Accounts, 241
5.8 Tax Concepts..................................... 242
5.9 Corporate Income Tax: Business Income 242
    5.9.1 Small Business Tax Credit, 246
    5.9.2 Manufacturing and Processing Profits Deduction, 246

    5.9.3 Determining Taxable Income, 247
    5.10 After-Tax Cash Flow 248
    5.11 Effect of Depreciation Method 251
    5.12 Effect of Interest on Borrowed Money................. 254
    5.13 Carry-back and Carry-forward Rules 258
    5.14 Capital Gains and Losses............................. 259
    5.15 Tax Treatment of Capital Assets 260
    5.16 Recaptured Capital Cost Allowance 261
    5.17 Income Tax Incentives 263
    5.18 Lease-Buy Considerations 266
    5.19 Depletion 269
    5.20 Summary 270
    Problems... 271

Chapter 6. Economic Analysis of Projects in the Public Sector
    6.1 Introduction... 283
    6.2 The Nature of Public Projects 283
    6.3 Objectives in Project Evalution........................ 284
    6.4 Benefit-Cost Analysis 285
    6.5 Important Considerations in Evaluating Public Projects . . 298
        6.5.1 Point of View, 298
        6.5.2 Selection of the Interest Rate, 301
        6.5.3 Assessment of Benefit-Cost Factors, 305
        6.5.4 Overcounting, 308
        6.5.5 Unequal Lives, 309
        6.5.6 Tolls, Fees, and User Charges, 311
    6.6 Multiple-Use Projects.................................. 313
    6.7 Problems with the B/C Ratio 314
    6.8 Cost-Effectiveness Analysis 316
        6.8.1 The Standardized Approach, 317
    6.9 Summary 320
    Problems 320

Chapter 7. Break-Even, Sensitivity, and Risk Analyses
    7.1 Introduction 329
    7.2 Linear Break-Even Analysis 330

7.3 Nonlinear Break-Even Analysis 335
7.4 Sensitivity Analysis 342
7.5 Risk Analysis 351
    7.5.1 Distributions, 352
    7.5.2 Risk Aggregation, 354
7.6 Computer Simulation 363
7.7 Summary 370
Problems. 371

Chapter 8. Decision Models

8.1 Introduction 381
8.2 The Matrix Decision Model 385
8.3 Decisions under Assumed Certainty 388
8.4 Decisions under Risk 390
    8.4.1 Dominance, 391
    8.4.2 Expectation-Variance Principle, 392
    8.4.3 Most Probable Future Principle, 393
    8.4.4 Aspiration-Level Principle, 394
8.5 Decisions under Uncertainty. 395
    8.5.1 The Laplace Principle, 396
    8.5.2 Maximin and Minimax Principles, 396
    8.5.3 Maximax and Minimin Principles, 397
    8.5.4 Hurwicz Principle, 398
    8.5.5 Savage Principle (Minimax Regret), 401
8.6 Sequential Decisions 403
    8.6.1 Decision Trees, 404
    8.6.2 The Conditional Probability Theorem, 417
    8.6.3 The Value of Perfect Information, 421
    8.6.4 The Value of Imperfect Information, 422
    8.6.5 Sequential Decisions—Summary Comments, 435
8.7 Multiple Objectives. 435
    8.7.1 Classifying Objectives According to Importance, 436
    8.7.2 Ranking, 436
    8.7.3 Weighting Objectives, 438
    8.7.4 Determining the Value of Multiple Objectives, 442
8.8 Summary 444
Problems 446

Chapter 9. Accounting Principles

    9.1 Introduction..459
    9.2 Balance Sheet......................................459
    9.3 Income Statement..................................462
    9.4 Interpretation of Financial Statements...............462
    9.5 Cost Accounting....................................471
    Problems...477

Chapter 10. Fundamental Economic Concepts

    10.1 Introduction......................................483
    10.2 Supply and Demand.................................483
        10.2.1 Demand, 483
        10.2.2 Shift in the Demand Curve, 486
        10.2.3 Supply, 488
        10.2.4 Shift in the Supply Curve, 490
        10.2.5 Price, 490
        10.2.6 Elasticity of Demand, 496
        10.2.7 Total Expenditure, 499
        10.2.8 Elasticity of Supply, 502
        10.2.9 Price Control, 503
        10.2.10 Price Support, 506

    10.3 Production..508
        10.3.1 The Production Function, 508
        10.3.2 The Univariable Production Function, 509
        10.3.3 Multivariable Production Function, 513
        10.3.4 Cost of Production, 515
        10.3.5 The Expansion Path, 521

    10.4 Summary...524
    Problems...524

Appendices

    A. Discrete Compounding...............................529
    B. Continuous Compounding.............................553
    C. Answers to Even-Numbered Problems..................583
    D. Glossary of Technical Terms........................589
    Index..607

Preface to the Revised Canadian Edition

To the professional engineer who is daily required to make financial decisions, skill in economic analysis is essential. Development of this skill must therefore be an integral part of an engineer's training. As we have taught the subject to thousands of students over the past thirty years, we have found that the excellent American text by Professors White, Agee, and Case meets the demands of such a course. However, Canadian instructors also need a text that provides Canadian information and addresses economic topics specific to this country. The concept of producing a Canadian book, originally based on White, Agee, and Case, *Principles of Enginering Economic Analysis* emerged in response to this need.

In this revised edition of the first Canadian edition, several topics have been amended, updated and extended. The section on cost and cost estimation is extended to include learning curves as well as cost-capacity relations and price indices. The section on depreciation and income taxes are updated to include the latest information about the Canadian Income Tax Act. A new chapter is created to cover the fundamentals of financial and cost accounting. And all other sections of the book are corrected, simplified and amended. The book includes information, illustration and problems on all traditional topics of engineering economic analysis, and management and business decision-making tools.

Throughout the text and in the problems, all units and symbols are expressed in SI units (Système International d'Unités).

While responsibility for the content of this edition must be solely ours, many users of the American edition and others have contributed to changes through their encouragement, comments, and suggestions. We are indebted to M. L. Bilodeau, McGill University; Philip H. Byer and Scott J. Rogers, University of Toronto; P. S. Chisholm, University of Guelph; A. Clayton and K. McLachlan, University of Manitoba; V. Bruce Irvine, University of Saskatchewan; and C. G. Miller,

Queen's University; whose reviews of specific chapters were so helpful. Thanks are also due to G. Keri of the National Research Council.

A. J. Szonyi
R. G. Fenton,
Toronto, Ontario
July 1999

Chapter 1
Introduction

1.1 Background

Engineering decisions cover a wide variety of areas ranging from choosing airport locations to selecting production methods and determining budget requirements. Since all corporate and government decisions are influenced by financial considerations, one of the most effective and important tools available to engineers is economic analysis. Here are only a few of the spheres of activity to which economic analysis can be applied: examining design alternatives; deciding the location and size of production plants, railway lines, roads, bridges and tunnels; and installing airports and harbour facilities, dams and irrigation systems, communication networks, power generating facilities and power distribution systems. The engineer may also be involved in production decisions such as selecting production methods, assessing what quantities of raw materials should be purchased, what types of equipment and machinery should be acquired; determining operating and maintenance procedures, methods for the storage, packaging and delivery of goods; and setting down testing and quality-control methods for production. Managerial decisions that engineers may be required to make include determining capital budget requirements; selecting research and development projects; and establishing production targets. This book is designed to help the engineer learn the rationale and methods of economic analysis and their application to engineering decision making in both the private and public sectors.

Traditionally, the application of economic analysis techniques in the comparison of engineering alternatives has been referred to as *engineering economy, economic analysis, and economic decision analysis*, among other names. However, the emergence of a widespread interest in the economic analysis of engineering decisions in the public sector has brought about greater use of the more general term *economic analysis*.

1.2 The Problem-Solving Process

Economic analyses are typically performed as a part of the overall problem-solving process. In designating a new or improved product, a manufacturing process, or a system to provide a desired service, the "problem solver" is involved in performing the following five steps:

1. Formulation of the problem.
2. Analysis of the problem.
3. Search for alternative solutions to the problem.
4. Selection of the preferred solution.
5. Specification of the preferred solution.

In Step 4 the selection of the preferred solution is frequently based on the economic performance of the alternatives.

The *formulation of the problem* involves the establishment of its boundaries, and it is aided by taking a black box approach. An originating state of affairs (State A) and a desired state of affairs (State B) exist. A transformation must take place in going from State A to State B, as depicted in Figure 1.1. More than one method of performing the transformation from State A to State B exists, and there is unequal preferability of these methods. The solution to the problem is visualized as a black box of unknown, unspecified contents having input A and output B.

Fig. 1.1 Black box approach.

The *analysis of the problem* consists of a detailed phrasing of the characteristics of the problem, including restrictions and the criteria to be used in evaluating alternatives. Considerable fact gathering is involved, including the real restrictions on the problem, which must be satisfied. Consequently, any budget, quality, safety, personnel, environmental, and service-level constraints that may exist are identified.

The *search for alternative solutions to the problem* involves the use of the engineer's creativity in developing feasible solutions to the problem. One suggested approach is as follows:

1. Exert the necessary effort.
2. Not get bogged down in details too soon.
3. Make liberal use of the questioning attitude.
4. Seek many alternatives.
5. Avoid conservatism.
6. Avoid premature rejection.
7. Avoid premature acceptance.
8. Refer to analogous problems for ideas.
9. Consult others.
10. Attempt to divorce your thinking from the existing solution.

Unfortunately, the natural tendency is to go on to the evaluation of alternatives at the expense of searching for additional alternative solutions. As a result, the search process often terminates with the development of the first alternative that can be justified economically. By separating the search process from the selection process, we enhance the possibility of generating a number of alternatives that can be justified economically.

The *selection of the preferred solution* consists of evaluating the alternatives, using the appropriate criteria. The alternatives are examined in light of the constraints, and infeasible alternatives are eliminated. The benefits produced by the feasible alternatives are then compared. Among the criteria considered for choosing the best alternative is the economic performance of each alternative.

The *specification of the preferred solution* consists of a detailed description of the solution to be implemented. Predictions of the performance characteristics of the solution to the problem are included in the specification.

Example 1.1

As an illustration of the problem-solving procedure, Proxax Industries, a leading manufacturer of automotive brake drums was faced with the need to expand its distribution operations. A study team was formed to analyze the problem and develop a number of feasible alternative solutions. After analyzing the problem and projecting future distribution requirements, the following alternatives were reached:

1. Consolidate all distribution activities and expand the existing distribution centre, located in London, Ontario.
2. Consolidate all distribution activities and construct a new distribution centre, location to be determined.
3. Decentralize the distribution function and build several new distribution centres geographically dispersed.

After considering the pros and cons of each alternative, the president of the company directed the study team to pursue the second.

An extensive plant location study was performed. The location study resulted in five candidate locations being selected for final consideration. The criteria used to make the final selection included:

1. Land cost and availability.
2. Labour availability and cost.
3. Proximity to supply and distribution points (present and future).
4. Property taxes and insurance rates.
5. Transportation (access to rail and main highways).
6. Community attitudes.
7. Building costs.

Based on site visits to each location, the director of engineering of the company selected a site in Oshawa, Ontario, and directed the study team to develop alternatives for the material handling system to be used in the distribution centre.

Applying the problem-solving procedure, four alternatives were identified for evaluation. The first involved the use of pallet racks, lift trucks, and flow racks; the second included the use of an automated stacker crane system, lift trucks, conveyors, and flow racks; the third suggested narrow-aisle, guided picking machines, high-rise shelving, conveyors, driverless tractor trains, and lift trucks; the fourth consisted of an automated stacker crane system, an automatic guided vehicle system, a conveyor system, and high-rise, narrow-aisle lift trucks.

A planning period of 10 years was used in performing an economic analysis of each alternative. The fourth was the most eco-

Introduction 5

nomical and was recommended to management. Based on the detailed presentation and the economics involved, management approved a budget of $90 million to implement the recommendations of the study team.

As demonstrated by the preceding illustration, engineers must be prepared to defend their solutions to problems. Economic performance is among the criteria used to evaluate each alternative. Monetary considerations seldom can be ignored. If economics are not considered in the criteria used in the final evaluation of the alternatives, they are usually involved in an initial screening of them. In fact, cost can be a limiting factor on the alternatives that can be considered, as well as the basis for the final selection.

This textbook treats in detail Step 4 in the problem-solving process, the step that involves the selection of the preferred solution. The process of measuring cash flows and benefits, and the consideration of multiple objectives in selecting the preferred alternative are also treated.

In comparing alternatives, the *differences* will be emphasized. Consequently, the aspects of the alternatives that are the same normally will not be included in the analysis.

1.3 Cash Flow

Our approach to the subject of engineering economic analysis will be a *cash flow* approach. A cash flow occurs when money *actually* changes hands from one individual or organization to another. A single cash flow item is either positive, if money is received (revenue), or negative, if money is dispersed (expenditure). Companies, systems, projects, or any engineering undertaking have cash flows. The cash flow of a project, for example, includes all revenues received and expenditures incurred during the life of the project. If economic criteria are used for finding the preferred solution to an engineering problem, the selection is always based on comparing the cash flows.

Cash flows are frequently represented graphically as *cash flow diagrams* (see Figure 1.2). Time, measured in periods (years, months, or days), is shown on the horizontal axis of the diagram. On Figure 1.2, the time period is a year. The numbers indicate ends of time periods. For example, the end of the third period is labelled 3, and it indicates the end of Year 3. The cash flow items are represented by vertical arrows. Positive quantities (up arrows) are revenues, and negative quantities (down arrows) are expenditures. The cash flow diagram of Figure 1.2 shows an expenditure of $10 000 incurred at the beginning of the first year, followed by revenue of $8 000 received at the end of Year 1, and $14 000 received at the end of Year 5.

Cash flow expenditures at any given time should include all monies dispersed, including income tax payments made. If income tax payments are excluded, the cash flow is referred to as *before-tax cash flow*. *After-tax cash flow* includes income tax payments in addition to all other expenditures.

In the *private sector* (business enterprises), after-tax cash flow should normally be used for selecting the preferred alternative. However, if income tax payments equally affect all alternatives, before-tax cash flow analysis would suffice. Occasionally, when income tax payment prediction is difficult, before-tax cash flow analysis may be used for obtaining a preliminary solution for the preferred alternative.

The *public sector* (federal and provincial governments, municipalities, school boards, universities, hospitals, police, army, charitable organizations, and government institutions, etc.) does not pay income tax, therefore in the economic analysis of public sector projects, the cash flow obviously does not include income tax payments.

Sound engineering decisions based on economic criteria can only be reached if the cash flows for all alternatives are complete and accurate.

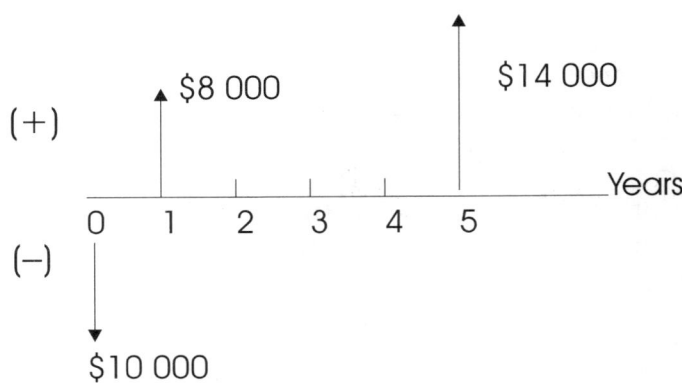

Fig. 1.2 A cash flow diagram.

1.4 A Fundamental Concept: The Time Value of Money

A fundamental concept underlies much of the material covered in the text: *money has a time value*. The *value* of a given sum of money depends on *when* the money is received.

Example 1.2

To illustrate the concept of the time value of money, suppose you received the following offer for an invention: you either receive $10 000 now or $X a year from now. Would you choose to receive the $10 000 now or the $X one year from now if $X equaled (1) $10 000, (2) $10 200, (3) $10 800, (4) $12 000?

In presenting this situation to numerous students, no students preferred Case 1, very few students preferred Case 2, most students preferred Case 3, and all students preferred Case 4. The point is that the value of $10 000 one year from now was perceived to be less than the value of $10 000 at present. For most students, the value of $10 200 one year from now was believed to be less than the value of $10 000 at present. Only a few students felt that $10 800 a year from now was less valuable than $10 000 at present. All students believed $12 000 a year from now was more valuable than $10 000 at present. Thus, for each individual student, some value (or range of values) of $X exists for which one would have no preference between receiving $10 000 now versus receiving $X a year from now. If, for example, one is indifferent for $X equal to $10 600, then we would conclude that $10 600 occurring one year from now has a *present value of $10 000 for that particular individual*.

Example 1.3

To continue our consideration of different cash flow situations, examine the two cash flow profiles given in Table 1.1. Both alternatives involve an investment of $10 000 in ventures that last for four years. Alternative A involves the purchase of a computer and software by a consulting engineer who is planning on providing computerized finite element analysis capability for clients. Since the engineer anticipates that competition will develop quickly, a declining revenue profile is anticipated.

Table 1.1 Cash Flow Profiles for Two Investment Alternatives

End of Year	CF A	CF B	CF A−B
0	− $10 000	− $10 000	$0
1	+ 7 000	+ 1 000	+ 6 000
2	+ 5 000	+ 3 000	+ 2 000
3	+ 3 000	+ 5 000	− 2 000
4	+ 1 000	+ 7 000	− 6 000

Alternative B involves an investment in a coal-mining venture by a group of individuals. Increasing quantities of coal are to be sold over a four-year periold. Consequently, an increasing revenue profile is anticipated.

The consulting engineer has available funds sufficient to undertake either investment, but not both. The cash flows shown are after taxes and other expenses have been deducted. Both investments result in $16 000 being received over the four-year period; hence, a net cash flow of $6 000 occurs in both cases.

Which would you prefer? If you prefer Alternative B, then you are not acting in a manner consistent with the concept that money has a time value. The $6 000 difference at the end of the first year is worth more than the $6 000 difference at the end of the fourth year. Likewise, the $2 000 difference at the end of the second year is worth more than the $2 000 difference at the end of the third year.

Example 1.4

As another illustration of the impact of the time value of money on the preference between investment alternatives, consider alternatives C and D, having the cash flow profiles depicted in Figure 1.3. The cash flow diagrams indicate that the positive cash flows for Alternative C are identical to those for Alternative D, except that the former occur one year sooner; both alternatives require an investment of $6000. *If only one of the alternatives must be selected,* then Alternative C would be preferred to Alternative D, based on the time value of money.

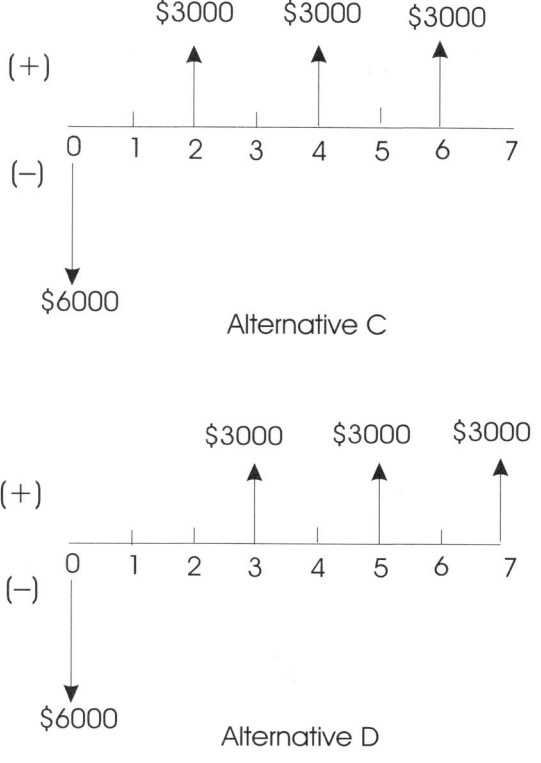

Fig. 1.3 Cash flow diagrams for Alternatives C and D.

Example 1.5

A third illustration of the effect of the time value of money on the selection of the preferred investment is presented in Figure 1.4. Either Alternative E or Alternative F must be selected; the only differences in the performance characteristics of the two alternatives are economic. As shown in Figure 1.4, the economic differences reduce to a situation in which the receipt of $100 is delayed in order to receive $200 a year later. For this illustration, we would conclude that most students polled earlier would prefer Alternative E to Alternative F.

While the previous examples provide perceptual impressions of the importance of the time value of money, there are specific mathematical conversions that may be used to determine exactly the differences between the above alternatives. These conversions, the time value of money operations, are presented in detail in Chapter 3.

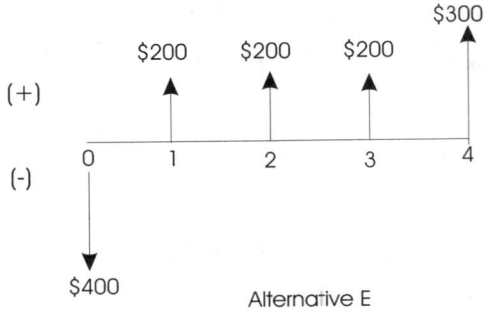

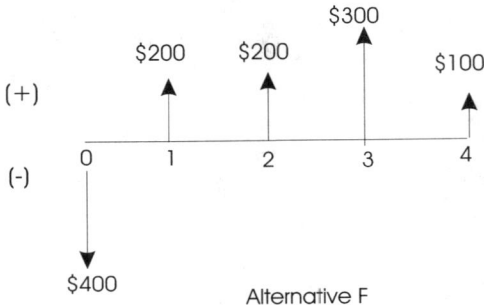

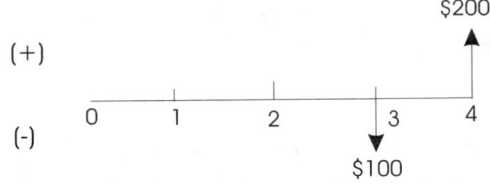

Difference between Alternatives E and F
Fig. 1.4 Cash flow diagrams for Alternatives E and F.

1.5 Economic Justification May Not Be Enough

Even though the text concentrates on the economic aspects of investment alternatives, note that in many instances economic justification may not be enough! Decisions can be quite different from recommendations. Managers typically have multiple criteria to consider in reaching a final decision about the alternative to be adopted. Among the factors to be considered are: quality, safety, environmental impact, community attitudes, labour-management relationships, cash flow position, risks, system reliability, system availability, system maintainability, system operability, system flexibility, impact on personnel levels, training requirements, comparisons with competitors, impact on the different units within the organization, ego, customers' preferences, capital requirements, and economic justification.

The list of criteria can be daunting. However, perhaps the most important ingredient that separates the selected alternative from the *runners-up* is the salesmanship of the individual who presents the alternative to the manager. It is not unusual for a manager to adopt a weaker alternative because of the persuasive powers of the person who presented it.

A tendency exists for problem solvers to spend too much time trying to develop the recommended solution to the problem and too little time in determining the best way to sell the solution to management. Effective communication is essential; in particular, it is important to communicate with management in a way that can be understood and accepted.

It is important to write a proposal that includes an *executive summary*. The executive summary normally reduces the performance characteristics of the recommendation to a few pages. Technical aspects of the recommended solution and the details of the economic analysis are usually provided in the main report for the manager who wishes additional information. A face-to-face presentation is also quite important in achieving effective communication.

Communication techniques such as the use of visual aids, voice control, dress, structure and clarity of expression are necessary to sell the proposed solution. However, it is probably more important to know your audience well and prepare your proposal using their language. Since the language of managers is often the language of finance, Chapter 9 introduces the financial terminology used by managers.

1.6 Non-Monetary Considerations

The principal emphasis of this textbook is on the use of logical methodology to choose an engineering project from among several available to the decision maker, where the criterion for choice will be some economic measure of effectiveness. However, the decision maker has several economic measures from which to select. For example, because of an increase in market demand, the management of a firm may face the choice between (1) going to a second shift of production using the same machinery, and (2) installing more highly automated machinery in order to meet the new production schedules. Clearly, a comparison of the costs for each of these alternatives over some planning period is in order, and one objective of management would no doubt be to meet production schedules at the lowest possible annual cost of production.

Although considerable investigation and study are required, many of the factors involved in a decision that may at first seem non-economic can be expressed in (or reduced to) monetary values. For instance, in the above example, maintenance personnel may have to be trained in the installation and repair of the new machinery but, most likely, these training costs can be determined and charged to the "automation" alternative. However, other factors involved are not so easily reduced to monetary values and, indeed, some factors involved are not so easily reduced at all. For example, the installation of the automated machinery would probably reduce the number of personnel required by the present production system, whereas the second shift alternative would result in additional employees. If persons are laid off or transferred to other jobs within the firm, their salaries could be considered annual savings accruing to the automated machinery alternative and, conversely, the salaries of additional employees integral to the second shift alternative would be annual costs. However, layoffs or transfers could have a deleterious effect on the morale of the remaining employees; on the other hand, adding new personnel could have a beneficial effect on morale. Assessing the cost or gain of these expected changes in morale is virtually impossible. Nevertheless, the potential effect on production that morale changes may have must be considered by management.

Factors that affect a decision but cannot be expressed in monetary terms are often called *intangibles* or *irreducibles.* Practically all real-world business decisions involve both monetary and intangible factors. In the above production example, assume that the alternative of a second shift results in a total annual cost of $100 000 for both shifts over a five-year planning period with no persons laid off or transferred or new employees hired. For the automated machinery alternative, there would be an estimated annual cost of $85 000 over the five-year planning period with six persons laid off. Thus, this oversimplified decision is reduced to two measures: a monetary annual cost and the single intangible factor of employee morale. If management chooses the second shift alternative, this would imply that management places a higher subjective utility value on "good employee morale" than on the $15 000 difference in annual costs between the two alternatives. That is, the objective of good employee morale is more important than the objective of lowest annual cost (at least for this difference of $15 000). Instances of decision situations where intangible considerations outweigh monetary ones are frequent, and the engineer should not become distraught at this fact. Within the hierarchy of management responsibilities for a firm, the higher the level of management, the more likely it is that intangible considerations will be given greater subjective weight. Engineering project proposals should therefore reflect a knowledge of intangibles that management will wish to consider.

Miller and Starr[1] have made interesting observations in regard to the decision-making process.

1. Being unable to satisfactorily describe goals in terms of one objective, *people customarily maintain various objectives.*

2. Multiple objectives are frequently in conflict with each other, and when they are, a *suboptimization* problem exists.

3. At best, we can only optimize as of *that time* when the decision is made. This will frequently produce a suboptimization when viewed in subsequent times.

4. Typically, decision problems are so complex that any attempt to discover *the* set of optimal actions is useless. Instead, people set their goals in terms of outcomes that are *good enough.*

5. Granted all the difficulties, human beings make every effort to be *rational* in resolving their decision problems.

Accepting the premise that multiple objectives are involved in a decision, it follows that different measures for these objectives may arise. For instance, two or more engineering project alternatives could each involve annual costs in dollars, mass in kilograms, repairability values on an arbitrary index from 1 to 10, and so forth. The decision problem would be greatly simplified if the various measures could be trans-

---

[1] David W. Miller and M. K. Starr, *Executive Decisions and Operations Research*, 2d. Ed., Prentice-Hall, 1969, pp. 52-53.

formed to a single measure, say, values on a "utility" scale. Then, each measure (dollars, kilograms, etc.) for a given alternative could be converted to a utility value, the new utility values could be weighted by their relative importance, and the weighted values could be aggregated by an appropriate functional form. A single utility value would thus result for each alternative, and a selection would be made by choosing the alternative having the maximum utility value. This area of study is the study of utility theory and value measurements, a controversial subject outside the scope of this book.

There will be some additional discussion on multiple objectives in Chapter 8, but otherwise it will be assumed throughout this book that only a single economic measure of effectiveness is relevant in comparing the alternative projects.

1.7 Overview of the Text

As an overview, we have organized the material in a manner consistent with the logical sequence of steps followed in performing an economic evaluation of investment alternatives. Chapter 2 provides a discussion of cost concepts and includes discussions of elements of *costs*, measurement of cash flows, incremental costs, future costs, sunk costs, intangibles, irreducibles, nonquantifiables, standard costs, direct costs, indirect costs, fixed costs, variable costs, average costs, marginal costs, and opportunity costs. Given a feel for how the data are to be obtained for an economic analysis, Chapter 3 presents some important fundamental concepts involving the *time value of money*. In fact, Chapter 3 provides the foundation for the remainder of the book; therefore the reader must understand the time value of money operations presented in this chapter.

Chapter 4 uses the time value of money operations in *comparing investment alternatives*. Measures of economic worth such as present worth, annual worth, future worth, payback period, rate of return, and savings/investment ratio are used in comparing mutually exclusive investment alternatives.

Chapter 5 addresses the issue of *income taxes* and their incorporation in economic analyses; Chapter 6 treats *benefit-cost analysis*. Those involved primarily in the private sector will probably choose to study Chapter 5; those involved primarily in the public sector will probably choose to study Chapter 6. When one is involved in both the public and the private sectors, both chapters are appropriate!

Chapter 7 presents supplementary analysis techniques. In particular, the effects of risk and uncertainty on the analysis of economic investment alternatives are considered. In modelling the effects of risk and uncertainty mathematically, two approaches are taken: a prescriptive (normative) approach and a descriptive approach. A prescriptive model prescribes the action that, in some sense, is optimal; a descriptive model describes the behaviour of the situation modelled. Chapter 7 provides a descriptive treatment of the effects of risk and uncertainty by presenting the subjects of *break-even, sensitivity,* and *risk analyses.* Prescriptive models are presented in Chapter 8. Subjects considered include *decision making* under risk and uncertainty.

Chapter 9 gives an introduction to financial and cost accounting. Chapter 10 covers the basic economic principles that are essential to the understanding of engineering economic analysis. The concept of supply and demand interaction and the theory of production are outlined in some detail, especially for the benefit of readers without any previous exposure to economic theory. Readers familiar with these fundamentals may omit Chapter 10. However, those who have no background in economic theory should read Chapter 10 first, before reading any other chapter of this book.

Chapter 2
Cost Concepts and Cost Estimation

2.1 Introduction

Engineering economic analysis is primarily concerned with comparing alternative projects on the basis of an economic measure of effectiveness. This comparison process has a variety of cost terminologies and cost concepts, and it will be helpful to present them prior to the discussion in Chapter 4 of the economic measures of effectiveness for comparing alternative projects. To implement the discussion of cost terminology, a typical production situation will now be described.

Let us assume that the business of a small manufacturing firm is job-shop machining. That is, the firm produces a variety of products and component parts according to customer order. Any given order may be for quantities of as few as five parts or as many as several hundred parts. The firm has periodically received orders to manufacture a part, which we will identify as Part No. 163H, for the Montreal Gear Works Ltd. The part has been manufactured in a four-stage production sequence consisting of (1) cutting bar stock to length on a horizontal band saw, (2) machining on an engine lathe, (3) drilling on a vertical drill press, and (4) packaging. The unit cost of producing Part No. 163H by this sequence is $30, where the unit cost consists of the cost of direct labour, materials, and overhead (prorated cost of administrative and management expenses, building and maintenance costs, utilities, etc.) The firm is now in the process of negotiations with the Montreal Gear Works to obtain a contract for producing 10 000 of these parts over a period of four years. A contract for this volume of parts is highly desirable but, in order to obtain the contract, the firm must lower the unit cost.

An engineer for the firm has been assigned to determine alternative production methods to lower the unit cost. After study, the engineer recommends the purchase of a turret lathe. With the turret lathe, the processing sequence of Part No. 163H would consist essentially of (1) machining bar stock on the turret lathe and (2) packaging. The estimated unit cost for Part No. 163H by this production method would only be $20. Furthermore, the production rate by the new method would be increased over the old method.

If the turret lathe is purchased, the saw, engine lathe, and drill press would not be sold but would be kept for other jobs for which the firm may receive orders. The turret lathe would be reserved for the production of Part No. 163H, and its excess production capacity could be devoted to other jobs.

The investment required to purchase the turret lathe and the new tooling required, as well as installing the machine, is $180 000. The physical life of the turret lathe is about 25 years, but the firm believes that the useful economic life of the machine will be only 10 years. The estimated salvage value of the turret lathe at the end of the 10-year period is $60 000. If the maximum unit price that the Montreal Gear Works will pay for Part No. 163H is $28, should the firm accept the contract for 10 000 parts and then purchase the turret lathe in order to execute the contract?

We will not answer the question in this chapter, but have cited the example to illustrate one type of decision with which this text is concerned. In fact, additional information and the methodology presented in Chapter 4 would have to be used to find a complete answer to the question. However, the reader can readily appreciate that certain cost figures in the above example must be determined or estimated before any rational decision can be reached as to the purchase of the turret lathe. The cost elements contributing to the $180 000 machine cost, the unit production cost of $20, and salvage value of $60 000 would be determined or estimated from information obtained from a variety of sources. These sources might typically be production records, accountants' records, manufacturers' catalogues, publications from Statistics Canada, or other government sources, etc. The engineer should therefore be familiar with cost terminology, cost factors, and cost concepts as used by different specialists if effective economic comparison and intelligent recommendations are to be made.

2.2 Cost Terminology

Because both cost definitions and cost concepts are included in this section, clarity will be achieved by the use of six categories:

1. Life-cycle costs.

2. Past and sunk costs.

3. Future and opportunity costs.

4. Direct, indirect, and overhead costs.

5. Fixed and variable costs.

6. Average and marginal costs.

2.2.1 Life-Cycle Cost

The *life-cycle cost* of an item is the sum of all expenditures associated with the item during its entire service life. The term *item* should be interpreted in the general sense as a machine, a unit of equipment, a building, a project, a system, and so forth. Life-cycle costs may include engineering design and development costs; building, manufacturing, fabrication, and testing costs; shipping and installation costs; operating and maintenance costs; and disposal costs. Life-cycle costs may also be expressed as the summation of acquisition, operating, maintenance, and disposal costs. Thus, life-cycle cost terminology may vary from author to author, but the basic meaning of the term is clear.

This textbook is predominantly concerned with the economic justification of engineering projects, the replacement of existing projects or capital assets, and the economic comparison of alternative projects. For the purpose of these types of analyses, we will define life-cycle costs as consisting of:

1. First cost (initial investment or capital cost).

2. Operating and maintenance costs.

3. Disposal costs.

The *first cost* of an item is the total initial investment required to get the item ready for service; such costs are nonrecurring during the life of the item. For example, for the purchase of a numerically controlled machine tool, the first cost may consist of the following major elements: (1) the basic machine cost; (2) installation cost (including: cost of foundation, vibration and noise insulation, temperature and humidity control, heat, light, compressed air, and power supply, and cable connection to computers, etc.); (3) cost of testing the machine; (4) cost of training personnel; (5) cost of special tooling, jigs, and fixtures; (6) cost of all supporting equipment (computer hardware, software, etc.). For some other item, a different set of first-cost elements may be appropriate, but the first cost of an item normally involves more cost elements than just the basic purchase price.

Operating and *maintenance costs* are recurring costs that are necessary to operate and maintain an item during its useful life. Operating costs usually consist of labour, material, and overhead items (fuel, electric power, insurance premiums, inventory charges, administrative and management expenses, etc.). It is usually assumed that these costs are annual, but maintenance costs may or may not be on a recurring, annual basis. That is, a regular annual schedule of minor or preventive maintenance may be followed, or it could be policy that maintenance is performed only when necessary, such as when an overhaul is required. In most cases, the maintenance policy would consist of both preventive maintenance and maintenance on an "as-needed" basis. In any case, repair and upkeep result in costs that must be recognized in the economic analysis of engineering projects.

When the life cycle of an item has ended, *disposal costs* usually result. Disposal costs may include labour and material costs for removal of the item, shipping costs, or special costs; an example of special costs is the disposal cost of radioactive waste materials. Although disposal costs may be incurred at the end of the life cycle, most items have some monetary value at the time of their disposal. This value is the *true* or *market value* (i.e., the actual selling price of the item). After deducting the cost of disposal from the market value, the net dollar worth of the item is obtained. This net dollar worth at the time of disposal is the *salvage value*.

Machinery, equipment, buildings, and other fixed assets gradually decrease in value through physical deterioration and obsolescence. This loss of value is an important cost element that must be considered when the operating costs are determined. The cost equivalent of the life-time loss of value of long-term assets are allocated to the yearly operating costs by the process of *depreciation.* Depreciation, according to the Canadian Institute of Chartered Accountants, is a "procedure in which the cost or other recorded value of a fixed asset less estimated salvage (if any) is distributed over its estimated useful life in a systematic, rational manner." The amount allocated in any one year from the lifetime loss of value of a fixed asset to the annual operating costs is the yearly *depreciation charge.* It is important to note that, although the yearly depreciation charge is part of the cost of operation, it does not involve an actual cash flow. A detailed discussion of depreciation will be given in Chapter 5.

The *book value* of an asset is the first cost minus accumulated depreciation charges. At the end of each accounting period a new book value is determined by reducing the book value at the end of the previous accounting period by the depreciation charge for the period. It should be noted that the book value is not related to and, therefore, not necessarily equal to the market value of the asset. The market value is determined by the supply and demand interaction at the marketplace; a detailed discussion of this topic is given in Chapter 10. The book value, on the other hand, is determined by an accounting process: initial investment cost minus accumulated depreciation.

Scrap value refers to the value of the material of which the item is made. For example, a four-year-old automobile may have a scrap value of $200, a market value of $3 500, and a book value of $5 200. Salvage value of this automobile is, therefore, $3 500 minus disposal costs (if any). A distinction must be made between these terms for evaluating potential investment projects.

The life cycle obviously involves a time period, and the end of an item's life may be judged from either a functional or economic point of view. The *economic life* of an item is generally shorter than its *functional life*. For example, an engine lathe may remain functionally useful for 40 years or more, but, because of periodic advancements in machine design technology, newer engine lathes will have higher production rates, and improved accuracy and reliability; therefore, the economically useful life of an engine lathe may be only 10 years. In this textbook, the life of an item will be its economical, not functional, life.

2.2.2 Past and Sunk Costs

Past costs are historical costs that have occurred for the item under consideration. *Sunk costs* are past costs that are unrecoverable. The distinction is perhaps best made through examples. Assume that an item of machinery is purchased for $40 000, salvage value at the end of five years is estimated as $10 000, and the annual depreciation charge is $6 000. The $6 000 annual cost of depreciation is a non-cash cost of production that should be allocated to the output of the machinery. After allocating this and other manufacturing costs, and general and administrative costs to each unit of production, the total unit cost is determined. A profit is then added to each unit of production in order to arrive at the unit selling price. Thus, when a unit is sold, a portion of each sales dollar returns a portion of the depreciation expense. In this illustration, it is assumed that the sales will return, or recover, the total estimated depreciation expense of $30 000 (first cost minus estimated salvage value) for the five-year period. However, if the machine is sold for only $2 000 at the end of five years, there is a loss of $8 000, ($10 000−$2 000). This loss results from underestimating the yearly depreciation charge, which should have been $\dfrac{\$40\,000 - \$2\,000}{5} = \$7\,600$, instead of \$6 000. The difference, \$1 600 yearly, was not added to the cost of production, was not recovered when the products were sold, and therefore was lost. Since this is a cost which occured during the past five years, it can no longer be recovered. This is a sunk cost. Of course, if the machine is kept, it is argued that the true value being kept is only \$2 000.

Sunk costs may qualify as capital losses and serve to offset capital gains or other taxable income and thus reduce income taxes paid. Examples in Chapter 5 will illustrate this point. Also, past costs and sunk costs provide information that can improve the accuracy of estimating future costs for similar items.

A sunk cost is a cost that cannot be recovered, and therefore is irrelevant for the future. For example, a manufacturing company has 1 000 defective valves that cost \$15 per unit to manufacture. The units can be sold as they are for \$8 each, or they can be rebuilt for an additional \$18 per unit and then sold for \$30. Should the company rebuild the valves or should it sell them in their present form? The original manufacturing cost of \$15 per unit is sunk cost and irrelevant in the decision affecting the future. The net return if the defective valves are sold is $1\,000(\$8) = \$8\,000$. The net return if the valves are rebuilt and sold is $1\,000(\$30 - \$18) = \$12\,000$. Based on the available information, it appears that rebuilding the valves is the better proposition.

2.2.3 Future and Opportunity Costs

All costs that may occur in the future are termed *future costs.* These future costs may be operating costs for labour and materials, maintenance costs, overhaul costs, and disposal costs. In any case, by virtue of occurring in the future, these costs are rarely known with certainty and therefore must be estimated. This is, of course, also true for future revenues or savings if these are involved in a given project. Estimates of future costs or revenues are uncertain and subject to error. The economic analysis is simplified if certainty of future costs, revenues, or savings is assumed. This assumption is made until Chapter 7, where concepts of probability are introduced.

The cost of foregoing the opportunity to earn interest or a return on investment funds is termed an *opportunity cost.* This concept is best explained by means of illustrations. For example, if a person has \$1 000 and stores this cash in a home safe, the person is foregoing the opportunity to earn interest on the money by establishing a savings account in a local bank that pays, for example, 4% annual interest. (Of course, investments other than savings accounts are possible.) For a one-year period, the person is foregoing the opportunity to earn $0.04(\$1\,000) = \$40$. The $40 amount is thus termed the opportunity cost associated with storing the $1 000 in the home safe.

As a similar illustration of an opportunity cost, assume that a person has $5 000 cash on hand. This amount is considered *equity capital* if the $5 000 was not borrowed (i.e., there is no debt obligation involved). The person has available secure investment opportunities such as a savings account in a bank or other financial instruments. From the available investment opportunities, suppose the optimum combination of risk and interest yield on the investment results in 6% annual interest. Thus, the investment of $5 000 would yield $0.06(\$5\,000) = \$300$ each year. If the person, instead of investing the $5 000, purchases an automobile for the same amount, the person will forego the opportunity to earn $300 interest per year. The $300 is an annual opportunity cost associated with purchasing the automobile.

The same logic applies in defining an annual opportunity cost for investments in engineering projects. The purchase of an item of production machinery with $600 000 of equity funds prevents this money from being invested elsewhere with greater security or higher profit potential. This concept of *opportunity cost* is fundamental to the study of engineering economic analysis and is a cost element that is included in virtually all methodologies for comparing alternative projects. In Chapter 4, the concept of opportunity costs will be discussed under the heading of *minimum attractive rate of return (MARR)*.

Some individuals define the *MARR* as the *cost of capital*. As used here, the term *cost of capital* refers to the cost of obtaining funds for financing projects through debt obligations. These funds are usually obtained from external sources by (1) borrowing money from banks or other financial organizations (e.g., trust companies, credit unions, insurance companies, and pension funds) and (2) issuing bonds. These debt obligations are normally long term, as opposed to short-term obligations for the purchase of supplies and raw materials. The debt obligations result in interest payments on, say, a monthly, quarterly, semiannual, or annual basis. The interest payments are thus a cost of borrowed capital. Financing projects through issuing bonds is a method of obtaining capital funds that is probably less known to the reader than borrowing money from a bank at a stated interest rate. Some elaboration of bonds is therefore appropriate.

Bonds are issued by various organizational units—partnerships, corporations (profit or nonprofit), governmental units (municipal, provincial, and federal), or other legal entities. The sale of bonds represents a legal debt incurred by the issuing organization. Bonds constitute a promise to pay a definite sum of money to the holder at a fixed date. Bonds may be owned and transferred during their lives by a number of people; therefore, bonds do not name the lender. Bonds can be secured or unsecured. Unsecured bonds are often called debentures. Only financially strong companies are able to market unsecured bonds. In any case, the purchaser of a bond has legal claim to the assets of the issuing unit but has no ownership privileges in the issuing unit. Purchasers of the common or preferred stock of an organizational unit do hold ownership status but may or may not have voting privileges, depending on the stipulations of the particular stock issue. In the sense that bonds are debt obligations and not ownership shares, bonds are considered a more secure investment than either common or preferred stock. This statement should not be taken as a universal truth, however, since the security level for a bond or a stock depends on many factors, economic and otherwise. However, the principal factor affecting the security level is the financial soundness of the issuing unit. Further details on interest payments on loans, bonds, and debentures are covered in Chapter 3.

Another method of financing engineering projects is through the use of *equity funds* (i.e., through ownership capital or cash that is debt free). The use of such funds incurs an opportunity cost, as mentioned previously, and will again be discussed in Chapter 4 where the minimum attractive rate of return (*MARR*) for an investment project is examined. However, the point is that the cost of capital is the cost of borrowed funds. Chapter 6 further discusses the cost of capital for financing public utility projects.

2.2.4 Direct, Indirect, and Overhead Costs

It will be helpful to provide definitions of direct, indirect, and overhead costs in the context of a manufacturing environment.

The *cost of goods sold*, as shown in Figure 2.1, is the total cost of a product (cost of goods manufactured plus general and administrative and selling costs). An amount of profit is then added to this total cost to arrive at a selling price. Such a cost structure is helpful in arriving at a

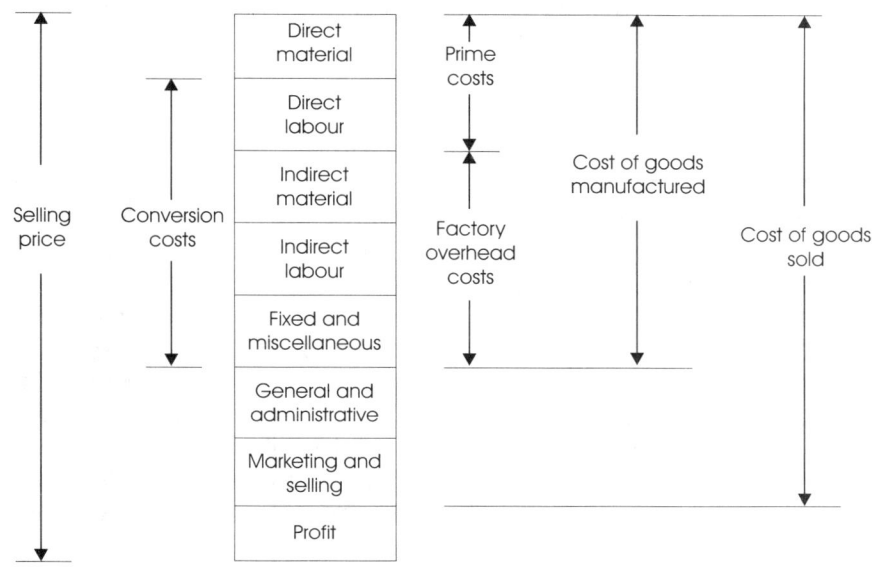

Fig. 2.1 A cost structure for manufacturing

unit cost, which is a primary objective of cost accounting. The term *cost of goods sold*, as used here for a manufacturing company, has a different meaning from the term *cost of goods sold*, used in general accounting practice for commercial companies, particularly for retail businesses. General accounting defines the cost of goods sold to be: beginning-of-year inventory plus purchases minus end-of-year inventory. Different meanings for the same terminology are unfortunate, but they do occur in the literature on accounting, and the reader is cautioned on this point. To simplify the treatment of the total cost of goods or services sold (as defined by Figure 2.1), the major cost elements can be defined as direct material, direct labour, and overhead costs.

Direct material and labour costs are the costs of material and labour that are easily measured and can be conveniently allocated to a specific operation, product, or project.

Indirect costs for both labour and material, on the other hand, are either very difficult or impossible to assign directly to a specific operation, product, or project. The expense of directly assigning such costs is prohibitive, and these costs are therefore considered to be indirect.

As an example of these different cost elements, suppose the raw material for a given part is a grey iron casting. The casting is machined on a vertical milling machine, the unmachined surfaces are painted, and four holes are drilled and tapped. The finished parts are stacked on wooden pallets and delivered to a customer.

In this example, the direct labour required per part to machine, paint, and package is probably readily determined. The labour required to receive the raw materials, handle parts between work stations, load pallets onto a truck, and deliver material to the customer is less easily identified and assigned to each part. This labour would be classified as indirect labour, especially if the labour in receiving, handling, shipping, and delivery is responsible for dealing with many different parts during the normal workday. The unit purchase price of the grey iron casting is an identifiable direct material cost. The cost of paint used per part may or may not be easily determined; if it is not, it is an example of indirect material cost. Also, any lubricating oils used during the machining processes would be an indirect material, not readily assigned on a cost per part basis.

Overhead costs consist of all costs other than direct material and direct labour. A given firm may identify different overhead categories, such as factory overhead, general and administrative overhead, and marketing and selling expenses. Furthermore, overhead amounts may be allocated to a total plant, departments within a plant, or even to a given item of equipment. Typical specific items of cost included in the category of factory overhead are indirect materials, indirect labour, taxes, insurance premiums, rent, maintenance and repairs, supervisory, technical, and management personnel, and utilities (water, electric power, etc.). Depreciation charge is usually included in the factory overhead; occasionally, however, it may be considered as part of the direct costs. It is the task of cost accounting to assign a propoprtionate amount of the above costs to various products manufactured or to services provided by an organization.

2.2.5 Fixed and Variable Costs

Fixed costs do not vary in proportion to the quantity of output. Administrative expenses, taxes and insurance, rent, building and equipment depreciation, and utilities are examples of cost items that are usually invariant with production volume and hence are termed *fixed costs*. Such costs may be fixed only over a given range of production; they then may change and be fixed for another range of production. *Vari-*

able costs vary in proportion to quantity of output. These costs are usually for direct material and direct labour.

Many cost items have both fixed and variable components. For example, a plant maintenance department may have a constant number of maintenance personnel at fixed salaries over a wide range of production output. However, the amount of maintenance work done and replacement parts required on equipment may vary in proportion to production output. Thus, total annual maintenance costs for a plant over several years would consist of both fixed and variable components. Indirect labour, equipment depreciation, and electrical power are other cost items that may consist of fixed and variable components. Determining the fixed and variable portions of such a cost item may not be possible, and where it is possible, the expense of establishing detailed measurement techniques and accounting records may be prohibitive.

Certain total costs (*TC*), then, can be expressed as the sum of fixed (*FC*) and variable costs (*VC*). As an example, the total annual cost for operating an automobile for a given year might be expressed as

$$TC(x) = FC + VC(x)$$

where *x* is the distance travelled yearly. Costs for insurance, registration fee, depreciation, certain maintenance costs, and interest on burrowed money if the car were financed are essentially fixed costs, independent of the distance travelled per year. Expenses for gasoline, oil, tire replacements, and certain maintenance items are proportional to the distance travelled. Arbitrarily assigning numerical values to the total cost function, assume that

$$TC(x) = \$2\,800 + \$0.28x$$

is a valid relationship for a given year in question (the expression is restricted to a given year, since actual depreciation expenses, and hence the fixed expenses, may vary from year to year). This relationship is linear in terms of *x*. However, the variable cost component is often a nonlinear function. Figure 2.2 graphically illustrates the total cost function.

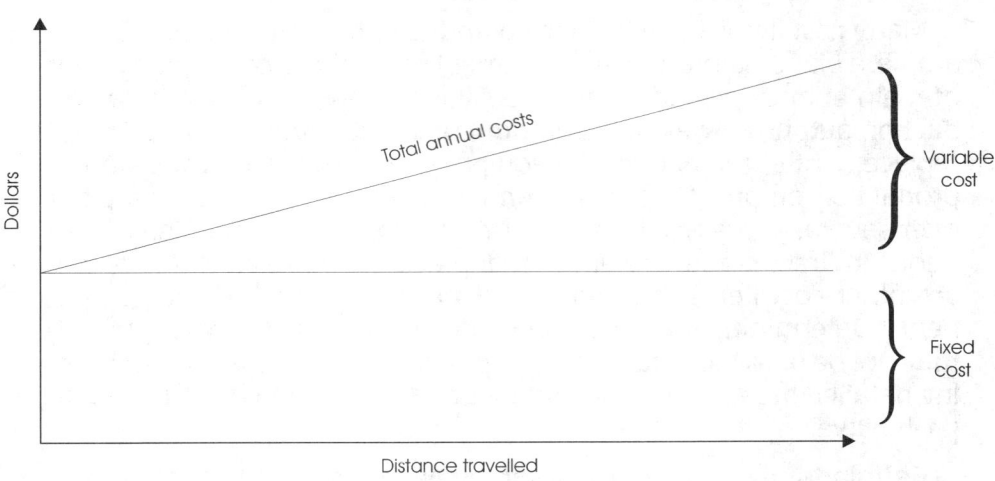

Fig. 2.2 Total annual costs as a linear function of distance travelled annually.

2.2.6 Average and Marginal Cost

The *average cost (AC)* of one unit of output (unit cost) is the ratio of total cost to quantity of output (or production volume)

$$AC(x) = \frac{TC(x)}{x}$$

where $TC(x)$ = total cost

x = output quantity

The average cost is usually a variable function of the output quantity and normally decreases with increasing quantity of output. For example, the average cost of operating a car, with total cost function of $(2800 + 0.28x)$, as depicted in Figure 2.2 is

Cost Concepts and Cost Estimation

$$AC(x) = \frac{2\,800 + 0.28x}{x}$$

$$= \$\left(\frac{2\,800}{x} + 0.28\right)\Big/\text{km}$$

If the car travels 10 000 km yearly, its average operating cost becomes

$$\$\left(\frac{2\,800}{10\,000} + 0.28\right)\Big/\text{km} = \$0.56/\text{km}$$

If, however, the car travels 20 000 km per year, its unit cost is only

$$\$\left(\frac{2\,800}{20\,000} + 0.28\right)\Big/\text{km} = \$0.42/\text{km}$$

Marginal cost is the additional (incremental) cost required to increase the quantity of output by one unit. It is the derivative of the cost function with respect to the output quantity. Marginal and average cost values corresponding to a specified output quantity are generally different. If the marginal cost is smaller than the average cost, an increase in output would result in a reduction of unit cost. If the total cost is a linear function of the output quantity, the marginal cost is constant. For example, for the car discussed above, the marginal cost is $0.28/km regardless of the distance travelled per year. If, however, the total cost is a nonlinear function of the output, the marginal cost is variable. The following example will illustrate the calculation of total, average, and marginal costs.

Example 2.1

Let us assume that the variable cost of producing electricity in a hydro-electric power plant is

$$VC(x) = \$(41.5x - 0.0001x^2)$$

where x is the yearly electricity output of the power plant measured in megawatt hours. The fixed cost of the plant is $2 500 000 per year. Calculate the total yearly cost of the power plant and the average and marginal costs of electricity produced. The total yearly electricity production is 150 000 MW·h.

The total cost is

$$TC(x) = \$(2\,500\,000 + 41.5x - 0.0001x^2)$$
$$= \$[2\,500\,000 + (41.5)(150\,000) - (0.0001)(150\,000)^2]$$
$$= \$6\,475\,000$$

The average cost of the electricity produced is

$$AC(x) = \frac{\$6\,475\,000}{150\,000 \text{ MW} \cdot \text{h}} = \$43.17/\text{MW} \cdot \text{h}$$

The marginal cost, however, is

$$\frac{d(TC)}{dx} = \$\frac{d}{dx}(2\,500\,000 + 41.5x - 0.001x^2)/\text{MW} \cdot \text{h}$$
$$= \$(41.5 - 0.0002x)/\text{MW} \cdot \text{h}$$
$$= \$[41.5 - (0.0002)(150\,000)]/\text{MW} \cdot \text{h}$$
$$= \$11.50/\text{MW} \cdot \text{h}$$

The computed marginal cost indicates that it would cost an additional $11.50 to increase the output of the power plant by 1 MW·h per year from the present level of 150 000 MW·h.

2.3 Estimation

The estimation of future events resulting from actions taken at the present is obviously a fact of life for every individual, group, or organization. Family budgeting, weather forecasts, market forecasts of demand for consumer products, and predicting the annual national revenues from income taxation are only a few examples of the almost limitless number and variety of estimates that are made in personal lives and business. In this textbook, we are concerned specifically with estimation based on factors relevant to comparing alternative engineering projects and making a selection from these projects. The annual revenues or savings, the initial and annual recurring costs, the life of a project, and the salvage value of buildings and equipment that may be associated with a given project are rarely, if ever, known with certainty.

Many different terms pertain to the general subject of estimation. We will not attempt to enumerate and explain all the terms exhaustively; selected terminology will be given throughout the book as needed to explain the topics of a given chapter or section. Furthermore, in-depth study of estimation procedures and the accuracy of such estimated values pertains to mathematical statistics and probability theory, about which a vast literature exists.

Chapters 3 to 6 are concerned with comparing alternative projects when the estimated values for relevant factors are single-valued or point estimates. In such cases, the single-valued estimates are considered certain to occur. Moving a step toward realism, an interval estimate could be made for the value of a given factor such as annual costs. That is, both a high and low value or range might be estimated for the factor. Chapter 7 discusses factors that contribute to uncertainty in estimated values and considers deviations from single-valued or point estimates. Chapter 8 continues the discussion on decision making with the topics of risk and uncertainty, defining these terms, and proposing methodology for selecting from a set of feasible alternatives, each of which has uncertain outcomes.

Although it is difficult to state precisely in quantitative terms, there is a relationship between the accuracy of an estimate and the cost of making the estimate. Intuitively, the more detailed the information that is obtained as the basis for an estimate and the more mathematical precision that is exercised in calculating the estimate, the more accurate the estimate should be. However, as the level of detail increases, the cost involved in making the estimate also increases. Ostwald has conceptualized this notion by the function

$$C_T = C_M(D) + C_E(D)$$

where

C_T = the total cost of making the estimate, dollars

$C_M(D)$ = the functional cost of making the estimate, dollars

$C_E(D)$ = the functional cost of errors in the estimate, dollars.

This relationship is depicted in Figure 2.3. As more detailed estimates are made the cost of making an estimate increases, but the cost of errors made in decisions based on incorrect estimates decreases. The total cost of making an estimate reaches a minimum value when the amount of detail is in the vicinity of value D_1. Quantitatively defining the amount of detail is at best difficult and may be a practical impossibility. However, this concept of the total cost of an estimate varying with the amount of detail involved in making it is realistic and is important to the general subject of estimation. In the abbreviated discussion on cost estimation techniques that follows, it will be noted that the in-

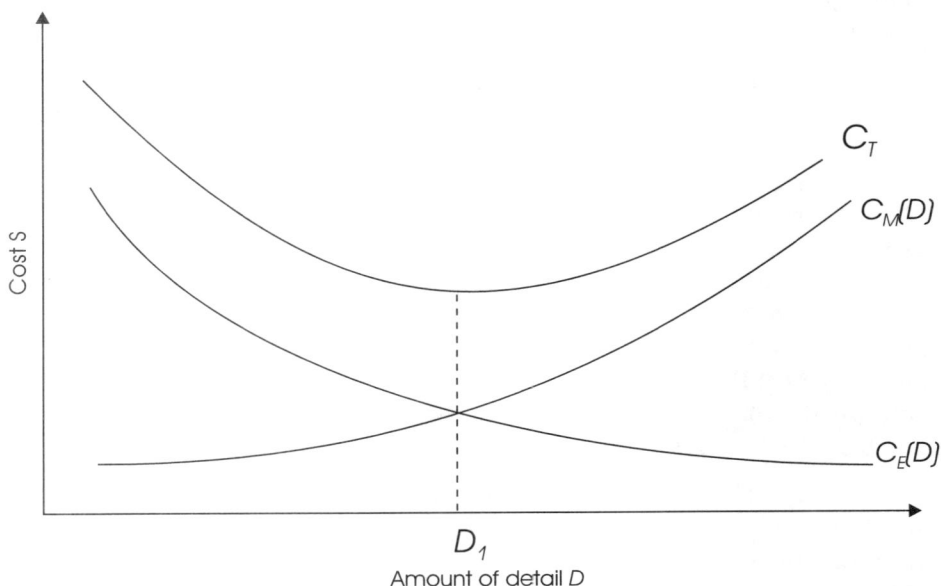

Fig. 2.3 Cost of increasing detail.

dividual techniques are based on varying amounts of detail with implied differences in the cost of making the estimate.

2.3.1 Project Estimation

For all categories of estimated items previously mentioned, three principal classes of estimates, based on accuracy and degree of detail, may be defined:

1. Order-of-magnitude estimates.
2. Preliminary estimates.
3. Detailed estimates.

Order-of-magnitude estimates are usually gross estimates based on experience and judgement and made without formal examination of the details involved. Accuracies of $\pm 50\%$, relative to actual cost, can be expected.

Preliminary estimates are also gross estimates, but more consideration is given to detail in making the estimate than for order-of-magnitude estimates. Certain sub-elements of the overall task are individually estimated, such as engineering specifications. Estimating the cost for manufacturing new components or products before designs and production plans are complete is an example of preliminary cost estimation. This type of estimate enables process engineering and product engineering groups to compare alternative designs or manufacturing methods and assess the economic impact of them. An accuracy level of about ±20% of actual cost can be expected with preliminary estimates.

Detailed estimates are expected to result in an accuracy level of ±5% of actual cost. In preparing the estimate, each sub-element of the overall task is considered, and an attempt is made to assign a realistic cost to the sub-element. Pricing a product or contract bidding usually involves detailed estimates of the costs.

Estimating the cost of a project may involve all of the above types of estimates. The time available to make the estimate, the information available, the experience and knowledge of the persons making the estimate, and the type and scope of the project are all factors that affect the desired accuracy of the project's cost estimate and the cost of making the estimate. When a comparison of alternative projects is based on future cash flows, the cost and revenue figures are necessarily estimates that will vary from actual cost and revenue figures.

In subsequent chapters, examples illustrating the principles of engineering economic analysis will primarily involve the following estimated items: first cost, operating costs, maintenance costs, revenues or savings, useful life, salvage value, disposal cost, income taxes, and cost of capital. These cost items were discussed previously in this chapter in the section on cost terminology (Section 2.2). Estimates of the functionally useful physical life of an item of equipment may be obtained from manufacturers and suppliers or, if a company repeatedly buys a particular item of equipment and records its life, an estimate of the functionally useful life of the item may be obtained from this record. For example, records may reveal that 10% of the items survive for four years, 20% survive for five years, 50% survive for six years, 15% survive for seven years, and the remaining 5% survive for eight years. The mean, or expected, life of this equipment is thus equal to $[(0.10)(4)+(0.20)(5)+(0.50)(6)+(0.15)(7)+(0.05)(8)]$ or 5.85 years.

In making an economic comparison of alternative projects, the analyst may be interested in only one particular period of time or planning period. This planning period may be (and often is) different from the functionally useful life of one or more of the projects under consideration. For example, a job contract may cover a five-year period when the functionally useful lives of the projects under consideration are 10 years. The planning period adopted might then be five years, and only the relevant cost and revenue figures for the five-year period are used in the comparison. Further discussion on this point is deferred until Chapter 4.

2.3.2 General Sources of Data

There are many sources of data, both internal and external, to make the various estimates required in comparing alternative investment projects. Examples of sources within a firm are sales records, production control records, inventory records, quality control records, purchasing department records, work measurement and other industrial engineering studies, maintenance records, and personnel records. Some or all of these records and perhaps, additional ones, are input to the accounting function of the firm, which then compiles various financial reports for management. A principal objective of the accounting system is to determine per-unit costs for direct materials, direct labour, and overhead involved in manufacturing a product or providing a service. Since overhead costs, by definition, cannot be allocated as direct charges to a given product or service, overhead costs are prorated among the various products or services by somewhat arbitrary methods, some of which are presented in Chapter 9. Thus, a caution is raised concerning the use of cost-accounting data for estimating the cost of overhead items associated with a project. Nevertheless, the accounting system can and usually does serve as an important, if not primary, internal source of detailed estimates on operating costs, maintenance costs, and material costs, among others.

Sources of data external to the firm may be grouped into two general classes: published information that is generally available and information (published or otherwise) available on request. Available published information includes the vast literature of trade journals, professional society journals, government publications, reference handbooks, other books, and technical directories. Information not generally available except by request includes many sources listed in the previous category. For instance, many professional societies and trade associations publish handbooks, other books, special reports, and research bulletins that are available on request. Manufacturers and distributors of equipment are excellent sources of technical data, and most will readily supply this information without charge. In addition, various government agencies, commercial banks (particularly holding companies involved in leasing buildings and equipment), and research organizations (commercial, governmental, industrial, and educational) may be sources of data to aid the estimating process.

2.4 Capital Cost Estimation

It is frequently necessary to estimate the costs of capital items such as machinery, equipment, plants, etc. In some instances, for example for bid preparation, detailed estimation is required. However, for many other purposes, for example for feasibility studies and for making a selection from alternative engineering or investment proposals, preliminary or order of magnitude estimates often suffice.

Past costs of capital items are often used to estimate their present costs for the purpose of preliminary or order of magnitude estimates. These past costs, however, have to be adjusted to take into consideration the effect of inflation and technological advancements. This adjustment is quickly accomplished with reasonable accuracy by using price indexes; for detailed discussion see Section 2.4.1.

For preliminary and order of magnitude estimates the cost-capacity relationship, discussed in Section 2.4.2, provides an easy and quick method.

2.4.1 Price Indexes

Cost data quickly become out of date because of inflation, and past cost items frequently have to be converted into present cost items for estimating purposes. This task is best accomplished with the help of cost indexes. The *cost index* is the relative cost of an item in terms of its cost in the base year. The base year index value is generally 100. For example, the base year of the Canadian Consumer Price Index, computed and published monthly by Statistics Canada, Prices Division, is 1992, with base year index value of 100. The Canadian Consumer Price Index stood, in July 1999, at 110.4, indicating that consumer goods which were purchased in January 1992 for $100 cost $110.4 in July 1999. The cost of an item in a particular year can be computed with the help of cost indexes using the equation:

$$C_X = C_B \frac{I_X}{I_B}$$

where

C_X = cost in year X
C_B = cost in year B
I_X = index value in year X
I_B = index value in year B

Cost indexes are calculated to reflect the change in average cost values, for example, the average cost of all rubber and plastic products. Specific cost values computed with the help of these cost indexes (e.g., the unit cost of 1.2 m wide, 6-ply cotton-nylon conveyor belt) are not accurate enough to be used in detailed estimates. There has always been a great deal of criticism of cost indexes because of their limited accuracy, but they still represent the fastest, most convenient means of obtaining preliminary estimates of costs.

The most frequently used indexes in Canada are the Consumer Price Index, the Industrial Production Price Index, and the Raw Materials Price Index, published monthly by Statistics Canada, Prices Division. A typical price index, illustrated in Figure 2.4, shows that the price of a

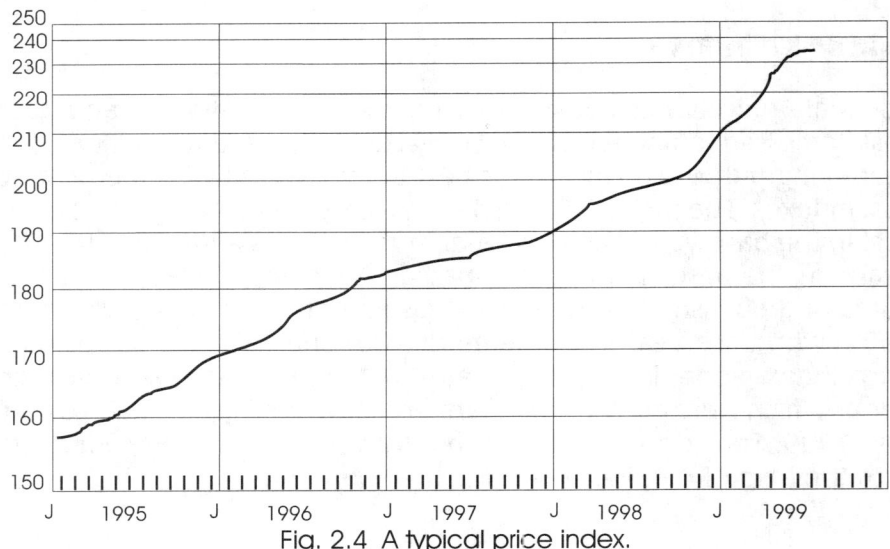

Fig. 2.4 A typical price index.

piece of equipment that cost $158 in January 1995 would be about $235 in August 1999.

Statistics Canada, Prices Division publishes separate price indexes for all major industrial products: Food, Rubber and Plastic Products, Leather, Wood, Paper, Primary Metal, Machinery, Electrical Products, Petroleum and Coal Products, and Chemical Products, etc. It also gives price indexes for industrial materials, raw materials, and electric power.

Other frequently used indexes are the Engineering News-Record Construction and Building Index, the Marshall and Swift's All-Industry Index, the Marshall and Swift's Process Industry Index, the Nelson Refinery Index, the Chemical Engineering Magazine Cost Index, and the Engelsman's General Construction Cost Guide. The Engineering News-Record Construction Index reflects the change in wage rates and material costs for a specified block of industrial construction. The Marshall and Swift's All-Industry Index is the arithmetic average of over forty different types of industrial indexes. The Marshall and Swift's Process Industry Index is the weighted average of eight industrial indexes. The Nelson Refinery Index shows the cost of refinery construction representing 40% material and 60% labour costs.

In the example below, a Construction Cost Index is used to estimate the cost values of constructing a factory building.

Example 2.2

A factory building was constructed in 1983 at a cost of $2 400 000. If a similar building is to be built in 2000, what would be its estimated cost?

Using the relevant portion of a Construction Cost Index shown in Table 2.1 and the formula $C_X = C_B \dfrac{I_X}{I_B}$, the estimated cost of the building in 2000 can be calculated as follows:

$$C = \$2\,400\,000 \dfrac{5.83}{1.79} = \$7\,816\,760$$

This value is, of course, only a preliminary estimate of the building cost. In order to get a more accurate value, a detailed estimate must be made.

Table 2.1 A General Construction Cost Index			
Year	Cost Index	Year	Cost Index
1981	1.67	1991	3.48
1982	1.73	1992	3.80
1983	1.79	1993	4.21
1984	1.92	1994	4.63
1985	2.11	1995	5.00
1986	2.23	1996	5.21
1987	2.53	1997	5.31
1988	2.79	1998	5.45
1989	3.03	1999	5.62
1990	3.21	2000	5.83

2.4.2 Cost-Capacity Relationship

A specific relationship may exist between the cost of a unit of equipment and its capacity (output). If the cost is plotted on a log-log diagram against the capacity (output) of a unit of equipment such as small steam turbines (see Figure 2.5), it is frequently found that the best curve fitted to the data is a straight line. Equipment cost values can be obtained directly from this diagram or statistical curve fitting methods (such as least square analysis) may be used to find an empirical *cost-capacity relationship* from the available data. The empirical cost-capacity relationship usually takes the form of

$$C_X = C_B \left(\frac{Q_X}{Q_B} \right)^M$$

where

C_X = cost of equipment of capacity Q_X
C_B = cost of equipment of capacity Q_B
M = cost capacity factor

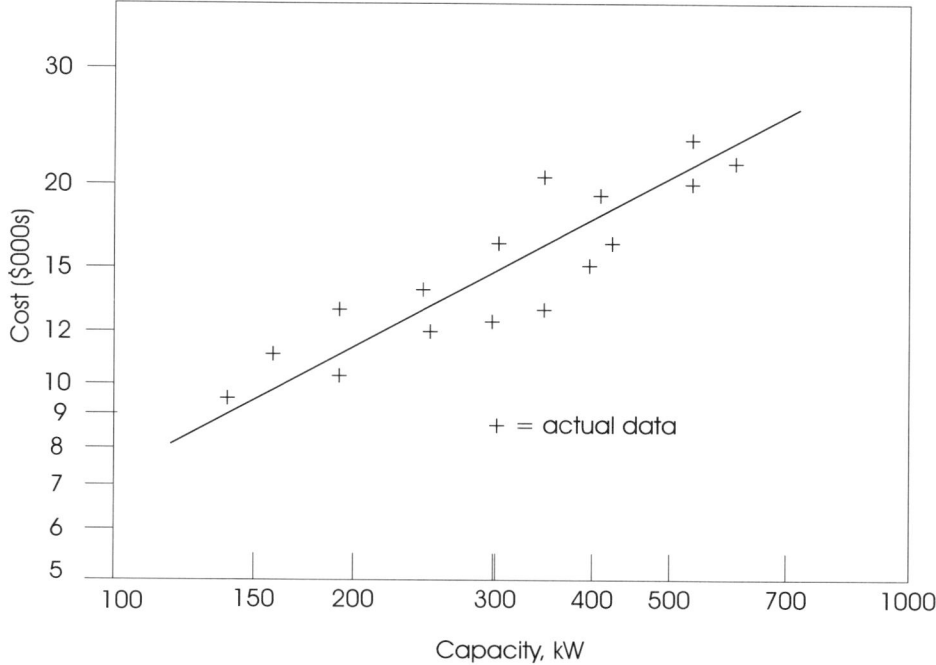

Fig 2.5 Cost capacity relationship for small steam turbines.

The value of the cost-capacity factor is generally in the interval of $0 \leq M \leq 1$, and its most frequently used value is 0.6. Preliminary estimates of the cost of a major item of equipment are often obtained using this empirical equation, provided that costs of similar equipment of different capacity are available. However, this method of estimation is generally not acceptable for the purpose of preparing detailed estimates, and of course this method of estimation is used only if more accurate estimates or bids cannot be obtained from suppliers or manufacturers.

The following example illustrates preliminary cost estimating with the help of the cost-capacity relationship.

Example 2.3

An automatic electric furnace is required for heat-treating 2 000 turbine blades per day. What is the estimated cost of this furnace if the cost of a recently installed similar furnace with a heat-treating capacity of 1 000 blades per day is $1 200 000, and the cost-capacity factor for this kind of installation is 0.6?

The estimated cost of the new furnace is

$$C = \$1\,200\,000 \left(\frac{2\,000}{1\,000}\right)^{0.6}$$

$$= \$1\,818\,860$$

It should be noted that this value is only a preliminary estimate. If more accurate information about the cost is required, a detailed estimate must be made using data other than that provided by this cost-capacity relationship.

2.5 The Learning Curve

The amount of direct labour hours required to make a product and the number of rejected parts produced (important components of the variable production cost) are normally reduced in time as:

1. Workers gain experience and learn to perform their tasks more efficiently.

2. Workers use their machinery more effectively.

3. Management improves the organization of the production process.

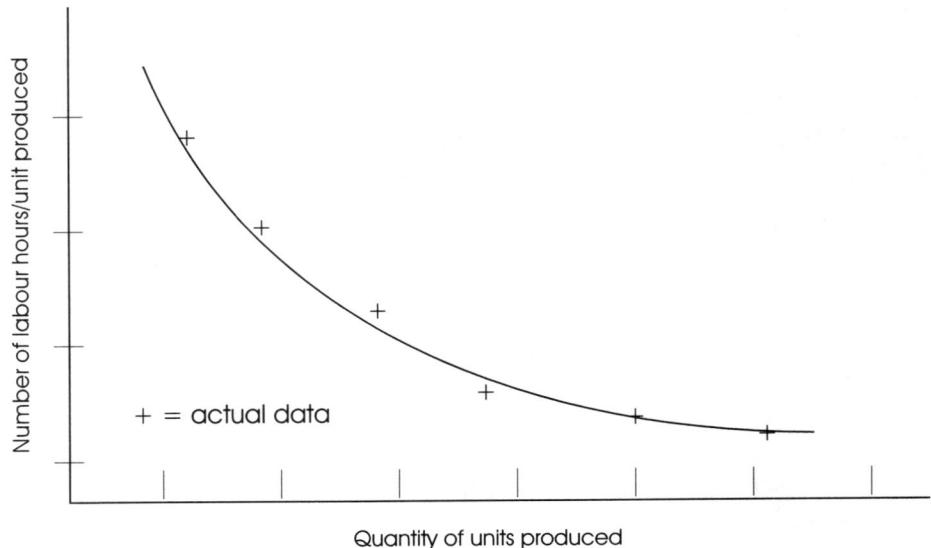

Fig 2.6 The learning curve.

Cost Concepts and Cost Estimation

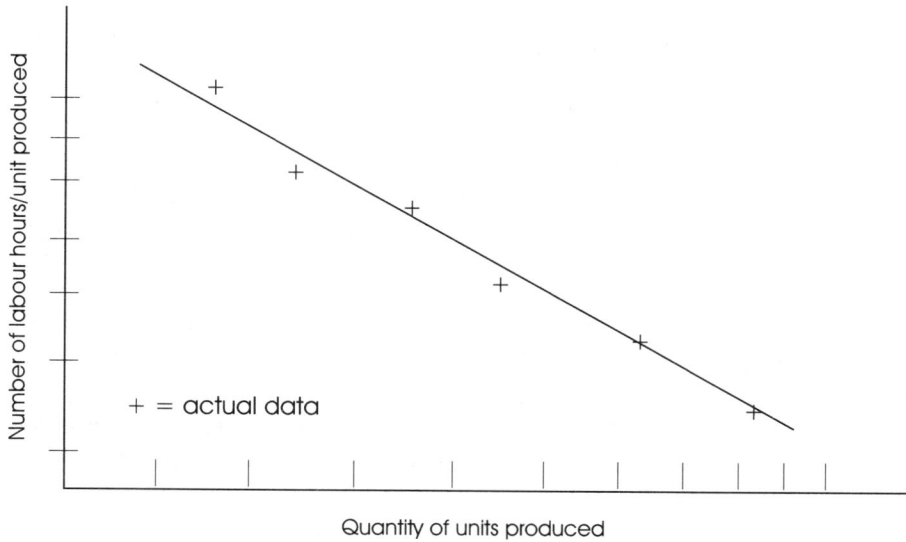

Fig 2.7 The learning curve on a log-log diagram.

4. Engineering introduces technological improvements (new tooling, process modifications, etc.).

Figure 2.6 shows the number of labour hours required to overhaul an aircraft jet engine. This diagram is typical for all repetitive manufacturing operations and reflects the effect of experience or learning. If the data of Figure 2.6 is plotted on a log-log diagram, it can be approximated with reasonable accuracy by a straight line, see Figure 2.7. The straight-line relationship on a log-log diagram indicates that an exponential relationship exists between the quantity of the product produced and the labour hours required to make it.

$$E_n = k n^s$$

where

E_n = labour hours required to make the nth product
k = labour hours required to make the first product
n = product number
s = learning curve experiment

The values of k and s are different for different production operations and their values have to be determined in each particular case. Once the values of k and s are known, the learning curve equation can be used to estimate future production costs quickly.

To determine the values of k and s for a given production process, at least two or possibly more values of data (corresponding values of n and E_n) are required. If only two values of data are available, k and s are simply determined by substituting the available data into the learning curve relation and thus obtaining two equations and solving the two equations for k and s. If more than two values of data are available, then there are two alternative methods to determine the learning-curve relation. Either the available data is plotted on a log-log diagram, a straight line is fitted between the points representing the data, and the equation of this straight line—the learning-curve equation—is determined; or, the equation of the best straight line representing the data can be determined by using statistical methods (the least square analysis). Obviously, if more data is available, a more accurate learning curve will be obtained.

Example 2.4

A company begins to manufacture newly designed cement cooling towers. The cooling towers consist of a vertical steel shell with outside cooling channels, a spiral rotating auger, a drive unit, cooling water and electrical connections, sensors, and a control system. Six prototype cooling towers are made before commercial production commences to test the product as well as the production process.

The labour hours required to make the third and the sixth cooling towers were 650 hours and 590 hours respectively.

Based on this information, determine the learning curve for manufacturing this cooling tower.

$$E_3 = 640 = k\,3^s$$

and

$$E_6 = 590 = k\,6^s$$

by division

$$\frac{640}{590} = \left(\frac{3}{6}\right)^s$$

Cost Concepts and Cost Estimation

and

$$\ln\left(\frac{640}{590}\right) = s \ln\left(\frac{3}{6}\right)$$

so

$$s = -0.1174$$

and

$$k = 728.1$$

Example 2.5

Based on the learning curve determined in Example 2.4, estimate the labour required to manufacture the 12th cooling tower.

$$E_{12} = k12^s$$

For this cement cooling tower manufacturing process $k = 718.07$ and $s = -0.1174$, see Example 2.4, therefore

$$E_{12} = (718.1)12^{-0.1174} = 543.9 \text{ hours}$$

Example 2.6

After making six prototype units, commercial production of the cooling towers described in Example 2.4 commences. Based on the learning curve for cooling towers determined in Example 2.4, estimate the labour hours required to manufacture the first 100 units (from $n = 7$ to $n = 106$).

$$E_T = E_7 + \cdots + E_{106} = \sum_{7}^{106} E_n = k \sum_{7}^{106} n^s$$

This sum can be approximated with the following integral

$$E_T \cong k \int_{6.5}^{106.5} n^s \, dn = k$$

For this cement cooling tower manufacturing process $k = 718.1$ and $s = -0.1174$, from Example 2.4, therefore

$$E_T = 728.1 \left[\frac{106.5^{-0.1174}}{-0.1174+1} - \frac{6.5^{-0.1174}}{-0.1174+1} \right]$$
$$= 46\,484 \text{ hours}$$

and

$$E_{ave} = \frac{46\,484}{100} = 464.8 \text{ hours.}$$

2.6 Standard Costs

Although the first task of cost accounting is to determine per-item or per-order costs, another major purpose of cost accounting is to interpret financial data so that management can (1) measure changes in production efficiency and (2) judge the adequacy of production performance. Establishing cost standards can be of great assistance in achieving these objectives. A standard-cost system involves—in advance of manufacture—(1) the preparation of standard rates for material, labour, and overhead and (2) the application of these rates to the standard quantities of material and labour required for a job order or for each production operation required to complete the job order.

Since a process-type manufacturing firm, such as an oil refinery, outputs the same product (or a few products) over a long time, cost standards are more readily determined for process firms than for job-shop firms, where the variety of output is large and varies with customer order. However, the number and type of production opertions required to complete various job orders are finite for a given manufacturing firm. Each job order is, of course, made up of single units. Thus, a standard amount of material can be determined for each unit, and standard labour times and machine times can be determined for each unit. It is usually the responsibility of the work measurement function within the firm to determine these standard quantities. Then, by applying standard unit material costs and standard labour rates, unit costs for material, labour, and overhead can be determined. The standard costs then serve as a basis for measuring production efficiency and performance over time. Deviations from standard costs may be caused by several factors, especially (1) raw material price variations and (2) actual quantities of material and labour used versus the standard amounts of these items. This latter factor is the primary concern in determining production efficiency and performance, measures of which provide information to management to aid in cost control.

Problems

1. The total yearly cost of operating a steel rolling mill is

    $$TC(x) = \$(3\,000\,000 + 50x)$$

    where x is the output of the mill measured in tonnes per year. The yearly output of the mill is 700 000 t.

    Determine:

    (a) the yearly fixed cost of the operation,
    (b) the variable unit cost,
    (c) the total cost (2.2.5),
    (d) the average cost (2.2.6).

2. The total annual cost of pumping drinking water for a small community is

    $$TC(x) = \$(200\,000 + 0.8x^{1.3})$$

    where x is water measured in cubic metres per year.

    Determine:

    (a) the average cost of water if the daily consumption is 4 000 m^3,
    (b) the marginal cost at a daily consumption of 4000 m^3,
    (c) the change in unit cost as the consumption increases from 4 000 m^3 to 5000 m^3 per day (2.2.6).

3. The cost of a numerically controlled machining centre is $240 000. The concrete foundation and the electrical connections for the machine cost $18 000. A number of special holding fixtures and tools are purchased for the machine for $36 500. A training course organized by the machine supplier for machine operators and programmers costs $12 500. Preventive and corrective maintenance costs are estimated to be $2 500 per year. Operating costs (labour) are expected to be $48 000 per year. The machine will be used for five years and then sold for $55 000.

    Determine the life-cycle cost of the machine (2.2.1).

4. Durex Industries manufactures a strapping device for packaging machines. The direct labour and material costs are $32.80 and $16.70 respectively per unit of product. Factory

overhead costs are calculated as 65% of the direct labour cost. General, administrative, marketing, and selling costs are determined as 120% of the prime costs. The profit is 25% of the selling price.

Determine:

(a) the cost of the strapping device,

(b) the conversion cost, and,

(c) the selling price (2.2.4).

5. Determine:

(a) the cost-capacity relationship for overhead cranes based on the following information:

Crane Capacity, tonnes	1	2	5	10
Cost, $000s	20	30	50	80

Estimate the cost of:

(b) a 4 t crane,

(c) a 20 t crane (2.3.3).

6. Estimate the April 2000 cost of a 50 MW turbine on the basis of the following information:

Capacity, MW	2	5	20	120
Cost, $10 000s (in May 1994)	15	23	45	110

The cost index values for turbines are 116 in May 1994 and 156 in April 2000.

7. The cost of a tundish used for transporting molten steel from the furnace to a continuous billet casting machine is a function of the capacity of the tundish. The company purchased five different sized tundishes in the past. The costs of these tundishes are given in the Table below.

Cost Concepts and Cost Estimation

Capacity, tonnes	10	20	40	50	100
Cost, $000s	125	203.5	330	385	426

    (a) Plot the data on a log-log diagram and determine the cost-capacity relation.

    (b) Use the least square method to determine the cost-capacity relation from the data given in the Table.

    (c) Use the cost-capacity relation determined above to estimate the cost of a 200-tonne capacity tundish (2.4.2).

8. The cost of a recently built 200 m long conveyor capable of transporting coal at the rate of 200 tonnes/hour is $240 000.

    (a) Estimate, based on the information given above, the cost of a similar 200 m long conveyor capable of transporting 300 tonnes of coal per hour.

    (b) Using additional information it was determined that the exponent of the cost-capacity relation for this type of conveyor is 0.68. Refine your previous estimate for the 300 tonnes/hour capacity conveyor (2.4.2).

9. Your company intends to purchase a 500-tonne capacity injection moulding machine. A similar machine was purchased in July 1992 for $89 500. The price index for heavy duty equipment for the plastic industry was 128.9 in July 1992 and it is 163.6 at present.

Estimate:

    (a) the current cost of the machine (2.4.1),

    (b) the current cost of a 2 000-tonne capacity machine (2.4.2).

10. Apex Industries starts producing 100 heavy duty inconnel turbine shafts on a horizontal boring mill. The labour requirements for the first and second shafts are 68.5 and 61.0 hours respectively.

    (a) Determine the learning-curve equation for this process.

    (b) Estimate the labour requirement for the 10th unit.

    (c) Estimate the total labour requirements for the first 100 units ($n=1$ to $n=100$) (2.5).

11. A special-purpose welding robot is manufactured by Robodex Ltd. The direct labour hours required to produce the first five units are given below.

Number	1	2	3	4	5
Labour hours	620	572	541	525	511

    (a) Plot the data on a log-log diagram and determine the learning-curve relation for this manufacturing process

    (b) Determine the learning-curve relation using least square analysis (2.5).

12. The learning-curve equation for direct labour for making a special-purpose welding maching was determined during prototype production of the first five units. It was found that $k = 240$ hours and $s = -0.09$. The labour cost is $22.00 per hour.

    Estimate the average direct labour cost of producing the next 300 welding machines (2.5).

13. Weldex Industries begins to manufacture a newly designed, small, spot welding machine. The labour hours required for making the second and third units are 28.6 and 28.2 hours.

    Estimate the labour required to manufacture the 50th unit (2.5).

Chapter 3
Time Value of Money Operations

3.1 Introduction

Engineering alternatives are normally compared by using a host of different criteria, including system performance and economic performance. The system performance characteristics of primary importance are quality, reliability, safety, and customer service. Economic performance characteristics normally considered are initial investment requirements, return on investment, and the cash flow profile. Since the cash flow profiles usually differ among the several alternatives, in order to compare the economic performances of the alternatives, one must compensate for the differences in the timing of cash flows.

As discussed in Chapters 1 and 2, the concept of the *time value of money* is fundamental in the comparison of the economic performances of alternatives. In this chapter we examine a number of mathematical operations that are based on the time value of money, with an emphasis on modelling cash flow profiles.

For the sake of ease and familiarity, we will look at the time value of money within the context of personal finance. However, all of the concepts examined in this chapter can also be applied to business.

3.2 Interest Calculations

In considering the time value of money, it is convenient to represent mathematically the relationship between the current or *present* value of a single sum of money and its *future* value. Letting time be measured in years, if a single sum of money has a current or *present* value of P, its value in n years would be equal to

$$F_n = P + I_n$$

where F_n is the accumulated value of P over n years, or the *future* value of P, and I_n is the increase in the value of P over n years. I_n is referred to as the accumulated *interest* in borrowing and lending transactions and is a function of *P, n,* and the *annual interest rate, i.* The annual interest rate is defined as the change in value for $1 over a one-year period.

Over the years, two approaches have emerged for computing the value of I_n. The first approach considers I_n to be a linear function of time. Since *i* is the rate of change over a one-year period, it is argued that *P* changes in value by an amount of *Pi* each year. Hence,

$$I_n = Pin$$

and

$$F = P(1+in)$$

This is called the *simple interest* approach.

The second approach used to compute the value of I_n is to interpret *i* as *the rate of change in the accumulated value of money.* Hence, it is argued that the following relation holds,

$$I_n = iF_{n-1}$$

and

$$F_n = F_{n-1}(1+i)$$

This approach is referred to as the *compound interest* approach.

The approach to be used in any particular situation depends on how the interest rate is defined. Since practically all monetary transactions are based currently on compound interest rates instead of on simple interest rates, we will assume compounding occurs unless otherwise stated.

A convenient method of representing the time value of money is to visualize positive and negative cash flows as though they were generated by a borrower and a lender. In particular, suppose you loaned $1 000 to an individual who agreed to pay you interest at a rate of 15% per year. The $1 000 is referred to as the *principal* amount. At the end of one year, you would receive $1 150 from the borrower. Thus, we might say that $1 150 one year from now is worth $1 000 today based on 15% interest rate or, conversely, we might say that $1 000 today has a value of $1 150 one year from now based on a 15% interest rate.

Whenever the interest charge for any time period is based only on the unpaid principal amount and not on any accumulated interest charges, simple interest calculations apply. The interest due for a given interest period does not convert to principal for the purpose of calculating the interest amount due in the subsequent interest period. In the preceding illustration, if the individual borrowed the $1 000 for two years, the amount owed at the end of two years would be the $1 000 principal plus the interest charges of (0.15)($1 000) = $150 each year, or $1 300 total.

If the individual borrowed $1 000 for two years and interest was compounded, the borrower would owe $1 150 at the end of the first year and $1 322.50 at the end of the second year, since the second yearly interest on the $1 150 owed at the end of the first year is (0.15)($1 150) = $172.50. The $1 322.50 for the compounding case compares with the $1 300 for the simple interest case. (Often the difference is much more dramatic!)

Compound interest involves the computation of interest charges during a time period based on the unpaid principal amount plus any accumulated interest charges up to the beginning of the time period. The interest amount due for a given interest period converts to principal for the purpose of calculating the interest amount due in the subsequent interest period. The relationship between the principal amount and the compounding of interest is given in Table 3.1.

Table 3.1 Illustrating the Effect of Compound Interest

End of Period	(A) Amount Owed	(B) Interest for Next Period	(C) = (A) + (B) Amount Owed for Next Period*
0	P	Pi	$P + Pi = P(1 + i)$
1	$P(1 + i)$	$P(1 + i)i$	$P(1 + i) + P(1 + i)i = P(1 + i)^2$
2	$P(1 + i)^2$	$P(1 + i)^2 i$	$P(1 + i)^2 + P(1 + i)^2 i = P(1 + i)^3$
3	$P(1 + i)^3$	$P(1 + i)^3 i$	$P(1 + i)^3 + P(1 + i)^3 i = P(1 + i)^4$
⋮	⋮	⋮	⋮
$n - 1$	$P(1 + i)^{n-1}$	$P(1 + i)^{n-1} i$	$P(1 + i)^{n-1} + P(1 + i)^{n-1} i = P(1 + i)^n$
n	$P(1 + i)^n$		

* Notice, the value in column (C) for the end of period $(n - 1)$ provides the value in column (A) for the end of period n.

Example 3.1

Person A borrows $4 000 from the Royal Bank and agrees to pay $1 000 plus accrued interest at the end of the first year and $3 000 plus the accrued interest at the end of the fourth year. What are the amounts for the two payments if 18% annual simple interest applies?

For the first year, the payment is

$$\$1\,000 + (0.18)(\$4\,000) = \$1\,720$$

For the fourth year, the payment is

$$\$3\,000 + (0.18)(\$4\,000 - \$1\,000)(3) = \$4\,620$$

3.3 Single Sums of Money

To illustrate the mathematical operations involved in modelling cash flow profiles using compound interest, first consider the investment of a single sum of money, P, in a savings account for n interest periods. Let the interest rate per interest period be denoted by i and let the accumulated total in the fund n periods in the future be denoted by F. As shown in Table 3.1, assuming no monies are withdrawn during the interim, the amount in the fund after n periods equals $P(1+i)^n$. As a convenience in computing values of F (the future worth) when given values of P (the present worth), the quantity $(1+i)^n$ is tabulated in Appendix A for various values of i and n. The quantity $(1+i)^n$ is referred to as the *single sum, future worth factor* and is denoted $(F|P\ i,n)$. The expression $(F|P\ i,n)$ is read as the F, given P factor at $i\%$ for n periods. The above discussion is summarized as follows:

Let P = the equivalent value of an amount of money at time zero, or present worth

F = the equivalent value of an amount of money at time n, or future worth

i = the interest rate per interest period

n = the number of interest periods

Thus, the future worth is related to the present worth as follows:

$$F = P(1+i)^n \qquad (3.1)$$

or equivalently,

$$F = P(F|P\ i,n) \qquad (3.2)$$

A cash flow diagram depicting the relationship between F and P is given in Figure 3.1. Remember that F occurs n periods after P. (Note that upward arrows represent cash inflows and downward arrows represent cash outflows.)

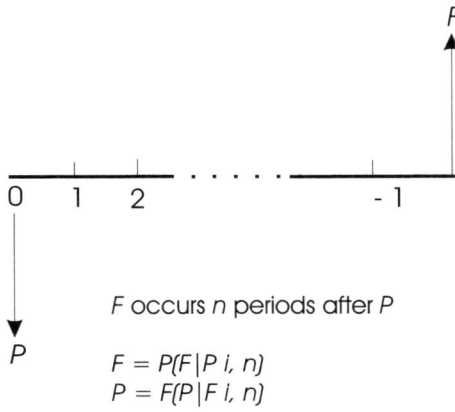

Fig. 3.1 Cash flow diagram of the time relationship between P and F.

Example 3.2

An individual borrows $1 000 at 6% compounded annually. The loan together with interest is paid back after five years. How much should be repaid?

Using the compound interest tables in Appendix A for 6% and five periods, the value of the $F = P(F|P\ 6,5)$ factor is found to be 1.3382. Thus,

$$\begin{aligned} F &= P(F|P\ 6,5) \\ &= \$1\,000(1.3382) \\ &= \$1\,338.20 \end{aligned}$$

The amount to be repaid equals $1 338.20. One could, of course, obtain the same result without using the table in Appendix A, simply by computing the repayment as

$$F = \$1\,000(1+0.06)^5 = \$1\,338.20,$$

or by using a financial or business calculator.

Since we are able to determine conveniently values of F when given values of P, i, and n, it is a simple matter to determine values of P when given values of F, i, and n. In particular, since

$$F = P(1+i)^n$$

on dividing both sides by $(1+i)^n$, we find that the present worth and future worth have the relation

$$\boxed{P = F(1+i)^{-n}} \quad (3.3)$$

or

$$\boxed{P = F(P|F\ i,n)} \quad (3.4)$$

where $(1+i)^n$ and $F(P|F\ i,n)$ are referred to as the *single sum, present worth factor*.

Example 3.3

To illustrate the computation of P given F, i, and n, suppose you wish to accumulate $2 000 in a savings account two years from now, and the account pays interest at a rate of 6% compounded annually. How much must be deposited today?

$$P = F(P|F\ 6,2)$$
$$= \$2\,000(0.8900)$$
$$= \$1\,780.00$$

3.4 Series of Cash Flows

Having considered the transformation of a single sum of money to a future worth equivalent when given a present worth amount and vice versa, we generalize that discussion to consider the conversion of a series of cash flows to present worth and future worth equivalents. In particular, let A_k denote the magnitude of a cash flow (receipt or disbursement) at the end of time period k. Using discrete compounding, the present worth equivalent for the cash flow series is equal to the sum of the present worth equivalents for the individual cash flows. Consequently,

$$P = A_1(1+i)^{-1} + A_2(1+i)^{-2} + \cdots + A_{n-1}(1+i)^{-(n-1)} + A_n(1+i)^{-n}$$

(3.5)

or, using the summation notation,

$$P = \sum_{k=1}^{n} A_k (1+i)^{-k}$$

(3.6)

or, equivalently,

$$P = \sum_{k=1}^{n} A_k (P|F\ i,k)$$

(3.7)

Example 3.4

Consider the series of cash flows depicted by the cash flow diagram given in Figure 3.2. Using an interest rate of 6% per interest period, the present worth equivalent is given by

$$P = \$300(P|F\ 6,1) - \$300(P|F\ 6,3) + \$200(P|F\ 6,4)$$
$$+ \$400(P|F\ 6,6) + \$200(P|F\ 6,8)$$
$$= \$300(0.9434) - \$300(0.8396) + \$200(0.7921)$$
$$+ \$400(0.7050) + \$200(0.6274)$$
$$= \$597.04$$

The future worth equivalent is equal to the sum of the future worth equivalents for the individual cash flows. Thus,

$$F = A_1(1+i)^{n-1} + A_2(1+i)^{n-2} + \cdots + A_{n-1}(1+i) + A_n$$

or, using the summation notation,

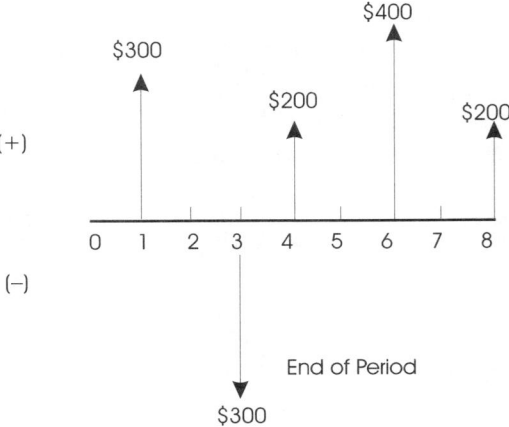

Fig. 3.2 Series of cash flows.

$$F = \sum_{k=1}^{n} A_k (1+i)^{n-k} \tag{3.9}$$

$$F = \sum_{k=1}^{n} A_k (F|P\,i, n-k) \tag{3.10}$$

Alternately, since we know the value of future worth is given by

$$F = P(1+i)^n \tag{3.11}$$

substituting Equation 3.6 into Equation 3.11 yields

$$F = (1+i)^n \sum_{k=1}^{n} A_k (1+i)^{-k}$$

Hence,

$$F = \sum_{k=1}^{n} A_k (1+i)^{n-k} \tag{3.12}$$

For the example under consideration, the worth of the series at time $t = 8$ is given by

$$F = \$300(F|P\ 6,7) - \$300(F|P\ 6,5) + \$200(F|P\ 6,4)$$
$$+ \$400(F|P\ 6,2) + \$200$$
$$= \$300(1.5036) - \$300(1.3382) + \$200(1.2625)$$
$$+ \$400(1.1236) + \$200$$
$$= \$951.56$$

Alternately, since we know the present worth is equal to $597.04

$$F = P(F|P\ 6,8)$$
$$= \$597.04(1.5938)$$
$$= \$951.56$$

Note that in computing the future worth of a series of cash flows, *the future worth amount obtained occurs at the end of time period n.* Thus, if a cash flow occurs at the end of time period n, it earns no interest in computing the future worth amount at the end of the nth period.

Obtaining the present worth and future worth equivalents of cash flow series by summing the individual present worths and future worths, respectively, can be quite time consuming if many cash flows are included in the series. However, with the availability of financial or business calculators, programmable pocket calculators, and desk computers, it is not uncommon to treat all series in the manner described above.

Nevertheless, it is possible to use more efficient solution procedures if the cash flow series have one of the following forms:

(i) Uniform Series of Cash Flows

$$A_k = A \qquad\qquad k = 1,\ldots, n$$

(ii) Gradient Series of Cash Flows

$$A_k = \begin{cases} 0 & k=1 \\ A_{k-1} + G & k = 2,\ldots,n \end{cases}$$

(iii) Geometric Series of Cash Flows

$$A_k = \begin{cases} A & k=1 \\ A_{k-1}(1+j) & k = 2,\ldots,n \end{cases}$$

3.4.1 Uniform Series of Cash Flows

A uniform series of cash flows exists when all of the cash flows in a series are equal. In the case of a uniform series, the present worth equivalent is given by

$$P = \sum_{k=1}^{n} A(1+i)^{-k} \tag{3.13}$$

where *A* is the magnitude of an individual cash flow in the series.

Letting $X = (1+i)^{-1}$ and bringing *A* outside the summation yields

$$P = A \sum_{k=1}^{n} X^k$$

$$= AX \sum_{k=1}^{n} X^{k-1}$$

Letting $h = k - 1$ gives the geometric series

$$P = AX \sum_{h=0}^{n-1} X^h \tag{3.14}$$

Since the summation in Equation 3.14 represents the first *n* terms of a geometric series, the closed form value for the summation is given by

$$\sum_{h=0}^{n-1} X^h = \frac{1 - X^n}{1 - X} \tag{3.15}$$

Hence, on substituting Equation 3.15 into Equation 3.14, we obtain

$$P = AX \left(\frac{1 - X^n}{1 - X} \right)$$

Replacing *X* with $(1+i)^{-1}$ yields the following relationship between *P* and *A*:

Time Value of Money Operations

$$P = A\left[\frac{(1+i)^n - 1}{i(1+i)^n}\right]$$

(3.16)

more commonly expressed as

$$P = A(P|A\ i,n)$$

(3.17)

where $(P|A\ i,n)$ is referred to as the *uniform series, present worth factor* and is tabulated in Appendix A for various values of i and n.

Example 3.5

An individual wishes to deposit a single sum of money in a savings account so that five equal annual withdrawals of $2 000 can be made before depleting the fund. If the first withdrawal is to occur one year after the deposit and the fund pays interest at a rate of 7% compounded annually, how much should be deposited?

Because of the relationship of P and A, as depicted in Figure 3.3, in which P occurs one period before the first A, we see that

$$P = A(P|A\ 7,5)$$
$$= \$2\ 000(4.1002)$$
$$= \$8\ 200.20$$

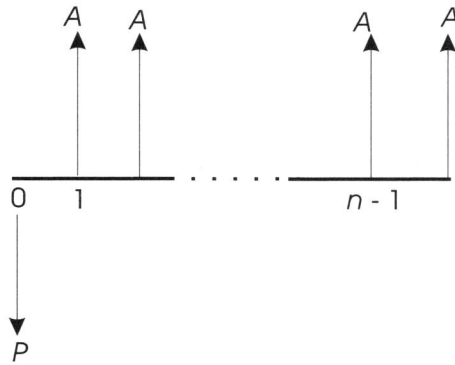

Fig. 3.3 Cash flow diagram of the relationship between P and A.

Thus, if $8 200.20 is deposited in a fund paying 7% compounded annually, then five equal annual withdrawals of $2 000 can be made.

Example 3.6

In Example 3.5, suppose that the first withdrawal will not occur until three years after the deposit.

As depicted in Figure 3.4, the value of P to be determined occurs at $t = 0$, whereas a straightforward application of the $(P|A\ 7,5)$ factor will yield a single sum equivalent at $t = 2$. Consequently, the value obtained at $t = 2$ must be moved backward in time to $t = 0$. The latter operation is easily performed using the $(P|F\ 7,2)$ factor. Therefore,

$$P = A(P|A\ 7,5)(P|F\ 7,2)$$
$$= \$2\ 000(4.1002)(0.8734)$$
$$= \$7\ 162.23$$

Deferring the first withdrawal for two years reduces the amount of the deposit by $8 200.40 − $7 162.23 = $1 038.17.

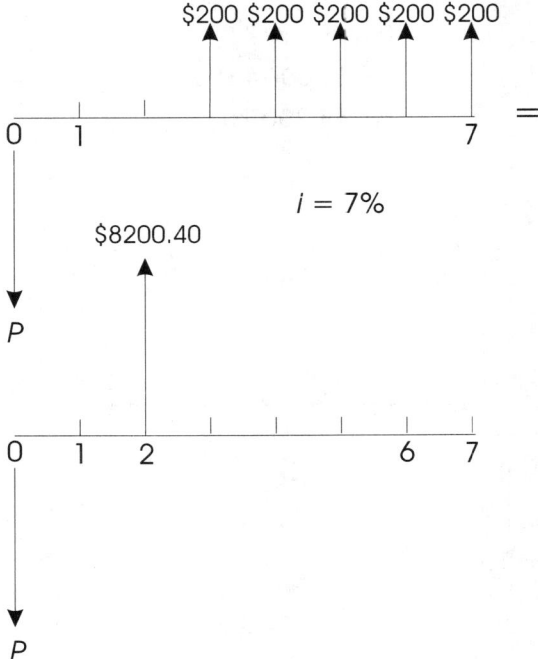

Fig. 3.4 Equivalent cash flow diagrams.

Time Value of Money Operations

The reciprocal relationship between P and A can be expressed as

$$A = P\left[\frac{i(1+i)^n}{(1+i)^n - 1}\right] \qquad (3.18)$$

or as

$$A = P(A|P\ i,n) \qquad (3.19)$$

The expression $(A|P\ i,n)$ is called the *capital recovery factor* for reasons that will become clear in Chapter 4. The $(A|P\ i,n)$ factor is used frequently in both personal financing and in comparing investment alternatives.

Example 3.7

Suppose $10 000 is deposited into an account that pays interest at a rate of 7% compounded annually. If 10 equal, annual withdrawals are made from the account, with the first withdrawal occurring one year after the deposit, how much can be withdrawn each year in order to deplete the fund with the last withdrawal?

Since we know that A and P are related by

$$A = P(A|P\ i,n)$$

then

$$A = P(A|P\ 7,10)$$
$$= \$10\,000(0.1424)$$
$$= \$1\,424$$

Example 3.8

Suppose that in Example 3.7 the first withdrawal is delayed for two years, as depicted in Figure 3.5. How much can be withdrawn each of the 10 years?

The amount in the fund at $t = 2$ equals

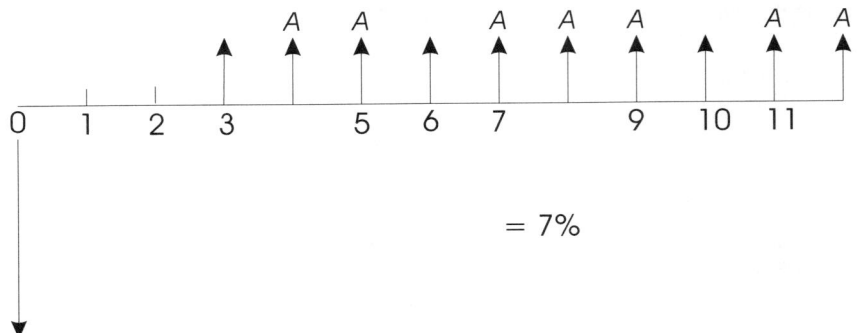

Fig. 3.5 Cash flow diagram of deferred payment example.

$$V_2 = P(F|P\ 7,2)$$
$$= \$1\,000(1.1449)$$
$$= \$11\,449$$

Therefore, the size of the equal annual withdrawals will be

$$A = V_2(A|P\ 7,10)$$
$$= \$11\,449(0.1424)$$
$$= \$1\,630.34$$

Thus, delaying the first withdrawal for two years increases the size of each withdrawal by $206.34.

The future worth of a uniform series is obtained by recalling that

$$F = P(1+i)^n \qquad (3.20)$$

By substituting Equation 3.16 into Equation 3.20 for P and reducing, one obtains

$$\boxed{F = A\left[\frac{(1+i)^n - 1}{i}\right]} \qquad (3.21)$$

or equivalently,

Time Value of Money Operations

$$F = A(F|A\ i,n) \qquad (3.22)$$

where $(F|A\ i,n)$ is referred to as the *uniform series, future worth factor*.

Example 3.9

If annual deposits of $1 000 are made into a savings account for 30 years, how much will be in the fund immediately after the last deposit if the fund pays interest at a rate of 8% compounded annually?

$$F = A(F|A\ 8,30)$$
$$= \$1\,000(113.2831)$$
$$= \$113\,283.10$$

The reciprocal relationship between A and F is easily obtained from Equation 3.21. Specifically, we find that

$$A = F\left[\frac{i}{(1+i)^n - 1}\right] \qquad (3.23)$$

or, equivalently,

$$A = F(A|F\ i,n) \qquad (3.24)$$

The expression $(A|F\ i,n)$ is referred to as the *sinking fund factor*, since the factor is used to determine the size of a deposit one should place (sink) in a fund in order to accumulate a desired future amount. As depicted in Figure 3.6, F occurs at the same time as the last A. Thus, the last A or deposit earns no interest.

Example 3.10

If $150 000 is to be accumulated in 35 years, how much must be deposited annually in a fund paying 8% compounded annually in order to accumulate the desired amount immediately after the last deposit?

$$A = F(A|F\ 8,35)$$
$$= \$150\,000(0.0058)$$
$$= \$870$$

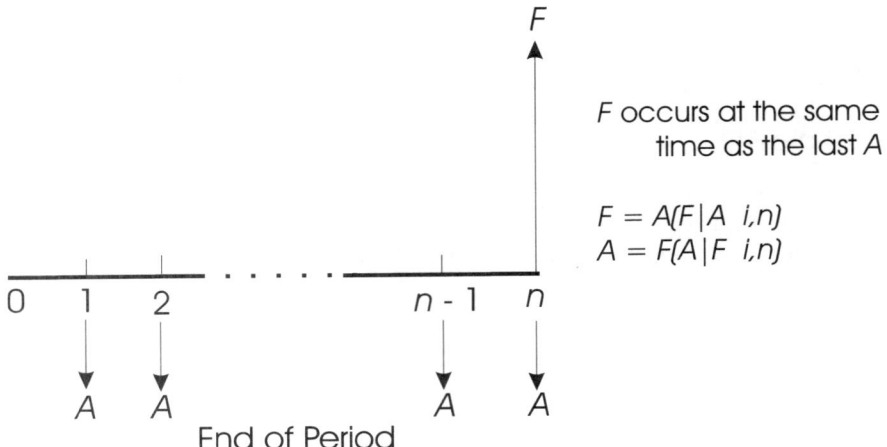

Fig. 3.6 Cash flow diagram of the relationship between A and F.

3.4.2 Gradient Series of Cash Flows

The gradient series of cash flows is depicted in Figure 3.7. The first cash flow occurs at the end of the second time period; each successive cash flow increases in magnitude by an amount G. A convenient representation of the size of the cash flow at the end of period k is given by

$$A_k = (k-1)G \qquad k = 1,\ldots,n \qquad (3.25)$$

The gradient series arises when the value of an individual cash flow differs by a constant, G, from the preceding cash flow. As an illustration, if an individual receives an annual bonus and the size of the bonus increases by $100 each year, then the series is a gradient series. Also, operating and maintenance costs tend to increase over time because of both inflation effects and a gradual deterioration of equipment; such costs are often approximated by a gradient series.

The present worth equivalent of a gradient series is obtained by recalling

$$P = \sum_{k=1}^{n} A_k (1+i)^{-k} \qquad (3.26)$$

Substituting Equation 3.25 into Equation 3.26 gives

Time Value of Money Operations

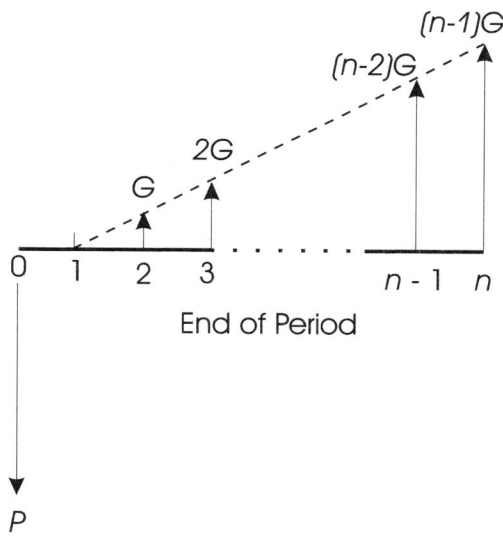

Fig. 3.7 Cash flow diagram of the gradient series.

$$P = \sum_{k=1}^{n}(k-1)G(1+i)^{-k} \qquad (3.27)$$

or, equivalently,

$$P = G\sum_{k=1}^{n}(k-1)(1+i)^{-k} \qquad (3.28)$$

As an exercise you may wish to show that the summation reduces to

$$P = G\left[\frac{1-(1+ni)(1+i)^{-n}}{i^2}\right] \qquad (3.29)$$

Expressing the term in brackets in terms of interest factors already treated yields

$$\boxed{P = G\left[\frac{(P|A\ i,n)-n(P|F\ i,n)}{i}\right]} \qquad (3.30)$$

or, equivalently,

$$P = G(P|G\ i,n) \quad (3.31)$$

where $(P|G\ i,n)$ is the *gradient series, present worth factor* and is tabulated in Appendix A.

A uniform series equivalent to the gradient series is obtained by multiplying the value of the gradient series present worth factor by the value of the $(A|P\ i,n)$ factor to obtain

$$A = G\left[\frac{1}{i} - \frac{n}{i}(A|F\ i,n)\right]$$

or, equivalently,

$$A = G(A|G\ i,n) \quad (3.32)$$

where the factor $(A|G\ i,n)$ is referred to as the *gradient-to-uniform series conversion factor* and is tabulated in Appendix A. To obtain the future worth equivalent of a gradient series at time n, multiply the value of the $(A|G\ i,n)$ factor by the value of the $(F|A\ i,n)$ factor.

It is not uncommon to encounter a cash flow series that is the sum or difference of a uniform series and a gradient series. To determine present worth and future worth equivalents of such a composite, one can deal with each special type of series separately.

Example 3.11

An individual deposits an annual bonus into a savings account that pays 6% compounded annually. The size of the bonus increases by $100 per year; the initial bonus was $300. Determine how much will be in the fund immediately after the fifth deposit.

A cash flow diagram for this example is given in Figure 3.8. Note that the cash flow series consists of the sum of a uniform series of $300 and a gradient series with G equal to $100. Converting the gradient series to a uniform series gives

$$A = G(A|G\ 6,5)$$
$$= \$100(1.8836)$$
$$= \$188.36$$

(Notice that n equals 5 even though only four positive cash flows are present in the gradient series.)

The cash flow series given in Figure 3.8 is equivalent to a uniform series having cash flows equal to $300 + $188.36 or $488.36. Converting the uniform series to a future worth equivalent:

$$F = A(F|A\ 6,5)$$
$$= \$488.36(5.6371)$$
$$= \$2\,752.93$$

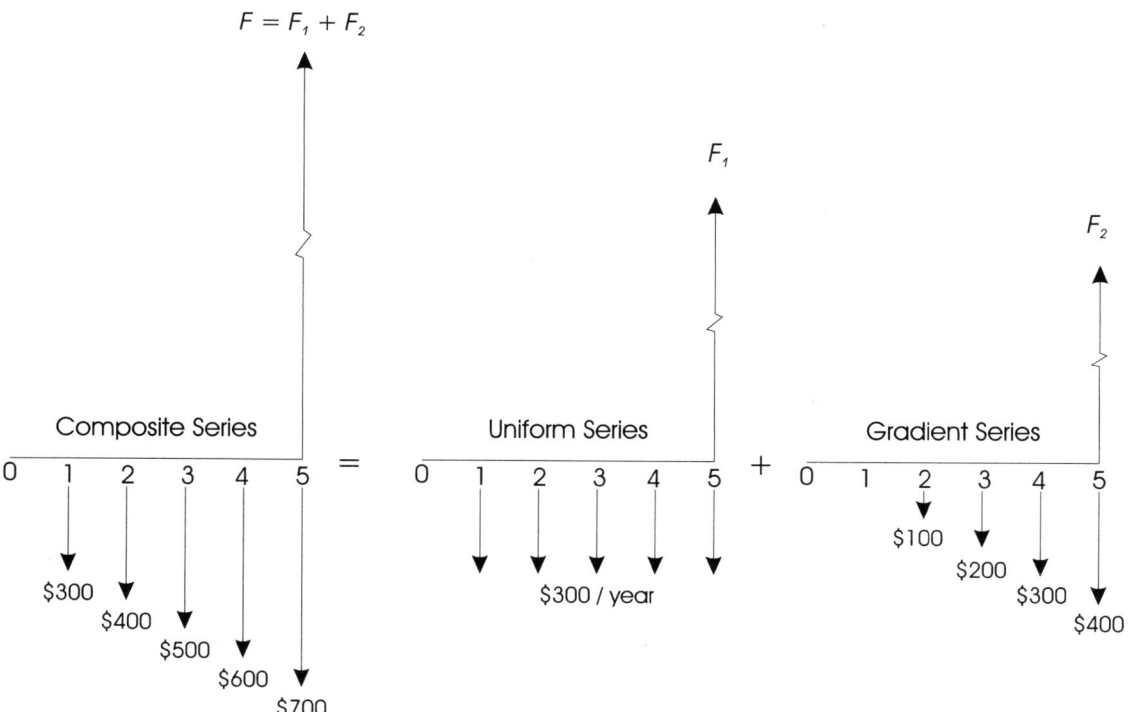

Fig. 3.8 Cash flow diagram for a gradient series example.

Thus, $2752.93 will be in the fund immediately after the fifth deposit.

Example 3.12

Five annual deposits are made into a fund that pays interest at a rate of 8% compounded annually. The first deposit equals $800; the second deposit equals $700; the third deposit equals $600; the fourth deposit equals $500; and the fifth deposit equals $400. Determine the amount in the fund immediately after the fifth deposit.

As depicted by the cash flow diagrams in Figure 3.9, the cash flow series can be represented by the difference in a uniform series of $800 and a gradient series of $100. The uniform series equivalent of the gradient series is given by

$$A = G(A|G\ 8,5)$$
$$= \$100(1.8465)$$
$$= \$184.65$$

Therefore, a uniform series having cash flows equal to $800 − $184.65, or $615.35, is equivalent to the original cash flow series. The future worth equivalent is found to be

$$F = A(F|A\ 8,5)$$
$$= \$615.35(5.8666)$$
$$= \$3\,610.01$$

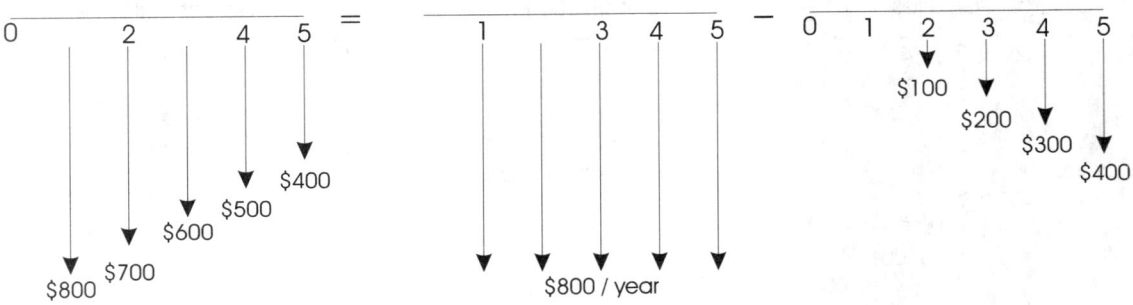

Fig. 3.9 Cash flow diagrams for the decreasing gradient series example.

3.4.3 Geometric Series of Cash Flows

The geometric cash flow series, as depicted in Figure 3.10, occurs when the size of a cash flow increases (decreases) by a fixed percent from one time period to the next. If j denotes the percent change in the size of a cash flow from one period to the next, the size of the kth cash flow can be given by

$$A_k = A_{k-1}(1+j) \qquad k=2,\ldots,n$$

or, more conveniently,

$$A_k = A_1(1+j)^{k-1} \qquad k=1,\ldots,n \qquad (3.33)$$

The geometric series is used most often to represent the growth (positive j) or decay (negative j) of costs and revenues due to inflation or recession. As an illustration, if labour costs increase by 10% a year, then the resulting series representation of labour costs will be a geometric series.

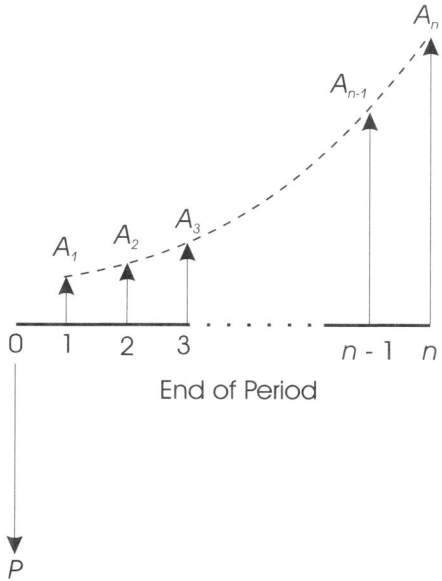

Fig. 3.10 Cash flow diagram of the geometric series.

The present worth equivalent of the cash flow series is obtained by substituting Equation 3.33 into Equation 3.6 to obtain

$$P = \sum_{k=1}^{n} A_1 (1+j)^{k-1} (1+i)^{-k} \qquad (3.34)$$

or

$$P = A_1 (1+j)^{-1} \sum_{k=1}^{n} \left(\frac{1+j}{1+i} \right)^k \qquad (3.35)$$

As an exercise you may wish to show that the following relationship results:

$$P = \begin{cases} A_1 \left[\dfrac{1-(1+j)^n (1+i)^{-n}}{i-j} \right] & i \neq j \\ \dfrac{nA_i}{1+i} & i = j \end{cases} \qquad (3.36)$$

or

$$P = A_1 (P|A_1 \; i, j, n) \qquad (3.37)$$

where $(P|A_1 \; i, j, n)$ is the *geometric series, present worth factor* and is tabulated in Appendix A for various values of $i, j,$ and n.

For the case of $j \geq 0$ and $i \neq j$, the relationship between P and A can be conveniently expressed in terms of compound interest factors previously considered.

$$P = A_1 \left[\frac{1-(F|P \; j,n)(P|F \; i,n)}{i-j} \right] \qquad i \neq j, \; j \geq 0 \qquad (3.38)$$

Example 3.13

Labour costs have been increasing at an annual rate of 8%. A firm wishes to set aside funds to cover labour costs for the next five years. Determine how much must be set aside today if the money will be invested and will earn interest at a rate of 10%. The labour cost next year will be $50 000.

For this example, $A_1 = \$50\,000$, $j = 8\%$, $i = 10\%$, and $n = 5$. The present worth equivalent will be

$$P = A_1(P|A_1\,10,8,5)$$
$$= \$50\,000(4.3831)$$
$$= \$219\,155$$

If labour costs increase at an annual rate of 10%,

$$P = \frac{nA_1}{1+i}$$
$$= \frac{5(\$50\,000)}{1.1}$$
$$= \$227\,227.73$$

The future worth equivalent of the geometric series is obtained by multiplying the value of the geometric series present worth factor and the $(F|P\,i,n)$ factor to obtain

$$F = \begin{cases} A_1\left[\dfrac{(1+i)^n - (1+j)^n}{i-j}\right] & i \neq j \\ nA_1(1+i)^{n-1} & i = j \end{cases}$$

(3.39)

or

$$F = A_1(F|A_1\,i,j,n)$$ (3.40)

where $(F|A_1\,i,n)$ is the *geometric series, future worth factor* and is tabulated in Appendix A. From Equation 3.39, notice that $(F|A_1\,i,j,n) = (F|A_1\,j,i,n)$.

Example 3.14

An individual receives an annual bonus and deposits it in a savings account that pays 6% compounded annually. The size of the bonus increases by 5% each year; the initial deposit was $500. Determine how much will be in the fund immediately after the tenth deposit.

In this case, $A_1 = \$500$, $j = 5\%$, $i = 6\%$, and $n = 10$. Thus, the value of F is given by

$$F = A_1(F|A_1\ 6,5,10) = A_1(F|A_1\ 5,6,10)$$
$$= \$500(16.1953)$$
$$= \$8\,097.65$$

The interest factors developed to this point are summarized in Table 3.2. Values of the factors given in Table 3.2 are provided in Appendix A.

Table 3.2 Summary of Discrete Compounding Interest Factors

To Find	Given	Factor	Symbol
P	F	$(1+i)^{-n}$	$(P\|F\ i,n)$
F	P	$(1+i)^n$	$(F\|P\ i,n)$
P	A	$\dfrac{(1+i)^n - 1}{i(1+i)^n}$	$(P\|A\ i,n)$
A	P	$\dfrac{i(1+i)^n}{(1+i)^n - 1}$	$(A\|P\ i,n)$
F	A	$\dfrac{(1+i)^n - 1}{i}$	$(F\|A\ i,n)$
A	F	$\dfrac{i}{(1+i)^n - 1}$	$(A\|F\ i,n)$
P	G	$\dfrac{1-(1+ni)(1+i)^{-n}}{i^2}$	$(P\|G\ i,n)$
A	G	$\dfrac{(1+i)^n - (1+ni)}{i[(1+i)^n - 1]}$	$(A\|G\ i,n)$
P	A_{1j}	$\dfrac{1-(1+j)^n(1+i)^{-n}}{i-j}$	$(P\|A_1\ i,j,n)$ *
F	A_{1j}	$\dfrac{(1+i)^n - (1+j)^n}{i-j}$	$(F\|A_1\ i,j,n)$ *

$*\ i \neq j$

3.5 Multiple Compounding Periods in a Year

Not all interest rates are stipulated as annual compounding rates. For example, if the interest rate is stipulated as 6% compounded quarterly, then the interest period is a three-month period with 1.5% interest rate/interest period. Thus, if $1 000 is borrowed at an interest rate of 6% compounded quarterly, then the amount owed at the end of five years or 20 interest periods is obtained as follows:

$$F = P(F|P\ 1.5,20)$$
$$= \$1\,000(1+0.015)^{20}$$
$$= \$1\,000(1.3469)$$
$$= \$1\,346.90$$

With quarterly compounding, $1 346.90 is to be repaid.

Example 3.15

As another illustration of multiple compounding periods within a year, suppose interest is stated as 8% compounded quarterly. If you borrow $1 000 for one year, how much must be repaid?

Using an interest rate of 2% per three-month period, after four interest periods the amount owed is given by:

$$F = P(F|P\ 2,4)$$
$$= \$1\,000(1.0824)$$
$$= \$1\,082.40$$

Thus, $1 082.40 would be owed. Note that if interest were stated as 8.24% compounded annually,

$$F = P(F|P\ 8.24,1)$$
$$= \$1\,000(1.0824)$$
$$= \$1\,082.40$$

and the same amount would be owed.

It can be concluded that 8% compounded quarterly is equivalent to 8.24% compounded annually. The rate of 8% is referred to as the *nominal interest rate*; the rate of 8.24% is referred to as the *effective annual interest rate* and the rate of 2% is referred to as the interest rate per compounding period. The effective interest rate is defined as the annual compounding rate that is equivalent to the stated interest rate.

Letting r denote the nominal annual interest rate, m denote the number of compounding periods per year, i denote the interest rate per interest period, and i_{eff} denote the effective interest rate per year,

$$i_{eff} = \left(1+\frac{r}{m}\right)^m - 1$$
$$= (1+i)^m - 1$$

$$\boxed{i_{eff} = \left(F|P\ \frac{r}{m}, m\right) - 1}$$

To illustrate with 8% compounded semiannually, r equals 0.08, m equals 2, and

$$i_{eff} = (1+0.04)^2 - 1$$
$$= (F|P\ 4,2) - 1$$
$$= 0.0816$$

Thus, 8% compounded semiannually is equivalent to 8.16% compounded annually (i.e., 8.16% is the effective annual interest rate when interest is 8% compounded semiannually).

Suppose interest is stated as 18% compounded monthly. The effective annual interest rate is given by

$$i_{eff} = (1+0.015)^{12} - 1$$
$$= (F|P\ 1.5,12) - 1$$
$$= 0.1956 \quad \text{or} \quad 19.56\%$$

Example 3.16

An individual borrowed $1 000 and paid off the loan with interest after 4.5 years. The amount paid was $1 250. What was the effective annual interest rate for this transaction? Letting the interest period be a six-month period, it is seen that the payment of $1 250 and the debt of $1 000 are related by the expression

$$F = P(F|P\ i,n)$$

Thus,

$$\$1\,250 = \$1\,000(F|P\ i,9)$$

or

$$\$1\,250 = \$1\,000(1+i)^9$$

Dividing both sides by $1 000 gives

$$1.25 = (1+i)^9$$

Taking the logarithm of both sides yields

$$\log 1.25 = 9\log(1+i)$$

Dividing both sides by nine and taking the antilog of the result provides the relation

$$(1+i) = 1.0251$$
$$i = 0.0251$$

Thus, the six-month interest rate is 2.51%, and computing the effective annual interest rate yields

$$i_{eff} = (1+i)^2 - 1$$
$$= (1+0.0251)^2 - 1$$
$$= 0.0508$$

The effective annual interest rate for the loan transaction was approximately 5.08%.

Instead of using logarithms, we could have searched the interest tables for a value of i that yielded a value of 1.25 for the $(F|P\ i,9)$ factor. Since i would be found to equal approximately 2.5%, the computation of the effective annual interest rate would have followed the same procedure used to determine the effective annual interest rate for 5% compounded semiannually.

Since both in personal and business financing the annual percentage rate might be computed differently among the alternative sources of funds, the effective annual interest rate is a very good basis for comparing alternative financing plans. Subsequently, we will find that the effective annual interest rate is also useful when the timing of cash flows does not coincide with the end of interest periods.

3.6 Continuous Compounding

In the discussion of the effective interest rate in the previous section it was noted that, as the frequency of compounding in a year increases, the effective interest rate increases. Since monetary transactions occur daily or hourly in most businesses and money is often "put to work" for the business as soon as it is received, compounding is occurring quite frequently. If one wishes to account explicitly for such rapid compounding, then *continuous compounding relations should be used*. Continuous compounding means that each year is divided into an infinite number of interest periods. Mathematically, the single payment compound amount factor under continuous compounding is given by

$$(1+i_{\text{eff}})^n = \lim_{m \to \infty}\left(1+\frac{r}{m}\right)^{mn} = e^{rn}$$

where n is the number of years, m is the number of interest periods per year, and r is the nominal annual interest rate. Given P, r, and n, the value of F can be computed using continuous compounding as follows:

$$F = Pe^{rn} \quad (3.42)$$

or

$$F = P(F|P\ r,n)_\infty \quad (3.43)$$

where $(F|P\ r,n)_\infty$ denotes the *continuous compounding, single sum, future worth factor*. The subscript ∞ is provided to denote that continuous compounding is being used. The interest tables for continuous compounding are given in Appendix B.

Example 3.17

If $2 000 is invested in a fund that pays interest at a rate of 6% compounded continuously, after five years the cumulative amount in the fund will total

$$F = P(F|P\ 6,5)_\infty$$
$$= \$2\,000(1.3499)$$
$$= \$2\,699.80$$

Thus, a withdrawal of $2 699.80 will deplete the fund after five years.

The effective interest rate under continuous compounding is easily obtained using the relation

$$\boxed{i_{eff} = e^r - 1} \qquad (3.44)$$

or

$$\boxed{i_{eff} = (F|P\ r,1)_\infty - 1} \qquad (3.45)$$

To illustrate, if interest is 8% compounded continuously, then the effective interest rate is given by

$$i_{eff} = (F|P\ 8,1)_\infty - 1$$
$$= 0.0833$$

Thus, 8.33% compounded annually is equivalent to 8% compounded continuously.

The inverse relationship between F and P indicates that

$$\boxed{P = F\,e^{-rn}} \qquad (3.46)$$

or

$$\boxed{P = F(P|F\ r,n)_\infty} \qquad (3.47)$$

where $(P|F\ r,n)_\infty$ is called the *continuous compounding, single sum, present worth factor*.

3.6.1 Discrete Flows

If it is assumed that cash flows are discretely spaced over time, then the continuous compound relations for the uniform, gradient, and geometric series can be obtained. Substituting e^{-rn} for $(1+i)^{-n}$, $e^r - 1$ for i, and e^{rn} for $(1+i)^n$ in the remaining discrete compounding formulas yields the continous compound interest factors summarized in Table 3.3. Values for these factors are provided in Appendix B.

Example 3.18

To illustrate the use of the continuous compound interest factors, suppose $1 000 is deposited each year into an account that pays interest at a rate of 6% compounded continuously. Determine both the amount in the account immediately after the tenth deposit and the present worth equivalent for ten deposits.

The amount in the fund immediately after the tenth deposit is given by the relation

$$F = \$1\,000(F|A\ 6,10)_\infty$$
$$= \$1\,000(13.2951)$$
$$= \$13\,295.10$$

The present worth equivalent for ten deposits is obtained using the relation

$$P = \$1\,000(P|A\ 6,10)_\infty$$
$$= \$1\,000(7.2965)$$
$$= \$7\,296.51$$

In the case of the geometric series the size of the kth cash flow will be assumed to be given by

$$A_k = A_{k-1}e^c \qquad k = 2,\ldots,n \qquad (3.48)$$

or, equivalently,

$$A_k = A_1 e^{(k-1)c} \qquad k = 1,\ldots,n \qquad (3.49)$$

where c is the nominal compound rate of increase in the size of the cash flow. The resulting expressions for the *continuous compounding, geometric series present worth factor* and the *continuous compounding, geometric series future worth factor* are given, respectively, by $(P|A_1,r,c,n)_\infty$ and $(F|A_1,r,c,n)_\infty$, as given in Table 3.3 for the case of

Table 3.3
Summary of Continuous Compounding Interest Factors

To Find	Given	Factor	Symbol
P	F	e^{-rn}	$(P\|F\ r,n)_\infty$
F	P	e^{rn}	$(F\|P\ r,n)_\infty$
F	A	$\dfrac{e^{rn}-1}{e^r-1}$	$(F\|A\ r,n)_\infty$
A	F	$\dfrac{e^r-1}{e^{rn}-1}$	$(A\|F\ r,n)_\infty$
P	A	$\dfrac{e^{rn}-1}{e^{rn}(e^r-1)}$	$(P\|A\ r,n)_\infty$
A	P	$\dfrac{e^{rn}(e^r-1)}{e^{rn}-1}$	$(A\|P\ r,n)_\infty$
P	G	$\dfrac{e^{rn}-1-n(e^r-1)}{e^{rn}(e^r-1)^2}$	$(P\|G\ r,n)_\infty$
A	G	$\dfrac{1}{e^r-1}-\dfrac{n}{e^{rn}-1}$	$(A\|G\ r,n)_\infty$
P	A_1,c	$\dfrac{1-e^{(c-r)n}}{e^r-e^c}$	$(P\|A_1\ r,c,n)_\infty,\quad r\ne c$
F	A_1,c	$\dfrac{e^{rn}-e^{cn}}{e^r-e^c}$	$(F\|A_1\ r,c,n)_\infty,\quad r\ne c$
P	$\bar{A}$	$\dfrac{e^{rn}-1}{re^{rn}}$	$(P\|\bar{A}\ r,n)$
$\bar{A}$	P	$\dfrac{re^{rn}}{e^{rn}-1}$	$(\bar{A}\|P\ r,n)$
F	$\bar{A}$	$\dfrac{e^{rn}-1}{r}$	$(F\|\bar{A}\ r,n)$
$\bar{A}$	F	$\dfrac{r}{e^{rn}-1}$	$(\bar{A}\|F\ r,n)$

$r \neq c$. As an exercise you may wish to derive the appropriate expressions when $r = c$.

Example 3.19

An individual receives an annual bonus and deposits it in a savings account that pays 5% compounded continuously. The size of the bonus increases each year at a rate of 6% compounded continuously; the initial deposit was $500. Determine how much will be in the fund immediately after the tenth deposit.

In this case, $A_1 = \$500, r = 5\%, c = 6\%, n = 10$. Thus, the value of F is given by

$$F = A_1(F|A_1\ 5,6,10)_\infty$$
$$= \$500(16.4117)$$
$$= \$8\ 205.85$$

The effect of continuous compounding increases the amount in the fund by $110.90.

If discrete flows occur during a year, it is necessary to define r consistent with the spacing of cash flows. To illustrate, suppose semiannual deposits are made into an account paying 6% compounded continuously. In this case the nominal semiannual rate would be 6% ÷ 2 time periods per year, or $r = 3\%$, and n would equal the number of semiannual periods involved.

Since the differences in discrete and continuous compounding are not great in most cases, it is not uncommon to see discrete compounding used when continuous compounding is more appropriate. The arguments given for this are that errors in estimating the cash flows will probably offset any attempts to be very precise by using continuous compounding and that the interest rate used in discrete compounding is actually the effective interest rate resulting from continuous compounding.

3.6.2 Continuous Flow

Thus far only discrete cash flows have been considered. It was assumed that cash flows occurred at, say, the end of the year. In some cases money is expended throughout the year on a somewhat uniform basis. (Costs for labour, carrying inventory, and operating and maintaining equipment are typical examples.) Consequently, as a mathematical convenience, instead of assuming that money flows in discrete increments at the end of monthly, weekly, daily, or hourly time periods, it is assumed that money flows continuously during the time period at a uniform rate. Instead of having a uniform series of discrete cash flows of magnitude A, it is assumed that a *total* of A dollars flows uniformly and continuously throughout a given time period. Such an approach to modeling cash flows is referred to as the *continuous flow* approach.

To illustrate the continuous flow concept, suppose you are to divide $1 000 into *k* equal amounts to be deposited at equally spaced points in time during a year. The interest rate per period is defined to be *r/k*, where *r* is the nominal rate. Thus, the present worth of the series of *k* equal amounts is

$$P = \frac{\$1000}{k}\left(P|A\ \frac{r}{k}, k\right)$$

or

$$P = \frac{\$1000}{k}\left\{\frac{[1+(r/k)]^k - 1}{(r/k)[1+(r/k)]^k}\right\}$$

which reduces to

$$P = \$1000\left\{\frac{1}{r} - \frac{1}{r[1+(r/k)]^k}\right\}$$

Taking the limit of *P* as *k* approaches infinity gives

$$\lim_{k \to \infty} P = \$1\,000\left(\frac{1}{r} - \frac{1}{re^r}\right)$$

or

$$P = \$1\,000\left(\frac{e^r - 1}{re^r}\right)$$

In general, for *n years,*

$$P = \overline{A}\left(\frac{e^{rn}-1}{re^{rn}}\right) \tag{3.50}$$

or

$$P = \overline{A}(P|\overline{A}\ r,n) \tag{3.51}$$

where $(P|\overline{A}\ r,n)$ is referred to as the *continuous flow, continuous compounding uniform series present worth factor* and is tabulated in Appendix B, and $\overline{A}$ is the amount of continuous flow per period given in units of dollars per period.

The remaining continuous flow, continuous compound interest factors are summarized in Table 3.3. Values of the factors are given in Appendix B for various values of r and n. With continuous flow and continuous compounding the continuous annual cash flow, $\overline{A}$, is equivalent to $\overline{A}\dfrac{r}{(e^r-1)}$, when discrete flow and continuous compounding is used. Hence, the discrete flow equivalent of $\overline{A}$ is given by

$$A = \frac{\overline{A}(e^r-1)}{r}$$

or

$$A = \overline{A}(F|\overline{A}\ r,1)$$

Example 3.20

What are the present worth and future worth equivalents of a uniform series of continuous cash flows totalling $1 000 per year for 10 years when interest is compounded continuously at a rate of 10% per year?

In this problem $\overline{A} = \$1\,000/\text{year}$, $r = 10\%$, $n = 10$, and the present worth equivalent is given by

$$P = \$1\,000(P|\overline{A}\ 10,10)$$
$$= \$1\,000(6.3212)$$
$$= \$6\,321.20$$

The future worth equivalent is given by

$$F = \$1\,000(F|\overline{A}\ 10,10)$$
$$= \$1\,000(17.1828)$$
$$= \$17\,182.80$$

Continuous flow and continuous compounding are concepts that are used to represent more closely the realities of certain business transactions. In fact, several of the computer programs developed by industries and governmental agencies and used to perform economic analyses incorporate both concepts. Despite such arguments for their use, the concepts have not been widely accepted among economic analysts. We have presented both concepts in anticipation that both will become more popular in the future.

3.7 Equivalence

Throughout the preceding discussion we have used the term *equivalence* without defining what was meant by the term. This was done intentionally in order to introduce subtly the notion that two cash flow series or profiles are *equivalent* at some specified interest rate, $k\%$, if their present worths are *equal* using an interest rate of $k\%$.

Example 3.21

> The two cash flow profiles shown in Figure 3.11 are equivalent at 6%, since each has a present worth of $565.66. You may wish to verify this claim.
>
> Of course, if the present worths of two cash flow profiles are equal at $k\%$, their "worths" at any given point in time will be equal when using a discount (interest) rate of $k\%$. Likewise, if two cash flow profiles are equivalent at $k\%$, their respective uniform series equivalents will be equal when expressed over the same time period. Hence we could state that the two cash flow profiles shown in Figure 3.11 are equivalent at 6%, since each has a future worth of $901.56 at the end of period eight; alternately, we could assert their equivalence at 6% on the basis that each can be represented by a uniform series of $91.09 per period over the interval [1, 8].

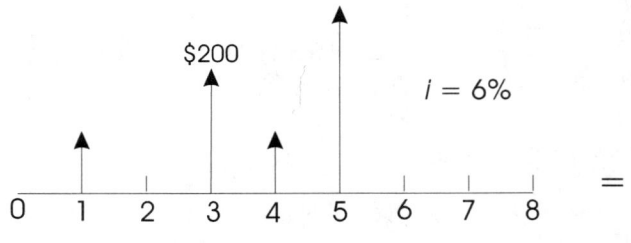

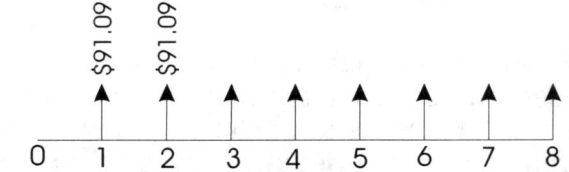

Fig. 3.11 Cash flow diagrams for the equivalence example.

Example 3.22

What single sum of money at $t = 6$ is equivalent to the cash flow profile shown in Figure 3.12 if $i = 5\%$?

The present worth of the cash flow profile is given by

$$P = -\$400(P|F\ 5,1) + \$100(P|A\ 5,3)(P|F\ 5,1)$$
$$+ \$100(P|A\ 5,3)(P|F\ 5,5)$$
$$= -\$400(0.9524) + \$100(2.7232)(0.9524)$$
$$+ \$100(2.7232)(0.7835)$$
$$= \$91.76$$

Moving $97.76 forward in time to $t = 6$ gives

$$F = \$91.76(F|P\ 5,6)$$
$$= \$91.76(1.3401)$$
$$= \$122.97$$

Thus, at 5%, a positive cash flow of \$122.97 at $t = 6$ is equivalent to the cash flow profile shown in Figure 3.12.

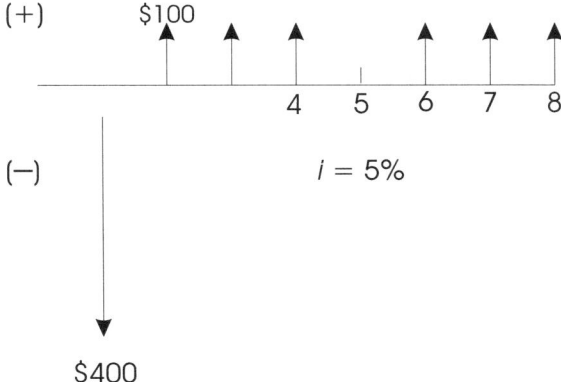

Fig. 3.12 Cash flow diagram for the equivalence example.

Example 3.23

Using an 8% discount rate (annual compounding yearly interest rate), what uniform series over five periods, [1, 5], is equivalent to the cash flow profile given in Figure 3.13?

The cash flow profile in Figure 3.13(a) consists of the difference in a uniform series of $500 and a gradient series, $G = \$100$. A uniform series equivalent of the cash flow profile (see Figure 3.13(b)) can be obtained for the interval [2, 6] as follows:

$$A = \$500 - \$100(A|G\,8,5)$$
$$= \$500 - \$100(1.8465)$$
$$= \$315.35$$

The uniform series of $315.35 over the interval [2, 6] must be converted to a uniform series over the interval [1, 5]. Thus, *each* of the five cash flows must be moved back in time one period. The discounted value of $315.35 over one time period, using an 8% discount rate is

$$P = \$315.35(P|F\,8,1)$$
$$= \$315.35(0.9259)$$
$$= \$291.98$$

Consequently, a uniform series of $291.98 over the interval [1, 5] (see Figure 3.13(c)) is equivalent to the cash flow profile given in Figure 3.13(a). If you have doubts concerning the equivalence, compare their present worths using an 8% interest rate.

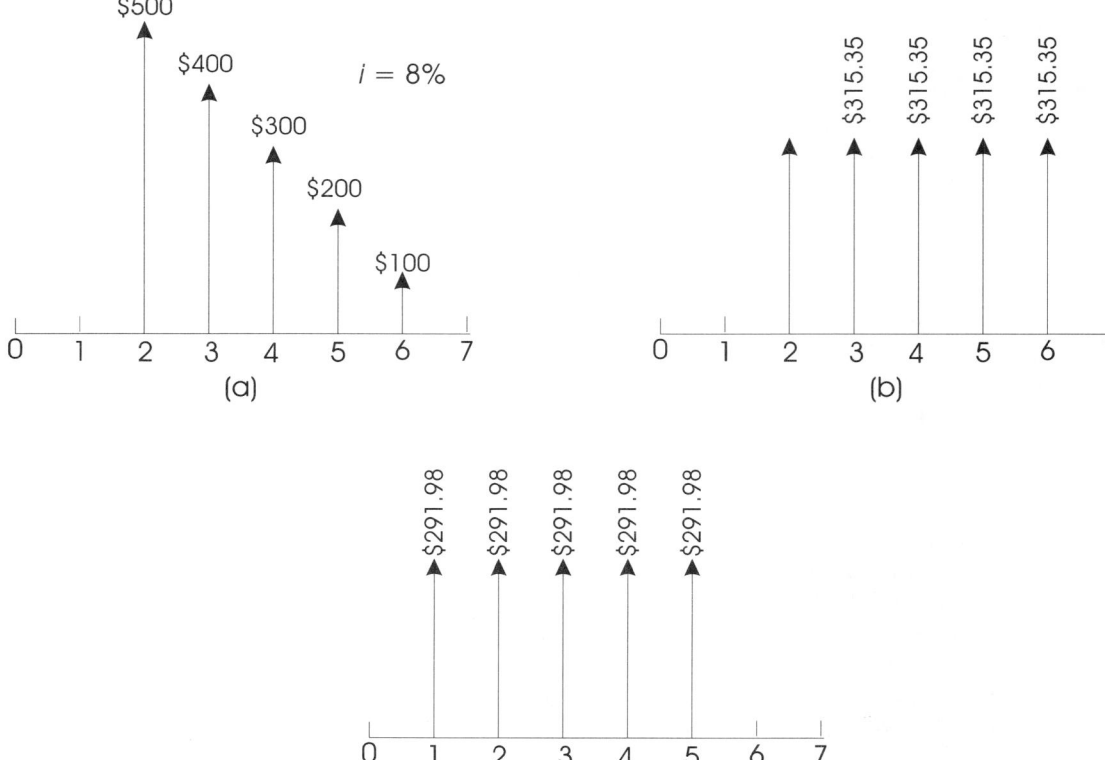

Fig. 3.13 Cash flow diagrams for the equivalence example.

Example 3.24

Determine the value of X that makes the two cash flows, as given in Figure 3.14, equivalent when a discount rate of 6% is used.

Equating the future worths of the two cash flow profiles at $t = 4$ gives

$$-\$200(F|A\ 6,4) - \$100(F|A\ 6,3) - \$100 = -[\$200 + X(A|G\ 6,4)](F|A\ 6,4)$$

Cancelling $[-\$200(F|A\ 6,4)]$ on both sides yields

$$\$100(3.1836) + \$100 = X(1.4272)(4.3746)$$

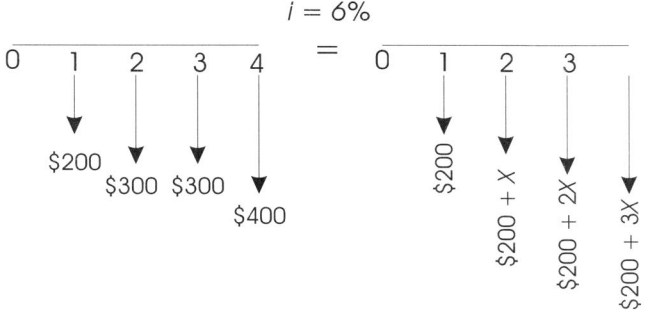

Fig. 3.14 Cash flow diagrams for the equivalence example.

Solving for X gives a value of $67.

Example 3.25

For what interest (discount) rate are the two cash flow profiles shown in Figure 3.15 equivalent?

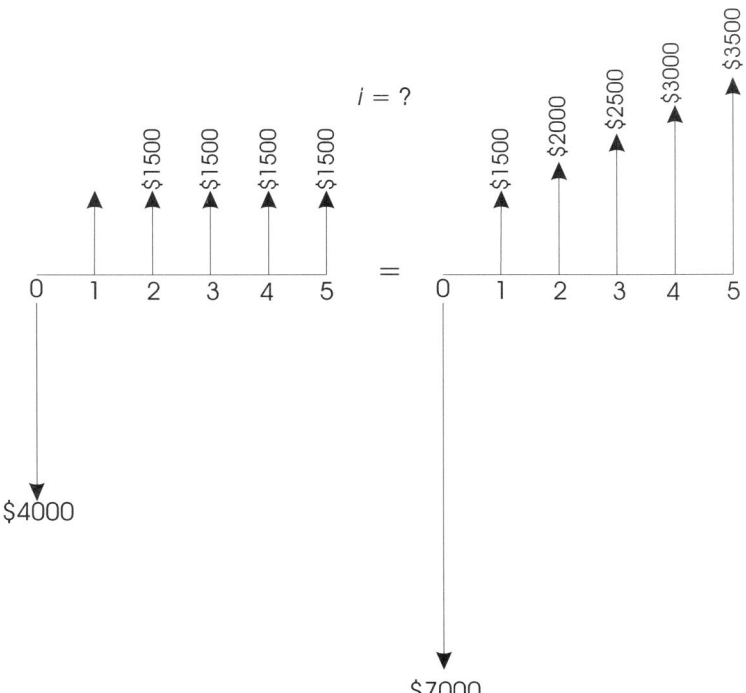

Fig. 3.15 Cash flow diagrams for the equivalence example.

Converting each cash flow profile to a uniform series over the interval [1, 5], gives

$$-\$4\,000(A|P\ i,5) + \$1\,500$$
$$= -\$7\,000(A|P\ i,5) + \$1\,500 + \$500(A|G\ i,5)$$

or

$$\$3\,000(A|P\,i,5) = \$500(A|G\,i,5)$$

which reduces to

$$(A|G\,i,5) = 6(A|P\,i,5)$$

On searching through the interest tables at $n = 5$, it is found that the $(A|G\,i,5)$ factor is six times the value of the $(A|P\ i,5)$ factor for an interest rate between 12 and 15%. Specifically, with a 12% interest rate,

$$(A|G\ 12,5) - 6(A|P\ 12,5) = 1.7746 - 6(0.2774) = 0.1102$$

and, using a 15% interest rate,

$$(A|G\ 15,5) - 6(A|P\ 15,5) = 1.7228 - 6(0.2983) = -0.0670$$

Interpolating for i gives

$$i = 0.12 + \frac{(0.15 - 0.12)(0.1102)}{(0.1102 + 0.0670)}$$

or

$$i = 0.1386$$

Therefore, using a discount rate of approximately 13.86% will establish an equivalence relationship between the cash flow profiles in Figure 3.15.

Example 3.26

Dortex Corporation purchases a machine for $30 000, keeps it for five years, and sells it for $6 000. During the time the machine was owned by the company, operating and maintenance costs totalled $8 000 the first year, $9 000 the second year, $10 000 the third year, $11 000 the fourth year, and $12 000 the fifth year. The firm uses a 15% interest rate in performing economic analyses. Determine the single sum of money occurring at (1) $t=0$ and (2) $t=5$, which is equivalent to the cash flow history for the machine. Also, determine the uniform series occurring over the interval [1, 5] that is equivalent to the cash flow profile for the machine.

The single sum equivalent at time zero for a cash flow profile is called the *present worth* or *present value* for the cash flow profile. We will use the term present worth and denote it as *PW*. For the example problem,

$$PW = -\$30\,000 - \$8\,000(P|A\ 15,5) + \$1\,000(P|G\ 15,5)$$
$$+ \$6\,000(P|F\ 15,5)$$
$$= -\$59\,609.57$$

Hence, a single expenditure of $59 609.57 at time zero would have been equivalent to the cash flows experienced during the ownership of the machine.

The single sum equivalent at the end of the life of a project is termed the *future worth* or *future value* for the project. We denote the future worth by *FW* and compute its value as follows:

$$FW = -\$30\,000(F|P\ 15,5)$$
$$-[\$8\,000 + \$1\,000(A|G\ 15,5)](F|A\ 15,5)$$
$$+ \$6\,000$$
$$= -\$119\,897$$

Alternately, the future worth can be obtained from the present worth:

$$FW = PW(F|P\ i,n)$$
$$= -\$59\,609.57(F|P\ 15,5)$$
$$= -\$119\,898.69$$

The difference of $1.69 is due to round-off error in the tables. Hence, a single expenditure of $119 897 at year five would have been equivalent to the cash flows associated with the machine.

A uniform series equivalent for a series of yearly cash flows is referred to as the *annual worth* or *equivalent uniform annual cost* for the project. The latter expression is most appropriate for the type of example under consideration, since the resulting uniform series is a cost, not an income, series. The annual worth designation is

AW; EUAC denotes the equivalent uniform annual cost. We will use both designations throughout the text.

For the example problem, the *EUAC* determination is performed as follows:

$$EUAC = \$30\,000(A|P\ 15,5) + [\$8\,000 + \$1\,000(A|G\ 15,5)]$$
$$- \$6\,000(A|F\ 15,5)$$
$$= \$17\,782.00$$

Hence, an annual expenditure of $17 782 per year for five years is equivalent to the cash flow profile associated with the machine investment. Alternately, the equivalent uniform annual cost can be obtained from either the present worth or the future worth.

$$EUAC = -PW(A|P\ i,n)$$
$$EUAC = -FW(A|F\ i,n)$$

Present worth, future worth, and annual worth computations are used often in comparing investment alternatives having different cash flow profiles. Consequently, we will have need for *PW, FW*, and *AW* calculations for the discussion of alternative comparisons in Chapter 4.

3.8 Loan Payments

When both personal and corporate investments are financed from borrowed funds, income taxes are affected by the amount of interest paid. The topic of income tax will be discussed in detail in Chapter 5. However, it is quite important to know how much of each payment is interest and how much is reducing the principal amount borrowed initially. To illustrate this situation, suppose you borrowed $10 000 and paid it back using four equal annual installments, with interest computed at 10% compounded annually. The payment size is computed to be

$$A = P(A|P\ 10,4)$$
$$= \$10\,000(0.3155)$$
$$= \$3\,155$$

The interest accumulation the first year is 0.10(10 000), or $1 000. Therefore, the first payment consists of a $1 000 interest payment and a $2 155 principal payment. The unpaid balance at the beginning of the second year is $10 000 − $2 155, or $7 845; consequently, the interest charge in the second year is 0.10($7 845), or $784.50. Thus, the second payment consists of a $784.50 interest payment and a principal payment of $3 155 − $784.50, or $2 370.50. The unpaid balance at the beginning of the third year is $7 845 − $2 370.50, or $5 474.50, so the interest charge in the third year is 0.10 ($5 474.50), or $547.45. Thus, the third payment consists of a $547.45 interest payment and a principal payment of $3 155 − $547.45, or $2 607.55. The unpaid balance at the beginning of the fourth year is $5 474.50 − $2 607.55, or $2 866.95; the interest charge in the fourth year is 0.10($2 866.95), or $286.70. Thus, the fourth payment consists of a $286.70 interest payment and a principal payment of $3 155 − $286.70, or $2 868.30. (The principal payment in the fourth payment should equal the unpaid balance of $2 866.95 at the beginning of the fourth year. The difference of $1.35 is due to errors made in rounding off when computing the payment size.)

The amount of principal remaining to be repaid immediately after making payment $(k-1)$ can be found by computing the present value of the remaining $n-k+1$ payments; letting U_{k-1} denote the unpaid principal after making payment $k-1$, we know that

$$U_{k-1} = A(P|A\ i, n-k+1) \tag{3.52}$$

or

$$U_{k-1} = A \sum_{j=1}^{n-k+1} (1+i)^{-j}$$

where A denotes the size of the individual payments and i represents the interest rate used to compute A.

The amount by which payment k reduces the unpaid principal will be designated E_k and is given by the relation, $E_k = U_{k-1} - U_k$. Hence,

$$E_k = A \sum_{j=1}^{n-k+1} (1+i)^{-j} - A \sum_{j=1}^{n-k} (1+i)^{-j} \tag{3.53}$$

or

$$E_k = A \left[\sum_{j=1}^{n-k} (1+i)^{-j} + (1+i)^{-(n-k+1)} \right] - A \sum_{j=1}^{n-k} (1+i)^{-j}$$

Thus,

$$E_k = A(1+i)^{-(n-k+1)} \tag{3.54}$$

or

$$\boxed{E_k = A(P|F\ i, n-k+1)} \tag{3.55}$$

Recalling how the payment size is determined, we express E_k as

$$\boxed{E_k = P(A|P\ i,n)(P|F\ i,n-k+1)} \tag{3.56}$$

where P is the original principal amount borrowed, and E_k is the amount of payment k, which is an *equity* payment (i.e., payment against principal). Letting I_k be the amount of payment k, which is an interest payment, it is seen that

$$I_k = A - E_k$$

$$\boxed{I_k = A[1-(P|F\ i,n-k+1)]} \tag{3.57}$$

Example 3.27

To illustrate the use of the formulas for computing the values of E_k and I_k, suppose $10\,000 is borrowed at 6% annual interest and repaid with five equal annual payments. The payment size is found to be

$$A = \$10\,000(A|P\ 6,5)$$
$$= \$10\,000(0.2374)$$
$$= \$2\,374$$

Thus

$$E_1 = \$2\,374(P|F\ 6,5) = \$1\,774, \quad I_1 = \$600$$
$$E_2 = \$2\,374(P|F\ 6,4) = \$1\,880, \quad I_2 = \$494$$
$$E_3 = \$2\,374(P|F\ 6,3) = \$1\,994, \quad I_3 = \$380$$
$$E_4 = \$2\,374(P|F\ 6,2) = \$2\,112, \quad I_4 = \$262$$
$$E_5 = \$2\,374(P|F\ 6,1) = \$2\,240, \quad I_5 = \$134$$

Example 3.28

An individual purchases a $50 000 house and makes a downpayment of $10 000. The remaining $40 000 is financed over a 25-year period at 7% compounded monthly. The monthly house payment is computed to be

$$A = \$40\,000\left(A|P\ \frac{7}{12},300\right)$$
$$= \$282$$

The individual keeps the house for five years and decides to sell it. How much equity is there in the house?

The amount of principal remaining to be paid can be determined by computing the present worth of the remaining 240 monthly payments using a $7/12$% interest rate, or

$$P = \$282\left(P|A\ \frac{7}{12},240\right)$$
$$= \$36\,378$$

Thus, the individual's equity equals ($40 000 − $36 378), or $3 622, plus the value of the downpayment, or $13 622. It is interesting (or perhaps depressing if you are a borrower) to determine that over the five-year period the individual made payments totalling $16 920, of which $13 298 was for interest, i.e. $13 298 = $16 920 − $3 622.

3.9 Special Topics

The previous discussion of compound interest methods and the concept of equivalence provides the foundation for the subsequent treatment of economic analysis. At this point, sufficient background material has been covered for an understanding of the discussion in Chapter 4. However, before leaving the subject of time value of money, a number of related topics will be covered. Methods for dealing with changing interest rates are treated; the situation where the spacing of cash flows and the compounding frequency differ is dealt with; methods are presented for analyzing situations involving an infinite number of cash flows; the use of compound interest methods in analyzing investments in bonds is discussed; methods are provided for coping with inflationary effects in compound interest calculations; and, finally, the concept of capital recovery cost is presented. Since each topic is treated independently, you may wish to study only those topics that are of particular interest.

3.9.1 Changing Interest Rates

The preceding discussion assumed that the interest rate did not change during the time period of concern. Recent experience indicates that such a situation is not likely if the time period of interest extends over several years (i.e., more than one interest rate may be applicable). Considering a single sum of money and discrete compounding, if i_k denotes the interest rate appropriate during time period k, the future worth equivalent for a single sum of money can be expressed as

$$F = P(1+i_1)(1+i_2)\cdots(1+i_{n-1})(1+i_n) \quad (3.58)$$

and the inverse relation

$$P = F(1+i_n)^{-1}(1+i_{n-1})^{-1}\cdots(1+i_2)^{-1}(1+i_1)^{-1} \quad (3.59)$$

Example 3.29

Consider the situation depicted in Figure 3.16 in which an individual deposited $1 000 in a savings account that paid interest at an annual compounding rate of 5% for the first three years, 6% for the next four years, and 7% for the next two years. How much was in the fund at the end of the ninth year?

Letting V_t denote the value of the account at the end of time period t, we see that

$$V_3 = \$1\,000(F|P\ 5,3)$$
$$= \$1\,000(1.1576)$$
$$= \$1\,157.60$$

Likewise,

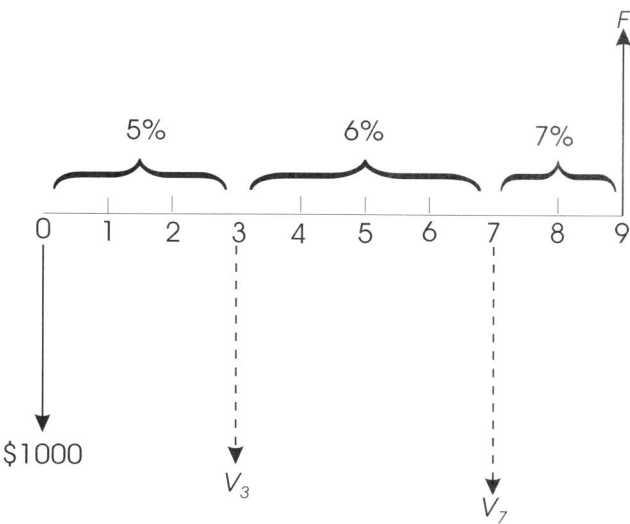

Fig. 3.16 Cash flow diagram for a changing interest example.

$$V_7 = \$1157.60(F|P\ 6,4)$$
$$= \$1157.60(1.2625)$$
$$= \$1461.47$$

Similarly,

$$F = V_9 = \$1461.47(F|P\ 7,2)$$
$$= \$1461.47(1.1449)$$
$$= \$1673.24$$

Alternately, the amount in the account at the end of nine years is given by

$$F = \$1000(1.05)(1.05)(1.05)(1.06)(1.06)(1.06)(1.06)(1.07)(1.07)$$
$$= \$1673.24$$

Extending the consideration of changing interest rates to series of cash flows, the present worth of a series of cash flows can be represented as

$$P = A_1(1+i_1)^{-1} + A_2(1+i_1)^{-1}(1+i_2)^{-1} + \cdots$$
$$+ A_n(1+i_1)^{-1}(1+i_2)^{-1}\cdots(1+i_n)^{-1} \qquad (3.60)$$

The future worth of a series of cash flows can be given by

$$F = A_n + A_{n-1}(1+i_n) + A_{n-2}(1+i_{n-1})(1+i_n) + \cdots$$
$$+ A_1(1+i_2)(1+i_3)\cdots(1+i_{n-1})(1+i_n) \qquad (3.61)$$

Example 3.30

Consider the cash flow diagram given in Figure 3.17 with the appropriate interest rates indicated. Determine the present worth, future worth, and uniform series equivalents for the cash flow series.

Computing the present worth gives

$$P = \$200(P|F\,4,1) - \$200(P|F\,4,1)(P|F\,4,1)$$
$$+ \$300(P|F\,4,1)(P|F\,4,1)(P|F\,5,1)$$
$$+ \$200(P|F\,4,1)(P|F\,4,1)(P|F\,5,1)(P|F\,5,1)(P|F\,6,1)$$
$$= \$200(P|F\,4,1) - \$200(P|F\,4,2)$$
$$+ \$300(P|F\,4,2)(P|F\,5,1) + \$200(P|F\,4,2)(P|F\,5,2)(P|F\,6,1)$$
$$= \$200(0.9615) - \$200(0.9246) + \$300(0.9246)(0.9524)$$
$$+ \$200(0.9246)(0.9070)(0.9434)$$
$$= \$429.79$$

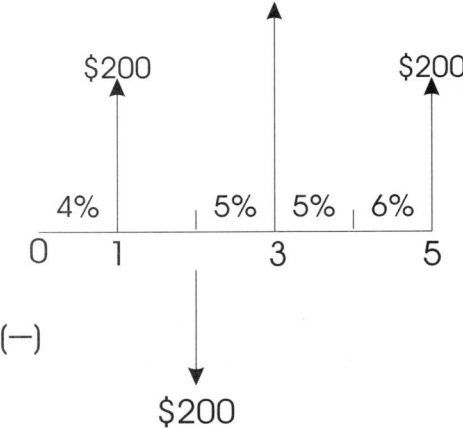

Fig. 3.17 Cash flow diagram for a changing interest rate example.

The future worth is given by

$$F = \$200 + \$300(F|P\,5,1)(F|P\,6,1) - \$200(F|P\,5,2)(F|P\,6,1)$$
$$+ \$200(F|P\,4,1)(F|P\,5,2)(F|P\,6,1)$$
$$= \$200 + \$300(1.05)(1.06) - \$200(1.1025)(1.06)$$
$$+ \$200(1.04)(1.1025)(1.06)$$
$$= \$543.25$$

The uniform series equivalent is obtained as follows:

$$P = A(P|F\,4,1) + A(P|F\,4,2) + A(P|F\,5,1)(P|F\,4,2)$$
$$+ A(P|F\,5,2)(P|F\,4,2) + A(P|F\,6,1)(P|F\,5,2)(P|F\,4,2)$$
$$\$429.79 = A[(0.9615) + (0.9246) + (0.9524)(0.9246)$$
$$+ (0.9070)(0.9246) + (0.9434)(0.9070)(0.9246)]$$
$$4.396\,A = \$429.79$$
$$A = \$97.76$$

Thus, $97.76 per time period for five time periods is equivalent to the original cash flow series. As an exercise one may wish to solve the example problem assuming continuous compounding.

3.9.2 End-of-Period Cash Flows and End-of-Period Compounding

Thus far we have emphasized end-of-period cash flows and end-of-period compounding as well as the treatment of continuous cash flows. Depending on the financial institution involved, accounts might not pay interest on deposits made in "the middle of a compounding period." Consequently, you should not expect answers obtained using the methods we describe to be exactly the same as those provided by the financial institution.

The beginning-of-period cash flows can be handled very easily by noting that the end of period k is the beginning of period $k + 1$. To illustrate, rental payments are usually made at the beginning of each month. However, one can think of the payment made at the beginning of, say, March as having been made at the end of February.

In Chapter 4, end-of-year cash flows are assumed unless otherwise noted. It is realized that monetary transactions take place during a calendar year, but it is convenient to ignore any compounding effects within a year and deal directly with end-of-year cash flows.

We previously mentioned the possibility of cash flows occurring at intervals that were not the same as compounding intervals. Two possibilities might arise: interest is compounded *more* frequently than the occurrence of cash flows and interest is compounded *less* frequently than the occurrence of cash flows.

Example 3.31

Suppose an individual makes monthly deposits into a fund that pays interest at a rate of 8% compounded quarterly. Depending on how the financial institution interprets the rate of "8% compounded quarterly," money deposited during a quarter might not earn any interest. For example, the institution may pay the 8% interest compounded quarterly on the minimum quarterly balance. We interpret 8% compounded quarterly to mean that deposits earn interest from the time the deposit is made at the rate equivalent to 8% compounded quarterly. Hence, the monthly interest rate to be applied to the monthly deposits should be equivalent to 8% compounded quarterly. Letting i be the monthly rate, we note that the effective interest rate for $i\%$ per month should be the same as for 8% compounded quarterly. Therefore,

$$i_{eff} = (1+i)^{12} - 1$$
$$= \left(1 + \frac{0.08}{4}\right)^4 - 1$$

or

$$(1+i)^{12} = (1.02)^4$$

or

$$(1+i)^3 = 1.02$$

Thus,

$$1+i = (1.02)^{1/3}$$
$$i = (1.02)^{1/3} - 1$$
$$i = 0.0066227$$

Consequently, a monthly interest rate of 0.66227% can be used to determine the balance in the fund at the end of any month.

Example 3.32

Suppose quarterly deposits are made into a fund that pays interest at a rate of 6% compounded monthly. By the same arguments used in the previous example, we know that

$$i_{eff} = (1+i)^4 - 1$$
$$= \left(1 + \frac{0.06}{12}\right)^{12} - 1$$

or

$$(1+i)^4 = (1.005)^{12}$$
$$i = (1.005)^3 - 1$$
$$= 0.015075$$

Thus, a quarterly interest rate of 1.5075% can be used to determine the balance in the fund at the end of any quarter.

3.9.3 Perpetuities and Capitalized Value

A specialized type of cash flow series is a perpetuity, a uniform series where the payments continue indefinitely. By virtue of being a special case, an infinite series of cash flows would be encountered much less frequently in the business world than a finite series of cash flows. However, for such very long-term investment projects as bridges, power plants, dams, mines, highways, forest harvesting, or the establishment of endowment funds where the estimated life is 50 years or more, an infinite cash flow series may be appropriate.

If a present value P is deposited into a fund at interest rate i per period so that a payment of size A may be withdrawn each and every period forever, then the following relation holds between P, A, and i:

$$Pi = A$$

Thus, as depicted in Figure 3.18, P is a present value that will pay out equal payments of size A indefinitely if the interest rate per period is i. The present value P is termed the *capitalized value* of A, the size of each of the perpetual payments.

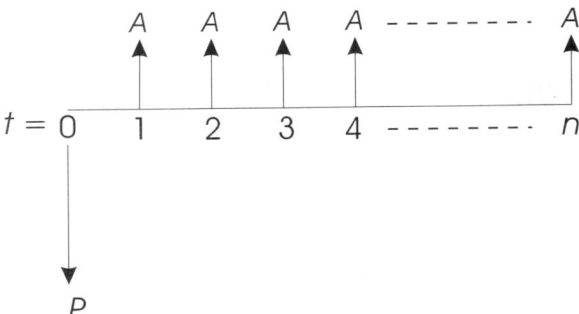

Fig. 3.18 An infinite cash flow series.

Example 3.33

What deposit at $t=0$ into a fund paying 15.5% annually is required in order to pay out $600 each year forever? The solution is straightforward:

$$P = \frac{A}{i} = \frac{\$600}{0.155} = \$3\,870.97$$

By means of the subsequent example, let us now broaden the meaning of capitalized value to be that present value that would pay repeatedly for the first cost of some project and provide for its perpetual maintenance at an interest rate i.

Example 3.34

Project MX107 consists of the following requirements:
1. A $20 000 first cost at $t=0$.
2. A $2 000 expense every year.
3. A $10 000 expense every third year forever, with the first expense occurring at $t=3$.

What is the capitalized cost of Project MX107 if $i=10\%$ annually?

It will be instructive to determine the capitalized cost of each requirement separately and then sum the results. First, the capitalized cost of "$10 000 every third year forever" may be determined from any of three points of view. One view is that a value P is required at the beginning of a three-year period such that, with interest compounded at 10% annually, a sum of $P + \$10\,000$ will accrue at the end of the three-year period. Thus, $10 000 would be withdrawn; thereby leaving the value P to repeat the cycle indefinitely each three-year period. The logic is illustrated in Figure 3.19.

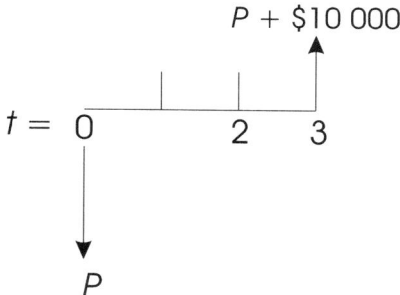

Fig. 3.19 Capitalized cost of $10 000 every third year—first view.

Thus,

$$P(F|P\ 10,3) = P + \$10\ 000$$

or

$$1.3310P = P + \$10\ 000$$

and

$$P = \$30\ 211.48$$

which is the capitalized cost of the requirement. A second view is to reason that, for each three-year period, three equal deposits of size A are required that will amount to $10 000 at the end of the third year. Then, if these payments of size A occurred every year, $10 000 would be available every third year forever. The present value P that yields the required payment A is the solution, as shown in Figure 3.20.

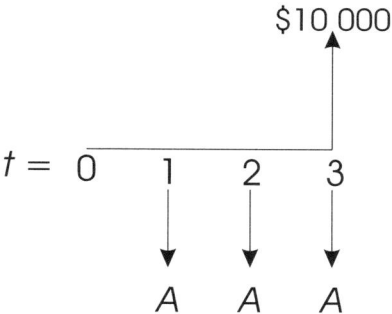

Fig. 3.20 Capitalized cost of $10 000 every third year—second view.

Furthermore,

$$A = F(A|F\ 10,3) = \$10\,000(0.3021)$$
$$= \$3\,021$$

Then,

$$P = \frac{A}{i} = \frac{\$3\,021}{0.10} = \$30\,210$$

which is approximately the same answer as from the first approach. The difference of $1.48 is due to rounding in the interest factor calculations.

A third point of view for the $10 000 requirement is to consider the infinite series depicted in Figure 3.21, such that $P = \frac{A}{i}$ would yield the desired result. However, the value of i required is the effective interest rate for a three-year period. Thus,

$$i = \left[(1+0.10)^3 - 1\right] = 0.3310$$

and

$$P = \frac{\$10\,000}{0.3310} = \$30\,211.48$$

The capitalized cost of Project MX107 is computed as:

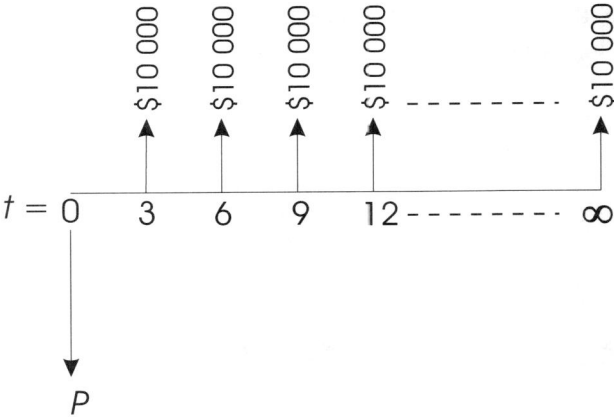

Fig. 3.21 Capitalized cost of $10 000 every third year—third view.

1. Capitalized cost of $20 000 first cost	=	$20 000
2. Capitalized cost of $2 000 every year = $2 000/0.10	=	$20 000
3. Capitalized cost of $10 000 every third year	=	$30 210
Total capitalized cost		$70 210

The capitalized value of $70 210 would provide

$$A = Pi = \$70\,210(0.10) = \$7\,021$$

every year forever.

3.9.4 Bond Problems

Although bonds are important financial instruments in the business world as investment opportunities, there are additional reasons for considering bond problems in an economic analysis textbook.

1. The issuance and sale of bonds is a mechanism by which capital may be raised to finance engineering projects.

2. Bond problems illustrate the notion of equivalence. That is, the purchase price of a bond is equivalent to (has the same present value as) the returns from the bond at an appropriate compound interest rate. The returns from the bond consist of periodic interest payments to the bondholder and the redemption value, or sales price, of the bond.

3. A bond problem to calculate the *yield* on the bond's purchase price is analogous to the calculation of the *internal rate of return* on an investment. The problems involving the calculation of bond yield will introduce the internal rate of return method, which is presented in Chapter 4 as a means of comparing alternative investment projects.

The latter two reasons are the primary ones for treating bond problems in this text. Subsequent discussion presents the appropriate terminology and illustrates the three types of problems that are possible.

An organizational unit desiring to raise capital may issue bonds totalling, say, $1 million, $5 million, $25 million, or more. A financial brokerage firm usually handles the issue on a commission basis and sells smaller amounts to other organizational units or individual investors. Individual bonds are normally issued in even denominations such as $500, $1 000, or $5 000. The stated value of the individual bond is termed the *face* or *par value*. The par value is to be repaid by the issuing organization at the end of a specified period of time, say 5, 10, 15, 20, or even 50 years. Thus, the issuing unit is obligated to *redeem* the bond at par value at *maturity*. Furthermore, the issuing unit is obligated to pay a stipulated *bond rate* on the face value during the interim between the date of issuance and date of redemption. This might be 5% payable quarterly, 5.125% payable semiannually, 5.25% payable annually, etc. For the purpose of the problems to follow, it is emphasized that the bond rate applies to the par value of the bond.

Example 3.35

A person purchases a $5 000, five-year bond on the date of issuance for $5 000. The stated bond rate is "8% seminannually," and the interest payments are received on schedule until the bond is redeemed at maturity for $5 000. The bond rate per interest period is 8%/2 = 4%. Thus, the bond holder receives (0.08/2)($5 000) = $200 payments every six months. A cash flow diagram for the duration of the investment is given in Figure 3.22, where time periods are six-month intervals. At the end of five years, in addition to the last interest payment, the face value of the bond is also repaid.

It is noted from Figure 3.22 that the $5 000 expenditure at $t = 0$ yields $200 each interest period for 10 periods and a $5 000 redemption value at $t = 10$. Thus, the $5 000 investment at $t = 0$ yielded the revenues from $t = 1$ through $t = 10$. What annual rate of return or interest rate did the $5 000 investment yield?

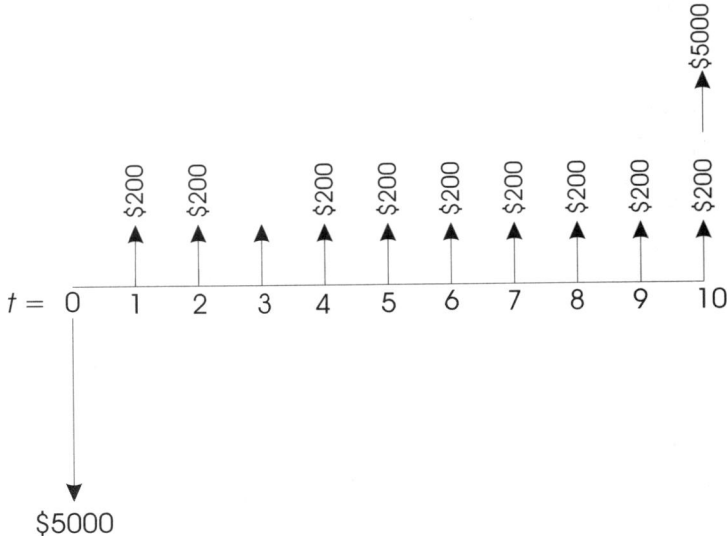

Fig. 3.22 Cash flow for a $5 000 bond.

One might intuitively answer the above question by stating that the $5 000 investment was exactly returned (no loss or gain in capital) at redemption and, since during the interim an interest rate of 8% paid semiannually was received, then the yield *must* be 8% semiannual. Accepting this argument for the moment, one might further pose certain hypotheses concerning the transaction. For instance, it is hypothesized that the $5 000 at $t=0$ is equivalent (has the same present worth) to the revenue cash flows if the time value of money is 8% compounded semiannually. That this hypothesis is true is shown by the relation

$$P = A(P|A\ 4,10) + F(P|F\ 4,10)$$

or

$$\$5\,000 = (0.04)(\$5\,000)(8.1109) + \$5\,000(0.6756)$$

and

$$\$5\,000 = \$5\,000$$

A second hypothesis concerning the transaction is that if a premium above par value is paid for the bond at $t=0$ and all revenue figures remain the same, then an annual yield (rate of return) less than the bond rate of 8% semiannual will be received on the investment. For example, if $5 500 were paid for the bond at $t=0$, the present worth of the $5 500 investment is not equal (equivalent) to the present worth of the revenues at an 8% semiannual yield. This is shown by

$$\$5\,500 \neq (0.04)(\$5\,000)(P|A\ 4,10) + \$5\,000(P|F\ 4,10)$$
$$\$5\,500 \neq \$200(8.1109) + \$5\,000(0.6756)$$
$$\$5\,500 \neq \$5\,000$$

This result now raises the pertinent question, what *is* the annual yield on the investment if $5 500 is paid for the bond and the revenues remain the same? Intuition may suggest that the yield will be less than 8% semiannual because the purchase value of $5 500 decreases to $5 000 on redemption, resulting in a loss of investment capital. Furthermore, the semiannual payments of $200 are not equal to 4% of the $5 500 purchase price. In order to answer the yield question more precisely, let us answer the alternate question of "what interest rate per period (or yield per period) will make the purchase price equivalent to the revenue cash flows?" That is,

Present worth of investment = Present worth of revenues

at what interest rate per period? Alternately, what interest rate satisfies the following equation?

$$\$5\,500 = (0.04)(\$5\,000)(P|A\ ?,10) + \$5\,000(P|F\ ?,10)$$

The solution to the above equation gives the answer to the original question, and the task is to solve for the positive roots of the polynomial. An accurate solution to this problem can be found by using an appropriate computer program. An approximate solution is by trial and error. An iterative procedure follows.

For $i = 4\%$ (8% semiannual),

$$\$5\,500 \neq (0.04)(\$5\,000)(P|A\ 4,10) + \$5\,000(P|F\ 4,10)$$
$$\$5\,500 \neq \$200(8.1109) + \$5\,000(0.6756)$$
$$\$5\,500 \neq \$5\,000$$

For $i = 3\%$ (6% semiannual),

$$\$5\,500 \neq (0.04)(\$5\,000)(P|A\ 3,10) + \$5\,000(P|F\ 3,10)$$
$$\$5\,500 \neq \$200(8.5302) + \$5\,000(0.7441)$$
$$\$5\,500 \neq \$5\,426.50$$

For $i = 2.5\%$ (5% semiannual),

$$\$5\,500 \neq (0.04)(\$5\,000)(P|A\ 2.5,10) + \$5\,000(P|F\ 2.5,10)$$
$$\$5\,500 \neq \$200(8.7521) + \$5\,000(0.7812)$$
$$\$5\,500 \neq \$5\,656.42$$

Table 3.4 Bond Yield Interpolation

For $i =$	0.03	X	0.025
PW of revenues =	$5426.50	$5500	$5652.42

From the last two trials (for $i = 3\%$ and $i = 2.5\%$), the equivalent present worth of $5 500 desired for revenues is bracketed, as shown in Table 3.4. Using the data of Table 3.4, we can solve for X by linear interpolation, or

$$\frac{0.03 - 0.025}{\$5\,426.50 - \$5\,652.42} = \frac{0.03 - X}{\$5\,426.50 - \$5\,500}$$

and

$$X = 0.02837 \text{ or } 2.83\%$$

Thus, the equivalent yield on the $5 500 investment is approximately 2.837% per period, or (2)(2.837%) = 5.674% semiannually, or an effective annual yield of

$$\left[(1+0.02837)^2 - 1\right](100\%) = 5.7545\%$$

These figures support the *a priori* intuition of an annual yield less than 8% semiannual, and the second hypothesis is accepted.

Bond problems often arise in economic analysis. Bonds trade daily through financial markets such as banks and securities brokerages. Thus, bonds may be purchased and sold for less than, greater than, or equal to par value, depending on the supply and demand. Furthermore, once purchased, bonds may be kept for a variable number of interest periods before being sold. A variety of situations can occur, but there are only three basic types of bond problems. These will be presented after formalizing the discussion thus far. We now employ the following notation:

P = the purchase price of a bond

F = the sales price (or redemption value) of a bond

V = the par or face value of a bond

r = the bond rate per interest period

i = the yield rate per interest period

n = the number of interest payments received by the bondholder

$A = Vr$ = a single interest payment

The general expression relating these is

$$P = Vr(P|A\ i,n) + F(P|F\ i,n) \qquad (3.62)$$

Now, the three types of bond problems follow:

1. Given P, r, n, V, and a desired i, find the sales price F.
2. Given F, r, n, V, and a desired i, find the purchase price P.

3. Given $P, F, r, n,$ and V, find the yield i that has been earned on the investment.

Each of these cases is illustrated in the following examples.

Example 3.36

A $1 000, 8% semiannual bond is purchased for $1 050 at $t = 0$. If the bond is sold at the end of three years and six interest payments, what is the selling price to earn 6% nominal? The cash flow for this example is given by Figure 3.23.

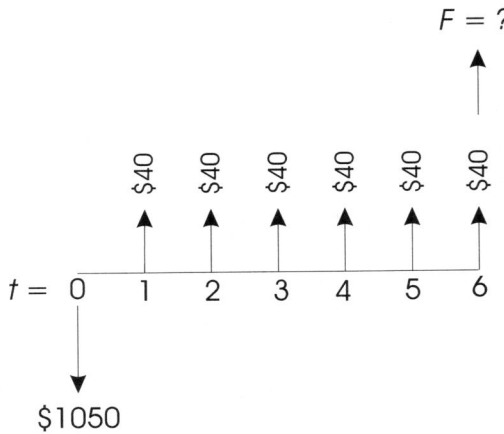

Fig. 3.23 Cash flow for a $1000 bond—determine sales price.

A mathematical statement of the cash flow depicted in Figure 3.23 follows.

$$P = Vr(P|A\ 3{,}6) + F(P|F\ 3{,}6)$$

or

$$\$1\,050 = (\$1\,000)(0.04)(5.4172) + F(0.8375)$$

Solving the above equation for F yields a value of $995.

Example 3.37

If a $1 000, 8% semiannual bond is purchased at $t = 0$, held for three years and six interest payments, and redeemed at par value, what must the purchase price have been in order to earn a nominal yield of 10%?

From the cash flow given in Figure 3.24,

$$P = Vr(P|A\ 5,6) + F(P|F\ 5,6)$$

or

$$P = (\$1\,000)(0.04)(5.0757) + \$1\,000(0.7462)$$

and

$$P = \$949.23$$

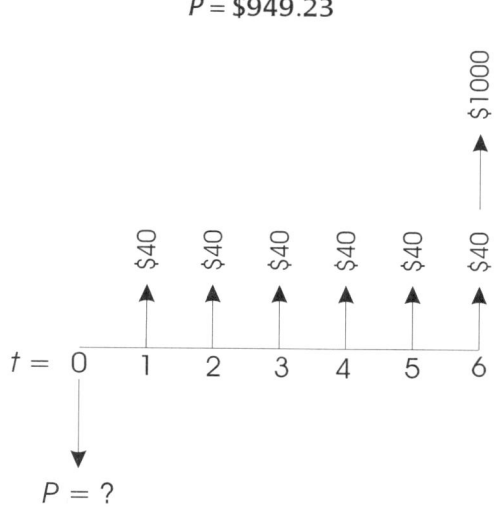

Fig. 3.24 Cash flow for a $1000 bond—determine purchase price.

Example 3.38

If a $1 000, 8% quarterly bond is purchased at $t = 0$ for $1 020 and sold three years later for $950, (1) what was the quarterly yield on the investment, and (2) what was the effective annual yield? From the cash flow diagram in Figure 3.25

$$\$1\,020 = (\$1\,000)(0.02)(P|A\ i,12) + \$950(P|F\ i,12)$$

It is then necessary to solve for the unknown i either analytically or by trial and error as follows:
For $i = 1\%$,

$$\$1\,020 \neq \$20(11.2551) + \$950(0.8874)$$
$$\$1\,020 \neq \$1\,068.13$$

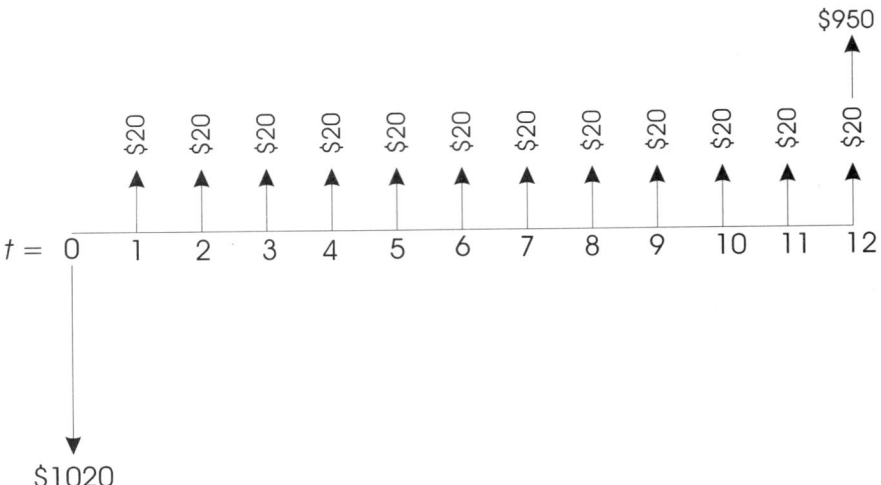

Fig. 3.25 Cash flow for a $1000 bond—determine yield.

For $i = 1.5\%$,

$$\$1\,020 \neq \$20(10.9075) + \$950(0.8364)$$
$$\$1\,020 \neq \$1\,012.73$$

Then, interpolating gives

$$\frac{0.01 - 0.015}{\$1\,068.13 - \$1\,012.73} = \frac{0.01 - X}{\$1\,068.13 - \$1\,020}$$

$$X = 0.01434 \quad \text{or} \quad 1.434\% \text{ per quarter}$$

The effective annual yield is

$$\left[(1+0.01434)^4 - 1\right]100\% = 5.861\%$$

3.9.5 Capital Recovery Cost

In the engineering economy literature there is frequent reference to the term *capital recovery cost*. If an investment of $P is made in an asset, the asset is used for n years, and disposed of for salvage value of $F, then the capital recovery cost, CR, is defined as

$$CR = P(A|P\,i,n) - F(A|F\,i,n) \qquad (3.68)$$

However, since

$$(A|Pi,n) = (A|Fi,n) + i \qquad (3.69)$$

then on substituting Equation 3.69 into Equation 3.68 we obtain

$$CR = (P - F)(A|Fi,n) + Pi \qquad (3.70)$$

As will be noted in Section 5.4.4, the first term on the right hand side of Equation 3.70 is the annual sinking fund deposit for sinking fund depreciation; the second term is the opportunity cost due to $P being tied up in the asset. On the basis of Equation 3.70, capital recovery cost is often defined as the cost of depreciation plus a minimum return on the investment.

Alternately, Equation 3.68 can be given as

$$CR = (P - F)(A|Pi,n) + Fi \qquad (3.71)$$

due to the relationship between the factors $(A|P\ i,n)$ and $(A|F\ i,n)$. Among the three methods of computing the capital recovery cost, Equation 3.71 appears to be the most popular. We tend to use Equation 3.68, since it is a direct application of the cash flow approach.

Example 3.39

Consider an investment of $10 000 in a unit of equipment which lasts for 10 years and is sold for $1 000. An interest rate of 10% is used. Determine the capital recovery cost using the three methods presented.

$$CR = \$10\,000(A|P\ 10\%,10) - \$1\,000(A|F\ 10\%,10)$$
$$= \$10\,000(0.1627) - \$1\,000(0.0627)$$
$$= \$1\,564.30 \quad \text{per year}$$

$$CR = (\$10\,000 - \$1\,000)(A|F 10\%,10) + \$10\,000(0.10)$$
$$= \$9\,000(0.0627) + \$1\,000$$
$$= \$1\,564.30 \quad \text{per year}$$

$$CR = (\$10\,000 - \$1\,000)(A|P\ 10\%,10) + \$1\,000(0.10)$$
$$= \$9\,000(0.1627) + \$100$$
$$= \$1\,564.30 \quad \text{per year}$$

3.10 Inflationary Effects

The dynamic nature of the economy in the past has focused considerable attention on inflation and its effect on economic decision making. Alternative approaches that are typically used to account for inflationary effects include:

1. Express all cash flows in terms of "then-current" dollar amounts and combine the inflation rate with the interest rate.
2. Express all cash flows in terms of "constant worth" dollar amounts and use an interest rate without an inflation rate component.

The latter approach appears to be the method most preferred by practitioners. However, it is not uncommon to encounter claims that the interest or discount rate used in time value of money computations includes a component for inflation. If care is not taken to insure a proper accounting for inflation, one might partially account for inflation, rather than account for it in both the discount rate used and the cash flow estimates.

Letting j be the inflation rate, C_k be the "constant worth" value of a cash flow at the end of period k, and T_k be the "then-current" value of a cash flow at the end of period k, the following relation holds:

$$T_k = C_k(1+j)^k \tag{3.63}$$

Thus, when the set of constant worth cash flows constitutes a *uniform series*, (i.e., $C_k = A$, $k = 1,\ldots,n$), the set of then-current cash flows constitutes a *geometric series*, i.e., $T_k = A_1(1+j)^{k-1}$, where $A_1 = A(1+j)$.

Using the first approach to account for inflationary effects, the present worth equivalent of a series of T_k cash flows is computed as follows:

Time Value of Money Operations

$$P = \sum_{k=0}^{n} T_k (1+j)^{-k} (1+i)^{-k}$$

$$= \sum_{k=0}^{n} T_k (1+j+i+ij)^{-k}$$

$$= \sum_{k=0}^{n} T_k (1+d)^{-k} \quad (3.64)$$

where d is a discount rate equal to $i+j+ij$. The second approach computes the present worth equivalent using constant worth dollar amounts by noting that

$$P = \sum_{k=0}^{n} T_k (1+j)^{-k} (1+i)^{-k}$$

can be given as

$$P = \sum_{k=0}^{n} C_k (1+i)^{-k} \quad (3.65)$$

In the preceding sections of the chapter we implicitly assumed that either the cash flows were in the form of constant worth dollars or the discount rate used with then-current dollars included a component for inflation.

Example 3.40

To illustrate the two approaches in dealing with inflation, suppose

$$j = 5\%$$
$$i = 10\%$$
$$T_0 = -\$10\,000$$
$$T_1 = \$1\,000$$
$$T_2 = \$3\,000$$
$$T_4 = \$7\,000$$

As shown in Table 3.5, the present worth equivalent for the situation is –$356. In this instance, $d = 0.05 + 0.10 + 0.005 = 0.155$, or 15.5%. If the present worth were computed using $d = 10\%$ and then-current dollar estimates, the present worth would have been $1 175 instead of the –$356 obtained considering inflation. Consequently, what appears to have been a profitable investment without considering inflation has a negative present worth when the effects of inflation are included in the analysis.

Table 3.5 Present Worth Calculations Under Inflation

k	T_k	$(1+d)^{-k}$	$T_k(1+d)^{-k}$	C_k^*	$(1+i)^{-k}$	$C_k(1+i)^{-k}$
0	−$10 000	1.0000	−$10 000	−$10 000	1.0000	−$10 000
1	1 000	0.8658	866	952	0.9091	866
2	3 000	0.7496	2 249	2 721	0.8264	2 249
3	4 000	0.6490	2 596	3 455	0.7513	2 596
4	7 000	0.5619	3 933	5 759	0.6830	3 933
			−356			−356

$^*C_k = T_k(1.05)^{-k}$

Although the second approach, involving constant worth dollar amounts, appears to be the simplest method of incorporating inflation in the analysis, estimating cash flows in terms of constant worth dollar amounts is not a trivial exercise. We mentioned previously that a uniform series of constant worth cash flows converts to a geometric series of then-current cash flows. You expect cash flows for, say, labour costs to increase over time; some portion, but not all, of the increase is due to inflation. Thus, it is difficult to factor out the portion of the increase that is due to inflation in order to provide inflation-free estimates of future cash flows.

Another aspect of inflation that tends to complicate the analysis is the differences that may exist in inflation rates for not only various types of cash flows, but also for different regions of the country and world. The firm that has numerous plants scattered throughout not only the same country, but also the world, must cope with the economic differences that exist among the locations, as well as the differences in inflation rates for labour, equipment, materials, utilities, and supplies.

As a further complication in dealing with inflation, the inflation rate tends to change from one time period to the next. If j_t denotes the inflation rate for period t, and i_t denotes the interest rate for period t, the present worth can be expressed in then-current dollars as

$$P = \sum_{k=0}^{n} T_k \prod_{t=0}^{k} (1+j_t)^{-1} (1+i_t)^{-1} \qquad (3.66)$$

in the case of discrete compounding and as

$$P = \sum_{k=0}^{n} T_k e^{-\sum_{t=0}^{k}(c_t + r_t)} \qquad (3.67)$$

in the case of continuous compounding, where c_t is the nominal inflation rate for period t, and r_t is the nominal interest rate for period t; in the above, c_0, r_0, j_0, and i_0 are defined to be equal to zero.

Inflation is a much discussed subject in the area of economic investment alternatives. Some argue that inflation effects can be ignored, since inflation will affect all investment alternatives in roughly the same way. Thus, it is argued, the relative differences in the alternatives will be approximately the same with or without inflation considered. Others argue that the inflation rate in the past has been so dynamic that an accurate prediction of the true inflation rate and its impact on future cash flows is not possible. Another argument for ignoring explicitly the effects of inflation is that it is accounted for implicitly, since cash flow estimates for the future are made by individuals conditioned by an inflationary economy. Thus, it is argued, any estimates of future cash flows probably incorporate implicitly inflationary effects. A final argument for ignoring inflation in comparing investment alternatives involving only negative cash flows is that an alternative that is preferred by ignoring inflation effects will be even more attractive when effects of inflation are incorporated in the analysis.

The above arguments are certainly valid in some instances. However, counter-arguments can be given for each. Consequently, before dismissing inflation effects as unnecessary in performing economic analysis, the individual situation should be considered closely.

3.11 Summary

In this chapter we developed the concept of the time value of money and defined a number of mathematical operations consistent with that concept. The principles introduced here from the viewpoint of personal financing will be applied in subsequent chapters to the study of investment alternatives from the viewpoint of an ongoing enterprise.

Problems

1. Person A sells Person B a used automobile for $2 000. Person B pays $500 cash down and gives Person A two personal notes for the remainder due. The principal of each note is therefore $750. One note is due at the end of the first year; the other note is due at the end of the second year. The annual simple interest rate agreed on is 8.5%. How much total interest will Person B pay Person A? (3.2)

2. A debt of $1 000 is incurred at $t = 0$. An annual simple interest rate of 8% on the unpaid balance is agreed on. Three equal payments of $388 each at $t = 1, 2, 3$ will pay off this debt and the relevant interest due. For *each* payment, what is (a) the payment on the principal and (b) the interest amount paid? (3.2)

3. If $5 000 is deposited at $t = 0$ into a fund paying 6% compounded per period, what sum will be accumulated at the end of eight periods, or $t = 8$? What would be the sum accumulated if the fund paid 5% compounded per period? (3.3)

4. If a deposit of $1 000 at $t = 0$ amounts to $4 300 at the end of the eighth compounding period, what value of i is involved? (3.3)

5. How long does it take a deposit in a 6% fund to triple in value? (3.3)

6. If a fund pays 6% compounded annually, what single deposit is required at $t = 0$ in order to accumulate $8 000 in the fund at the end of the tenth year ($t = 10$)? (3.3)

7. How much money today is equivalent to $1 000 in five years, with interest at 7% compounded annually? (3.3)

8. A person deposits $500, $1 200, and $2 000 at $t = 0, 1, 2$, respectively. If the fund pays 6% compounded per period, what sum will be accumulated in the fund at (a) $t = 2$ and (b) $t = 3$? (3.4)

9. How much should be deposited at $t = 0$, into a fund paying 10% compounded per period, in order to withdraw $700 at

$t = 1$, $1 500 at $t = 3$, and $2 000 at $t = 7$ and the fund be depleted?

10. A person deposits $4 000 in a savings account that pays 6% compounded annually. Three years later he deposits $4 500. Two years after the $4 500 deposit $2 500 is deposited. Four years after the $2 500 deposit, half of the accumulated funds is transferred to a fund that pays 7% compounded annually. How much money will be in each fund six years after the transfer? (3.4)

11. A woman annually deposits $$A_k$ in an account at time k, $k = 1, \ldots, 20$, where $A_k = 10k(1.05)^{k-1}$. If the fund pays 5%, how much is in the fund immediately after the seventeenth deposit? (3.4)

Note:

$$\sum_{k=1}^{n} k = \frac{n(n+1)}{2}$$

$$\sum_{k=1}^{n} kx^k = \frac{(x-1)(n+1)x^{n+1} - x^{n+2} + x}{(x-1)^2}$$

12. What equal, annual deposits must be made at $t = 1, 2, 3, 4, 5, 6$ in order to accumulate $10 000 at $t = 6$ if money is "worth" 10% compounded annually? (3.4.1)

13. A debt of $1 000 is incurred at $t = 0$. What is the amount of three equal payments at $t = 1, 2, 3$ that will repay the debt if money is "worth" 8% compounded per period? (3.4.1)

14. Five deposits of $300 each are made at $t = 1, 2, 3, 4, 5$ into a fund paying 8% compounded per period. How much will be accumulated in the fund at (a) $t = 5$ and (b) $t = 9$? (3.4.1)

15. If you know the values of the $(F|P\ 6, n)$ factor for $n = 1, 2, \ldots, 10$, show how you would determine the value for the $(A|P\ 6, 10)$ factor. (3.4.1)

16. A deposit of $$X$ is placed into a fund paying 10% compounded annually at $t = 0$. If withdrawals of $3 154.67 can be made at $t = 1, 2, 3$, and 4 such that the fund is depleted with the last withdrawal, show that the value of X is $10 000 by (a)

use of the interest factor $(P|A\ 10,4)$ and (b) use of the interest factors: $(P|F\ 10,1)$, $(P|F\ 10,2)$, $(P|F\ 10,3)$, and $(P|F\ 10,4)$. (3.4.1)

17. A person deposits $1 000 in a savings account; five years after the deposit, half of the account balance is withdrawn; $2 000 is deposited annually for five more years, with the total balance withdrawn at the end of the fifteenth year. If the account earns interest at a rate of 5%, how much is withdrawn (a) at the end of five years and (b) at the end of 15 years? (3.4.1)

18. With 5% interest compounded annually, how much money will be accumulated in a fund immediately after the fifth deposit, if equal deposits of $521.08 are made annually? (3.4.1)

19. A person borrows $10 000 at 8% compounded annually and wishes to pay the loan back over a five-year period with annual payment. However, the second payment is to be $500 greater than the first payment; the third payment is to be $500 greater than the second payment; the fourth payment is to be $500 greater than the third payment; and the fifth payment is to be $500 greater than the fourth payment. Determine the size of the first payment. (3.3)

20. A debt of $X is incurred at $t=0$ (purchase of land). It is agreed that payments of $5 000, $4 000, $3 000, and $2 000 at $t=4$, 5, 6, and 7, respectively, will satisfy the debt if 10% compounded per period is the interest rate. Determine the amount of the debt, $X. (3.4.2)

21. Determine the future worth, F, at $t=15$ of the following deposits: $1 000 at $t=8$, $900 at $t=9$, $800 at $t=10$, and $700 at $t=11$. Assume the fund pays 20% compounded per period. (3.4.2)

22. Mr. Jones receives an annual bonus from his employer. He wishes to deposit the bonus in a fund that pays interest at a rate of 6% compounded continuously. His first bonus is $1 000. The size of his bonus is expected to increase at a rate of 4% per year. How much money will be in the fund immediately *before* the tenth deposit? (3.6)

23. An individual works for a company that pays an annual bonus, the size of which is based on experience with the

company. After one year with the company, the bonus equals $500. The size of the bonus thereafter compounds at an annual rate of 4%. The individual decides to place half the bonus in a fund that pays 5% compounded annually.

(a) How much money will be in the fund immediately prior to the sixth deposit?

(b) What is the answer to (a) if the fund compounds at 4% annually? (3.4.3)

24. A man invests $10 000 in a venture that returns him $500(0.80)^{k-1}$ at the end of year k for k = 1, ..., 20. With an interest rate of 10%, what is the equivalent uniform annual profit (cost) for the venture? (3.4.3)

25. A man places $1 000 in a fund at the end of 1999. He places end-of-year deposits in the fund until the end of 2018, when his last deposit is made. The fund pays 5% compounded annually. If the size of a deposit, A_k, at the end of year k, equals $0.90 A_{k-1}$, how much will be in the fund immediately after the last deposit? (3.4.3)

26. Ms. Smith deposits $1 000 in a savings account in her local bank. The bank pays interest at a rate of 6% compounded annually. Three years after making the single deposit she withdraws half the accumulated money in her account. Five years after the initial deposit, she withdraws all of the accumulated money remaining in the account. How much does she withdraw five years after her initial deposit? (3.5)

27. A person wishes to make a single deposit P at t = 0 into a fund paying 8% compounded quarterly such that $1 000 payments are received at t = 1, 2, 3, and 4 (periods are three-month intervals) and a single payment of $5 000 is received at t = 12. What single deposit is required? (3.5)

28. A man borrows $20 000 at 8% compounded quarterly. He wishes to repay the money with 10 equal semiannual installments. What must be the size of the payment if the first payment is made one year after obtaining the $20 000? (3.5)

29. A woman borrows $2 000 at 8% per year compounded monthly. She wishes to repay the loan with 12 end-of-month

payments. She wishes to make her first payment three months after receiving the $2 000. She also wishes that, after the first payment, the size of her payment be 10% greater than the previous payment. What is the size of her sixth payment? (3.5)

30. Monthly deposits of $100 are made into an account paying 4% compounded quarterly. Ten monthly deposits are made. Determine how much will be accumulated in the account two months after the last deposit. (3.8.3)

31. A person makes four consecutive semiannual deposits of $1 000 in a savings account that pays interest at a rate of 6% compounded semiannually. How much money will be in the account two years after the last deposit? (3.5)

32. An individual borrows $5 000 at an interest rate of 8% per year compounded semiannually and desires to repay the money with five equal end-of-year payments, with the first payment made two years after receiving the $5 000. What should be the size of the annual payment? (3.5)

33. What is the effective interest rate for 6% compounded monthly? (3.5)

34. What monthly deposits must be made in order to accumulate $4 000 in five years if a fund pays 9% compounded monthly? (Assume the first deposit is made at the end of the first month, and the final deposit is made at the end of the fifth year.) (3.5)

35. What monthly payments are required in order to pay off a $20 000 debt in five years if the *nominal* rate of interest charged is 9% compounded annually? (3.8.3)

36. A man borrows $20 000 at 8% compounded semiannually. He pays back the loan with four equal semiannual payments, with the first payments made one year after receiving the $20 000. What should be the size of each of the four payments? (3.5)

37. A man borrows $1 000 and pays the loan off, with the interest, after two years. He pays back $1 150. What is the effective interest rate for this transaction? (3.5)

38. A woman borrows $5 000 at 1% per month. She desires to repay the money using equal end-of-month payments for 10 months. The woman makes four such payments and decides to pay off the remaining debt with one lump sum payment at the time for the fifth payment. What should the size of this payment be if interest is truly compounded at a rate of 1% per month?

39. Mr. Wright borrows $8 000 from a bank that charges interest at 6% compounded semiannually. Mr. Wright is to pay the money back with six equal payments. However, the first payment is to be made immediately on receipt of the $8 000. Successive payments are spaced one year apart.

   (a) Determine the size of the equal annual payment.

   (b) At the time of the fourth payment, suppose Mr. Wright decides to pay off the loan with one lump sum payment. How much should be paid? Include the fourth payment. (3.5)

40. Approximately how long will it take a deposit to triple in value if money is worth 8% compounded semiannually? (3.5)

41. Assume a person deposits $2 000 now, $1 000 two years from now, and $5 000 five years from now into a fund paying 8% compounded semiannually.

   (a) What sum of money will have accumulated in the fund at the end of the sixth year?

   (b) What equal deposits of size A, made every six months (with the first deposit at $t = 0$ and the last deposit at the end of the sixth year), are equivalent to the three deposits stated above? (3.5)

42. A man borrows $1 000 from the Shady Deal Finance Company. He is told the interest rate is merely 1.7% per month, and his payment is computed as follows:

   Payback period = 30 months
   Interest = 30(0.017)($1 000) = $510
   Credit investigation and insurance = $20
   Total amount owed = $1 530
   Payment size = $1 530/30 = $51 per month

What is the approximate interest rate for this transaction? (3.5)

43. Operating and maintenance costs for a production machine occur continuously during the year. If the total annual operating and maintenance cost is $8 000 and money is worth 15% compounded continuously, what single sum of money at the present is equivalent to five years of operating and maintenance costs? (3.6)

44. Labour costs occur continuously during the year. At the end of each year a new labour contract becomes effective for the following year. Let $\bar{A}_j$ denote the cumulative labour cost occurring uniformly during year j, where $\bar{A}_j = 1.05\bar{A}_{j-1}$. If money is worth 10% compounded continuously and $\bar{A}_1 = \$25\,000$, determine the present worth equivalent for five years of labour costs. (3.6)

45. A person borrows $10 000 and wishes to pay it back with 10 equal annual payments. What will the size of the payments be if the interest charge is 10% compounded (a) annually, (b) semiannually, and (c) continuously? (3.6.1)

46. Operating and maintenance costs for year k are given as $C_k = \left(e^{0.05}\right)C_{k-1}$, $k = 2, 3, \ldots, 15$, with $C_1 = \$1\,000$. Determine the equivalent uniform annual operating and maintenance cost based on continuous compounding with a nominal interest rate of 10%. (3.6.1)

47. Four equal, quarterly deposits of $1 000 each are made at $t = 0, 1, 2$, and 3 (time periods are three-month intervals) into a fund that pays 8% compounded *continuously*. Then, at $t = 7$ and $t = 10$, withdrawals of size $A are made so that the fund is depleted at $t = 10$. What is the size of the withdrawals? (3.6.1)

48. Semiannual deposits of $500 are made into a fund paying 8% compounded continuously. What is the accumulated value in the fund after 10 such deposits? (3.6.1)

49. A firm buys a new computer system that costs $100 000. It may either pay cash now or pay $25 000 down and $10 000 per year for 10 years. If the firm can earn 5% on investments, which would you suggest? (3.7)

50. An individual is considering two investment alternatives. Alternative A requires an initial investment of $10 000; it will yield incomes of $2 500, $3 000, $3 500, and $4 000 over its 4-year life. Alternative B requires an initial investment of $12 000; it is anticipated that the revenue received will increase at a compound rate of 6% per year. Based on an interest rate of 6% compounded annually, what must be the revenue the first year for B in order for A and B to be equivalent? (3.7)

51. It is desired to determine the size of the uniform series over the time period [2, 6] that is equivalent to the cash flow profile shown below using an interest rate of 10%. (3.7)

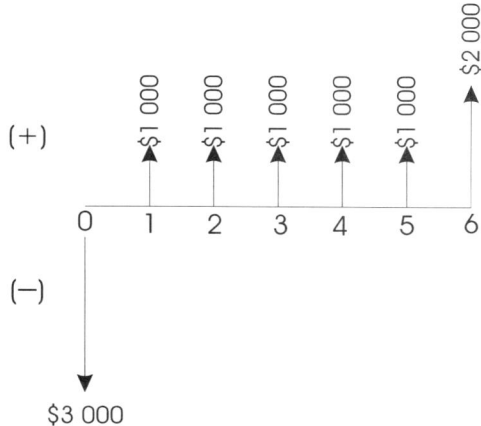

52. Given the following cash flows, what single sum at $t = 6$ is equivalent to the given data. Assume $i = 8\%$. (3.7)

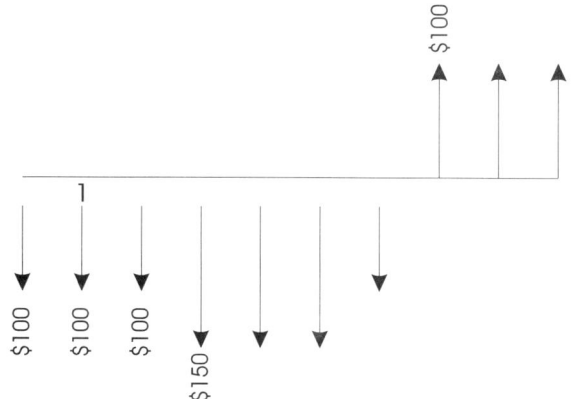

53. A machine is purchased at $t = 0$ for $20 000 (including installation costs). Net annual revenues resulting from operating the machine are $6 000. The machine is sold at the end of 10 years, $t = 10$, for $2 000. The cash flow series for the machine is equivalent to what single sum, $X, at $t = 6$ if money is worth 10% compounded continuously? (3.7)

54. Examine the following two investment plans.

   (a) Purchase for $4 386.68 (a negative cash flow) and receive $400 at the end of each six months for four years $(t = 1,2,\ldots,8)$, and a single payment of $2 000 at the end of the fourth year.

   (b) Purchase for $4 091.14 and receive $900 at the *beginning* of each year for the four-year period, and a single payment of $X at the end of the fourth year.

   If money is worth 6% compounded semiannually, what is the value of X so that the two investments are equivalent? (3.7)

55. A man invests $5 000 and receives $500 each year for 10 years, at which time he sells out for $1 000. With a 10% interest rate, what equal annual cost (profit) is equivalent to the venture? (3.7)

56. Consider the following cash flow series.

EOY	CF
0	− $10 000
1	3 000
2	3 500
3	4 000
4	4 500
5	5 000
6	5 500
7	6 000
8	6 500

At 10% annual compound interest, what uniform annual cash flow is equivalent to the above cash flow series? (3.7)

57. A person borrows $1 000 from a bank at $t=0$ at 8% simple interest for two years. He pays the total interest due for the two-year period at $t=0$ and thus receives $840 at $t=0$. If he pays back $1 000 at $t=2$, the person is, in effect, paying an interest rate of X% compounded annually. Solve for X. (3.7)

58. Assume payments of $2 000, $5 000, and $3 000 are received at $t=3$, 4, and 5, respectively. What five equal payments occurring at $t=1$, 2, 3, 4, and 5, respectively, are equivalent if i=10% compounded per period? (3.7)

59. What single deposit of size $X into a fund paying 10% compounded annually is required at $t=0$ in order to make withdrawals of $500 each at $t=4$, 5, 6, and 7 and a single withdrawal of $1 000 at $t=20$?

    If the above withdrawals are immediately placed into another fund paying 6% compounded annually, what amount will be accumulated in this fund at $t=20$? (3.7)

60. A college student borrows $4 000 and repays the loan with four quarterly payments of $400 during the first year and four quarterly payments of $1 000 during the second year after receiving the $4 000 loan. Determine the effective interest rate for the loan transaction. (3.7)

61. Quarterly deposits of $500 are made at $t=1$, 2, 3, 4, 5, 6, 7. Then withdrawals of size A are made at $t=12$, 13, 14, 15 and the fund is depleted with the last withdrawal. If the fund pays 6% compounded quarterly, what is the value of A? (3.7)

62. Determine the value of X so that the following cash flow series are equivalent at 8% interest. (3.7)

EOY	CF(A)	CF(B)
0	−$8 000	−$15 000
1	6 000	4 000
2	5 000	3 000+X
3	4 000	2 000+2X
4	5 000	3 000+3X
5	6 000	4 000+4X
6	5 000	3 000+5X

63. Given the cash flow diagram shown below and an interest rate of 8% per period, solve for the value of an equivalent amount at (a) $t = 5$, (b) $t = 12$, and (c) $t = 15$. (Remember that upward arrows represent cash inflows and downward arrows represent cash outflows.)

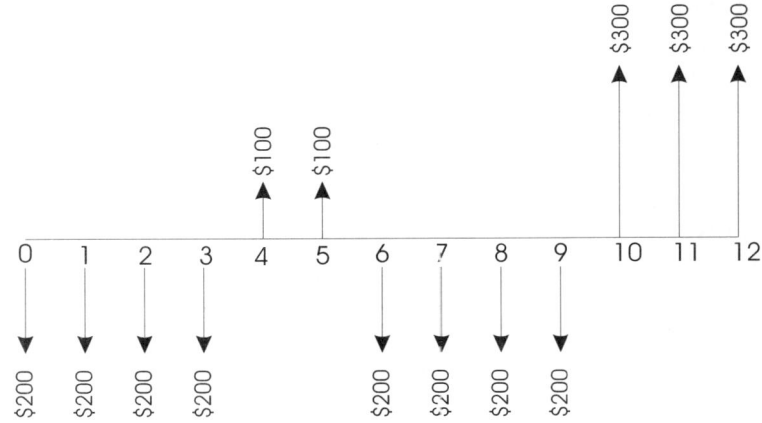

64. Given the cash flow profiles shown below, determine the value of X so that the two cash flow profiles are equivalent at a 10% interest rate. (3.7)

EOY	CF(A)	CF(B)
1	−$12 000	−$10 000
2	1 000	7 000
3	3 000	6 000+0.5X
4	5 000	5 000+1.0X
5	7 000	4 000+1.5X
6	9 000	

65. What single sum of money at $t = 4$ is equivalent to the cash flow profile shown below? Use a 6% interest rate in your analysis. (3.7)

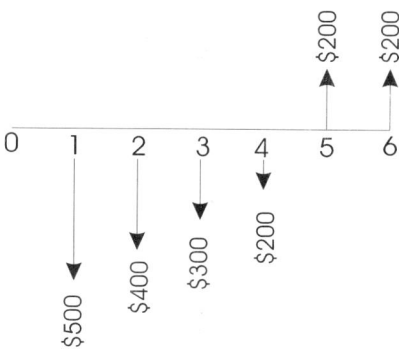

66. Given the two cash flow profiles shown below, for what value of X are the two series equivalent using an interest rate of 10%? (3.7)

EOY	CF(A)	CF(B)
0	− $200 000	− $140 000
1	24 000	16 000
2	32 000	16 000
3	40 000	16 000+X
4	48 000	16 000+2X
5	56 000	16 000+3X
6	64 000	16 000+4X
7	72 000	16 000+5X
8	80 000	16 000+6X

67. At what interest rate is $1 000 today equivalent to $1 967 in 10 years? (3.7)

68. Given the cash flow profiles shown below, determine the value of X such that the two cash flow profiles are equivalent at 8% compounded annually. (3.7)

EOY	CF(A)	CF(B)
1	– $12 000	$ – X
2	1 000	7 000
3	4 000	9 000
4	6 000	10 000
5	7 000	10 000
6	5 000	7 000

69. If a machine costs $10 000 and lasts for 10 years, at which time it is sold for $2 000, what equal annual cost over its life is equivalent to these two cash flows with a 10% annual interest rate? (3.7)

70. Given the two cash flow profiles shown below, for what value of X are the two series equivalent using an interest rate of 10%? (3.7)

EOY	CF(A)	CF(B)
0	– $100 000	– $70 000
1	12 000	8000
2	16 000	8000
3	20 000	8000
4	24 000	8000+X
5	28 000	8000+2X
6	32 000	8000+3X
7	36 000	8000+4X
8	40 000	8000+5X

71. Given the three cash flow profiles shown below, determine the values of X and Y so that all three cash flow profiles are equivalent at an annual interest rate of 8%. Show all work.

EOY	CF(A)	CF(B)	CF(C)
0	−$1 000	−$2 500	$ Y
1	X	3 000	Y
2	1.5X	2 500	Y
3	2.0X	2 000	2Y
4	2.5X	1 500	2Y
5	3.0X	1 000	2Y

72. An investment opportunity offered by commercial banks is guaranteed investment certificates. For example, a guaranteed investment certificate (GIC) having a four-year maturity date may have a stated interest rate of "7.5%, payable quarterly." This statement means that every three months, the issuing bank would pay the GIC holder an amount of (0.075)(1/4) times the face value (purchase price) of the GIC. Assuming a $5 000, four-year, 7.5% GIC were purchased at $t = 0$, quarterly payments of [(0.075)(0.25)($5 000)] = $93.75 would be received at $t = 1, 2, 3, \ldots, 16$, and at $t = 16$, the GIC would be redeemed by the bank for the original purchase price, or $5000. (This payment series is an application of simple interest on a quarterly basis.)

It is also possible for the purchaser of the GIC to arrange with the bank for each interest payment to be deposited into a regular savings account that pays, say, 5.5% compounded quarterly. Let it be assumed that this is done; the first deposit of $93.75 occurs at $t = 1$, and the last deposit is made at $t = 16$.

(a) For the initial investment of $5 000, what total sum of money will be received by the purchaser at the end of the fourth year (GIC redemption value plus savings account balance)?

(b) Now consider only the initial investment, $P = \$5\ 000$, and the F value calculated from (a). What effective annual compound interest rate relates the two single sums of money?

(c) For P and F from (b) above, what annual simple interest rate relates the two single sums of money? (3.7)

73. Assume the following two investment plans:

   Plan A—Purchase for $4 386 and receive the following:
   (1) $400 at the end of each six months for four years.
   (2) A single payment of $2 000 at the end of the fourth year.

   Plan B—Purchase for $3 302 and receive the following:
   (1) $900 at the beginning of each year for four years.
   (2) A single payment of $X at the end of the fourth year.

   If money is worth 6% compounded semiannually, what is the value of X such that the two plans are equivalent? (3.7)

74. A person borrows $5 000 at 6% compounded annually. Five equal annual payments are used to repay the loan, with the first payment occurring one year after receiving the $5 000. Determine the interest amount included in each payment. (3.8)

75. An individual borrows $8 000 at 7% compounded annually. Six equal annual payments are used to repay the loan, with the first payment occurring two years after receiving the $8 000. Determine the payment on principal included in each payment. (3.8)

76. A man borrows $10 000 at 8% compounded monthly. He is to pay off the loan with 60 monthly payments. One month after making the thirtieth payment, he elects to pay off the unpaid balance on the note. How much should he repay? (3.8)

77. An individual makes five annual deposits of $1 000 in a savings account that pays interest at a rate of 4% compounded annually. One year after making the last deposit, the interest rate changes to 5% compounded annually. Five years after the last deposit the accumulated money is withdrawn from the account. How much is withdrawn? (3.9.1)

78. A person deposits $10 000 in a savings account paying 5% compounded annually for the first two years and 6% compounded annually for the next two years. Four annual withdrawals are made from the savings account. The size of the withdrawal increases by $1 000 per year, with the first withdrawal occurring one year after the deposit. Determine the size of the last withdrawal, which depletes the balance of the account. (3.9.1)

79. Assume that six equal deposits of $500 are made at $t = 1, 2, 3, 4, 5,$ and 6 (three-month periods) into a fund paying 6% compounded quarterly. This interest applies until $t = 16$. The accumulated sum is withdrawn at this time and immediately deposited into a fund paying 8% compounded continuously. Beginning with $t = 17$, determine the amount of three equal quarterly payments, A, that may be withdrawn such that the fund is depleted at $t = 19$. (3.9.1)

80. A man borrows $10 000 and repays the loan with four equal annual payments. The interest rate for the first two years of the loan is 5% compounded continuously, and for the third and fourth years of the loan it is 6% compounded continuously. Determine the size of the annual payment. (3.9.1)

81. A person deposits $5 000 in a savings account. One year after the initial deposit, $1 000 is withdrawn. Two years after the first withdrawal, $4 000 is deposited in the account. Three years after the second deposit, $2 000 is withdrawn from the account. Four years after the second withdrawal, all funds are withdrawn from the account. During the period of time the savings account was in use, the bank paid 4% compounded continuously for the first two years, 5% compounded annually for the next four years, and 6% compounded quarterly for the remainder of the time the account was in use. Determine the amount of the final withdrawal. (3.9.1)

82. A man borrows $5 000 and repays the loan with three equal annual payments. The interest rate for the first year of the loan is 5% compounded annually, for the second year of the loan is 6% compounded annually, and for the third year of the loan is 7% compounded annually. Determine the size of the equal annual payment. (3.9.1)

83. A person deposits $1 000 in a fund that pays interest at a continuous compound rate of r_t for the tth year after the initial deposit, where

$$r_t = \begin{cases} 0.04 + 0.005t & t = 1, 2, 3, 4, 5, 6 \\ 0.07 & t = 7, 8, 9, 10, \ldots \end{cases}$$

Thus, a continuous compound rate of 4.5% is earned during the first year, 5% is earned during the second year, 5.5% is earned during the third year, etc. A maximum of 7% compounded continuously is earned during the sixth and each successive year. If the person withdraws $500 two years after the initial deposit, how much money will be in the fund six years after the initial deposit? (3.9.1)

84. A person deposits $2 500 in a savings account that pays interest at a rate of 5% compounded annually. Two years after the deposit, the savings account begins paying interest at a rate of 5% compounded continuously. Five years after the deposit, the savings account begins paying interest at a rate of 6% compounded semiannually.

    (a) How much money should be in the savings account 10 years after the initial deposit?

    (b) What annual compound interest rate is equivalent to the interest pattern of the savings account over the 10-year period? (3.9.1)

85. A man deposits $1 000 in a fund each year for a 10-year period. The fund initially pays 4% compounded annually. Immediately after the man makes his sixth deposit, the fund begins paying 5% compounded annually. The man removes his money from the fund three years after his last deposit. How much should he be able to withdraw at that time? (3.9.1)

86. Based on discrete cash flows and discrete compounding, compute the present worth and annual worth for the following situations: (3.9.1)

EOP t	CF	Interest Rate During Period
0	−$10 000	0
1	2 000	0.05
2	4 000	0.06
3	6 000	0.07
4	8 000	0.08
5	10 000	0.09

87. Solve problem 86 for the case of continuous compounding with the interest rates shown interpreted as nominal rates. (3.9.1)

88. A person deposits $2 000 in a savings account that pays 7% compounded annually; two years after the deposit, the interest rate increases to 8% compounded annually. A second deposit of $2 000 is made immediately after the interest rate changes to 8%. How much will be in the fund five years after the second deposit? (3.9.1)

89. An individual makes monthly deposits of $100 in a savings account that pays interest at a rate equivalent to 6% compounded quarterly. How much money should be in the account immediately after the sixtieth deposit? If no interest is earned on money deposited during a quarter and the first deposit coincides with the beginning of a quarter, what will be the account balance immediately after the sixtieth deposit? (3.9.2)

90. An individual makes semiannual deposits of $500 into an account that pays interest equivalent to 7% compounded quarterly. Determine the account balance immediately *before* the tenth deposit. (3.9.2)

91. A rental contract consists of annual charges payable in advance. The first charge is $1 000 at $t = 0$ and then decreases by $200 each year. If five annual payments are made and money is worth 15% compounded annually, what is the present worth at $t = 0$ of all payments by using the uniform gradient series formula? (3.9.2)

92. If a fund pays 6% compounded annually, what deposit is required today such that $1 000 can be withdrawn every five years forever? (Ignore any tax considerations.) (3.9.3)

93. Maintenance on a reservoir is cyclical with the following *costs* occurring over a five-year period: $3 000, $2 000, $5 000, $0, and $1 000. It is anticipated that the sequence will repeat itself every five years forever. Determine the capitalized cost for the maintenance costs based on a time value of money of 8%. (3.9.3)

94. A bond is purchased for $900 and kept for 10 years, at which time it matures at a face value of $1 000. During the 10-year period $60 is received every six months (i.e., 20 receipts of $60 each). What is the rate of return for the investment? (3.9.4)

95. A person buys a $2 000 bond for $1 800. The bond has a bond rate of 8% with bond premiums paid annually. If the bond is kept for 10 years and sold for par value, determine the equivalent annual return (rate of return) for the bond investment. (3.9.4)

96. A person is considering purchasing a bond having a face value of $2 500 and a bond rate of 8% payable semiannually. The bond has a remaining life of 15 years. How much should be paid for the bond in order to earn a rate of return of 10% compounded semiannually? Assume the bond will be redeemed for face value. (3.9.4)

97. A person wishes to sell a bond that has a face value of $2 000. The bond has a bond rate of 8% with bond premiums paid annually. Four years ago, $1800 was paid for the bond. At least a 10% return on investment is desired. What must be the minimum selling price for the bond in order to make the desired return on investment? (3.9.4)

98. A $5 000, 10-year, 8% semiannual bond is purchased at $t=0$ for par value. After receiving the twelfth dividend, the bond is sold at a price to yield a 6% annual nominal rate of return on this original purchase price.

    (a) What was the selling price?

    (b) If the purchaser keeps the bond until maturity and redeems it for par value, what approximate annual nominal rate of return on the bond investment will be earned? (3.9.4)

99. The following labour *costs* are anticipated over a five-year period: $7 000, $8 000, $10 000, $12 000, and $15 000. It is estimated that a 6% inflation rate will apply over the time period in question. The labour costs given above are expressed in then-current dollars. The time value of money, excluding inflation, is estimated to be 5%. Determine the present worth equivalent for labour cost. (3.10)

100. Labour costs over a four-year period have been forecast in then-current dollars as follows: $10 000, $12 000, $15 000, and $17 500. The inflation rates for the four years are forecast to be 8, 9, 10, and 10%; the interest rate (excluding inflation) is anticipated to be 6, 6, 5, and 5% over the four-year period. Determine the present worth equivalent for labour cost. (3.10)

101. Determine the capital recovery cost for an investment of $100 000 over a 10-year period, with salvage value of $20 000. Use an interest rate of 20%. (3.9.5)

102. Determine the capital recovery cost for an investment of $75 000 over a five-year period, with salvage value of −$25 000. Use an interest rate of 10%. (3.9.5)

Chapter 4
Comparison of Alternatives

4.1 Introduction

Chapter 1 recommended that engineers solve problems by formulating and analyzing the problem, generating a number of feasible solutions (alternatives) to the problem, comparing the alternatives, selecting the preferred solution, and implementing the solution. The process of evaluating the alternatives and selecting the preferred solution is the subject of this chapter.

Here we apply the time value of money concept to the comparison of economic investment alternatives. Although multiple objectives are often involved in performing a comparison of alternatives, for now we concentrate on the comparison of *mutually exclusive* alternatives on the basis of monetary considerations alone. The term, mutually exclusive alternatives, means that no more than one alternative can be chosen. The adage of "not being able to have one's cake and eat it too" illustrates the notion of mutually exclusive alternatives.

A systematic approach that can be used in selection of investment alternatives is summarized as follows:

1. Define the set of feasible, mutually exclusive economic investment alternatives to be compared.
2. Define the planning period to be used in the economic study.
3. Develop the cash flow profiles for each alternative.
4. Specify the time value of money to be used.
5. Specify the measure(s) of merit or effectiveness to be used.
6. Compare the alternatives using the measure(s) of merit or effectiveness.
7. Perform supplementary analyses.
8. Select the preferred alternative.

The procedures for comparing investment alternatives outlined in this chapter are intended to aid in making better measurements of the *quantitative aspects* of investment alternatives. *It cannot be too strongly emphasized that no economic evaluation can replace the sound judgement of experienced managers concerned with both the quantitative and non-quantitative aspects of investment alternatives.* Typical of the aspects of alternatives not considered in this chapter are safety, personnel considerations, product quality, customers' satisfaction, environmental effects, and engineering and construction capability. Such factors are relevant and often control decisions on capital expenditures. However, our concern is with the monetary aspects of the alternatives.

We interpret our role to be the development of logical approaches to be used in *analyzing* investment alternatives and *recommending* action to be taken, based on economic considerations alone. The process of *deciding* which alternative to choose for implementation involves the consideration of monetary and non-monetary factors. Approaches that can be used to assimilate multiple objectives are treated in Chapter 8.

4.2 Defining Mutually Exclusive Alternatives

An individual alternative selected from a set of mutually exclusive alternatives can be made up of several *investment proposals*. Investment alternatives are decision options; investment proposals are single projects or undertakings that are being considered as investment possibilities.

Example 4.1

As an illustration of the distinction between investment proposals and investment alternatives, consider a distribution centre that receives pallet loads of product, stores the product, and ships pallet loads of product to various customer locations. A new distribution centre is to be constructed, and the following proposals have been made:

1. Method of moving materials from receiving to storage and from storage to shipping.
    a. Conventional lift trucks for operating in 4-m aisles.
    b. Narrow-aisle lift trucks for operating in 2-m aisles.
    c. Automatically guided vehicle system.
    d. Belt conveyor system.
    e. Pallet conveyor system.

Comparison of Alternatives

2. Method of placing materials in and removing materials from storage.
   a. Conventional lift trucks for operating in 4-m aisles.
   b. Narrow-aisle lift trucks for operating in 2-m aisles.
   c. Narrow-aisle, operator-driven, rail-guided storage/retrieval vehicle.
   d. Narrow-aisle, automated, rail-guided storage/retrieval vehicle.
3. Method of storing materials.
   a. Stacking pallet loads of material (3 m high, 4-m aisles).
   b. Conventional pallet rack (6 m high, 4-m aisles).
   c. Flow rack (6 m high, 4-m aisles).
   d. Narrow-aisle, pallet rack (6 m high, 2-m aisles).
   e. Flow rack (6 m high, 2-m aisles).
   f. Medium height, pallet rack (12 m high, 2-m aisles).
   g. High rise, pallet rack (24 m high, 2-m aisles).

Given the set of proposals, alternative designs for the material handling system can be obtained by combining a proposed method of moving materials from receiving to storage, a proposed method placing materials in storage, a proposed method of storage, a proposed method of removing materials from storage, and a proposed method of transporting materials from storage to shipping. Some of the combinations of proposals will be eliminated because of their incompatibility. For example, lift trucks requiring 4-m aisles cannot be used to place materials in and remove materials from storage when 2-m aisles are used. Other combinations might be eliminated because of budget limitations; a desire to minimize the variation in types of equipment due to maintainability, availability, reliability, flexibility, and operability considerations; ceiling height limitations; physical characteristics of the product (crushable product might require the use of storage racks); and a host of other considerations. Characteristically, experience and judgement are used to trim the list of possible combinations to a manageable number.

Example 4.2

To illustrate the formation of mutually exclusive investment alternatives from a set of investment proposals, consider a situation involving m investment proposals, let x_j be defined to be 0 if proposal j is not included in an alternative and let x_j be defined to be 1 if proposal j is included in an alternative. Using the binary variable x_j we can form 2^m mutually exclusive alternatives. Thus, if there are three investment proposals, we can form eight mutually exclusive investment alternatives, as depicted in Table 4.1.

Table 4.1 Developing Mutually Exclusive Investment Alternatives from Investment Proposals

Alternative	x_1	x_2	x_3	Explanation
1	0	0	0	Do nothing (proposals 1, 2, and 3 not included)
2	0	0	1	Accept proposal 3 only
3	0	1	0	Accept proposal 2 only
4	0	1	1	Accept proposals 2 and 3 only
5	1	0	0	Accept proposal 1 only
6	1	0	1	Accept proposals 1 and 3 only
7	1	1	0	Accept proposals 1 and 2 only
8	1	1	1	Accept all three proposals

As pointed out in the discussion of the design of a material handling system for the distribution centre, among the alternatives formed some might not be feasible, depending on the restrictions or constraints placed on the problem. To illustrate, there might be a budget limitation that precludes the possibility of combining all three proposals; thus, Alternative 8 would be eliminated. Additionally, some of the proposals might be *mutually exclusive proposals*. For example, Proposals 1 and 2 might be alternative computer systems and only one is to be selected; in this case Alternative 7 would be eliminated from consideration. Other proposals might be *contingent proposals* so that one proposal cannot be selected unless another proposal is also selected. As an illustration of a contingent proposal, Proposal 3 might involve the procurement of computer terminals, which depend on the selection of the computer system associated with Proposal 2. In such a situation, Alternatives 2 and 6 would be infeasible. Thus, depending on the restrictions present, the number of feasible mutually exclusive alternatives that result can be considerably less than 2^m.

In many organizations there is a rather formalized hierarchy for determining how the organization will invest its funds. Typically, the entry point in this hierarchy involves an engineer who is given an assignment to solve a problem; the problem may be one requiring the design of a new product, the improvement of an existing manufacturing process, or the development of an improved system for performing a service. The engineer performs the steps involved in the problem-solving procedure and recommends the preferred solution to the problem. In arriving at the preferred solution, a number of alternative solutions are normally compared; hopefully, the eight-step approach for comparing investment alternatives was followed!

The preferred solution is usually forwarded to the next level of the hierarchy for approval. In fact, one would expect many preferred solutions to various problems to be forwarded. Each preferred solution becomes an investment *proposal*, the resulting set of mutually exclusive investment alternatives are formed, and the process of comparing economic investment alternatives is repeated. This sequence of operations is usually performed in various forms at each level of the hierarchy until, ultimately, the preferred solution by the individual analyst or engineer is accepted or rejected. In this textbook we concentrate on the process of comparing investment alternatives at the first level of the hierarchy; however, the need for such comparisons at many levels of the organization should be kept in mind.

4.3 The Planning Period

In comparing investment alternatives, it is important to compare them over a common period of time. We define that period of time to be the *planning period*. In the case of investments in, say, equipment to perform a required service, the period of time over which the service is required might be used as the planning period. Likewise, in one-shot investment alternatives, the period of time over which receipts continue to occur might define the planning period.

In a sense, the planning period defines the width of a "window" that is used to view the cash flows generated by an alternative. In order to make an objective evaluation, the same window must be used in viewing each alternative.

In some cases the planning period is easily determined; in other cases the duration of one or more projects is sufficiently uncertain to cause concern over the time period to use. Some commonly used methods for determining the planning period to use in economic analysis include:

1. Least common multiple of lives for the set of feasible, mutually exclusive alternatives, denoted $\hat{T}$.

2. Shortest project life among the alternatives, denoted T_s.

3. Longest project life among the alternatives, denoted T_l.

4. Some other period of time.

In the economic analysis literature, the most commonly used method of selecting the planning period appears to be the least common multiple of lives. In most cases, such a selection is made implicitly, not explicitly. When three alternatives are being considered and the individual lives are six years, seven years, and eight years, using such a procedure yields a planning period of $\hat{T} = 168$ years.

If the lives had been either six years, six years, and eight years or six years, eight years, and eight years, $\hat{T} = 24$ years. Clearly, strict reliance on $\hat{T}$ as the planning period is not advisable.

If the shortest project life is used to define the planning period, estimates are required for the values of the unused portions of the lives of the remaining alternatives. Thus, for the situation considered above, with $T_s = 6$ years, the salvage or residual values at the end of six years' use must be assessed for the other two alternatives.

If the longest project life, T_l, is used in determining the planning period, some difficult decisions must be made concerning the period of time between T_s and T_l. If the alternative selected is to provide a necessary service, that service must continue throughout the planning period, regardless of the alternative selected. Consequently, when the shortest life alternative reaches the end of its project life, it must be replaced with some other asset capable of performing the required service. However, since technological developments will probably take place during the period of time T_s, new and improved candidates will be available for selection at time T_s. Thus, the specification of the cash flows for the shortest life alternative during the period of time from T_s to T_l is a difficult undertaking. As a result, T_l is seldom used as the planning period.

A number of organizations have adopted a standard planning period for all economic alternatives. Letting T denote the planning period specified by the organization, different approaches are recommended, depending on whether $T < \hat{T}$ or $T \geq \hat{T}$. If the planning period selected is less than the least common multiple of lives, the cash flows for each alternative must be provided for a period of time equal to the planning period. When the planning period is greater than or equal to the least common multiple of lives, it is recommended that the eco-

Comparison of Alternatives

nomic analysis be based on a period of time equal to the least common multiple of lives. The reason for the latter recommendation is that at time $\hat{T}$ a new economic analysis can be performed based on the alternatives available at that time. After $\hat{T}$ years new alternatives might be available; furthermore, one can more accurately estimate the values of cash flows occurring after $\hat{T}$ if one waits until nearer time $\hat{T}$ to make the estimates.

Example 4.3

To illustrate the difficulties associated with the selection of the planning period, consider the two cash flow profiles given in Table 4.2. Alternatives A and B have anticipated lives of four years and six years, respectively.

Table 4.2 Cash Flow Profiles for Two Mutually Exclusive Investment Alternatives Having Unequal Lives

Alternative A		Alternative B	
EOY	CF(A)	EOY	CF(B)
0	− $5 000	0	− $6 000
1-4	− 3 000	1-6	− 2 000

Using a least common multiple of lives approach, a planning period of $\hat{T} = 12$ years would be used. Using a 12-year planning period requires answers to the following questions. What cash flows are anticipated for years 5 to 12 if Alternative A is selected? What will be the cash flows for Alternative B for years 7 to 12?

As shown in Table 4.3, the traditional approach is to assume that Alternative A will be repeated twice and Alternative B will be repeated once and that identical cash flows occur during these repeating life cycles. Inflation effects, as well as the possibility of technological improvements, tend to invalidate such assumptions.

If the shortest life approach is used, a planning period of four years would be used. In such a case, an estimate of the salvage value of Alternative B should be indicated at the end of year 4, as denoted in Table 4.3 by an asterisk.

Using the longest life approach yields a six-year planning period. In this instance a decision must be made concerning the cash flows in years 5 and 6 for Alternative A. If an initial investment must be made to provide the required service for years 5 and 6, it would occur at the end of year 4. The assumption made in Table

4.3 is that Alternative A will be repeated with identical cash flows, and a terminal salvage value of $2500 will apply after two years of use.

Suppose a standard planning period of 10 years must be used. The same questions that arose in the cases of $\hat{T}, T_s,$ or T_l being the planning period apply when a 10-year planning period is used. As depicted in Table 4.3, one might assume that identical life cycles will be repeated until the end of the planning period and provide estimates of terminal salvage values at that time.

Table 4.3 Cash Flow Profiles for Various Planning Periods

	$T = \hat{T} = 12$			$T = T_s = 4$			$T = 10$	
EOY	CF(A)	CF(B)	EOY	CF(A)	CF(B)	EOY	CF(A)	CF(B)
0	–$5000	–$6000	0	–$5000	–$6000	0	–$5000	–$6000
1	– 3000	– 2000	1	– 3000	– 2000	1	– 3000	– 2000
2	– 3000	– 2000	2	– 3000	– 2000	2	– 3000	– 2000
3	– 3000	– 2000	3	– 3000	– 2000	3	– 3000	– 2000
4	– 8000	– 2000	4	– 3000	– 2000	4	– 8000	– 2000
5	– 3000	– 2000			+2000*	5	– 3000	– 2000
6	– 3000	– 8000				6	– 3000	– 8000
7	– 3000	– 2000		$T = T_l = 6$		7	– 3000	– 2000
8	– 8000	– 2000	EOY	CF(A)	CF(B)	8	– 8000	– 2000
9	– 3000	– 2000	0	–$5000	–$6000	9	– 3000	– 2000
10	– 3000	– 2000	1	– 3000	– 2000	10	– 3000	– 2000
11	– 3000	– 2000	2	– 3000	– 2000		+2500*	+2000*
12	– 3000	– 2000	3	– 3000	– 2000			
			4	– 8000	– 2000			
			5	– 3000	– 2000			
			6	– 3000	– 2000			
				+2500*				

Although the use of a standard planning period has the benefit of a consistent approach in comparing investment alternatives, there are also some dangers that should be recognized. In some cases, the major benefits associated with an alternative might occur in the later stages of its project life. If the planning period is less than the project life, such alternatives would seldom be accepted. Just such a practice caused one major textile firm to lose its strong position in the industry. A major modernization of the processing departments had been pro-

Comparison of Alternatives

posed, but its benefits would not be realized until the bugs had been worked out of the new system, all personnel were trained under the new system, and the marketing people had regained the lost customers. Unfortunately, the planning period specified by the firm was too short in duration, and the modernization plan was not approved.

For the example, deciding which planning period to use should depend on the particular situation instead of on the duration of the individual alternatives. If, for instance, Alternatives A and B are two different lift truck designs, and it is anticipated that the material handling function to be performed by the lift truck will continue for at least 12 years, the least common multiple of lives approach might make sense. Likewise, since the lift truck industry is quite dynamic and technological improvements are quite likely to occur in the future, we might prefer to use the shortest life as the planning period.

Example 4.4

As a second illustration of the planning period selection process, consider the two cash flow diagrams given in Figure 4.1. The two alternatives are mutually exclusive, one-shot investment alternatives. We are unable to predict what investment alternatives will be available in the future, but we do anticipate that recovered capital can be reinvested and earn a 10% return.

For this type of situation, a six-year planning period is suggested, with zero cash flows occurring in years 5 and 6 with Alternative 1. At the end of six years, the net future worths for the two alternatives will be:

$$FW_1(10\%) = \$4\,500(F|P10,2) + \$3\,500(P|A10,3)(F|P10,6)$$
$$-\$4\,000(F|P10,6)$$
$$= \$13\,778.87$$

$$FW_2(10\%) = \$1\,000(F|A10,6) + \$1\,000(A|G10,6)(F|A10,6)$$
$$-\$5\,000(F|P10,6)$$
$$= \$16\,014.01$$

Thus, we would recommend Alternative 2.

If one did not give careful thought to the situation involved and blindly assumed a least common multiple of lives planning period, with identical cash flows in repeating life cycles, then Alternative 1 would be recommended. Hence, it is important to consider the particu-

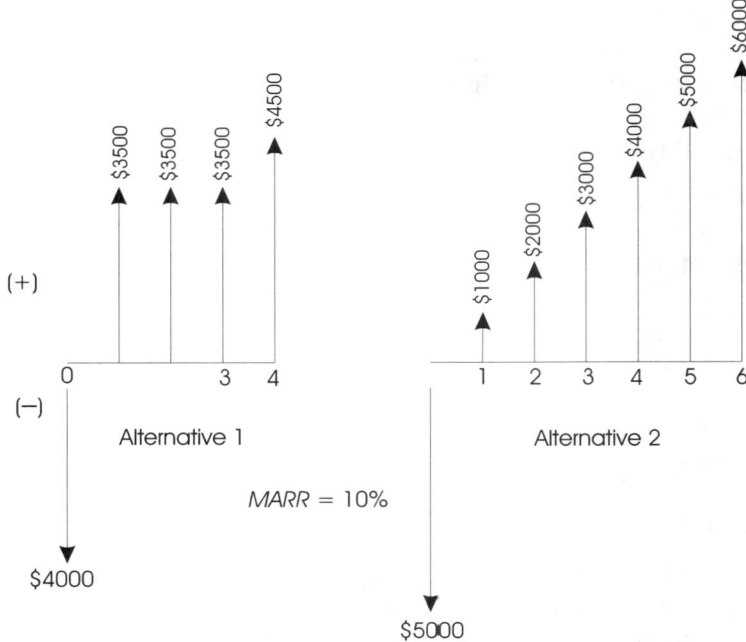

Fig. 4.1 Cash flow diagram for the example problem.

lar situation involved and specify the planning period instead of employing a rule of thumb for establishing a planning period that does not consider the nature of the investments.

It appears that the preferred approach would be to have a "flexible" standard planning period. All of the "routine" economic analyses would be based on the standard planning period of, say, five to ten years; nonroutine economic analyses would be based on a planning period that was appropriate for the situation.

4.4 Developing Cash Flow Profiles

Once the set of mutually exclusive alternatives has been specified and the planning period has been made, the cash flow profiles can be developed for the alternatives. As has been emphasized, the cash flow profiles should be developed by giving careful consideration to *future* conditions instead of relying completely on *past* cash flows. The cash flows for an investment alternative are obtained by aggregating the

Comparison of Alternatives

cash flows for all investment proposals included in the investment alternative.

Example 4.5

To illustrate the approach to be taken, suppose a planning period of five years is used and there are three investment proposals. Cash flow profiles for the proposals are given in Table 4.4. A budget limitation of $50 000 is available for investment among the proposals. Proposal 2 is contingent on Proposal 1, and Proposals 1 and 3 are mutually exclusive. Based on the restrictions associated with the combinations of proposals, only four investment alternatives are to be considered. Alternative A is the do-nothing alternative; Alternative B is Proposal 1 alone; Alternative C involves a combination of Proposals 1 and 2; and Alternative D is Proposal 3 alone. The cash flow profiles for the four alternatives are given in Table 4.5.

Table 4.4 Cash Flow Profiles for Three Investment Proposals

EOY	CF(1)	CF(2)	CF(3)
0	− $20 000	− $30 000	− $50 000
1	− 4 000	+ 4 000	− 5 000
2	+ 2 000	+ 6 000	+ 10 000
3	+ 8 000	+ 8 000	+ 25 000
4	+ 14 000	+ 10 000	+ 40 000
5	+ 25 000	+ 20 000	+ 10 000

Table 4.5 Cash Flow Profiles for Four Mutually Exclusive Investment Alternatives

EOY	CF(A)	CF(B)	CF(C)	CF(D)
0	0	− $20 000	− $50 000	− $50 000
1	0	− 4 000	0	− 5 000
2	0	+ 2 000	+ 8 000	+ 10 000
3	0	+ 8 000	+ 16 000	+ 25 000
4	0	+ 14 000	+ 24 000	+ 40 000
5	0	+ 25 000	+ 45 000	+ 10 000

The do-nothing alternative is the status quo condition and serves as the base against which other alternatives are considered. In some cases, the do-nothing alternative is not feasible (e.g., failing to comply with pollution standards). Moreover, the do-nothing alternative does not necessarily have zero cash flows associated with it. In principle, one should forecast the cash flows that will result if the present method is continued and compare the cash flows with those associated with other alternatives.

In most, if not all, economic evaluations, it is not necessary to develop a detailed forecast of all items of cost, revenue, and investment associated with an alternative. Costs and revenues that will be the same regardless of the alternative selected can be omitted. If a cost reduction alternative will not affect sales revenues, no forecast of such revenues need be developed. Attention is focused on the items of cost and revenue that will be affected by the alternative selected.

Since economic analyses are to be performed to judge the merits of investment alternatives for the future, the data required differ from those normally provided by the accounting system. The fundamental and traditional purpose of accounting is to maintain a consistent *historical* record of the financial results of the operations of the organization. Accounting figures are based on definitions derived consistent with this objective. Accounting methods are not designed to determine the economic worth of alternative courses of action.

4.5 Specifying the Time Value of Money

An important step in evaluating investment alternatives involves the specification of the *interest* or *discount rate* to be used. Even though a project may be financed entirely from internal sources of funds, an interest rate is recommended in evaluating investment alternatives. One reason for doing so is to reflect the cost of investing money in a particular project instead of investing it elsewhere and earning a return on the investment. The cost of foregoing other investment opportunities is referred to as the *opportunity cost,* as discussed in Chapter 2.

Except where other intangible benefits are involved, the discount rate should be greater than the cost of securing additional capital. Indeed, it should be greater than the cost of capital by an amount that will cover unprofitable investments that a firm must make for nonmonetary reasons. Examples of the latter would include investments

Comparison of Alternatives

in antipollution equipment, safety devices, and recreational facilities for employees.

The discount rate that is specified establishes the firm's *minimum attractive rate of return (MARR)* in order for an investment alternative to be justified. If the present worth for an investment alternative were negative, indicating that a negative cash flow was equivalent to the investment alternative, it would not be recommended for adoption.

Some firms establish a standard discount rate or minimum attractive rate of return to be used in all economic analysis; others maintain a flexible posture. For example,

> The XYZ Company return on investment (ratio or earnings to gross investment) has been, on the average over the past five years, approximately equal to a rate-of-return of 15% per year. Accordingly, 15% per year is being established as a tentative *minimum requirement for investment alternatives* whose results are primarily measurable in quantitative dollar terms. The principle being applied is that such alternatives should be expected to maintain or to improve overall return-on-investment performance for the XYZ Company. The minimum requirement is based on overall XYZ financial results (as opposed to results for various parts of the Company) in order to avoid situations in which investment alternatives of a given level of attractiveness are unknowingly undertaken in one part of the Company and rejected in another.
>
> The 15% minimum attractive rate-of-return standard is intended as a guide, rather than as a hard-and-fast decision rule. Furthermore, it is intended to apply to alternatives having risks of the kind usually associated with investments which are primarily in plant and facilities. For alternatives involving expenditures with substantially lower risks, such as those solely for inventories or those in buy-or-lease alterntives, a lower minimum requirement may apply.

Other approaches that are used to establish the *MARR* include:

1. Add a fixed percentage to the firm's cost of capital.

2. Average rate of return over the past five years is used as this year's *MARR*.

3. Use different *MARR* for different planning periods.

4. Use different *MARR* for different magnitudes of initial investment.

5. Use different *MARR* for new ventures rather than for cost improvement projects.

6. Use as a management tool to stimulate or discourage capital investments, depending on the overall economic condition of the firm.

7. Use the average shareholder's return on investment for all companies in the same industry group.

There are a number of different approaches used by companies in establishing the discount rate to be used in performing economic analyses. The issue is not a simple one.

The proper determination of the discount rate has been the subject of considerable controversy in the economic analysis literature for many years. In some ways we are no closer to an agreement today than we were 40 years ago. For this reason, among others, step 7 (perform supplementary analyses) is incorporated in the eight-step procedure for comparing economic alternatives. In many cases, a particular alternative will be preferred over a range of possible discount rates; in other cases, the alternative preferred will be quite sensitive to the discount rate used. Hence, depending on the situation under study, it might not be necessary to specify a particular value for the discount rate—a range of possible values might suffice. We examine this situation in more depth in Chapter 7.

Subsequently, it will be convenient to refer to the interest rate or discount rate used as the minimum attractive rate of return (*MARR*) and to interpret its value using the opportunity cost concept. The argument will be made that money should not be invested in an alternative if it cannot earn a return at least as great as the *MARR*, since it is reasoned that other opportunities for investment exist that will yield returns equal to the *MARR*.

In the case of the public sector, a different interpretation is required in determining the discount rate to use. Since this chapter emphasizes economic analysis in the private sector, Chapter 6 discusses establishing the discount rate for the public sector.

4.6 The Measures of Merit

As noted in Chapter 3 in discussing equivalence, investment alternatives can be compared in a number of ways. Present worth (*PW*) and annual worth (*AW*) comparisons are two commonly used approaches. Among the several methods of comparing investment alternatives are:

1. Present worth method.
2. Annual worth method.
3. Future worth method.
4. Payback period method.
5. Rate of return method.
6. Savings/investment ratio method.

Each of the above measures of merit or measures of effectiveness has been used numerous times in comparing real-world investment alternatives. They may be described briefly as follows:

1. Present worth method converts all cash flows to a single sum equivalent at time zero.
2. Annual worth method converts all cash flows to an equivalent uniform annual series of cash flows over the planning period.
3. Future worth method converts all cash flows to a single sum equivalent at the end of the planning period.
4. Payback period method determines how long at a zero interest rate it will take to recover the initial investment.
5. Rate of return method determines the interest rate that yields a future worth of zero.
6. Savings/investment ratio method determines the ratio of the present worth of savings to the present worth of the investment.

With the exception of the payback period method, all of the measures of merit listed are equivalent methods of comparing investment alternatives. Hence, applying each of the measures of merit to the same set of investment alternatives will yield the same recommendation (with the possible exception of the payback period method).

Since the present worth, annual worth, future worth, rate of return, and savings/investment ratio methods are equivalent, why does more than one of the methods exist? The primary reason for having different, but equivalent, measures of effectiveness for economic alternatives appears to be the differences in preferences among managers. Some individuals (and firms) prefer to express the net economic worth of an investment alternative as a single sum amount; hence, either the present worth method or the future worth method is used. Other individuals prefer to see the net economic worth spread out uniformly over the planning period so the annual worth method is used by them. Yet another group of individuals wishes to express the net economic worth as a rate or percentage; consequently, the rate of return method would be preferred. Finally, some individuals prefer to see the net economic worth expressed as a percentage of the investment required; the savings/investment ratio is one method of providing such information.

Since many organizations have established procedures for performing economic analyses, it seems worthwhile to consider in this chapter the more popular measures of merit that are used. Among those listed, it appears that the present worth, rate of return, and payback period methods are currently the most popular. However, a number of governmental agencies have recently adopted some version of the savings/investment (or benefit/cost) ratio method for purposes of comparing investment alternatives; hence, it is gaining in popularity.

4.6.1 Present Worth Method

The present worth of Alternative j can be represented as

$$PW_j(i) = \sum_{t=0}^{n} A_{jt}(1+i)^{-t} \tag{4.1}$$

with

$PW_j(i)$ = present worth of Alternative j using MARR of $i\%$
n = planning period
A_{jt} = cash flow for Alternative j at the end of period t
i = MARR

Comparison of Alternatives

The alternative having the greatest present worth is the alternative recommended using the present worth method.

Example 4.6

A pressure vessel is purchased for $10 000, kept for five years, and sold for $2 000. Annual operating and maintenance costs were $2 500. Using a 10% minimum attractive rate of return, what was the present worth for the investment?

$$PW_1(10\%) = -\$10\,000 - \$2\,500(P|A10,5) + \$2\,000(P|F10,5)$$
$$= -\$18\,235.20$$

Thus, a single expenditure of $18 235.20 at time zero is equivalent to the cash flow profile for the investment alternative.

4.6.2 Annual Worth Method

The annual worth of Alternative j can be computed as

$$AW_j(i) = \left[\sum_{t=o}^{n} A_{jt}(P|Fi,t) \right](A|P\ i,n)$$

or

$$AW_j(i) = PW_j(i)(A|P\ i,n) \qquad (4.2)$$

where $AW_j(i)$ denotes the annual worth of Alternative j using $i = MARR$. The alternative having the greatest annual worth is selected using the annual worth method.

Example 4.7

Determine the annual worth for an anticipated investment of $10 000 in a computer that will last for eight years and have zero salvage value at the time. Operating and maintenance costs are projected to be $1 000 the first four years and $1 500 the last four years. The minimum attractive rate of return is specified to be 10%. One method of determining the annual worth is:

$$AW_1(10\%) = -\$10\,000(A|P\ 10,8) - \$1\,000$$
$$- \$500(F|A\ 10,4)(A|F\ 10,8)$$
$$= -\$3\,077 \text{ per year}$$

4.6.3 Future Worth Method

The future worth of Alternative *j* can be determined using the relationship

$$FW_j(i) = \sum_{t=0}^{n} A_{jt}(1+i)^{n-t}$$

(4.3)

where $FW_j(i)$ is defined as the future worth of Alternative *j* using a MARR of *i*%. The future worth method is equivalent to the present worth method and the annual worth method, since the ratio of $FW_j(i)$ and $PW_j(i)$ equals a constant, $(F|P\ i,n)$ and the ratio of $FW_j(i)$ and $AW_j(i)$ equals a constant, $(F|A\ i,n)$. The alternative having the greatest future worth is the preferred alternative when the future worth method is used.

Example 4.8

For the previous example, the future worth is given by

$$FW_1(10\%) = \$10\,000(F|P\ 10,8) - \$1\,000(F|A\ 10,8)$$
$$- \$500(F|A\ 10,4)$$
$$= -\$35\,192.40$$

or

$$FW_1(10\%) = AW_1(10\%)(F|A\ 10,8)$$
$$= -\$3\,077(11.4359)$$
$$= -\$35\,188.26$$

Note: The difference in $35 192.40 and $35 188.26 is due to round-off errors.

4.6.4 Payback Period Method

The payback period method involves the determination of the length of time required to recover the initial investment based on a zero interest rate. Letting C_{0j} denote the initial investment for Alternative j and R_{jt} denote the net revenue received from Alternative j during period t, if we assume no other negative cash flows occur, then the smallest value of m_j such that

$$\sum_{t=1}^{m_j} R_{jt} \geq C_{0j}$$

defines the payback period for Alternative j. The alternative having the smallest payback period is the preferred alternative using the payback period method.

Example 4.9

Based on the data given in Table 4.6 on page 158,

$$\sum_{t=1}^{3} R_t = \$2\,000 + \$2\,500 + \$3\,130 < C_0 = \$10\,000$$

and

$$\sum_{t=1}^{4} R_t = \$2\,000 + \$2\,500 + \$3\,130 + \$3\,510 > C_0 = \$10\,000$$

Consequently, $m = 4$, indicating that four years are required to pay back the original investment.

A number of variations of the payback period method have been used by different organizations. However, the basic deficiencies of the payback method we have described are present in the variations with which we are familiar; the *timing* of cash flows and the *duration* of the project are ignored.

To illustrate the deficiencies of the payback period, consider the alternatives depicted in Figure 4.2. In the first case, if the payback period is used to compare Alternatives A and B, then Alternative B would be preferred. In the second case, Alternatives C and D would be equally preferred using the payback period method.

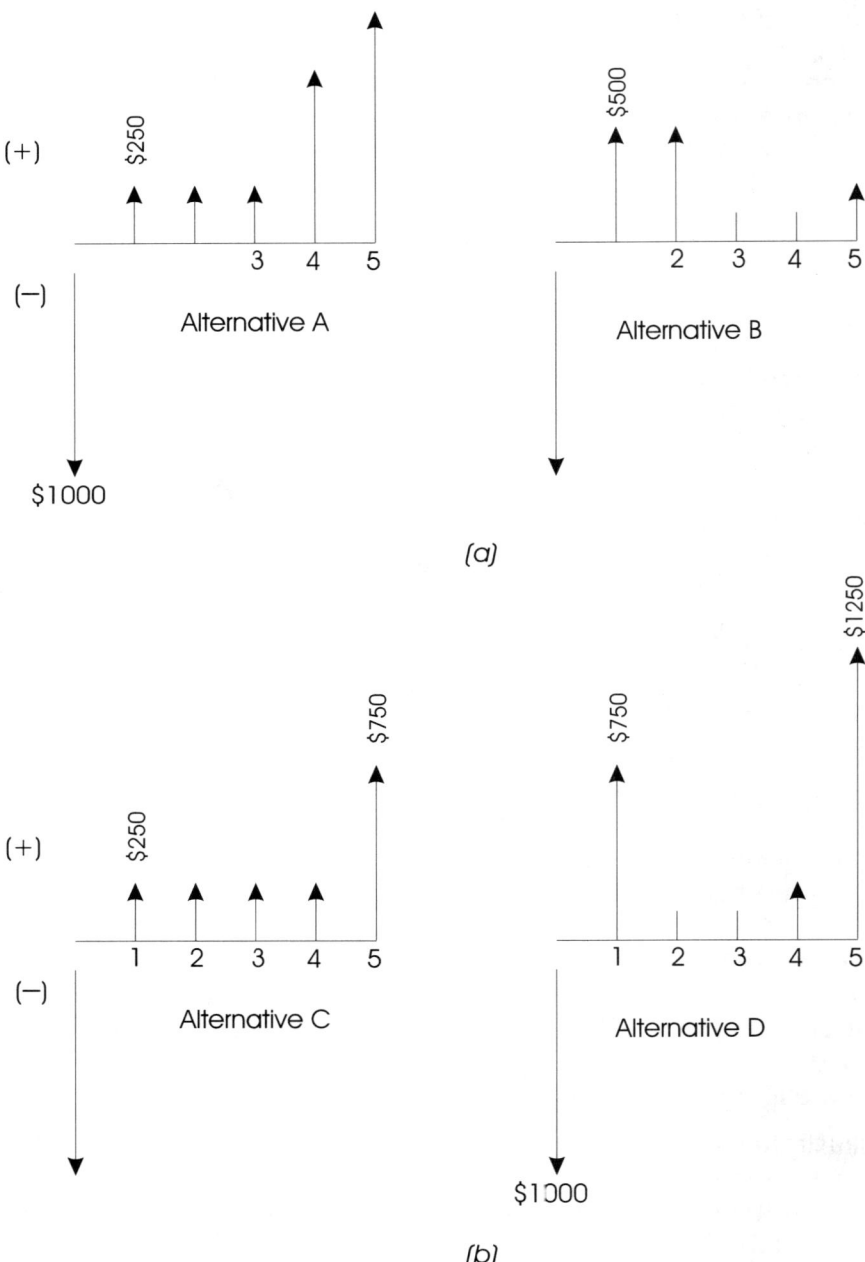

Fig. 4.2 Cash flow diagrams for two mutually exclusive situations, each involving two mutually exclusive investment alternatives.

Despite its obvious deficiencies, the payback period method continues to be one of the most popular methods of judging the desirability of investing in a project. The reasons for its popularity include the following:

1. No interest rate calculations are required.
2. No decision is required concerning the discount rate (*MARR*) to use.
3. It is easily explained and understood.
4. It reflects a manager's attitudes when investment capital is limited.
5. It hedges against uncertainty of future cash flows.
6. It provides a rough measure of the liquidity of an investment.

The payback period method is recommended as a supplementary method of evaluating investment alternatives. In particular, it is suggested that the payback period method be used after the set of feasible alternatives has been evaluated using methods based on the "time vaue of money." The top two or three alternatives obtained using, say, the present worth method could be compared using the payback period method in making the final selection. An alternate approach is to treat the minimization of payback period and the maximization of worth to be two of many objectives and employ the technique presented in Chapter 8 for dealing with multiple objectives.

4.6.5 Rate of Return Method

The rate of return for an alternative can be defined as the interest rate that equates the future worth to zero. Letting i_j^* denote the rate of return for Alternative *j*,

$$0 = \sum_{t=0}^{n} A_{jt}\left(1+i_j^*\right)^{n-1} \tag{4.4}$$

This method of defining the rate of return is referred to in the economic analysis literature as the discounted *cash flow rate of return, internal rate of return,* and the *true rate of return.* We prefer the term *internal rate of return (IRR).*

Table 4.6 Data Illustrating the Meaning of the Internal Rate of Return

t	A_t	B_t	I_t^a	E_t
0	−$10 000	—	—	− $10 000
1	+ 2 000	−$10 000	−$2 000	− 10 000
2	+ 2 500	− 10 000	− 2 000	− 9 500
3	+ 3 130	− 9 500	− 1 900	− 8 270
4	+ 3 510	− 8 270	− 1 654	− 6 414
5	+ 4 030	− 6 414	− 1 283	− 3 667
6	+ 4 400	− 3 667	− 733	− 0

*Based on $i = 0.20$.

Note that the present worth and the annual worth can be obtained by multiplying both sides of Equation 4.3 by the appropriate interest factor [i.e., $(P|Fi_j^*,n)$ and $(A|Fi_j^*,n)$]. Hence, the internal rate of return can also be defined to be the interest rate that yields either a present worth or an annual worth of zero. Depending on the form of a particular cash flow profile, it might be more convenient to use a present worth or an annual worth formulation to determine the internal rate of return for an alternative.

It is important to understand the definition of rate of return inherent in the use of the *IRR* method. In particular, the internal rate of return on an investment can be defined as *the rate of interest earned on the unrecovered balance of an investment*. It is illustrated in Table 4.6, where $10 000 is invested to obtain the receipts shown over a six-year period. A_t denotes the cash flow at the end of period t, B_t represents the unrecovered balance at the *beginning* of period t, E_t is the unrecovered balance at the *end* of period t, and I_t is defined as the interest on the unrecovered balance during period t. The following relationships exist:

$$E_0 = A_0$$
$$B_t = E_{t-1} \qquad t = 1,\ldots,n$$
$$I_t = B_t i \qquad t = 1,\ldots,n$$
$$E_t = A_t + B_t + I_t \qquad t = 1,\ldots,n$$

If i is the internal rate of return, then E_n will equal zero. As indicated in Table 4.6, if i is 20%, E_n is approximately zero. Consequently, i^* is approximately 20%. The equivalence of E_n being zero and the future worth being zero is easily understood by recognizing that E_n is actually the future worth of the cash flow profile. To see why this is true, notice that

$$E_n = A_n + B_n + I_n$$

Employing the definition of I_n,

$$E_n = A_n + B_n(1+i)$$

By the relationship between B_n and E_{n-1}, it is seen that

$$E_n = A_n + E_{n-1}(1+i)$$

Since a similar relationship exists between E_{n-1} and E_{n-2}, we note that

$$E_n = A_n + A_{n-1}(1+i) + E_{n-2}(1+i)^2$$

Generalizing, the recursive relationship between E_t and E_{t-1} gives

$$E_n = A_n + A_{n-1}(1+i) + A_{n-2}(1+i)^2 + \cdots + A_0(1+i)^n$$

Hence, we see that

$$E_n = FW(i\%)$$

as anticipated.

The above example illustrates that the time value of money operations involved in the *IRR* method are equivalent to assuming that all monies received are reinvested and earn interest at a rate equal to the internal rate of return. In particular, if the net cash flow in period t is negative, it is denoted by C_t; if the net cash flow in period t is positive, it is denoted R_t. Letting r_t be the reinvestment rate for positive cash flows occurring in period t and i' be the rate of return for negative cash flows, then the following relationship can be defined:

$$\boxed{\sum_{t=0}^{n} R_t(1+r_t)^{n-t} = \sum_{t=0}^{n} C_t(1+i')^{n-t}}$$

(4.5)

The future worth of reinvested monies received must equal the future worth of investments.

If r_t equals i', Equation 4.4 becomes

$$0 = \sum_{t=0}^{n}(R_t - C_t)(1+i')^{n-t}$$
(4.6)

Letting A_t equal $R_t - C_t$ defines the *IRR* method given by Equation 4.4. Hence, we see that the rate of return obtained using the *IRR* method can be interpreted as the reinvestment rate for all recovered funds.

Since the determination of the rate of return involves solving Equation 4.4 for i_j^*, it is seen that (for a given alternative *j*), it is necessary to determine the values of *x* that satisfy the following *n*-degree polynomial: $0 = A_0 x^n + A_1 x^{n-1} + \cdots + A_{n-1} x + A_n$, where $x = (1 + i^*)$. In general, there can exist *n* distinct roots (values of *x*) for an *n*-degree polynomial; however, most cash flow profiles encountered in practice will have a unique root (rate of return).

The number of real positive roots of an *n*-degree polynomial with real coefficients is less than or equal to the number of changes of sign in the sequence of cash flows, $A_0, A_1, \ldots, A_{n-1}, A_n$. Since the typical cash flow pattern begins with a negative cash flow, followed by positive cash flows, a unique root will normally exist.

Example 4.10

As an illustration of a cash flow profile having multiple roots, consider the data given in Table 4.7. The future worth of the cash flow series will be zero using either a 20% or 50% interest rate.

Table 4.7 Cash Flow Profile	
EOY	CF
0	– $1 000
1	4 200
2	– 5 850
3	2 700

$$FW(20\%) = -\$1\,000(1.2)^3 + \$4\,200(1.2)^2 - \$5\,850(1.2) + \$2\,700 = 0$$
$$FW(50\%) = -\$1\,000(1.5)^3 + \$4\,200(1.5)^2 - \$5\,850(1.5) + \$2\,700 = 0$$

A plot of the future worth for this example is given in Figure 4.3. The future worth polynomial is a third-degree polynomial and there are three changes of sign in the ordered sequence of cash flows (−,+,−,+); however, there are only two unique roots, corresponding to $i = 0.20$ and $i = 0.50$. In this case, there is a repeated root corresponding to $i = 0.50$, since the future worth polynomial can be written as

$$FW(i\%) = \$1\,000(1.2 - x)(1.5 - x)^2$$

where $x = (1 + i)$.

The possibility of multiple roots occurring in the internal rate of return calculation, coupled with the reinvestment assumptions concerning recovered funds, has led to the development of an alternative rate of return method, called the *external rate of return method*. The external rate of return (*ERR*) method consists of the determination of the value of i' that satisfies Equation 4.7.

$$\boxed{\sum_{t=0}^{n} R_t (1+r_t)^{n-t} = \sum_{t=0}^{n} C_t (1+i')^{n-t}}$$

(4.7)

where, as defined above,

A_t = net cash flow in period t

$$C_t = \begin{cases} A_t, & \text{if } A_t < 0 \\ 0, & \text{otherwise} \end{cases}$$

$$R_t = \begin{cases} A_t, & \text{if } A_t \geq 0 \\ 0, & \text{otherwise} \end{cases}$$

r_t = reinvestment rate for funds recovered in period t

i' = external rate of return

Normally, r_t equals the minimum attractive rate of return, since the *MARR* reflects the opportunity cost for money available for investment. The preferred alternative is the alternative with the greatest investment such that each increment of investment has a return at least equal to the *MARR*.

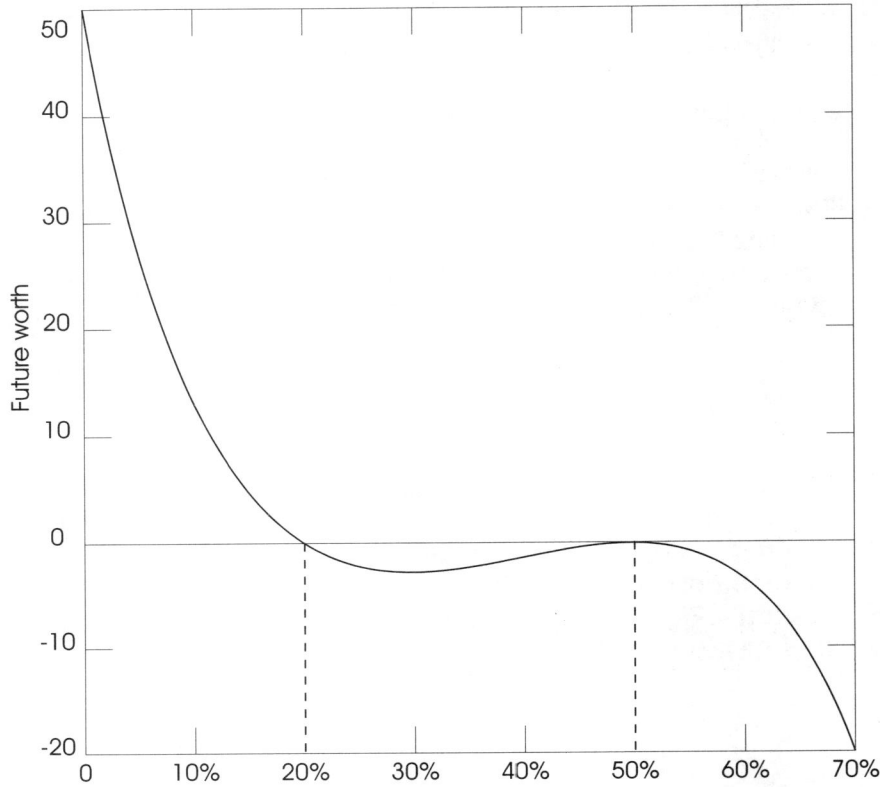

Fig. 4.3 Plot of future worth for the example problem.

Example 4.11

For the data provided in Table 4.6, suppose money received from the intitial investment is reinvested and earns 10% interest. At the end of the sixth year the reinvested funds total:

$$\$2\,000(F|P\ 10,5) + \$2\,500(F|P\ 10,4) + \$3\,131(F|P\ 10,3)$$
$$+ \$3\,510(F|P\ 10,2) + \$4\,030(F|P\ 10,1) + \$4\,400 = \$24\,127.38$$

Consequently, the *external rate of return* is defined as the interest rate such that the future worth of the \$10 000 investment equals \$24 127.38. Thus,

$$\$10\,000(1+i')^6 = \$24\,127.38$$

Comparison of Alternatives

Taking the logarithm and solving for i' yields a value of 15.8% as the rate of return.

As another example, if recovered funds are reinvested at 20%, the future worth will be

$$\$2\,000(F|P\ 20,5) + \$2\,500(F|P\ 20,4) + \$3\,131(F|P\ 20,3)$$
$$+\$3\,510(F|P\ 20,2) + \$4\,030(F|P\ 20,1) + \$4\,400 = \$29\,845.24$$

Setting the future worth of the $10 000 investment equal to $29 845.24 yields

$$\$10\,000(1+i')^6 = \$29\,845.24$$

Solving for i' gives a value of about 20% as the rate of return, as anticipated.

4.6.6 Savings/Investment Ratio Method

The *savings/investment ratio method* can be defined in various ways; two typical approaches will be described. First, the *savings/investment ratio* can be given as

$$SIR_j(i) = \frac{\sum_{t=1}^{n} A_{jt}(1+i)^{-t}}{C_{0j} - F_j(1+i)^{-n}}$$

(4.8)

where $SIR_j(i)$ is the savings investment ratio for Alternative j based on an *MARR* of i%, A_{jt} is the cash flow for Alternative j for period t, C_{0j} is the initial investment for Alternative j, and F_j is the terminal salvage value for Alternative j. An alternative formulation of the savings/investment ratio will be employed in this chapter; the *SIR* is defined as follows:

$$SIR_j(i) = \frac{\sum_{t=1}^{n} R_{jt}(1+i)^{-t}}{\sum_{t=0}^{n} C_{jt}(1+i)^{-t}}$$

(4.9)

where R_{jt} denotes the magnitude of the net positive cash flow for Alternative j occurring in period t and C_{jt} denotes the magnitude of the net negative cash flow for Alternative j occurring in period t.

Example 4.12

Consider the cash flows given in Table 4.6 and let the MARR be 10%. The present worth of the cash flows during periods 1 to 6 totals

$$\$2\,000(P|F\,10,1) + \$2\,500(P|F\,10,2) + \$3\,130(P|F\,10,3)$$
$$+\$3\,510(P|F\,10,4) + \$4\,030(P|F\,10,5) + \$4\,400(P|F\,10,6) = \$13\,619.13$$

Thus, the savings/investment ratio is given by

$$SIR_1(10\%) = \frac{\$13\,619.13}{\$10\,000.00} = 1.361913$$

Any savings/investment ratio greater than one indicates the alternative is economically desirable. An alternative label for the savings/investment ratio is the *benefit-cost ratio*. Savings are interpreted as the benefits derived from a venture and the difference in the initial investment, and the present worth of the salvage value is denoted as the cost of the venture. We explore benefit-cost analysis more fully in Chapter 6.

4.7 Comparing the Investment Alternatives

A comparison of investment alternatives, to be complete, requires a knowledge of all differences in cash flows among the investment alternatives. In a number of public sector applications the quantification of benefits in economic units is not a trivial undertaking. Consequently, we will postpone until Chapter 6 a discussion of the comparison of alternatives involving benefits which are not normally measured in dollars and cents.

Additionally, our discussion in this chapter will be based on an assumption that either the cash flows are after-tax cash flows or a before-tax study is desired. In Chapter 5, we address the subject of income taxes and their effect on the preference among investment alternatives.

Example 4.13

To illustrate the use of the various methods of comparing investment alternatives, consider the three mutually exclusive alternatives having the cash flow profiles given in Table 4.8 for a planning period of five years. A minimum attractive rate of return of 10% is to be used in the analysis.

Table 4.8 Cash Flow Profiles for Three Mutually Exclusive Investment Alternatives

T	A_{1t}	A_{2t}	A_{3t}
0	0	−$10 000	−$15 000
1	$4 000	7 000	7 000
2	4 000	7 000	8 000
3	4 000	7 000	9 000
4	4 000	7 000	10 000
5	4 000	7 000	11 000

4.7.1 Ranking Approaches

In comparing mutually exclusive investment alternatives using either present worth, annual worth, future worth, or payback period as the measure of merit, one can simply compute the value of the measure of merit for each alternative and rank the alternatives on the basis of the value obtained. The use of the ranking approach will be demonstrated for each of the four measures of merit by comparing the alternatives depicted in Table 4.8.

Present Worth Method. Using the present worth method gives:

$$PW_1(10\%) = \$4\,000(P|A\ 10,5)$$
$$= \$15\,163.20$$

$$PW_2(10\%) = -\$10\,000 + \$7\,000(P|A\ 10,5)$$
$$= \$16\,535.60$$

$$PW_3(10\%) = -\$15\,000 + \$7\,000(P|A\ 10,5)$$
$$+ \$1\,000(P|G\ 10,5)$$
$$= \$18\,397.33$$

Since Alternative 3 has the greatest present worth, it is recommended.

Annual Worth Method. The annual worths for the three alternatives are:

$$AW_1(10\%) = \$4\,000$$

$$AW_2(10\%) = -\$10\,000(A|P\ 10,5) + \$7\,000$$
$$= \$4\,362$$

$$AW_3(10\%) = \$15\,000(A|P\ 10,5) + \$7\,000$$
$$+\$1\,000(A|G\ 10,5)$$
$$= \$4\,853$$

(Alternatively, the annual worth values could have been obtained directly from the present worth values.) Since Alternative 3 has the greatest annual worth, it is recommended.

Future Worth Method. The future worths for the three alternatives are computed as follows:

$$FW_1(10\%) = \$4\,000(F|A\ 10,5)$$
$$= \$24\,420.40$$

$$FW_2(10\%) = -\$10\,000(F|P\ 10,5) + \$7\,000(F|A\ 10,5)$$
$$= \$26\,630.70$$

$$FW_3(10\%) = -\$15\,000(F|P\ 10,5) + \$7\,000(F|A\ 10,5)$$
$$+\$1\,000(A|G\ 10,5)(F|A\ 10,5)$$
$$= 29\,629.04$$

(Alternatively, the future worth values could have been obtained directly from the present worth and from the annual worth values.) Since Alternative 3 has the greatest future worth, it is recommended.

Payback Period Method. The payback period is computed based on the difference in the cash flows for Alternatives 2 and 3 and the cash flows for Alternative 1 (do nothing alternative). By investing $10 000, annual savings of $3 000 are provided by Alternative 2; by investing $15 000, savings of $3 000, $4 000, $5 000, $6 000, and $7 000 are produced for years 1 to 5 respectively.

For Alternative 2, the payback period is found to be four years, since

$$\sum_{t=1}^{3} R_{2t} = \$3\,000 + \$3\,000 + \$3\,000 < \$10\,000 = C_{02}$$

and

$$\sum_{t=1}^{4} R_{2t} = \$3\,000 + \$3\,000 + \$3\,000 + \$3\,000 > \$10\,000 = C_{02}$$

For Alternative 3, the payback period is also found to be four years, since

$$\sum_{t=1}^{3} R_{3t} = \$3\,000 + \$4\,000 + \$5\,000 < \$15\,000 = C_{03}$$

and

$$\sum_{t=1}^{4} R_{3t} = \$3\,000 + \$4\,000 + \$5\,000 + \$6\,000 > \$15\,000 = C_{03}$$

Using the payback period method, Alternatives 2 and 3 are equally preferred.

4.7.2 Incremental Approaches

In comparing mutually exclusive investment alternatives using either the rate of return or the savings/investment method, an incremental approach must be taken. The comparisons are based on the *differences* in the cash flows for combinations of investment alternatives.

It is also possible to use incremental methods to compare alternatives using present worth, annual worth, and future worth methods. A step-by-step representation of the incremental procedure for present worth, annual worth, and future worth comparisons is given below:

Step 1. Order the feasible alternatives according to the size of the initial investment. Go to step 2.

Step 2. Compute the value of the measure of merit for the feasible alternative having the smallest initial investment. Go to step 3.

Step 3. Obtain the cash flow profile for the differences in cash flows for the two remaining feasible alternatives that require the smallest investments. Go to step 4.

Step 4. Compute the value of the measure of merit for the cash flow profile obtained in step 3. If the value of the measure of merit is positive (negative), eliminate from further consideration the alternative that has the smallest (largest) initial investment. Go to step 5.

Step 5. If only one alternative remains, it is the preferred alternative; if more than one alternative remains, go to step 3.

In order to illustrate the step-by-step procedure, an annual worth comparison will be performed for the example problem, given in Table 4.8.

Step 1. Order 1, 2, 3.

Step 2. $AW_1(10\%) = \$4\,000$.

Step 3.

t	$A_{2t} - A_{1t}$
0	−$10 000
1	3 000
2	3 000
3	3 000
4	3 000
5	3 000

Step 4. $AW_{2-1}(10\%) = -\$10\,000(A|P\ 10,5) + \$3\,000$
$$= \$362$$

$AW_{2-1}(10\%) > 0$; therefore, eliminate Alternative 1.

Step 5. More than one alternative (2, 3) remains; therefore, go to step 3.

...

Comparison of Alternatives

Step 3.

t	$A_{3t} - A_{2t}$
0	−$5000
1	0
2	1000
3	2000
4	3000
5	4000

Step 4. $AW_{3-2}(10\%) = -\$5\,000(A|P\ 10,5) + \$1\,000(A|G\ 10,5)$
$$= \$491$$
$AW_{3-2}(10\%) > 0$; therefore, eliminate Alternative 2.

Step 5. Only one alternative remains; therefore, Alternative 3 is the preferred alternative.

As exercises, you may wish to solve the example problem using the incremental procedure in conjunction with the present worth and future worth methods. Additionally, you may wish to develop an incremental procedure for the payback period method. Similar, but slightly different, incremental procedures are employed when either the rate of return or the savings/investment ratio methods are used.

Rate of Return Method. The rate of return method includes the same sequence of comparisons used in comparing the differences in alternatives using present worth, annual worth, and future worth methods. The steps to be performed using the rate of return method in comparing investment alternatives are:

Step 1. Order the alternatives according to the size of the initial investment. Go to step 2.

Step 2. Compute the rate of return for the alternative having the smallest investment. Go to step 3.

Step 3. If the rate of return obtained in step 2 is less than the minimum attractive rate of return, eliminate the alternative from further consideration and repeat step 1. If the rate of return obtained in step 2 is greater than or equal to the MARR, go to step 4.

Step 4. Compute the rate of return for the difference in cash flows for the two alternatives having the smallest initial investments. Go to step 5.

Step 5. If the rate of return based on incremental investment obtained in step 4 is less than (greater than or equal to) the MARR, eliminate from further consideration the alternative having the larger (smaller) investment. Go to step 6.

Step 6. If only one alternative remains, it is the preferred alternative; if more than one alternative remains, go to step 4.

Applying the sequence of steps to the example problem given in Table 4.8, using the internal rate-of-return method yields the following results:

Step 1. Order 1, 2, 3.

Step 2. $i_1^* = \infty$ since no "initial investment" is required to obtain income of $4 000/year.

Step 3. $i_1^* >$ MARR, go to step 4.

Step 4. Compute rate of return on difference in cash flows for Alternatives 1 and 2. Using a present worth formulation yields

$$-\$10\,000 + \$3\,000(P|A\ i,5) = 0$$

and solving for i yields $i_{2-1}^* = 15.26\%$

Step 5. $i_{2-1}^* >$ MARR $= 10\%$; eliminate Alternative 1 from further consideration.

Step 6. More than one alternative remains; go to step 4.

...

Step 4. Compute rate of return on difference in cash flows for Alternatives 2 and 3. That is, $-\$5\,000 + \$1\,000(P|G\ i,5) = 0$ and solving for i yields $i_{3-2}^* = 19.48\%$.

Step 5. $i_{3-2}^* >$ MARR; eliminate Alternative 2 from further consideration.

Step 6. Only one alternative remains; thus, Alternative 3 is preferred.

It should be noted that the procedure for the *IRR* method given above depends on at least one of the alternatives having a rate of return greater than the *MARR*. However, in a number of investment situations, it is not uncommon to find that all of the alternatives involve only negative cash flows. In such a situation, steps 2 and 3 are eliminated from the procedure.

Applying the *ERR* method in comparing the investment alternatives given in Table 4.8 yields the following results:

Step 1. Order 1, 2, 3.

Step 2. $i'_1 = \infty$.

Step 3. $i'_1 > MARR$; go to step 4.

Step 4. $-\$10\,000(F|P\,i',5) + \$3\,000(F|A\,10,5) = 0$, and $i'_{2-1} = 12.87\%$.

Step 5. $i'_{2-1} > MARR$; eliminate Alternative 1 from further consideration.

Step 6. More than one alternative remains; go to step 4.

...

Step 4. $-\$5\,000(F|P\,i',5) + \$1\,000(A|G\,10,5)(F|A\,10,5) = 0$, and $i'_{3-2} = 17.19\%$.

Step 5. $i'_{3-2} > MARR$; eliminate Alternative 2 from further consideration.

Step 6. Only one alternative remains; thus, Alternative 3 is preferred.

Savings/Investment Ratio Method. Consider again the problem given in Table 4.8. The savings/investment ratio is computed by comparing the cash flows of Alternatives 1 and 2. Investing $10 000 yields annual savings of $3 000 per year. The present worth of the savings equals $3 000(P|A 10,5)$, or $11 372.40. Consequently, the *SIR* value is

$$SIR_{2-1}(10\%) = \frac{\$11\,372.40}{\$10\,000.00} = 1.137$$

Since $SIR_{2-1}(10\%) > 1.00$, Alternative 2 is preferred to Alternative 1.

Next we compare Alternatives 2 and 3. An incremental investment of $5 000 in Alternative 3 yields savings of $1 000, $2 000, $3 000, and $4 000 in years 2 to 5. The present worth of the savings equals $6 861.73. Thus, the *SIR* value is

$$SIR_{3-2}(10\%) = \frac{\$6\,861.73}{\$5\,000.00} = 1.372$$

Since $SIR_{3-2}(10\%) > 1.00$, Alternative 3 is preferred to Alternative 2.

4.8 Supplementary Analyses

The seventh step in performing an economic evaluation of investment alternatives is the performance of supplementary analyses. A sensitivity analysis consists of an exploration of the behaviour of the measure of merit when changes occur in the values of the parameters for an investment alternative. The parameters subject to change might include the planning period, the discount rate, and any or all of the cash flows. This step recognizes that the process of providing estimates of cash flows and decisions concerning the planning period and discount rate is not a precise, errorless process. An examintion of sensitivity analysis as it relates to economic analyses is reserved for Chapter 7. For now, we assume that the vaues assigned to the parameters are neither inaccurate nor subject to change.

4.9 Selecting the Preferred Alternative

The final step in performing a comparison of investment alternatives is the selection of the preferred alternative. Our discussion in this chapter has concentrated solely on the economic factor; we have been concerned with determining the most economical alternative. The final decision may be based on a host of criteria instead of on the single criterion of economics. In Chapter 8, we examine the decision process in the face of multiple objectives.

The selection process is complicated not only by the presence of multiple objectives, but also by the risks and uncertainties associated with the future. The analysis of Chapter 7 is extended in Chapter 8 to include decision making in the face of risk or uncertainty.

The selection or rejection of the recommended solution is heavily dependent on the sales ability of the individual presenting the recommendation to management. Since the corporate decision makers are normally presented with many more investment alternatives than can be funded, it is important to communicate effectively in order to compete favourably for the company's limited capital.

The decision makers' perspective is broad. The proposal should be developed accordingly. The project's capital requirements should be related to previous and estimated future capital requirements on similar investments. The proposal should be shown to fit in with long-range corporate plans and support short-range objectives. Comparison should be made with similar investments by competition and competititive advantage (if any) shown. Ancillary marketing, public relations and/or political benefits that the company will derive from the investment should be pointed out. The effect that the investment will have on other functional activities of the company should be noted and overall benefits stressed. In other words, the investment proposal should be related to the well-being of the total enterprise.

This investment proposal is only one of several that the decision makers may review; not all proposals will be accepted. It is important to know how the decision makers classify investments and the relative strengths and weaknesses of the investment proposal vis-à-vis competing investment proposals. If the proposed investment is relatively risk-free, that point should be made strongly to possibly offset less desirable aspects of the proposal.

The engineer should know and use the measure of merit that decision makers prefer, and support it with any other measures of merit that may be necessary or helpful. The engineer should become familiar with the values that the decision makers attach to different aspects of investment proposals—particularly economic uncertainty. Low risk should be emphasized in proposals to decision makers who tend to be risk-avoiders; potential economic gain should be highlighted for the risk-taking decision maker. This, of course, does not suggest that the engineer should be less than honest, but that the communication of the investment proposal should be developed with the decision maker in mind.

Decision makers are primarily interested in the economic aspects of the proposals. The engineer must resist the temptation to overstate the complicated and sophisticated technology that might underlie a proposal. The less decision makers understand about the technical/engineering aspects of a proposal, the more uneasy they become. This apprehension becomes part of the uncertainty which the decision makers subjectively assign to the proposal and the result could be its rejection in favour of a proposal with which they are more familiar, or at least more comfortable. The engineer should keep in mind the background and interests of the decision makers and use technical and commercial terms with which they are familiar.

4.10 Analyzing Alternatives with No Positive Cash Flows

The previous analysis of an investment decision given in Table 4.8 involved three mutually exclusive alternatives, including the do-nothing alternative, which involved positive valued cash flows. However, there exist situations in which no positive valued cash flows are present.

Example 4.14

Consider a cost reduction case in which two cost reduction alternatives have been proposed. The present method, which we refer to as the do-nothing alternative, is also a feasible alternative. Thus, three mutually exclusive alternatives are considered. Cash flow profiles for the alternatives are given in Table 4.9.

Table 4.9 Cash Flow for Three Mutually Exclusive Investment Alternatives with No Positive Cash Flow

t	A_{1t}	A_{2t}	A_{3t}
0	0	−$10 000	−$15 000
1	−$12 000	− 9 000	− 9 000
2	− 12 000	− 9 000	− 8 000
3	− 12 000	− 9 000	− 7 000
4	− 12 000	− 9 000	− 6 000
5	− 12 000	− 9 000	− 5 000

Alternative 1 is the do-nothing alternative in which the present expenditure of $12 000 per year is continued. Alternative 2 involves an initial investment of $10 000 in order to reduce the annual expenditures by $3 000 over the five-year period. Alternative 3 requires an initial investment of $15 000 in order to obtain decreasing annual expenditures. The minimum attractive rate of return is specified to be 10%.

The procedures we employed in analyzing alternatives having positive-valued cash flows can be applied to the present situation as well. As an illustration, consider the use of the incremental approach in conjunction with the annual worth procedure.

Step 1. Order 1, 2, 3.

Step 2. $AW_1(10\%) = -\$12\,000$.
Step 3.

t	$A_{2t} - A_{1t}$
0	−$10 000
1	3 000
2	3 000
3	3 000
4	3 000
5	3 000

Step 4. $AW_{2-1}(10\%) = \$362 > 0$; therfore eliminate Alternative 1.
Step 5. Alternatives 2 and 3 remain.

...

Step 3.

t	$A_{3t} - A_{2t}$
0	−$5 000
1	0
2	1 000
3	2 000
4	3 000
5	4 000

Step 4. Consider differences in cash flows for Alternatives 2 and 3. $AW_{3-2}(10\%) = \$491 > 0$; therefore, eliminate Alternative 2.
Step 5. Alternative 3 is the only remaining alternative; hence, it is the preferred alternative.

You may wish to verify on your own that the present worth, future worth, payback period, and savings/investment methods can be applied in the case of nonpositive valued cash flows. We find it instructive, however, to consider the rate of return method explicitly.

Since the sum of the cash flows (ignoring the time value of money) is negative for each alternative, positive-valued rates of return among the three alternatives do not exist. Consequently, you may ask, "How can the rate of return method be used to compare alternatives that do not have rates of return?" Admittedly, Alternative 2 does not have a rate of return by itself, but in comparison with Alternative 1, it is clear that the investment of the $10 000 yields annual savings of $3 000 in annual expenditures. Thus, on an incremental basis, the $10 000 incremental investment produces positive-valued cash flows of $3 000 per year for each of the five years.

The return on the incremental investment, using the internal rate of return method, is found to be approximately 15.26%. Since we would do better to invest the $10 000 in Alternative 2 and earn 15.26% than to choose Alternative 1 and only earn the minimum attractive rate of return on the remaining money available for investment, Alternative 2 is preferred to Alternative 1.

Of course, we have $15 000 to invest (otherwise Alternative 3 would be infeasible). Thus, at this point we are willing to invest $10 000 in Alternative 2 and earn 15.26% and invest the remaining $5 000 in some other opportunity and earn the minimum attractive rate of return of 10%. The question now considered is, "Should we use the additional $5 000 and pool it with the $10 000 to invest in Alternative 3?" The rate of return on the incremental investment required to obtain Alternative 3 can be calculated to be approximately 19.48%.

At this point, the question is "Should we invest $10 000 in Alternative 2 and invest $5 000 elsewhere at the *MARR*, or should we invest $15 000 in Alternative 3?" Investing in Alternative 3 yields a return of 15.26% on the $10 000 increment and 19.48% on the $5 000 increment. Consequently, Alternative 3 would be preferred to the investment of $10 000 in Alternative 2 and investing the remaining $5 000 to earn only 10%.

The rate of return philosophy is to continue investing so long as each increment of investment is justified. Since the application of this philosophy in evaluating investment alternatives is often difficult to grasp, we will consider yet another example to illustrate this important concept.

Example 4.15

Consider the four alternatives given in Table 4.10. A six-year planning period is used, along with a 10% minimum attractive rate of return. The alternatives are ranked in order of increasing investment. By following the step-by-step procedure outlined previously for the rate of return method, one obtains the following results:

Table 4.10 Data for a Rate of Return Example Problem

t	A_{1t}	A_{2t}	A_{3t}	A_{4t}
0	−$5 000	−$8 000	−$10 000	−$14 000
1-6	1 504	2 026	2 720	3 585

Step 1. Order 1, 2, 3, 4.

Step 2. Compute rate of return for Alternative 1:

$$0 = -\$5\,000(A|P\ i,6) + \$1\,504$$
$$(A|P\ i,6) = 0.3008$$
$$i_1^* = 20\%$$

Step 3. $i_1^* > MARR$

Step 4. Compute rate of return on the $3 000 incremental investment required to go from Alternative 1 to Alternative 2.

$$0 = -\$3\,000(A|P\ i,6) + \$522$$
$$(A|P\ i,6) = 0.1740$$
$$i_{2-1}^* = 1.25\%$$

Step 5. $i_{2-1}^* < MARR$, eliminate Alternative 2.

Step 6. Alternatives 1, 3, and 4 remain.

...

Step 4. Compute rate of return on the $5 000 incremental investment required to go from Alternative 1 to Alternative 3.

$$0 = -\$5\,000(A|P\ i,6) + \$1\,216$$
$$(A|P\ i,6) = 0.2432$$
$$I_{3-1}^* = 12\%$$

Step 5. $I_{3-1}^* > MARR$, eliminate Alternative 1.

Step 6. Alternatives 3 and 4 remain.

...

Step 4. Compute rate of return on the $4 000 incremental investment required to go from Alternative 3 to Alternative 4.

$$0 = \$4\,000(A|P\ i,6) + \$865$$
$$(A|P\ i,6) = 0.21625$$
$$I_{4-3}^* = 8\%$$

Step 5. $I_{4-3}^* < MARR$, eliminate Alternative 4.

Step 6. Alternative 3 is preferred.

The conclusion we draw is that Alternative 3 yields a return of 20% on the first $5 000 increment and 12% on the next $5 000 increment. Unused monies earn 10%. Letting the rates of return on individual alternatives be designated as *gross* rates of return, the following gross rates of return are obtained.

$$i_1 = 20\%, \quad i_2 = 13.5\%, \quad i_3 = 16.1\%, \quad i_4 = 13.85\%$$

Ranking alternatives on the basis of gross rate of return indicates that Alternative 1 is preferred. However, using the incremental rate of return approach, Alternative 3 is preferred even though it does not have the highest gross rate of return. Again, the reason for choosing Alternative 3 over Alternative 1 is that the second increment of $5000 will earn 12% if Alternative 3 is chosen; investing in Alternative 1 caused the remaining $5000 to be invested and earn *only* the *MARR* of 10%.

As long as the *IRR* method is used to evaluate increments of investment instead of ranking alternatives on the basis of gross rate of return, the recommended alternative will be the same as obtained using the annual worth, present worth, future worth, and savings/investment ratio methods. Likewise, if the *ERR* method employs a reinvestment rate equal to the minimum attractive rate of return, then the recommendation will be consistent with that obtained by using the *IRR* method.

Example 4.16

To illustrate the use of the external rate of return method, suppose a 10% reinvestment rate is used in the previous illustration. Following the step-by-step procedure gives the following results:

Step 1. Order 1, 2, 3, 4.

Step 2. $1\,504(F|A\ 10{,}6) = \$5\,000(F|P\ i_1',6)$

$(F|P\ i_1',6) = 2.32085$

$i_1' = 15.06\%$

Step 3. $i_1' > MARR$

Step 4. $\$522(F|A\ 10{,}6) = \$3\,000(F|P\ i_{2-1}',6)$

$(F|P\ i_{2-1}',6) = 1.34251$

$i_{2-1}' = 5.03\%$

Step 5. $i_{2-1}' < MARR$

Step 6. Alternatives 1, 3, and 4 remain.

...

Step 4. $\$1\,216(F|A\ 10{,}6) = \$5\,000(F|P\ i_{3-1}',6)$

$(F|P\ i_{3-1}',6) = 11.06\%$

Step 5. $i_{3-1}' > MARR$

Step 6. Alternatives 3 and 4 remain.

...

Step 4. $\$865(F|A\ 10,6) = \$4\,000(F|P\ i'_{4-3},6)$
$(F|P\ i'_{4-3},6) = 1.6685$
$i'_{4-3} = 8.91\%$

Step 5. $i'_{4-3} < MARR$

Step 6. Alternative 3 is preferred.

4.11 Classical Method of Dealing with Unequal Lives

The method that we presented of dealing with investment alternatives having unequal lives is not a widely used approach in the economic analysis literature. Recall that it was recommended that a planning period be specified, cash flows over the planning period be given explicitly, and the evaluation be performed over the common planning period. The classical method of dealing with unequal lives is to use a least common multiple of lives planning period and assume identical cash flow profiles in repeating life cycles. It is not uncommon to see alternatives having unequal lives compared on the basis of the annual worths for individual life cycles.

Example 4.17

To illustrate the classical approach, consider two mutually exclusive alternatives having cash flow profiles as depicted in Figure 4.4.

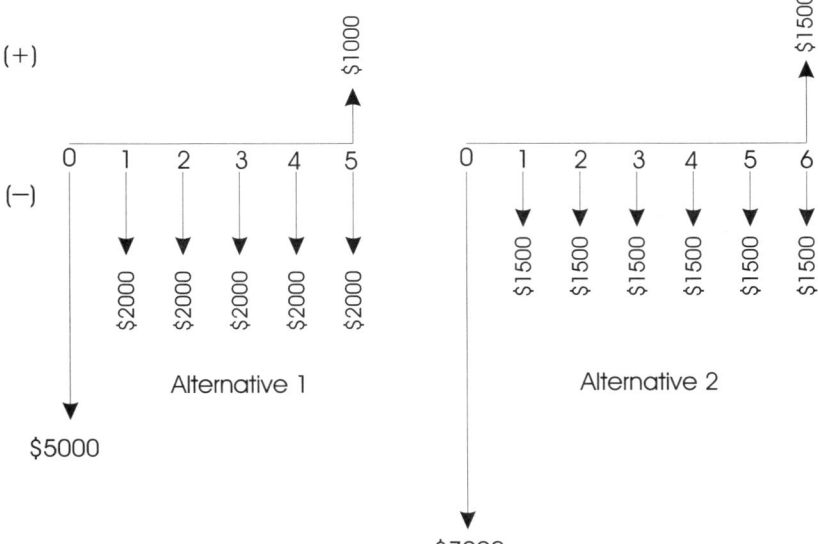

Fig. 4.4 Cash flow diagrams for two mutually alternatives.

Using a minimum attractive rate of return of 15%, the following annual worths are obtained:

$$AW_1(15\%) = -\$5\,000(A|P\ 15,5) - \$2\,000 + \$1\,000(A|F\ 15,5)$$
$$= -\$3\,343.20$$
$$AW_2(15\%) = -\$7\,000(A|P\ 15,6) - \$1\,500 + \$1\,500(A|F\ 15,6)$$
$$= -\$3\,178.10$$

Implicit in the comparison of the alternatives using annual worths based on individual life cycles is the assumption that a 30-year planning period is being used. As depicted in Table 4.11, if the present worths are computed for a 30-year period, the following values will be obtained:

$$PW_1(15\%) = -\$21\,952.10$$
$$PW_2(15\%) = -\$20\,868.30$$

Converting the present worths to annual worths yields the values obtained using individual life cycles.

An alternative assumption that could be made is that a five-year planning period is being used and the salvage value for Alternative 2 is such that an annual worth of $-\$3\,178.10$ will still be obtained. Hence, S_5, salvage value at the end of the fifth year for Alternative 2, must be:

$$AW_2(15\%) = -\$7\,000(A|P\ 15,5) - \$1\,500 - S_5(A|F\ 15,5)$$
$$-\$3\,178.10 = -\$3\,588.10 - 0.1483 S_5$$

or

$$S_5 = \$2764.67$$

(It is interesting to note that the salvage value obtained in this case is the book value at the end of five years of service using the sinking fund depreciation method described in Chapter 5.)

Example 4.18

To illustrate the shortcoming associated with an assumption that the salvage value for unused portions of an asset's life will be such that the annual worth will be unchanged, consider the alternatives depicted in Figure 4.5.

Table 4.11 Cash Flows for Two Mutually Exclusive Alternatives, with a 30-Year Planning Period

EOY	CF(1)	CF(2)
0	− 5 000	− $7 000
1-4	− 2 000	− 1 500
5	− 6 000	− 1 500
6	− 2 000	− 7 000
7-9	− 2 000	− 1 500
10	− 6 000	− 1 500
11	− 2 000	− 1 500
12	− 2 000	− 7 000
13-14	− 2 000	− 1 500
15	− 6 000	− 1 500
16-17	− 2 000	− 1 500
18	− 2 000	− 7 000
19	− 2 000	− 1 500
20	− 6 000	− 1 500
21-23	− 2 000	− 1 500
24	− 2 000	− 7 000
25	− 6 000	− 1 500
26-29	− 2 000	− 1 500
30	− 1 000	0
Present worth	− $21 952.10	− $20 868.30

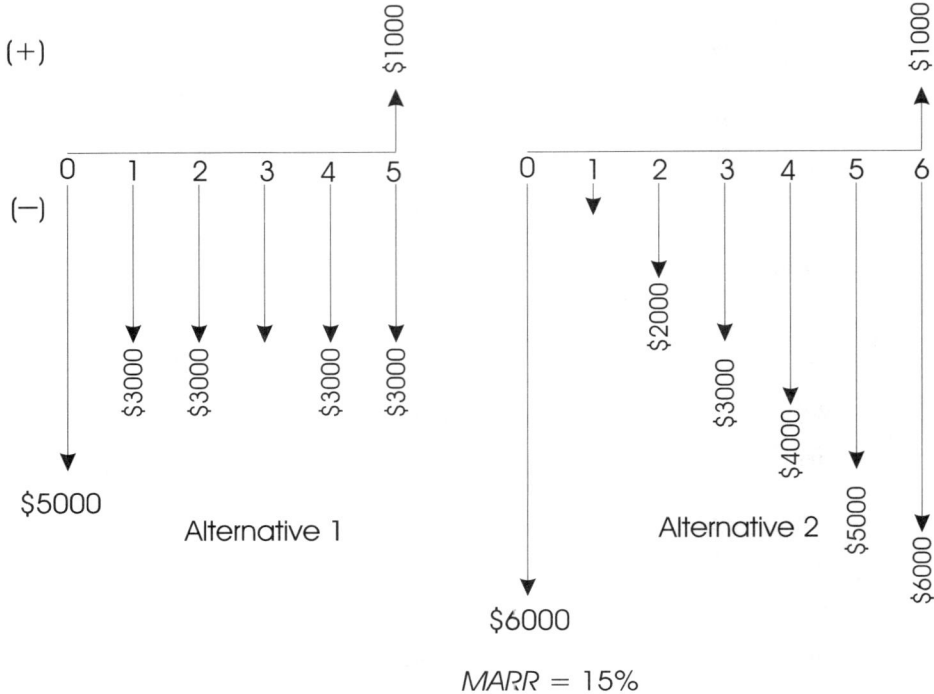

Fig. 4.5 Cash flow diagrams for two mutually exclusive alternatives.

The annual worths for individual life cycles are found to be:

$$AW_1(15\%) = -\$5\,000(A|P\ 15,5) - \$3\,000 + \$1\,000(A|F\ 15,5)$$
$$= -\$4\,343.20$$
$$AW_2(15\%) = -\$6\,000(A|P\ 15,6) - \$1\,000 + \$1\,000(A|G\ 15,6)$$
$$+ \$1\,000(A|F\ 15,6)$$
$$= -\$4\,568.20$$

If a five-year planning period is used, to obtain an annual worth for Alternative 2 of $-\$4\,568.20$ requires that salvage value at the end of the fifth year, S_5, be such that:

$$-\$4\,568.20 = -\$6\,000(A|P\ 15,5) - \$1\,000 - \$1\,000(A|G\ 15,5)$$
$$+ S_5(A|F\ 15,5)$$

or $S_5 = -\$374.92$. Thus, instead of having a $1 000 salvage value at the end of the sixth year, salvage value of $-\$374.92$ must exist to yield the annual worth obtained by treating individual life cycles.

We do not recommend that alternatives having unequal lives be blindly compared on the basis of the annual worths for individual life cycles. Such an approach assumes *implicitly* that either a least common multiple of lives planning period is appropriate or salvage values for unused portions of an asset's life are such that the annual worth is unchanged. We prefer to make explicit all assumptions concerning the planning period and salvage values.

4.12 Replacement Analysis

Among the various types of investment alternatives that are encountered in the real world, the *replacement* alternative deserves special consideration. Typically, in replacement analyses, one of the feasible alternatives involves maintaining the status quo; the remaining alternatives provide various replacement options that are available.

Whether you are considering replacing your automobile or your television set, or the machinery in a factory, there are definitely economic considerations involved. The need for considering replacement can be due to a number of factors, including inadequacy, excessive maintenance, declining efficiency, obsolescence (both functional and economic), physical impairment, and rental or lease possibilities. For example, if you are confronted with the need for a major overhaul on your car, you begin to consider replacing the car.

Changing demands can render a present asset inadequate to meet the production levels required. As an asset ages, it tends to deteriorate gradually over time and either operate less efficiently, less accurately, or experience increasing demands for maintenance. A decrease in demand might result in the present asset being obsolete functionally, since new assets having lower capacity are available to accommodate the new demand level. New and improved designs can render an existing asset obsolete economically. The rapid development of new models of computers is an illustration of the possibility of an asset becoming obsolete economically. When an asset fails and the service it renders is still required, a major repair is made or a replacement is acquired. The development of equipment rental firms has caused some homeowners to change from buying to leasing role to some expensive, but infrequently used, tools; the same is true for businesses.

4.12.1 Cash Flow Approach

The replacement decision can be analyzed using the approaches outlined previously in this chapter. Also, we continue to recommend you use a cash flow approach and ask of each alternative, "How much money will be spent and received if I adopt this alternative?" Past costs should be viewed from the proper perspective; as pointed out in Chapter 2, unrecoverable past costs are *sunk costs* and are not to be included in economic analyses that deal with the future, except as these sunk costs may affect income taxes if a present asset is disposed of. The past is beneficial in that it allows the identification of estimation errors previously made and assists in making better estimates of the future.

We will be concerned with defining the planning period based on the remaining life of the asset in current use and the lives of the alternative candidates for replacement. Likewise, if we replace the current asset we will want to know how much money will be received for it (i.e., its salvage value). Another consideration involves income taxes, which are treated in Chapter 5.

Example 4.19

To illustrate a replacement alternative, consider a situation involving a chemical plant that owns a filter press that was purchased three years ago for $10 000. Actual operating and maintenance (O & M) expenses (excluding labour) for the press have been $2 000, $2 500, and $3 000 each of the past three years, as depicted in Table 4.12. It is anticipated that the filter press can be used for five more years and salvaged for $1 000 at that time. The current market value for the used filter press is $4 000. If the old filter press is retained, annual operating and maintenance costs are anticipated to be as shown in Table 4.12.

A new filter press is available and can be purchased for $18 000. The new filter press is expected to have annual operating and maintenance costs as depicted in Table 4.12.

Table 4.12 Data for a Replacement Alternative

Alternative 1		Alternative 2		
t	O&M	t	O&M	S_t
−3	—	0	—	$18 000
−2	−$2 000	1	—	15 000
−1	− 2 500	2	−$ 500	12 000
0	− 3 000	3	− 1 000	9 900
1	− 3 500	4	− 1 500	7 800
2	− 4 000	5	− 2 000	6 000
3	− 4 500	6	− 2 500	4 500
4	− 5 000	7	− 3 000	3 300
5	− 5 500	8	− 3 500	2 400
		9	− 4 000	1 800
		10	− 4 500	1 500

The new filter press has an anticipated useful life of 10 years. Based on historical data concerning salvage values of filter presses, estimated salvage values (S_t) for the new press are given in Table 4.12.

Since the "old" press can only be used for five more years, a planning period of five years is specified; salvage value for the "new" press is estimated to be $6 000 in five years.

Alternative 1 is defined to be "keep the old press," Alternative 2 is defined to be "replace the old press with the new press." Cash flows for each alternative are given in Table 4.13.

Table 4.13 Cash Flows for a Replacement Alternative

t	A_{1t}	A_{2t}	$A_{2t} - A_{1t}$
0	0	−$18 000 + 4 000	−$14 000
1	−$3 500	0	3 500
2	− 4 000	− 500	3 500
3	− 4 500	− 1 000	3 500
4	− 5 000	− 1 500	3 500
5	−5 500+1 000	− 2 000 + 6 000	8 500

Computing the annual worths for each alternative using a minimum attractive rate of return of 10% yields the following results:

$$AW_1(10\%) = -\$3\,500 - \$500(A|G\ 10,5) + \$1\,000(A|F\ 10,5)$$
$$= -\$4\,241.25$$

$$AW_2(10\%) = -\$14\,000(A|P\ 10,5) - \$500(A|G\ 10,5) + \$6\,000(A|F\ 10,5)$$
$$= -\$3\,615.45$$

On the basis of the annual worth comparison, it is recommended that the new filter press be purchased and the old press be sold.

Note that, in carrying out the replacement analysis when the cash flows were listed for each alternative in Table 4.13, we ignored the operating and maintenance costs that will occur after the fifth year if the new filter press is purchased. The argument against including the operating and maintenance costs is based on the specification of a planning period of five years. The five-year planning period was based on the maximum useful life for the old filter press. If the old press is retained, then it will have to be replaced in five years (if not before, because of the possible development of attractive replacement alternatives in the future). Consequently, in five years we might have available an alternative that will yield even greater operating and maintenance savings than the filter press currently being considered. Thus, it is not fair to include savings that might occur after five years for one alternative without including such anticipated savings for the other alternative.

The above interpretation defines the planning period as a window through which can be seen only the cash flows that occur during the planning period, *with the exception of the terminal value for the alternative.* At the end of the planning period, an estimate is provided for the terminal value for each alternative, even though the alternative might not be physically replaced at that time. The end of the planning period defines a point at which another replacement study is planned. At that time, the future savings and costs will be compared against other available replacement candidates.

Example 4.20

If, in Example 4.19, a 10-year planning period is desired, we recommend that consideration be given to the replacement of the "old" filter press in five years. Based on a forecast of the growth of filter press technology, suppose we anticipate that at the end of five years a filter press will be available at a cost of $15 000; net operating and maintenance costs, and a terminal salvage value are anticipated to be as depicted in Table 4.14. A computation establishes that the difference in annual worths for the two alternatives is $38.02, with Alternative 1 being most economic. Thus, on the basis of technological forecasts of filter press alternatives in five years, it would appear advantageous to postpone the replacement. However, the degree of uncertainty in our forecast of the cash flows for a projected future replacement candidate would cause us to question the merits of postponing the replacement because of a difference of $38.02 per year. These kinds of considerations are explored more fully in Chapters 7 and 8

Table 4.14 Cash Flows for a Replacement Alternative Using a 10-Year Planning Period

t	A_{1t}	A_{2t}	$A_{2t} - A_{1t}$
0	0	−$18 000 + 4 000	−$14 000
1	−$3 500	0	3 500
2	− 4 000	− 500	3 500
3	− 4 500	− 1 000	3 500
4	− 5 000	− 1 500	3 500
5	−5 500+1 000 − 15 000	− 2 000	17 500
6	0	− 2 500	− 2 500
7	− 500	− 3 000	− 2 500
8	− 1 000	− 3 500	− 2 500
9	− 1 500	− 4 000	− 2 500
10	− 2 000 + 7 500	− 4 500 + 1 500	− 8 500

4.12.2 Classical Approach

As with the treatment of unequal lives, the classical approach in analyzing replacement decisions differs from the cash flow approach. The classical approach considers the salvage value of the old asset (defender) to be the investment cost for the defender if it is retained in service. Such an approach is consistent with the opportunity cost concept described in Chapter 2. Since the retention of the defender is equivalent to a decision to forego the receipt of its salvage value, then an opportunity cost is assigned to the defender.

Example 4.21

In order to illustrate the classical approach, consider Example 4.19. Applying the classical approach, the "cash flows" would be as depicted in Table 4.15. As can be seen from comparing Tables 4.13 and 4.15, the differences in the alternatives are the same; hence, the cash flow approach and classical approach are equivalent approaches.

While the concept underlying the classical approach is sound, there are some potential pitfalls in its application which should be avoided. Such pitfalls arise when either the defender has multiple trade-in values or unequal lives exist. Both of these situations will be illustrated with example problems.

Table 4.15 "Cash Flows" Using the Classical Approach

t	A_{1t}	A_{2t}	$A_{2t} - A_{1t}$
0	−$4 000	−$18 000	−$14 000
1	− 3 500	0	3 500
2	− 4 000	− 500	3 500
3	− 4 500	− 1 000	3 500
4	− 5 000	− 1 500	3 500
5	−5 500+1 000	− 2 000 + 6 000	8 500

Example 4.22

In the previous example, suppose a second replacement alternative is available. A filter press which sells for $20 000 is being considered; annual operating and maintenance costs are expected to equal $500 the first year and increase by 20% thereafter. Salvage value of $10 000 is anticipated in five years. If this new filter press is

Comparison of Alternatives

purchased, then a $5 000 trade-in allowance will be provided for the old filter press. A cash flow approach yields the data presented in Table 4.16.

Table 4.16 Cash Flows for Example 4.22

t	A_{1t}	A_{2t}	A_{3t}
0	0	$-$18\ 000 + 4\ 000$	$-$20\ 000 + 5\ 000$
1	$-$3\ 500$	0	$- 500$
2	$- 4\ 000$	$- 500$	$- 600$
3	$- 4\ 500$	$- 1\ 000$	$- 720$
4	$- 5\ 000$	$- 1\ 500$	$- 864$
5	$-5\ 500+1\ 000$	$- 2\ 000 + 6\ 000$	$- 1\ 037 + 10\ 000$

Alternatively, if the classical approach is used, then a decision must be made concerning the appropriate investment cost for the defender. Should it be $4 000 or $5 000? For either investment cost, the relative differences in investment costs among the three alternatives should remain the same. Thus, if $4 000 is used, then the cash flows at $t = 0$ will be –$4 000, –$18 000, and –$19 000 for the three alternatives; if $5000 is used, then the cash flows at $t = 0$ will be –$5 000, –$19 000, and –$20 000, respectively. Since the classical approach is based on the opportunity cost concept, it seems appropriate for a $5 000 investment cost to be assigned to the defender.

Example 4.23

As in Example 4.20, suppose a planning period of 10 years is used. If the cash flow approach is used, then, as was shown in Table 4.14, an explicit decision must be made concerning the replacement of the defender after five years, whereas the classical approach can result in such a decision being suppressed. In particular, with the classical approach the equivalent uniform annual cost of the defender is often compared with the equivalent uniform annual cost of the challenger. The dangers of such an approach were elaborated in Section 4.11; in the case of replacement decisions the dangers are magnified.

Using the classical approach, the *EUAC* for the defender and the challenger can be determined to be:

$$EUAC_1(10\%) = \$4\,000(A|P\ 10,5) + \$3\,500 + \$500(A|G\ 10,5)$$
$$-\$1\,000(A|F\,10,5)$$
$$= \$5\,296.45$$

$$EUAC_2(10\%) = \$18\,000(A|P\ 10,10) + \$500(A|G\ 10,10)$$
$$-\$1\,500(A|F\ 10,10)$$
$$= \$4\,697.30$$

However, in order for the equivalent uniform annual costs to be accurate over the 10-year planning period, it is assumed that the defender will be replaced with an asset having a five-year life and a cash flow profile identical to that of the defender. Such assumptions do not seem reasonable.

Recall that the cash flow approach requires the cash flows to be stated explicitly for each and every year during the planning period. Consequently, unequal trade-in values and unequal lives pose no difficulties. If the cash flow approach of Table 4.14 were modified to treat the trade-in value of the defender as an investment cost, then the only changes required are to let $A_{1t} = -\$4\,000$ for $t = 0$ and $A_{2t} = -\$18\,000$ for $t = 0$.

Although logical arguments can be given for treating a replacement decision as just another economic investment alternative, many firms fail to subject their existing equipment to careful scrutiny on a periodic basis to determine if replacement is required. Despite the fact that replacement studies can yield significant reductions in costs, many firms postpone replacing assets beyond the "optimum" time for replacement, perhaps because a decision to replace an asset involves a change, and resistance to change is inherent in most individuals. For example, an engineer who two years ago successfully argued that compressor Z should be replaced by compressor Y may now find that compressor X is more economical than compressor Y. If X is now championed, it may be viewed by management as an admission that the wrong compressor was selected as a replacement for Z.

Some reasons for delaying the replacement of assets beyond the economic replacement time are:

1. The firm is making a profit with its present equipment.

2. The present equipment is operational and is producing a product of acceptable quality.

3. There is risk or uncertainty associated with predicting the expenses of a new machine, whereas one is relatively certain about the expenses of the current machine.

4. A decision to replace equipment is a stronger commitment for a period of time into the future than keeping the existing equipment.

5. Management tends to be conservative in decisions regarding the replacement of costly equipment.

6. There may be a limitation on funds available for purchasing new equipment, but no limitation on funds for maintaining existing equipment.

7. There may be considerable uncertainty concerning the future demand for the services of the equipment in question.

8. Sunk costs psychologically affect decisions to replace equipment.

9. An anticipation that technological improvements in the future might render obsolete equipment available currently; a wait-and-see attitude prevails.

10. Reluctance to be a pioneer in adopting new technology; instead of replacing now, wait for the competition to act.

As equipment ages, operating and maintenance costs increase. At the same time, the capital recovery cost decreases with prolonged use of the equipment. The combination of decreasing capital recovery costs and increasing annual operating and maintenance costs results in the equivalent uniform annual cost taking on a form similar to that depicted in Figure 4.6.

By forecasting the operating and maintenance costs for each year of service, as well as anticipated salvage values for various replacement ages, one can determine the replacement interval for equipment.

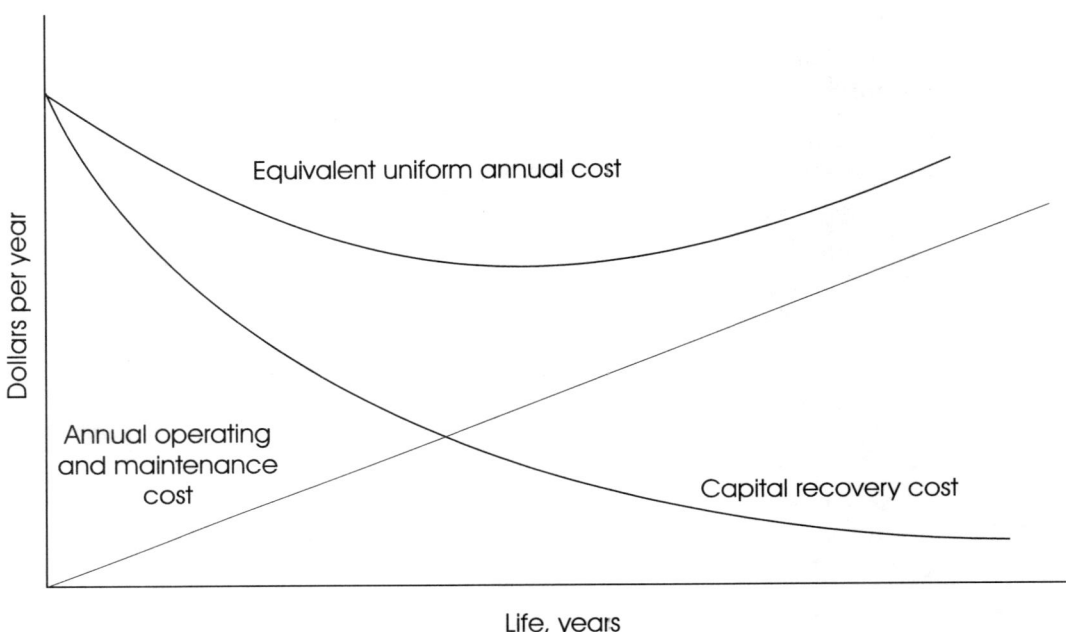

Fig. 4.6 Portrayal of components of equivalent annual cost.

Example 4.24

Suppose a small compressor can be purchased for $1 000; salvage value for the compressor is assumed to be negligible, regardless of the replacement interval. Annual operating and maintenance costs are expected to increase by $75 per year, with the first year's cost anticipated to be $150. Using a minimum attractive rate of return of 8%, the following equivalent uniform annual costs are obtained:

$$EUAC_{n=1} = \$1\,000(A|P\ 8,1)+\$150+\$75(A|G\ 8,1)$$
$$= \$1\,230$$

$$EUAC_{n=2} = \$1\,000(A|P\ 8,2)+\$150+\$75(A|G\ 8,2)$$
$$= \$746.52$$

$$EUAC_{n=3} = \$1\,000(A|P\ 8,3)+\$150+\$75(A|G\ 8,3)$$
$$= \$608.25$$

$$EUAC_{n=4} = \$1\,000(A|P\ 8,4)+\$150+\$75(A|G\ 8,4)$$
$$= \$555.49$$

$$EUAC_{n=5} = \$1\,000(A|P\ 8,5)+\$150+\$75(A|G\ 8,5)$$
$$= \$536.26$$

$$EUAC_{n=6} = \$1\,000(A|P\ 8,6)+\$150+\$75(A|G\ 8,6)$$
$$= \$533.07$$

$$EUAC_{n=7} = \$1\,000(A|P\ 8,7)+\$150+\$75(A|G\ 8,7)$$
$$= \$538.72$$

$$EUAC_{n=8} = \$1\,000(A|P\ 8,8)+\$150+\$75(A|G\ 8,8)$$
$$= \$549.34$$

$$EUAC_{n=9} = \$1\,000(A|P\ 8,9)+\$150+\$75(A|G\ 8,9)$$
$$= \$563.03$$

$$EUAC_{n=10} = \$1\,000(A|P\ 8,10)+\$150+\$75(A|G\ 8,10)$$
$$= \$578.41$$

As n increases beyond six years, the equivalent uniform annual cost increases. Hence, for this example, a replacement interval of six years is indicated.

The assumptions inherent in the determination of the optimum replacement interval should be considered closely. In particular, the analysis is based on a planning period that is a whole multiple of the replacement interval selected. (Recall our discussion of the use of the annual worth based on one life cycle and a planning period of the least common multiple of lives.) Second, we have assumed that each time the compressor is replaced, it will be replaced with a compressor having an identical cash flow profile. If neither assumption is valid, then the approach we used is not valid.

Example 4.25

Suppose that in the previous case a nine-year planning period is appropriate. For simplicity, we will continue to assume that replacements will have identical cash flow profiles. If the original compressor is kept for nine years, the present worth equivalent will be:

$$PW_{p=9} = -\$563.03(P|A\ 8,9)$$
$$= -\$3\,517.19$$

If the compressor is to be replaced at some intermediate point during the planning period, say after k years, and the replacement is to be kept until the end of the planning period, the following present worth calculations result for $k = 5$, 6, or 7:

$$PW_{k=5} = -\$536.26(P|A\ 8,5) - \$555.49(P|A\ 8,4)(P|F\ 8,5)$$
$$= -\$3\,393.32$$

$$PW_{k=6} = -\$533.07(P|A\ 8,6) - \$608.25(P|A\ 8,3)(P|F\ 8,6)$$
$$= -\$3\,452.18$$

$$PW_{k=7} = -\$538.72(P|A\ 8,7) - \$746.52(P|A\ 8,2)(P|F\ 8,7)$$
$$= -\$3\,581.59$$

Thus, the compressor should be replaced after five years and the replacement should be kept until the end of the planning period. [Why is it unnecessary to consider the remaining values of k (i.e., $k = 4, 3, 2$, and 1)? In order to answer this, consider the case of $k = 4$ and compare the cash flow profiles over the nine-year planning period using $k = 4$ and $k = 5$. As shown in Figure 4.7, a consideration of the time value of money eliminates the possibility of $k = 4$ having a lower present worth than $k = 5$. Similar comparisons can be used to eliminate the cases of $k = 1, 2$, and 3.

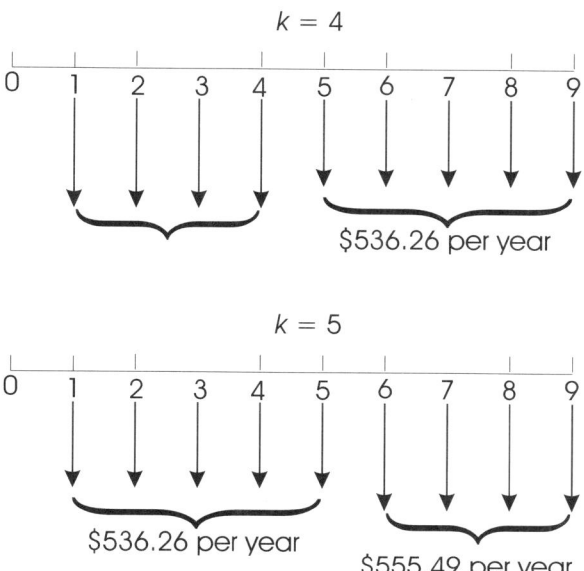

Fig. 4.7 Comparison of replacement strategies involving replacement at the end of year 4 versus the end of year 5.

For this particular example, failure to replace the equipment at precisely the optimum time will not result in a significantly lower annual *worth* or present *worth*. This is often the case with many replacement situations; the resulting measure of effectiveness is relatively insensitive to deviations from the optimum strategy. Hence, the firm might establish an operating policy of reviewing the *actual* operating and maintenance costs for the equipment at the anticipated "optimum" time of replacement. Then, perform a replacement study at that time, using the eight-step procedure for comparing investment alternatives.

4.13 Computer Applications

Financial and corporate computer-based planning models are common in many organizations. Regardless of which financial planning package is being considered, the use will typically find a basic set of features and capabilities. Packages differ primarily in terms of how they are designed and implemented, as well as their capacity for supporting large and complex financial planning models.

A typical planning package contains an extensive collection of functions, including:
- depreciation by multiple methods
- capital cost allowance
- rate of return (discrete or continuous)
- payback computations
- present value
- compounding–future value
- loan payment and amortization computations
- capital budgeting
- inventory management
- forecasting
- risk simulation (Monte Carlo analysis)
- sensitivity analysis
- profit and loss analysis
- cost accounting

The use of such programs for engineering economic analysis is now widespread in the mining, manufacturing, and processing industries.

4.14 Summary

The process of comparing investment alternatives was described in this chapter. A cash flow approach was recommended, as was a planning period approach. Replacement problems were treated as alternative comparison problems. Measures of merit examined included the present worth, annual worth, payback period, rate of return, and savings/investment ratio. The discussion of the comparison of investment alternatives did not consider either the effects of income taxes, the public-sector requirements for measuring benefits in economic units, or the effects of risk and uncertainty; such considerations are reserved for the following chapters.

Problems

1. Three investment proposals have been collected by the capital budget committee. Proposals A and C are mutually exclusive; Proposal A is contingent on Proposal B; and Proposal B is contingent on Proposal C. A budget limitation of $100 000 exists. The cash flow profiles for the three investment proposals are given below for the five-year planning period.

EOY	CF(A)	CF(B)	CF(C)
0	− $60 000	− $40 000	− $80 000
1-5	20 000	10 000	15 000
5	−	20 000	20 000

Using a minimum attractive rate of return of 5%, compare the set of feasible, mutually exclusive alternatives and specify the preferred alternative. Use all measures of merit described in the text. Assume a payback period of three years applies for the payback period method. (4.2, 4.7)

2. Four investment proposals, A, B, C, and D, are available. Proposals A and B are mutually exclusive and Proposals A and C are mutually exclusive. Proposal D is contingent on either Proposal A or B. Funds available for investment are limited to $50 000. Using a MARR of 15% and a present worth analysis, determine the recommended investment program. Justify the elimination of any mutually exclusive alternatives. (4.2, 4.7)

EOY	CF(A)	CF(B)	CF(C)	CF(C)
0	− $20 000	− $30 000	− $25 000	− $20 000
1-4	7 000	9 000	8 000	8 500

3. Two mutually exclusive proposals, each with a life of five years, are under consideration. Each proposal has the following cash flow profile:

EOY	CF(A)	CF(B)
0	−$20 000	−$34 000
1	4 878	4 762
2	4 878	4 762
3	4 878	4 762
4	4 878	4 762
5	4 878	4 762

(a) Specify clearly the mutually exclusive *alternatives* available to the decision maker. (4.2)

(b) Using a rate of return analysis, which *alternative* should the decision maker select? Assume $MARR = 10\%$. Use both the internal rate of return method and the external rate of return method. (4.7)

4. A firm has available four proposals, A, B, C, and D. Proposal A is contingent on acceptance of either Proposal C or Proposal D. The firm has a budget limitation of $200 000. In addition, Proposal C is contingent on Proposal D, while Proposal D is contingent on either Proposal A or Proposal B. Using a *MARR* of 10%, determine the preferred alternative using the savings/investment ratio method. (4.2, 4.7)

EOY	CF(A)	CF(B)	CF(C)	CF(D)
0	−$100 000	−$140 000	−$20 000	−$15 000
1	10 000	7 500	3 000	1 000
2	10 000	8 500	3 000	1 500
3	10 000	9 500	3 000	2 000
4	10 000	10 500	3 000	2 500
5	20 000	30 000	3 000	3 000

5. A firm has available three investment proposals, A, B, and C, having the cash flow profiles shown below. Proposals B and C are mutually exclusive and Proposal C is contingent on Proposal A being chosen. The firm has a *MARR* of 20%.

Comparison of Alternatives

	CF(A)	CF(B)	CF(C)
Initial investment	$200 000	$300 000	$150 000
Life	8 years	12 years	8 years
Annual receipts	$160 000	$190 000	$200 000
Annual disbursements	$115 000	$120 000	$150 000
Salvage value	$ 50 000	$100 000	$ 50 000

The firm is willing to use a planning period of 24 years and accepts the assumption of identical cash flow profiles for successive life cycles of a proposal.

  (a) Determine the set of feasible alternatives and give the cash flow profile for each over the planning period. (4.2, 4)

  (b) Determine the preferred alternative, based on an annual worth analysis. (4.7)

6. Four investment proposals, W, X, Y, and Z, are being considered by the Ajax Corporation. Proposals X and Z are mutually exclusive. Proposal Y is contingent on either X or Z. Proposals W and Y are mutually exclusive. A budget limitation of $200 000 exists. Either Proposal X or Proposal Z must be included in the alternative selected. Using a *MARR* of 20%, determine the annual worth for each *feasible investment alternative*. (4.2, 4.7)

EOY	CF(W)	CF(X)	CF(Y)	CF(Z)
0	− $100 000	− $125 000	− $90 000	− $100 000
1-8	30 000	50 000	25 000	20 000
8	40 000	10 000	120 000	100 000

7. A firm is faced with three investment proposals, A, B, and C, having the cash flow profiles shown below over the planning period of five years. Proposals B and C are mutually exclusive, and Proposal C is contingent on Proposal A being selected. A budgetary limitation of $125 000 exists on the amount that can be invested at the end-of-year zero. A *MARR* of 10% is to be used in performing a present worth analysis.

  (a) Specify the cash flow profiles for all mutually exclusive, feasible investment alterntives. (4.4)

(b) Determine which investment alternative yields the maximum present worth. (4.6)

EOY	CF(A)	CF(B)	CF(C)
0	− $100 000	− $50 000	− $20 000
1	20 000	20 000	5 000
2	30 000	20 000	5 000
3	40 000	20 000	5 000
4	50 000	20 000	5 000
5	60 000	20 000	15 000

8. Ms. Brown invested $64 000 in a business venture with the following cash flow results:

EOY	CF	EOY	CF	EOY	CF
0	− $64 000	3	$9 000	6	$9 000
1	5 000	4	11 000	7	7 000
2	7 000	5	11 000	8	5 000

To the nearest percent, what was her internal rate of return on the venture? (4.6)

9. Mr. Smith invests $10 000 in an investment fund. One year after making the investment he receives $1 000 and continues to receive $1 000 annually until 10 such amounts are received. He receives nothing further until 15 years after the initial investment, at which time he receives $10 000. Over the 15-year period, what was the man's internal rate of return? (4.6)

10. A firm purchases a pressure vessel for $20 000. Half of the purchase price is borrowed from a bank at 6% compounded annually. The loan is to be paid back with equal annual payments over a five-year period. The pressure vessel is expected to last 10 years, at which time it will have salvage value of $4 000. Over the 10-year period the operating and maintenance costs are anticipated to equal $4 000 per year.

The firm expects to earn 10% on its investments. What is the equivalent uniform annual cost for the investment? (4.6)

11. A firm purchases a computer for $10 000; the computer is used for five years and is then sold for $2 000. Annual disbursement for operating and maintenance costs equalled $4 000 per year over the five-year period. On the basis of a 20% MARR, what was the equivalent annual worth for the investment? (4.6)

12. A firm purchases a gravity settling tank by borrowing the $10 000 purchase price. The loan is to be repaid with six equal annual payments at an annual compound rate of 7%. It is anticipated that the tank will be used for nine years and then be sold for $1 000. Annual operating and maintenance expenses are estimated to be $5 000 per year. The firm uses a MARR of 15% for its economic analyses. Determine the equivalent uniform annual cost for the piece of equipment. (4.6)

13. A firm borrows $50 000 to purchase a new numerically controlled milling machine and pays the loan back over a five-year period with equal payments. Interest on the loan is 10% compounded annually. The machine is estimated to have annual operating and maintenance costs of $12 000 per year and have a life of 10 years. Salvage value is estimated to be $15 000. The firm has a MARR of 8%. Determine the equivalent uniform annual cost over the 10-year period. (4.6)

14. The Belt Company is considering four investment proposals, A, B, C, and D. Proposals A and B are mutually exclusive, and Proposal C is contingent on Proposal D. Proposals A and D are mutually exclusive. The cash flow data for the investments over a 10-year planning period are given below. The Belt Company has a budget limit of $750 000 for investments of the type being considered currently.

	CF(A)	CF(B)	CF(C)	CF(D)
Initial investment	$450 000	$600 000	$350 000	$300 000
Life	10 years	10 years	10 years	10 years
Salvage value	$50 000	$100 000	$50 000	$50 000
Annual receipts	$300 000	$450 000	$200 000	$350 000
Annual disbursements	$100 000	$200 000	$ 50 000	$200 000

MARR = 25%

(a) State all mutually exclusive investment alternatives available. (4.2)

(b) Develop the cash flow profiles for the mutually exclusive alternatives. (4.4)

(c) Compare the alternatives using the present worth method. (4.7)

15. The Acme Manufacturing Company purchased an automatic transfer machine for $15 000 installed. At the end of 10 years the company paid $2 000 to have the machine removed and junked. The operating and maintenance cost history was as follows:

EOY	O & M Cost
1	$1000
2	1000
3	1000
4	1000
5	1500
6	1500
7	1500
8	1500
9	2000
10	2000

The company has a 10% MARR. Supply the values of U, V, W, X, and Y in the following equation for computing the equivalent uniform annual cost for the machine. (4.6)

$$EUAC = (\$15\,000 - U)(A|F\ 10\%, 10) + V(0.10) + W$$
$$+ \$500[(F|A\ 10\%, 6) + (F|A\ 10\%, X)](Y, 10\%, 10)$$

16. A distillation column is purchased for $200 000. Operating and maintenance costs for the first year are $20 000. Thereafter, operating and maintenance costs increase by 10% per year over the previous year's costs. At the end of 10 years the column is sold for $50 000. During the life of investment revenue was produced that could be related directly to the investment in the column. The revenue the first year was $50 000. Thereafter, revenue increased by $4 000 over the previous year's revenue. Using a MARR of 15%, determine the equivalent present worth for the investment. (4.6)

17. Machine A initially costs $40 000. Its resale value at the end of the kth year of service equals $40 000 − $3 000k for k = 1, 2, ..., 12. Operating and maintenance expenses equal $8 000 per year. Machine B initially costs $18 000. Its resale value at the end of the kth year, S_k, equals $18\,000(0.80)^k$ for k = 1, 2, ..., 8. Operating and maintenance expenses equal $10 000 per year. What is the equivalent uniform annual cost for each machine, based on a planning period of (a) 24 years? (b) 5 years? Use a MARR of 10%. Assume the recommended machine will be kept until the end of the planning period. (4.6)

18. The Ajax Manufacturing Company wishes to choose one of the following machines:

	Machine A	Machine B	Machine C
First cost	$10 000	$14 000	$18 000
Planning period	8 years	8 years	8 years
Salvage value	—	$2 000	$3 000
Operating and maintenance costs for year k, k = 1,...,n	$600(1.10)^{k-1}$	$400(1.08)^{k-1}$	$200 + 100k$

What is the equivalent uniform annual cost for each machine, based on a MARR of (a) 0%, (b) 8%, (c) 10%? Assume machines are not replaced during the planning period. (4.6)

19. Mr. Shrewd invested $100 000 in a business venture with the following cash flow results. Determine his rate of return using the internal rate of return method. (4.6)

EOY	CF	EOY	CF	EOY	CF
0	− $100 000	7	$14 000	14	$7 000
1	20 000	8	13 000	15	6 000
2	19 000	9	12 000	16	5 000
3	18 000	10	11 000	17	4 000
4	17 000	11	10 000	18	3 000
5	16 000	12	9 000	19	2 000
6	15 000	13	8 000	20	1 000

20. The Ajax Manufacturing Company wishes to choose one of the following machines:

	Machine A	Machine B
First cost	$10 000	$16 000
Planning period	6 years	6 years
Salvage value	$1 000	$1 000
Operating and maintenance costs for year k	500+80k	200+100k

What are the equivalent uniform annual costs for each machine, based on a 10% MARR? (4.6)

21. Owners of a nationwide motel chain are considering building a new 200-unit motel in Calgary, Alberta. The present worth cost of building the motel is $2 250 000; the firm estimates furnishings for the motel will cost an additional $700 000 and will require replacement every five years. Annual operating and maintenance costs for the facility are estimated to be $60 000. The average rate for a unit is anticipated to be $14 per day. A 15-year planning period is used by the firm in evaluating new ventures of this type; a terminal salvage value of 15% of the original building cost is anticipated; furnishings

are estimated to have no salvage value at the end of each five-year replacement interval. Assuming an average daily occupancy percentage of 70%, 80%, and 90%, a MARR of 20%, and 365 operating days per year, should the motel be built? Ignore the cost of the land. (4.7)

22. A consulting engineer is designing an irrigation system and is considering two pumps to meet a pumping demand of 15 000 L/min at 17.3 m total dynamic head. The specific gravity of the fertilizer mixture being pumped is 1.25. Pump A operates at 78% efficiency and costs $9 450; Pump B operates at 86% efficiency and costs $11 250. Power costs $0.06 / kW · h. Continuous pumping for 200 days per year is required (i.e., 24 hours per day). Using a MARR of 10%, a planning period of five years, and equal salvage values for the two pumps, which should be selected? (Note: Dynamic head × litres per minute × specific gravity ÷ 60 000 = kilowatts.) (4.7)

23. The motor on a numerically controlled machine tool must be replaced. Two different 15 kW electric motors are being considered. Motor U sells for $850 and has an efficiency rating of 90%; Motor V sells for $625 and has a rating of 84%. The cost of electricity is $0.06 / kW · h. An eight-year planning period is used, and zero salvage values are assumed for both motors. Annual usage of the motor averages approximately 4 000h. A MARR of 20% is to be used. Assume the motor selected will be loaded to capacity. Compare the alternatives using present worth, savings/investment ratio, internal rate of return, and external rate of return methods. (4.7)

24. A chemical plant is considering the installation of a storage tank for water. The tank is estimated to have an initial cost of $142 000; annual costs for maintenance are estimated to be $1 200 per year. As an alternative, a holding pond can be provided some distance away at an initial cost of $90 000 for the pond, plus $15 000 for pumps and piping; annual operating and maintenance costs for the pumps and holding pond are estimated to be $4 500. Based on a 20-year planning period, zero salvage values, and a MARR of 10%, which alternative is preferred on the basis of present worth? (4.7)

25. An investment of $16 000 for a new condenser is being considered. Estimated salvage value of the condenser is $5 000

at the end of an estimated life of eight years. Annual gross income each year for the eight years is $4 500. Annual operating expenses are $1 500. Assume money is worth 25% compounded annually.

  (a) Using the annual worth method, should the investment in the condenser be made?
  (b) Is the internal rate of return on this investment greater than 25% or less than 25%? (4.7)

26. An individual is considering two mutually exclusive investment alternatives, each requiring an initial investment of $1 000. Alternative 1 returns $200 after one year and $1 200 after two years; Alternative 2 returns $150 per year for the first three years and $1 150 after four years. The individual has established a *MARR* of 8%. Which alternative should be selected using internal rate of return, external rate of return, present worth, and annual worth methods? (4.7)

27. A city near the Red River in Manitoba has a history of flooding during heavy rainfalls. Two improvement alternatives are under consideration by the city engineer:

  A. Leave the existing 1-m corrugated steel culvert in place and install another of the same size alongside at an installed cost of $4 200.

  B. Remove the culvert currently in place and replace with a single 2-m culvert at an installed cost of $6 400.

  If Alternative A is adopted, the existing 1-m culvert will need to be replaced in 10 years. It is assumed that a replacement at that time will cost $5 000. A newly installed culvert will last for 20 years; replacement prior to the end of 20 years will result in zero salvage value. Use a planning period of 20 years and a 10% *MARR* to determine the preferred alternative. (4.7)

28. Insulation is to be installed on a steam pipe. Either 1-cm or 2-cm insulation is to be selected. The annual heat loss from the pipe at present is estimated to be $2.25 per metre of pipe. The 1-cm insulation costs $0.60 per metre and will reduce the heat loss by 86%; the 2-cm insulation costs $1.15 per metre and will reduce the heat loss by 90%. If insulation lasts 10 years with no salvage value and an 8% *MARR* is used, which size (if any) should be selected? (4.7)

Comparison of Alternatives

29. The provincial ambulance service wishes to establish a communications network to cover the province. A specified minimum signal strength at all points in the province is desired. Two alternatives have been selected for detailed consideration. Design I involves the installation of 5 transmitting stations of low power. The investment at each installation is estimated to be $40 000 in structure and $25 000 in equipment. Design II involves the installation of two transmitting stations of much higher power. Investment in structure will be approximately $50 000 per installation; equipment at each installation will cost $200 000. Annual operating and maintenance costs per installation are anticipated to be $15 000 for Design I and $18 000 for Design II. Structures are anticipated to last for 20 years; equipment life is estimated to be 10 years, with replacements assumed to be identical in costs and performance. Using a planning period of 20 years, salvage values of zero at $t = 0$, and a *MARR* of 10%, which design should be recommended on the basis of a present worth analysis? (4.7)

30. Two compressors are being considered by the Ajax Company. Compressor A can be purchased for $5 500; annual operating and maintenance costs are estimated to be $1 950. Alternatively, Compressor B can be purchased for $4 150; annual operating and maintenance costs are estimated to be $2 325. An eight-year planning period is to be used; salvage values are estimated to be 15% of the original purchase price; a *MARR* of 25% is to be used. Compare the alternatives using both internal and external rate of return methods, as well as the savings/investment ratio. (4.7)

31. The manager of the distribution centre for the Western Canadian territory of a major pharmaceutical manufacturer is contemplating installing an improved material handling system linking receiving and storage, as well as storage and shipping. Two designs are being considered. The first consists of a conveyor system that is tied into an automated storage/retrieval system. Such a system is estimated to cost $750 000 initially, have annual operating and maintenance costs of $48 000, and salvage value of $75 000 at the end of the 15-year planning period.

The second design consists of manually operated narrow-aisle, high-stacking lift trucks. To provide service comparable to that provided by the alternative design, an initial investment of $325 000 is required. Annual operating and maintenance costs of $103 000 are anticipated. An estimated salvage value of $20 000 is expected at the end of the planning period.

Using a *MARR* of 20%, compare the alternatives using annual worth, savings/investment ratio, and rate of return methods. (4.7)

32. A manufacturing plant in Halifax has been contracting snow removal at a cost of $250 per day. The past four years have yielded unusually heavy snowfalls, resulting in the cost of snow removal being of concern to the plant manager. The plant engineer has found that a snow-removal machine can be purchased for $38 000; it is estimated to have a useful life of 10 years, and zero salvage value at that time. Annual costs for operating and maintaining the equipment are estimated to be $12 000. Based on a *MARR* of (i) 0%, (ii) 10%, (iii) 20%, and (iv) 30%, and an estimated demand for the equipment equal to (a) 40, (b) 80, and (c) 150 days per year, determine the preferred alternative using an annual worth analysis. (4.7)

33. Given the following data, which pump is more economical, using the annual cost comparison? Assume a 10-year planning period and a *MARR* of 12%. (4.7)

	Pump A	Pump B
First cost	$6 000	$10 000
Salvage value	0	2 000
Annual operating cost	500	750
Life	5 years	10 years

34. A firm is considering two compressors X and Y. One must be chosen. Based on the following data, compare the present worths of each alternative and recommend the compressor having the largest present worth. Use a *MARR* = 10% and a planning period of 42 years. (4.11)

Comparison of Alternatives

	Compressor X	Compressor Y
Initial Investment	$10 000	$20 000
Life	6 years	7 years
Salvage value	$1 000	$5 000
Annual disbursments	$5 000	$3 000

35. A firm has available two mutually exclusive investment proposals, A and B. Based on the cash flow profiles shown below and using a savings/investment ratio analysis, recommend the preferred alternative. Use a MARR of 10%. (4.7)

EOY	CF(A)	CF(B)
0	– $40 000	– $30 000
1	6 000	6 500
2	6 000	6 000
3	6 000	5 500
4	6 000	5 000
5	6 000	4 500
6	6 000	4 000
7	6 000	3 500
8	6 000	3 000
9	6 000	2 500
10	6 000	2 000

36. Two mutually exclusive alternatives are to be evaluated and the least cost alternative specified. The cash flow profiles for the alternatives are shown below.

EOY	CF(A)	CF(B)
0	– $10 000	– $15 000
1	1 000	1 500
2	1 500	2 000
3	—	500
4	2 500	3 000
5	3 500	4 000
6	—	500
7	4 500	5 000
8	6 000	11 500

Using a 12% before-tax *MARR* and assuming one of the alternatives must be chosen, compare the alternatives by employing an internal rate of return approach. (4.7

37. Consider the following investment decision.

	Machine A	Machine B
First cost	$15 000	$20 000
Estimated life	5 years	10 years
Estimated annual revenues	$9 000	$11 066
Estimated annual operating costs	$6 000	$8 000
Estimated salvage value at end of life	$1 500	$3 000

(a) If the *MARR* is 10% and a 10-year planning period is assumed (two cycles of A), which of the machines should be purchased, if either? Solve by using the annual worth method.

(b) What are the approximate internal and external rates of return for the incremental investment in Machine B?

(c) What would the salvage value of Machine A have to be in order for this investment to yield an internal and an external rate of return equal to 10%? (4.7)

38. Using the rate of return method, compare the three mutually exclusive investment alternatives having the cash flow profiles shown below. Base your analysis on a 30% *MARR*. Use both future worth and savings/investment methods. (4.7)

EOY	CF(A)	CF(B)	CF(C)
0	− $8 000	− $5 000	0
1	3 400	1 000	0
2	3 400	2 000	0
3	3 400	3 000	0
4	3 400	4 000	0
5	3 400	5 000	0

39. Two numerically controlled drill presses are being considered by the production department of a major corporation; one

must be selected. Both machines meet the quality and safety standards of the firm. Comparative data are as follows:

	Drill Press X	Drill Press Y
Initial investment	$20 000	$30 000
Estimated life	10 years	10 years
Salvage value	$5 000	$7 000
Annual operating costs	$12 000	$6 000
Annual maintenance costs	$2 000	$4 000

Using a 10% interest rate and a present worth comparison, which machine is preferred? (4.7)

40. A firm is considering either leasing or buying a small computer system. If purchased, the initial cost will be $200 000; annual operating and maintenance costs will be $80 000. Based on a five-year planning period, it is anticipated the computer will have salvage value of $50 000 at that time. If the computer is leased, annual operating and maintenance costs in excess of the annual lease payment will be $60 000. Based on an interest rate of 10%, what annual end-of-year lease payment will make the firm be indifferent between leasing and buying, based on economic considerations alone? (4.7)

41. Company W is considering investing $10 000 in a heat exchanger. The heat exchanger will last eight years, at which time it will be sold for $2 000. Maintenance costs for the exchanger are estimated to increase by $200 per year over its life. The maintenance cost for the first year is estimated to be $1 000. As an alternative, the company may lease the equipment for $X per year, including maintenance. For what value of X should the company lease the heat exchanger? The company expects to earn 10% on its investments. Assume end-of-year lease payments. (4.7)

42. A firm is faced with four investment proposals, A, B, C, and D, having the cash flow profiles shown below. Proposals A and C are mutually exclusive, and Proposal D is contingent on Proposal B being chosen. Currently, $400 000 is available for

investment, and the firm has stipulated a *MARR* of 20%. Determine the preferred alternative. (4.7)

	CF(A)	CF(B)	CF(C)	CF(D)
Initial investment	$200 000	$200 000	$300 000	$150 000
Planning period	10 years	10 years	10 years	10 years
Annual receipts	$140 000	$160 000	$200 000	$200 000
Annual disbursements	$110 000	$125 000	$120 000	$150 000
Salvage value	$50 000	$50 000	$100 000	$50 000

43. A firm will either lease an office copier at an end-of-year cost of $10 000 for a 10-year period, or they will purchase the copier at an initial cost of $66 117. If purchased, the copier will have zero salvage value at the end of its 10-year life. No other costs are to be considered. Using rate-of-return methods, should the firm lease or buy if their *MARR* is (a) 6% or (b) 10%? Justify your answer on the basis of an internal rate-of-return analysis. (4.7)

44. A company can construct a new warehouse for $700 000, or it can lease an equivalent building for $75 000 per year for 25 years with the option of purchasing the building for $100 000 at the end of the 25-year period. Lease payments are due at the *beginning* of each year. The company can earn 12% per year before taxes on its invested capital. Indicate the preferred alternative. (4.7)

45. In Problem 44, for what (a) beginning-of-year, (b) end-of-year lease payment will the company be indifferent between leasing and buying?

46. The XYZ Company must decide whether they should purchase or lease a computer. The computer costs $30 000 initially and will last five years, when it will have salvage value of $5 000. If the computer is purchased, all maintenance costs must be paid by the XYZ Company. Maintenance costs are $2 000 per year over the life of the equipment. The XYZ Company uses an interest rate of 10% in evaluating investment alternatives. For what end-of-year annual leasing charge is the

Comparison of Alternatives

firm indifferent between purchasing and leasing over the five-year period? (4.7)

47. A firm is considering replacing its material handling system and either purchasing or leasing a new system. The old system has an annual operating and maintenance cost of $12 000, a remaining life of 10 years, and an estimated salvage value of $5 000 in 10 years.

A new system can be purchased for $100 000; it will be worth $20 000 in 10 years; and it will have annual operating and maintenance costs of $8 000 per year. If the new system is purchased, the old system can be sold for $20 000.

Leasing a new system will cost $5 000 per year, payable at the beginning of the year, plus operating costs of $5 000 per year, payable at the end of the year. If the new system is leased, the old system will be scrapped at no value.

Use a MARR of 10% to compare the annual worths of keeping the old system, buying a new system, and leasing a new system. (4.12)

48. A firm is considering replacing a compressor that was purchased four years ago for $50 000. Currently, the compressor has a book value of $30 000, based on straight-line depreciation. If the compressor is retained it will probably be used for four years more, at which time it is estimated to have salvage value of $10 000. If the old compressor is retained for eight years more, it is estimated that its salvage value will be negligible at that time. Operating and maintenance costs for the compressor have been increasing at a rate of $1000 per year, with the cost during the past year being $9000.

A new compressor can be purchased for $60 000. It is estimated to have uniform annual operating and maintenance costs of $8000 per year. Salvage value for the compressor is estimated to be $30 000 after four years and $15 000 after eight years. If a new compressor is purchased, the old compressor will be traded in for $20 000.

Using a before-tax analysis, a *MARR* of 6%, and an annual worth comparison, determine the preferred alternative using a planning period of (a) four years, and (b) eight years. (4.12)

49. A firm is considering replacing a computer system they purchased three years ago for $40 000. It will have salvage value of $2 500 in five years. Operating and maintenance costs have been $7 500 per year. Currently the computer has a trade-in value of $25 000 toward a new system that costs $60 000 and has a life of five years, with salvage value of $25 000 at that time. The new computer will have annual operating and maintenance costs of $8 000.

If the current system is retained, another computer will have to be purchased in order to provide the required computing capacity. The additional computer will cost $30 000, has salvage value of $5 000 in five years, and has annual operating and maintenance costs of $5 000.

Using an annual worth comparison before taxes, with a *MARR* of 30%, determine the preferred course of action. (4.12)

50. A firm owns a pump that it is contemplating replacing. The old pump has annual operating and maintenance costs of $5 000 per year, it can be kept for five years more and will have zero salvage value at that time.

The old pump can be traded in on a new pump. The trade-in value is $2 500, with the purchase price for the new pump being $12 000. The new pump will have a value of $5 000 in five years and will have annual operating and maintenance costs of $2 000 per year.

Using a *MARR* of 20%, evaluate the investment alternative using the present worth method. (4.12)

51. A company owns a five-year old turret lathe that has a book value of $10 000. The present market value for the lathe is $8 000. The expected decline in market value is $1 000 per year to a minimum market value of $1 000. Maintenance plus operating costs for the lathe equal $2 200 per year.

A new turret lathe can be purchased for $20 000 and will have an expected life of 12 years. The market value for the turret lathe is expected to equal $20\,000(0.80)^k$ at the end of year k. Annual maintenance and operating cost is expected to equal $600.

Based on a 10% before-tax *MARR*, should the old lathe be replaced now? Use an equivalent uniform annual cost comparison. (4.12)

52. The ABC Company has an overhead crane that has an estimated remaining life of 10 years. The crane can be sold for $6 000. If the crane is kept in service, it must be overhauled immediately at a cost of $3 000. Operating and maintenance costs will be $2 000 per year after the crane is overhauled. After overhauling it, the crane will have a zero salvage at the end of the 10-year period. A new crane will cost $16 000, will last for 10 years, and will have $3 000 salvage value at that time. Operating and maintenance costs are $1 000 for the new crane. The company uses an interest rate of 10% in evaluating investment alternatives. Should the company buy the new crane? (4.12)

53. The Ajax Specialty Items Corporation has received a five-year contract to produce a new product. To do the necessary machining operations, the company is considering two alternatives.

    Alternative A involves continued use of the currently owned lathe. The lathe was purchased five years ago for $12 000. Today, the lathe is worth $5 000 on the used machinery market. If this lathe is to be used, special attachments must be purchased at a cost of $2 000. At the end of the five-year contract, the lathe (with attachments) can be sold for $1 000. Operating and maintenance costs will be $4 000 per year if the old lathe is used.

    Alternative B is to sell the currently owned lathe and buy a new lathe at a cost of $15 000. At the end of the five-year contract, the new lathe will have salvage value of $8 000. Operating and maintenance costs will be $2 500 per year for the new lathe. Using present worth analysis, should the firm use the currently owned lathe or buy a new lathe? Base your analysis on a minimum attractive rate of return of 20%. (4.12)

54. A corporation purchased a numerically controlled production machine five years ago for $450 000. The machine currently has trade-in value of $100 000. If the machine continues to be used, another machine, X, must be purchased to supplement the old machine. Machine X costs $300 000, has annual operating and maintenance costs of $50 000, and will have salvage value of $50 000 in 10 years. If the old machine is retained, it will have annual operating and maintenance costs of $80 000 and will have salvage value of $20 000 in 10 years.

As an alternative to retaining the old machine, it can be replaced with Machine Y. Machine Y costs $600 000, has anticipated annual operating and maintenance costs of $100 000, and has salvage value of $200 000 in 10 years.

Using a *MARR* of 20% and an annual worth comparison, determine the preferred economic alternative. (4.12)

55. A small foundry is considering the replacement of a No. 1 Whiting cupola furnace that is capable of melting gray iron only with a reverberatory-type furnace that has gray iron and nonferrous metals melting capability. Both furnaces have approximately the same melting rates for gray iron in kilograms per hour. The foundry company plans to use the reverberatory furnace, if purchased, primarily for melting gray iron, and the total quantity melted is estimated to be about the same with either furnace. Annual raw material costs would therefore be the same for each furnace. Available information and cost estimates for each furnace are given below.

Cupola furnace. Purchased used and installed eight years ago for a cost of $5 000. The present market value is determined to be $2 000. Estimated remaining life is somewhat uncertain but, with repairs, the furnace should remain functional for seven more years. If kept seven more years, salvage value is estimated as $500, and average annual expenses expected are:

Fuel	$10 000
Labour (including maintenance)	$12 000
Direct benefits	10% of direct labour costs
Taxes and insurance on furnace	1% of purchase price
Other	$ 6 000

Reverberatory furnace. This furnace costs $8 000. Expenses to remove the cupola and install the reverberatory furnace are about $600. The new furnace has estimated salvage value of $800 after seven years of use, and annual expenses are estimated as:

Fuel	$7 500
Labour (operating)	$9 000
Direct benefits	10% of direct labour costs
Taxes and insurance on furnace	1% of purchase price
Other	$ 6 000

In addition, the furnace must be relined every two years at a cost of $1 000 per occurrence. If the foundry presently earns an average of 20% on invested capital before income taxes, should the cupola furnace be replaced by the reverbatory furnace? (4.12)

56. A firm has an automatic chemical mixer that it has been using for the past four years. The mixer originally cost $18 000. Today the mixer can be sold for $9 000. The mixer can be used for 10 more years and will have salvage value of $2 000 at that time. The annual operating and maintenance costs for the mixer equal $5 000 per year.

    Because of an increase in business, a new mixer must be purchased. If the old mixer is retained, a new mixer will be purchased at a cost of $16 000 and have salvage value of $2 000 in 10 years. This new mixer will have annual operating and maintenance costs equal to $4 000 per year.

    The old mixer can be sold and a new mixer of larger capacity purchased for $28 000. This mixer will have a $3 000 salvage value in 10 years and will have annual operating and maintenance costs equal to $8 000.

    Based on a *MARR* of 15%, what do you recommend? (4.12)

57. A firm is presently using a machine that has a market value of $8 000 to do a specialized production job. The requirement for this operation is expected to last only six more years, after which it will no longer be done. The predicted costs and salvage values for the present machine are:

Year	1	2	3	4	5	6
Operating cost	$1000	$1200	$1400	$1800	$2300	$3000
Salvage value	$5000	$4500	$4000	$3300	$2500	$1400

A new machine has been developed that can be purchased for $12 000 and has the following predicted cost performance.

Year	1	2	3	4	5	6
Operating cost	$ 500	$ 700	$ 700	$1 200	$1 500	$1 900
Salvage value	$11 000	$10 500	$10 000	$9 500	$8 500	$7 500

If interest is at 0%, when should the new machine be purchased? (4.12)

58. A firm has received a production contract for a new product. The contract lasts for five years. To do the necessary machining operations, the firm can use one of its own lathes, which was purchased four years ago at a cost of $11 000. Today the lathe can be sold for $5 000. In five years the lathe will have zero salvage value. Annual operating and maintenance costs for the lathe are $4 000. If the firm uses its own lathe, it must also purchase an additional lathe at a cost of $7 000; its value in five years will be $1 000. The new lathe will have annual operating and maintenance costs of $2 500.

As an alternative the presently owned lathe can be sold and a new lathe of larger capacity purchased for a cost of $16 000; its value in five years is estimated to be $5 000, and its annual operating and maintenance costs will be $5 000.

An additional alternative is to sell the presently owned lathe and subcontract the work to another firm. Company X has agreed to do the work for the five-year period at an annual cost of $8 000 per end-of-year.

Using a 10% interest rate, determine the least cost alternative for performing the required production operations. (4.12)

59. A machine was purchased five years ago for $8 000. At that time, its estimated life was 10 years with estimated end-of-life salvage value of $800. The average annual operating and

maintenance costs have been $12 000 and are expected to continue at this rate for the next five years. However, average annual revenues have been and are expected to be $15 000. Now, the firm can trade in the old machine for a new machine for $3 000. The new machine has a list price of $10 000, an estimated life of 10 years, annual operating plus maintenance costs of $5 000, annual revenues of $9 000, and salvage values at the end of the *j*th year according to

$$S_j = \$10000 - \$1000j, \quad \text{for } j = 0, 1, 2, \ldots, 10$$

Determine whether to replace or not by the annual equivalent method using a MARR equal to 10% compounded annually. (Use a five-year planning period.) (4.12)

60. An automatic car wash has been experiencing difficulties in keeping its equipment operational. The owner is faced with the alternative of overhauling the present equipment or replacing it with new equipment. The cost of overhauling the present equipment is $5 000. The present equipment has annual operating and maintenance costs of $5 000. If it is overhauled, the present equipment will last for five years more and be scrapped at zero value. If it is not overhauled, it has a trade-in value of $2 000 toward the new equipment.

New equipment can be purchased for $20 000. At the end of five years the new equipment will have a resale value of $8 000. Annual operating and maintenance costs for the new equipment will be $2 000.

Using a MARR of 8%, what is your recommendation to the owner of the car wash? Base your recommendation on a present worth comparison. (4.12)

61. A highway contractor must decide whether to overhaul a tractor and scraper or replace it. The old equipment was purchased five years ago for $70 000; it had a projected life of 15 years, with salvage value of $5 000 at that time. If traded in on a new tractor and scraper, it can be sold for $30 000. Overhauling the old equipment will cost $10 000. If overhauled, operating and maintenance costs will be $8 000 per year, and the overhauled equipment will have projected salvage value of $4 000 in 10 years.

A new tractor and scraper can be purchased for $80 000; it will have annual operating and maintenance costs of $5 000, and will have salvage value of $40 000 in 10 years.

Using an annual worth comparison with a *MARR* of 10%, should the equipment be overhauled or replaced? (4.12)

62. A highway construction firm purchased earth-moving equipment three years ago for $50 000. Salvage value at the end of 10 years was estimated to be 34% of first cost. The firm earns an average annual gross revenue of $45 000 with the equipment, and the average annual operating costs have been and are expected to be $26 000.

The firm now has the opportunity to sell the equipment for $38 000 and subcontract the work normally done by the equipment over the next seven years. If the subcontracting is done, the average annual gross revenue will remain $45 000, but the subcontractor charges $35 000 per end-of-year for these services.

If a 25% rate of return before taxes is desired, determine whether the firm should subcontract or not by the annual worth method. (4.12)

63. A building supplies distributor purchased a gasoline-powered forklift truck five years ago for $8 000. At that time, the estimated useful life was 10 years with salvage value of $800 at the end of this time. The truck can now be sold for $2 500. For this truck, average annual operating expenses for year j have been

$$C_j = \$2\,000 + \$200(j-1)$$

Now the distributor is considering the purchase of a smaller battery-powered truck for $6 500. The estimated life is 10 years, with salvage value decreasing by $600 each year. Average annual operating expenses are expected to be $1 600. If a *MARR* = 10% is assumed and a five-year planning period is adopted, should the replacement be made now? (4.12)

64. A particular unit of production equipment has been used by a firm for a period of time sufficient to establish very accurate estimates of its operating and maintenance costs.

Comparison of Alternatives

Replacements can be expected to have identical cash flow profiles in successive life cycles if constant worth dollar estimates are used. The appropriate discount rate is 30%. Operating and maintenance costs for a unit of equipment in its tth year of service, denoted by C_t, are as follows:

t	C_t	t	C_t
1	$2 000	6	$5 000
2	2 500	7	5 750
3	3 050	8	6 550
4	3 650	9	7 400
5	4 300	10	8 300

Each unit of equipment costs $15 000 initially. Because of its special design, the equipment cannot be disposed of at positive salvage value following its purchase; hence zero salvage value exists, regardless of the replacement interval used.

(a) Determine the optimum replacement interval assuming an infinite planning period. (Maximum feasible interval = 10 years.)

(b) Determine the optimum replacement interval assuming a finite planning period of 15 years, with

$$C_{t+1} = C_t + \$500 + 50(t-2) \text{ for } t = 10, 11, \ldots$$

(c) Solve parts (a) and (b) using a discount rate of 0%.

(d) Based on the results obtained, what can you conclude concerning the effect the discount rate has on the optimum replacement interval? (4.12)

65. Given an infinite planning period, identical cash flow profiles for successive life cycles, and the following functional relationships for C_t, the operating and maintenance cost for the tth year of service for the unit of equipment in current use, and F_n, the salvage value at the end of n years of service:

$$C_t = \$1\,000(1.25)^t \quad t = 1, 2, \ldots, 12$$

$$F_n = \$11\,000(0.75)^n \quad n = 0, 1, 2, \ldots, 12$$

Determine the optimum replacement interval assuming a MARR of (a) 0%, (b) 10%. (Maximum life = 12 years.) (4.12)

66. Solve problem 65 given the following functional relationships: (4.12)

$$C_t = \$1\,000(1.10)^t \quad t = 1, 2, \ldots, 12$$

$$F_n = \$11\,000(0.50)^n \quad n = 0, 1, 2, \ldots, 12$$

Chapter 5

Depreciation and Income Tax Considerations

5.1 Introduction

Depreciation and taxes are particularly important in engineering economic analyses. Although depreciation allowances are not actual cash flows, their magnitude and timing do affect taxes. Tax dollars are real cash flows and are therefore just as important as dollars spent on wages, utilities, and raw materials. The care taken to use the most favourable depreciation and tax methods allowed by law can result in substantial savings by companies which otherwise would have been paid out in taxes to the government.

This chapter presents the basic depreciation methods commonly used in both Canada and the United States and shows their effects on corporate after-tax cash flow profiles. In addition, the importance of the depreciation write-off period, borrowed funds, capital gains and losses, and tax credits will be illustrated. No attempt will be made to discuss the minute aspects of depreciation or Canadian and American tax law. Experts in these areas devote entire careers to keeping abreast of the latest laws and judgements. Instead, the objective here will be to highlight the importance of depreciation and taxes so that you will perform economic analyses on an after-tax basis and seek professional legal and accounting assistance as required. It is very important to recognize that decisions concerning the acquisition or retirement of fixed assets should be made only by a person who has a thorough understanding of depreciation accounting.

5.2 The Meaning of Depreciation

Fixed assets, the physical plant and equipment (with the exception of land), are of use to the company for only a limited number of years. *Depreciation* is the systematic allocation of the cost of each asset over its economic life. It may be viewed as the gradual conversion of the *cost* of an asset into *expense*; thus, depreciation is a noncash expense. When a capital asset is purchased, in effect, a quantity of usefulness that will contribute to production throughout the life of the asset is acquired. Depreciation is the expiration of a capital asset's quantity of usefulness, and the recording of depreciation is a process of *allocating* and *charging* the cost of this usefulness to the accounting period that benefits from the asset's use.

Capital Cost Allowance (CCA) is the term used to describe the rate of depreciation of a capital asset allowed by the Canadian Income Tax Act. The CCA specifies the maximum percentage of a capital asset's cost that is allowed to be deducted in a given year from income for the purpose of determining taxable income. The CCA therefore does not reflect an actual cash flow, but is in fact a means of allocating the expense of an asset over time. Depreciable fixed assets of similar nature are grouped into separate pools or classes and are treated on a pool, not an individual, basis. Under normal circumstances the capital cost allowance is a *percentage of the original cost of the average item* in the pool. From one fiscal period to the next, this predetermined percentage remains constant. As capital cost allowances are deducted from the assets in the pool, the balance remaining declines and the deduction at the subsequent fiscal year end is made on that smaller balance; hence, the term *declining balance method.*

Most capital assets depreciate. Buildings, equipment, machines, and even property such as patents, franchises, or licences, which are used for only a limited period, may be subject to depreciation. The capital cost allowance system is designed to ensure that a firm will be allowed to recover the full capital cost of depreciable properties such as these. Some major causes of the reduction in value of an asset are:

1. *Physical deterioration* due to ordinary wear and tear.

2. *Technological deterioration or obsolescence*, which occurs because of advancements in the design and technology of the product or process.

3. *Functional deterioration*, which occurs when the usefulness of an asset is reduced because the rapid growth of a company renders it inadequate. In this case the asset would have to be replaced with a larger unit, even though it might be in good physical condition and not technologically obsolete.

While the capital cost allowance regulations recognize that fixed assets do undergo these kinds of deterioration, the deductions permitted by these regulations more closely reflect government policy than the actual decline in the useful life of the asset. (One purpose of permitting the rapid write-off of fixed assets is to encourage new investment in plant and equipment.) It should be remembered that depreciation is *not* a process of valuation. Accounting records and financial statements do not purport to show the constantly fluctuating market values of the fixed assets. Although the market value of a building may rise substantially over a period of years, its depreciation continues as a percentage of its original purchase price regardless of the increase in its market value. It is *actual cost*, not apportioned value.

Canadian corporate income tax is based on net income that results from deducting all expenses from gross income. For tax purposes, an investment is treated as a prepaid expense, and the capital cost allowance allocates that expense over time. It is important to remember that depreciation (CCA) is not an actual cash flow, but is merely treatable as an expense for income tax purposes. A larger depreciation allowance in a year decreases net taxable income and, hence, income taxes payable, making more money available for the firm. Because of the time value of money, it is generally desirable to take larger depreciation allowances in the early years and lesser allowances in the latter years of an asset's life. This is true if the firm is, in fact, earning money from which to take the depreciation deduction. All these deductions must, of course, be made within the limits prescribed by the Canadian Income Tax Act.

5.3 Factors Used to Determine Depreciation

Depreciation can be taken on any property that is used in a firm's trade or business or held for the production of income. Depreciation cannot be taken on inventories held for sale to customers or on land. A useful life and a cost basis (investment) must be ascertained before depreciation can be taken. The *cost basis (P)* is essentially the taxpayer's investment. In most cases, this is the cost of the property plus the cost of capital additions to that property, including installation cost. The *useful life (n)* depends on the use that the taxpayer intends for the asset. As pointed out in the previous section, this life may have little relationship to the inherent physical life of the asset. For example, obsolescence because of technological improvements and foreseeable economic changes may render an asset useless long before it is physically worthless. *Salvage value (F)* is an estimate of the market value at the end of an asset's useful life. The salvage value enters explicitly into the calculation of some depreciation methods and not at all in others. In no case may total depreciation exceed the cost basis less the estimated salvage value. Although we shall use the term *book value (B)* to describe the cost of an asset minus the accumulated depreciation, this amount is more accurately described as *undepreciated cost.*

5.4 Methods of Depreciation

Since the purpose of depreciation is to offer a realistic view of profit from both taxation and company viewpoints, the company usually keeps two different sets of records to satisfy these two different objectives. The capital cost allowance (depreciation) rates allowed by the Canadian Income Tax Act may well be different from those that the company perceives to be appropriate. Therefore, depreciation may be recorded on the books on a straight-line basis even though the fixed-percentage-on-the-declining-balance method is used for income tax purposes. In order to understand financial statements, then, it is important that we become familiar with the most common depreciation methods used in industry. Each depreciation method has unique features that appeal to different management philosophies and are based on different measures of merit.

For taxation purposes in Canada, the *fixed-percentage-on-the-declining-balance* method *must be used* as prescribed by the Income Tax Act. Special provisions apply, however, to certain situations where methods of accelerated depreciation or straight-line depreciation may be used. In the United States several types of depreciation procedures are acceptable to the Treasury Department. The straight-line, declining balance, and sum of the years' digits methods are specifically mentioned in the Internal Revenue Code. Canadian engineers need to have an understanding of these methods, since in multinational companies, Canadian projects frequently compete with projects in the United States. In order to compare projects, the Canadian engineer must be familiar enough with accepted US methods of calculating depreciation to be able to determine the benefits and after-tax cash flows, etc., of a competing US project.

5.4.1 Straight-Line Depreciation

The *straight-line method* provides for the uniform write-off of an asset. The depreciation allowed at the end of each year (D_t) is equal throughout the asset's useful life and is given by:

$$D_t = \frac{P-F}{n} \tag{5.1}$$

The undepreciated or book value at the end of each year (B_t) is given by:

$$B_t = P - \left(\frac{P-F}{n}\right)t \tag{5.2}$$

Example 5.1

We have just purchased a computer at a cost of $10 500 with an estimated salvage value of $500 and a projected useful life of six years. The depreciation and book value for each year are given in Table 5.1.

Table 5.1 Straight-Line Depreciation and Book Value

End of Year, t	Depreciation, D_t	Book Value, B_t
0	—	$10 500.00
1	$1 666.67	8 833.33
2	1 666.67	7 166.67
3	1 666.67	5 500.00
4	1 666.67	3 833.33
5	1 666.67	2 166.67
6	1 666.67	500.00

According to the Canadian Income Tax Act, properties such as patents, copyrights, franchises or licences used for a limited period may be depreciated using the straight-line method.

5.4.2 Sum of the Years' Digits Depreciation

The *sum of the years' digits method* of depreciation is known for its accelerated write-off of assets. That is, it provides relatively high depreciation allowances in the early years and lower allowances throughout the rest of an asset's useful life. The name "sum of the years' digits" comes from the fact that the sum

$$1 + 2 + \cdots + (n-1) + n = \frac{n(n+1)}{2}$$

is used directly in the calculation of depreciation.

The formula is composed of a denominator, which is the sum of the years' digits, and the numerator is the present year of the asset's life in inverse order. The depreciation allowance during any year t is expressed as

$$\boxed{D_t = \frac{n-(t-1)}{n(n+1)/2}(P-F)} \tag{5.3}$$

The book value at the end of each year t is given by

$$B_t = P - \sum_{j=1}^{t} \frac{n-(j-1)}{n(n+1)/2}(P-F)$$

which reduces to

$$\boxed{B_t = (P-F)\frac{(n-t)(n-t+1)}{n(n+1)} + F} \tag{5.4}$$

Example 5.2

Returning to our computer in Example 5.1 where $P = \$10\,500$, $F = \$500$, and $n = 6$, sum of the years' digits depreciation results in the allowances and book values summarized in Table 5.2.

Table 5.2 Sum of the Years' Digits Depreciation and Book Value

End of Year, t	Value of $\frac{n-(t-1)}{n(n+1)/2}$	Depreciation, D_t	Book Value, B_t
0	—	—	$10 500.00
1	6/21	$2 857.14	7 642.86
2	5/21	2 380.95	5 261.90
3	4/21	1 904.76	3 357.14
4	3/21	1 428.57	1 928.57
5	2/21	952.38	976.19
6	1/21	476.19	500.00

Sum of the years' digits depreciation may be used in the United States for federal income tax purposes only for tangible property having a useful life of three years or more (special rules apply to depreciable realty) where the taxpayer was the original user of the property. In Canada it is used only in special cases.

5.4.3 Declining Balance Depreciation

The *declining balance method,* like sum of the years' digits depreciation, is known for its accelerated write-off of assets. According to this method, the depreciation allowed at the end of each year t is a constant fraction p of the book value at the end of the previous year. That is,

$$D_t = pB_{t-1} \tag{5.5}$$

The book value at the end of each year t is given by

$$B_t = P(1-p)^t \tag{5.6}$$

Substituting Equation 5.6 into Equation 5.5 allows us to calculate the year t depreciation directly as

$$D_t = pP(1-p)^{t-1} \tag{5.7}$$

Note that in the declining balance method of depreciation the estimated salvage value need not come into play in figuring the deduction. However, the asset may not be depreciated beyond its salvage value.

Example 5.3

Assuming declining balance depreciation and a rate of $p = 0.3$, let us again calculate depreciation and book value for the computer example. The results are listed in Table 5.3.

Table 5.3 Declining Balance Depreciation and Book Value

End of Year, t	Depreciation, D_t	Book Value, B_t
0	—	$10 500.00
1	$3 150.00	7 350.00
2	2 205.00	5 145.00
3	1 543.50	3 599.50
4	1 079.85	2 519.65
5	755.90	1 763.75
6	528.90	1 234.85

With the declining balance method the book value of an asset never reaches zero. Consequently, when the asset is sold, exchanged, or scrapped, any remaining book value is used in determining the gain or loss on disposal. This is discussed in more detail in Section 5.14 (Capital Gains and Losses) and Section 5.16 (Recaptured Capital Cost Allowance). In Example 5.3, the remaining book value of $1 234.85 can be treated as a *terminal loss* only if there are no assets remaining in an asset class (pool).

5.4.4 Capital Cost Allowance

For Canadian income tax purposes the deductible yearly depreciation or capital cost allowance is determined using a declining balance method similar to the one we have described in Section 5.4.3. As mentioned in Section 5.2, the CCA is fixed by government regulation, the allowable maximum rate of depreciation being different for each class of assets. The taxpayer, of course, can claim less than the allowable maximum in any given year, but not more. Table 5.4 sets out the most commonly used classifications (there are about 30 classes in total).

Table 5.4 Capital Cost Allowance Rates by Class

Class	Rate (%)	Description
1	4	Most buildings bought after 1987, including components such as wiring, plumbing, heating, and cooling systems.
3	5	Most buildings, including components, bought after 1978 and before 1988. In some cases including part of the cost of additions made after 1987 in Class 1.
6	10	Frame, log, stucco on frame, galvanized iron, or corrugated metal buildings that do not have any footings below the ground. Also includes fences and greenhouses.
8	20	Property not included in any other Class. Some examples are fixtures, furniture, machinery, photocopiers, refrigeration equipment, telephones, and tools costing $200 or more.
9	25	Aircraft, including furniture or equipment attached to the aircraft, and spare parts.
10	30	Automobiles, except those used as a taxi or in a daily rental business, including vans, trucks, tractors, wagons, and trailers. Also inlcudes computer hardware and systems software.
12	100	Dies, jigs, moulds, cutting or shaping parts of a machine, tools and medical or dental instruments costing under $200, computer software (except systems software), and video cassettes.
13		Leasehold interest. (Maximum rate varies.)
14		Patents, franchises, concessions, or licences for a limited period. (Maximum rate varies.)
17	8	Roads, parking lots, sidewalks, airplane runways, storage areas, or similar surface construction.
22	50	Most power-operated, movable equipment bought before 1988 that is used for excavating, moving, placing, or compacting earth, rock, concrete, or asphalt.
38	30	Most power-operated, movable equipment bought after 1987 and used for excavating, moving, placing, or compacting earth, rock, concrete, or asphalt.

Depreciable assets are grouped in asset pools according to their nature with a capital cost allowance write-off prescribed for each pool. As noted earlier, the prescribed rates for each asset class are applied to the undepreciated capital cost balance remaining in the class. Any proceeds of sale up to the amount of the original cost must be applied to reduce the balance in the class. If the proceeds exceed the balance remaining in the class as of the end of the year, the excess is considered to be capital cost allowance recapture and must be treated as income for that year. If all assets in a class have been disposed of and a balance still remains in the class, then that balance may be written off as an expense in that year (terminal loss).

Capital cost allowance is available at one-half of the normal rate in the year that a qualifying asset is acquired, whether or not it is used before the end of that year. A taxpayer is considered to have acquired an asset on the date on which title is obtained or the date on which possession is taken, whichever is earlier, even if legal title remains with the vendor as security for the purchase price. The capital cost of depreciable property includes the purchase price and all other costs and expenses related to the acquisition, including federal and provincial sales taxes, customs duties, legal, accounting, engineering and other fees, installation costs, etc.

5.4.5 Double Declining Depreciation

According to United States tax law, twice the straight-line rate, or $2/n$, is the maximum constant fraction permissable for depreciation, and then only under certain conditions. When $p = 2/n$, as it most frequently does, this method of depreciation is known as the *double declining balance method*.

Example 5.4

Assuming that double declining balance depreciation is acceptable, let us again calculate depreciation and book value for the computer example. The rate $p = \dfrac{2}{6} = 0.333$. The results are tabulated in Table 5.5.

Table 5.5
Double Declining Balance Depreciation and Book Value

End of Year, t	Depreciation, D_t	Book Value, B_t
0	—	$10 500.00
1	$3 500.00	7 000.00
2	2 333.33	4 666.67
3	1 555.56	3 111.11
4	1 037.04	2 074.07
5	691.36	1 382.72
6	460.91	921.82

In this example, the book value at the end of year 6 was $921.81. Since the declining balance rate does not require the estimate of salvage value, the book value in the last year need not be the same as for the other methods. In the event that the resale value is different from the book value, compensating adjustments of a capital or business income nature must be made at the time of asset disposal. (Capital and ordinary income are discussed later in this chapter.)

In the United States, the Internal Revenue Service also allows switching from the double declining balance method to straight-line depreciation.[1] This may be desirable in order to present a greater depreciation charge, resulting in lower taxes in the current year, deferring taxes until later years, and thus providing a present worth tax advantage. In this case, the switch should take place whenever straight-line depreciation on the undepreciated portion of the asset exceeds the double declining balance allowance. That is, we should switch to straight-line at the first year for which

$$\boxed{\frac{B_{t-1} - F}{n - (t - 1)} > pB_{t-1}}$$

(5.8)

The estimated salvage value will be used in determining the straight-line depreciation component, even though it is neglected in the dou-

---

[1] Switching from sum of the years' digits to straight-line depreciation is also allowed, but normally cannot be justified economically.

ble declining balance method. Switching to straight-line depreciation will never be desirable if the estimated salvage value F exceeds the double declining balance book value for the last year B_n.

Another approach to declining balance depreciation that can be used in the United States is to select the value of the rate p for which the last year's book value equals an estimated salvage value. This is the only time that estimated salvage is used explicitly in declining balance calculations. The value of p required to make the last year's book value equal to the salvage value can be found by setting $B_n = F$ and solving for p in the book value expression $B_t = P(1-p)^t$ where $t = n$. That is,

$$B_n = F = P(1-p)^n \qquad (5.9)$$

Consequently,

$$p = 1 - \sqrt[n]{\frac{F}{P}} \qquad (5.10)$$

Example 5.5

In the examples used to this point in the chapter, $P = \$10\,500$, $F = \$500$, and $n = 6$. We now want to calculate the fixed rate p that will result in a book value of $500 after six years. Solving for p,

$$p = 1 - \sqrt[6]{\frac{\$500}{\$10\,500}} = 0.39795$$

Unfortunately, this value of p exceeds the maximum allowable rate in the USA, $2/n = 0.333$, and we revert to the solutions tabulated previously for the double declining balance method. The same would be true any time p exceeds $2/n$, such as when $F = 0$.

Example 5.6

Suppose in our example that $F = \$1\,000$ instead of $500. Now calculate the depreciation and the book value using a value of p that will result in a book value of $1\,000$ after six years. Solving for p,

$$p = 1 - \sqrt[6]{\frac{\$1000}{\$10\,500}} = 0.324$$

The results using a fixed declining balance rate of $p = 0.324$ are illustrated in Table 5.6.

Table 5.6 Declining Balance with Calculated p, Depreciation and Book Value

End of Year, t	Depreciation, D_t	Book Value, B_t
0	—	$10 500.00
1	$3 404.37	7 095.63
2	2 300.59	4 795.05
3	1 554.68	3 240.37
4	1 050.61	2 189.76
5	709.98	1 479.78
6	479.78	1 000.00

5.4.6 Sinking Fund Depreciation

Sinking fund depreciation for long-lived assets was used in certain industries and by both the Canadian and American governments until a few years ago. It is now of less importance in accounting practice, but since it is still referred to in some situations, we will give the sinking fund method a brief treatment here. The sinking fund model assumes that the asset depreciates at an increasing rate. If sum of the years' digits and declining balance depreciation are considered to be accelerated methods, sinking fund depreciation is a decelerated method.

Sinking fund depreciation can be determined by first imagining a bank account (call it a sinking fund) into which we deposit an equal amount at the end of each year. The sinking fund pays interest at a rate of $i\%$ and will have a balance equal to the total amount to be depreciated, $P - F$, after n years. From Chapter 3, we know that the equal annual deposit must be

$$A = (P - F)(A|F\ i, n) \qquad (5.11)$$

The depreciation allowance for any year t is then considered to be the sum of the deposit, A, plus the interest earned on the account in that year. That is, the first year's depreciation is A, the second year's is

$A(1+i)$, and the tth year's is $A(1+i)^{t-1}$, which equals $A(F|P\ i, t-1)$. Sinking fund depreciation for any year t may then be expressed as:

$$D_t = (P-F)(A|F\ i,n)(F|P\ i,t-1) \tag{5.12}$$

Since total depreciation taken at any time t is just the amount in our imaginary bank account, the book value at the end of year t equals first cost minus the size of the sinking fund at that time. That is,

$$B_t = P - A(F|A\ i, t)$$

or

$$B_t = P - (P-F)(A|F\ i,n)(F|A\ i,t) \tag{5.13}$$

Sinking fund depreciation has an interesting property in that the depreciation allowance for each year (D_t) plus interest charged on the undepreciated balance at the beginning of year (iB_{t-1}) is equal during each of the n years. This is sometimes called *capital recovered* (through depreciation) *plus return* (interest) *on the unrecovered capital* (book value). It is calculated for sinking fund depreciation as

$$D_t + iB_{t-1} = A(F|P\ i, t-1) + Pi - Ai(F|A\ i, t-1)$$
$$= A(1+i)^{t-1} + Pi - A\left[(1+i)^{t-1} - 1\right]$$

Hence,

$$D_t + iB_{t-1} = Pi + A \tag{5.14}$$

which does not depend on the year t in question. It is interesting to note that this is equal to the equivalent uniform annual cost (*EUAC*) of an asset, regardless of its method of depreciation. This is seen as follows:

Depreciation and Income Tax Considerations

$$EUAC = Pi + A$$
$$= Pi + (P-F)(A|F\ i,n)$$
$$= Pi + (P-F)\frac{i}{(1+i)^n - 1}$$
$$= \frac{Pi(1+i)^n - Fi}{(1+i)^n - 1}$$
$$= \frac{Pi(1+i)^n}{(1+i)^n - 1} - \frac{Fi}{(1+i)^n - 1}$$

or

$$\boxed{EUAC = P(A|P\ i,n) - F(A|F\ i,n)} \quad (5.15)$$

Example 5.7

Reconsidering our computer example, where $P = \$10\,500$, $F = \$500$, and $n = 6$, let us calculate not only depreciation and book value, but also illustrate that capital recovery plus return on the unrecovered capital is equal each year. Letting $i = 10\%$, the results are illustrated in Table 5.7.

Table 5.7 Sinking Fund Depreciation, Book Value, and Capital Recovery Plus Return

End of Year, t	Depreciation (Capital Recovered), D_t	Book Value (Capital Unrecovered), B_t	Return on Capital Unrecovered, iB_{t-1}	Capital Recovered Plus Return, $D_t + iB_{t-1}$
0	—	$10 500.00	—	—
1	$1 296.07	9 203.93	$1 050.00	$2 346.07
2	1 425.68	7 778.25	920.39	2 346.07
3	1 568.25	6 210.00	777.82	2 346.07
4	1 725.07	4 484.92	621.00	2 346.07
5	1 897.58	2 587.34	448.49	2 346.07
6	2 087.34	500.00	258.73	2 346.07

We can see how the depreciation allowances increase as time progresses, contrary to the depreciation methods seen previously.

Also, we see that the capital recovery plus return remains constant during each year as the item is depreciated.

Now we calculate the uniform annual cost of an asset where $P = \$10\,500$, $F = \$500$, $n = 6$, and $i = 10\%$. Having performed similar calculations in previous chapters, we know that

$$EUAC = P(A|P\ 10,6) - F(A|F\,10,6)$$
$$= \$10\,500(0.2296) - \$500(0.1296)$$
$$= \$2\,346$$

The annual cost equals the capital recovery plus return for each year of the sinking fund depreciation method. Since *EUAC* is calculated here without respect to depreciation, it should be apparent that the uniform annual cost of an asset before taxes remains the same, regardless of the depreciation method used.

5.5 Comparison of Depreciation Methods

Straight-line (*A*), sum of the years' digits (*B*), double declining balance (*C*), declining balance (*D*), and sinking fund (*E*) depreciation may be compared using value-time curves, as in Figure 5.1. The data displayed are those from several of the previous examples. Note that the method of depreciation determines the path of book values.

We can see that the book value, particularly in the intermediate years, depends largely on the method of depreciation chosen. In the United States, this will have a strong bearing on taxes paid during a given year and on the resulting after-tax cash flow profile. In Canada, where the method of depreciation for tax purposes is predetermined, companies cannot alter their after-tax cash flow profiles in this way.

5.6 Special Provisions for Accelerated Depreciation

In order to expand Canada's industrial capacity and to provide employment opportunities, the government is empowered to allow special depreciation on certain kinds of property, including property that is considered to be helpful to the defence effort. This special depreciation is in addition to the normal allowance and requires special certification. These and other special provisions will be discussed in more detail in Section 5.17 (Income Tax Incentives).

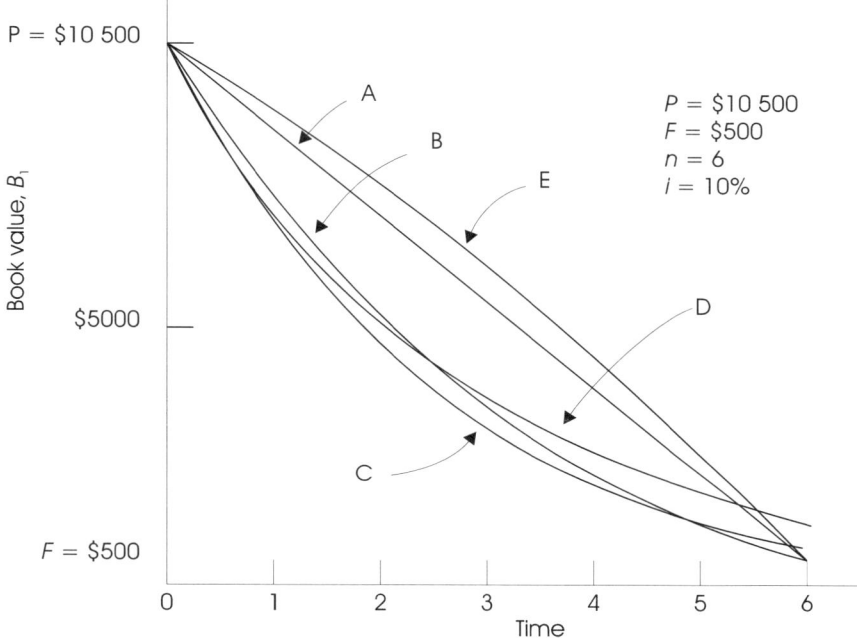

Fig. 5.1 Value-time curves of five different depreciation methods.

Example 5.8

In our earlier examples, we had a computer with investment cost basis $P = \$10\,500$, $F = \$500$, and $n = 6$ years. If our $10 500 investment qualifies for a special (50%) first-year allowance, that allowance (D_f) is

$$D_f = 0.50(\$10\,500)$$
$$= \$5\,250$$

Adjusting the investment cost basis, we have

$$P = P_{old} - D_f$$
$$= \$10\,500 - \$5\,250$$
$$= \$5\,250$$

A depreciation schedule would then be set up on the basis of $P = \$5\,250$ and $F = \$500$, with the option of writing this amount off during the second year.

Additional first-year depreciation or accelerated depreciation (in special cases) will not be illustrated in the remaining examples of the chapter. Although it is an easily used allowance for both Canadian and US income tax calculations, its use may mask some of the points to be made by subsequent examples. Because of the complexity of additional first-year depreciation, it would be wise to seek accounting assistance to determine the current allowance should this be a factor in an economic analysis.

5.7 Other Methods of Depreciation

We are not limited to the methods of depreciation treated to this point. Any other consistent method may be used to determine the depreciation allowance for property similar to that for which the declining balance method is applicable. The major limitation is that total allowances for income tax purposes at the end of each year do not exceed the maximum capital cost allowance permitted by the Canadian Income Tax Act. Some recognized methods are briefly presented below.

5.7.1 Units of Production Method

This procedure allows equal depreciation for each unit of output, regardless of the lapse of time involved. The allowance for year t is equal to the total depreciable amount $(P-F)$ times the ratio of units produced during the year (U_t) to the total units that may be produced during the useful life of the asset (U). That is,

$$D_t = (P-F)\frac{U_t}{U}$$

(5.16)

5.7.2 Operating Day Method

This is similar to the previous method in that year t depreciation is based on the ratio of days used during the year (Q_t) to total days expected in a useful life (Q). Depreciation is expressed as:

$$D_t = (P-F)\frac{Q_t}{Q}$$

(5.17)

5.7.3 Income Forecast Method

This method is applicable to depreciate the cost of rented property. The ratio of year t rental income (R_t) to the total useful life income (R) is multiplied by the total lifetime depreciation, or:

$$D_t = (P - F)\frac{R_t}{R}$$

(5.18)

5.7.4 Multiple-Asset Accounts

Until now each asset has been considered separately, having its own depreciation method, useful life, and salvage value. With multiple-asset accounts, a number of items can be combined and a single depreciation scheme applied to the entire account.

Through statistical analysis and past history, survivor curves can be determined that indicate the proportion of items in the group that will survive to any age. It should not be surprising that this is possible for property—life insurance companies base their entire business future on survivor curves for human beings. Depreciation is charged against the entire group of assets instead of against just a single item, using any reasonable method. If the useful life is taken as the average life of the grouped assets, it is logical to expect some items to last less time and some to last more. To retire some items early is considered "normal," and accounting for the capital loss on short-lived items is eliminated.

Three types of accounts often used are *group accounts, composite accounts,* and *classified accounts.* A group account contains similar assets with nearly the same useful lives, such as typewriters, lathes, passenger cars, or filing cabinets. Composite accounts include assets of dissimilar character and useful lives, such as machinery, office equipment, *and* furniture in the same account. Classified accounts have items of homogeneous character without regard to useful life, such as transportation equipment, office equipment, *or* machinery. Classified accounts are used frequently in the manufacturing industries with item accounts used mainly for large special assets, buildings, and structures.

5.8 Tax Concepts

The taxes paid by a corporation represent a real cost of doing business and, consequently, affect the cash flow profile. For this reason, it is wise to perform economic analyses on an *after-tax* basis. After-tax analysis procedures are identical to the before-tax evaluation procedures studied already; however, the cash flows are adjusted for taxes paid or saved.

There are numerous kinds of taxes including *ad valorem* (property), *goods and services, excise* (a tax or duty on the manufacture, sale, or consumption of various commodities), and *income taxes.* Income taxes are usually the only significant taxes to be considered in an engineering economic analysis. Property taxes, if considered, are normally treated as annual disbursements. Goods and services and excise taxes are included in the first cost of goods and services purchased. Income taxes are assessed on gross income less certain allowable deductions, incurred both in the normal course of business and on capital gains resulting from the disposal of property.

Federal and provincial income tax regulations are not only detailed and intricate, but they are subject to change over time. During periods of recession and inflation, there is a tendency for the tax laws to be changed in order to improve the state of the economy. For this reason, only the general concepts and procedures for calculating after-tax cash flow profiles and performing after-tax analyses are treated here.

5.9 Corporate Income Tax: Business Income

Income taxes may have an important impact on investment decisions, because they usually affect the estimates of future cash flows and, consequently, the internal rate of return on an investment project. All estimates of future cash flows should therefore be examined on an *after-tax* basis before discounted cash flow techniques are applied.

In these examples of economic analysis, it will be assumed that the effective rate of combined federal and provincial tax on income of most Canadian corporations is 50%, and since capital cost allowances are deductible expenses for income tax purposes, they will reduce the amount of income tax payable by 50%. Normally the full amount that is spent on the acquisition of a depreciable fixed asset is chargeable to taxable income as an expense over a number of years. However, because different rates of capital cost allowances apply to different classes of assets and because such allowances usually result in the faster write-off of assets for tax purposes, capital investment in depreciable fixed assets has a long-term, decreasing effect on the amount of tax saving from year to year.

The effect of capital cost allowances on the net amount of capital investment is accentuated when the time value of money is taken into account. For instance, the present value of the tax saving on a capital expenditure of $1 000 with a capital cost allowance of 30% will amount to only $375 when discounted at the rate of 10%. This means that for a company with a tax rate of 50%, the after-tax cost of an investment project will amount to approximately $625 for every $1 000 invested, as set out in Table 5.8. We should note here that an expenditure that is classified as a present *expense item* (as opposed to a capital asset) is *fully deductible* from income for tax purposes and has an after-tax cost to the company of $500 when taxes are paid at the 50% rate.

Table 5.8 After-Tax Cost of $1 000 Capital Investment

Year	Undepreciated Capital Cost $	Capital Cost Allowance @ 30%	Income Tax @ 50% $	Present Worth Factor $i=10\%$ $	Present Value of Tax Credit $
1	700	300	150	0.9091	136
2	490	210	105	0.8264	87
3	343	147	73	0.7513	55
4	240	103	52	0.6830	36
5	168	72	36	0.6209	22
6	118	50	25	0.5645	14
7	83	35	18	0.5132	9
8	58	25	13	0.4665	6
9	41	17	8	0.4241	3
10	29	12	6	0.3855	2
1 to 10		971	486		370
Future years		29	14		5
TOTAL		1000	500		375

The detailed calculations shown in Table 5.8 can be simplified by the use of a formula which results in expressing the after-tax present value of an investment's cost as a factor (ratio) called Capital Cost Tax Factor (*CCTF*). In the example above, *CCTF* would be 0.625 per dollar of capital invested. The Capital Cost Tax Factor will vary depending upon the applicable tax rate, capital cost allowance rate, and interest (discount) rate. The formula which can be used to incorporate any variation in these rates to determine the applicable capital cost tax factor is as follows:[2]

$$\text{Capital Cost Tax Factor} = 1 - \frac{td}{i+d}$$

(5.19)

where t = tax rate
d = capital cost allowance
i = interest rate

substituting for the data given above:

$$\text{CCTF} = 1 - \frac{0.50 \times 0.30}{0.10 + 0.30} = 1 - \frac{0.15}{0.40} = 0.625$$

Similar formulae can be derived for other methods of depreciation.

Corporations pay tax on their income, whether or not it is distributed to shareholders. For tax purposes, Canadian companies are divided into two main categories: "public corporations" and "private corporations."

A *public corporation* is generally a corporation resident in Canada whose shares are listed on a Canadian stock exchange. However, a corporation not listed on a Canadian stock exchange can be treateed as a public corporation if it meets certain conditions. A subsidiary of a public corporation is also treated as a public corporation. A *private corporation* is a corporation resident in Canada (i.e., carries on its business in Canada), which is neither a public corporation (as previously defined) nor controlled in any manner by one or more public corporations.

A *witholding tax* (usually limited to 10% to 15% by treaty) is imposed on a wide variety of payments from Canadian residents to nonresidents, both corporations and individuals. These incude interest, dividends, rents, royalties, maintenance payments, and certain management or administration fees.

---

[2] A more detailed description of the development and use of the Capital Cost Tax Factor may be found in the excellent text, *A Practical Approach to the Appraisal of Capital Expenditures* by C. G. Edge and V. B. Irvin, The Society of Management Accounts of Canada, 1981.

The *basic rate* of federal corporation income tax is 38%. The basic federal corporate income tax rate is determined by legislation, and it is subject to change from time to time. A temporary surtax may also be added.

All provinces and territories impose taxes on corporate income and to compensate for these provincial taxes the Income Tax Act permits corporations to deduct from the federal tax otherwise payable an amount equal to 10% of the corporation's taxable income earned during the tax year in a given province. The deduction, which is customarily referred to as a *provincial tax credit,* or *abatement,* is taken independently of any tax actually paid or payable to any province. The effective basic federal tax rate is thus 28%; that is, the basic rate of 38% less 10% provincial abatement.

The rates of corporation income tax levied by the provinces are shown in Table 5.9. The Quebec and Ontario tax applies to the taxable income of corporations as computed under the Quebec Taxation Act and the Ontario Corporations Tax Act, respectively. In all other provinces and the territories the tax applies to the taxable income of corporations as computed under the federal Income Tax Act. Except in the cases of Quebec and Ontario, these taxes are collected by the federal government on behalf of the province or territory.

Table 5.9 Provincial Corporation Income Tax Rates

	Small Business Rate %	Other %
Newfoundland	5	14
Prince Edward Island	7.5	15
Nova Scotia	5	16
New Brunswick	7	17
Quebec	3	13
Ontario	10, 13	14
Manitoba	9	17
Saskatchewan	8	17
Alberta	5	11
British Columbia	9	16.5
Nothwest Territories	5	14
Yukon Territory	6	15

5.9.1 Small Business Tax Credit

Canadian controlled private corporations enjoy a 16 percentage point tax rate reduction from the top basic rate of 38% (28% after the provincial abatement) on active business income up to $200 000 annually. Any income in excess of the small business limits is taxed at the regular rate. The term *active business* is defined very broadly to include income from virtually any business.

5.9.2 Manufacturing and Processing Profits Deduction

All corporations resident in Canada are entitled to a reduced rate of federal tax on profits derived from manufacturing and processing. This reduction amounts to 6% of eligible income. Investment income, business income from activities other than manufacturing and processing, income from operations outside Canada, and income eligible for the small business deduction do not qualify for the lower federal tax rate.

Manufacturing or processing excludes:

1. farming or fishing
2. logging
3. construction
4. operating an oil or gas well
5. extracting minerals or processing ore from a mineral resource
6. producing or processing electrical energy or steam for sale
7. processing of gas by a public utility

The lower rates apply only to profits derived from and allocated to actual manufacturing and processing activities and not to profits resulting from such activities as merchandising, sales, and distribution.

Example 5.9

> In this example we will show how federal corporate taxes are computed under different circumstances depending on eligibility for small business deductions and/or manufacturing and processing credits. Note that in this example the federal surtax mentioned at the beginning of this section is not included in the basic tax rate.

Effective Federal Corporation Income Tax Rate
Illustrations Year Ended July 31, 1999

I. Non-manufacturing company
  Taxable business income $120 000
  Basic tax   (38%−10%)×$120 000 $33 600
  Less: small business deduction   16%×$120 000   (19 200)
  Net federal income tax $14 400

II. Non-manufacturing company
  Taxable business income $420 000
  Basic tax   (38%−10%)×$420 000 $117 600
  Less: small business deduction   16%×$200 000   (32 000)
  Net federal income tax $85 600

III. Manufacturing company
  Total taxable business income $420 000
  Manufacturing and processing profit $160 000
  Basic tax   (38%−10%)×$420 000 $117 600
  Less: small business deduction   16%×$200 000   (32 000)
  Net federal tax payable $85 600

IV. Manufacturing company
  Total taxable business income $420 000
  Manufacturing and processing profit $360 000
  Basic tax   (38%-10%)×$420 000 $117 600
  Less: small business deduction   16%×$200 000   (32 000)
  Less: manufacturing credit   6%×($360 000−$200 000)   (9 600)
  Net federal tax payable $76 000

5.9.3 Determining Taxable Income

Taxable income must first be determined before any tax rate can be applied. Basically, *taxable income is gross income less allowable deductions. Gross income* is income in a general sense less any monies specifically exempt from tax liability. Corporate deductions are subtracted from gross income, and commonly include items such as salaries, wages, repairs, rent, bad debts, taxes (other than income), charitable contributions, casualty losses, interest, and depreciation. Interest and depreciation are of particular interest, since we can control them to

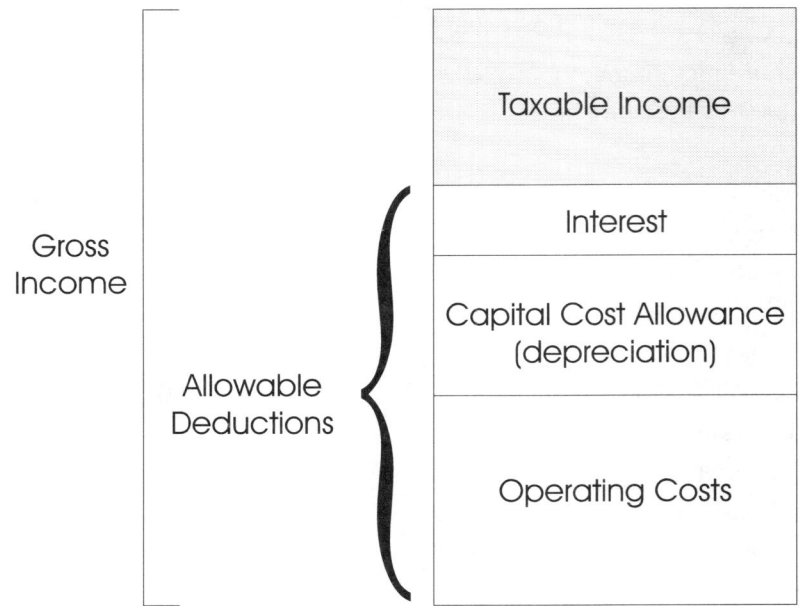

Fig. 5.2 Pictorial representation of taxable income.

some extent through financing arrangements and accounting procedures.

Taxable income is represented pictorially in Figure 5.2, which shows that taxable income for any year is what is left after operating costs and deductions, including interest on borrowed money and capital cost allowance (depreciation), are subtracted from gross income. These components are not all cash flows, since depreciation is simply treated as an expense in determining taxable income.

5.10 After-Tax Cash Flow

We have now looked at the basic elements needed to calculate after-tax cash flows. These elements are summarized in Figure 5.3. This shows that the after-tax cash flow is the amount remaining after operating costs, income taxes and deductions, including interest but excluding depreciation, are subtracted from gross income.

In many of the tables to follow, we simplify our terminology by speaking of before-tax cash flows. The term *before-tax cash flow* is used

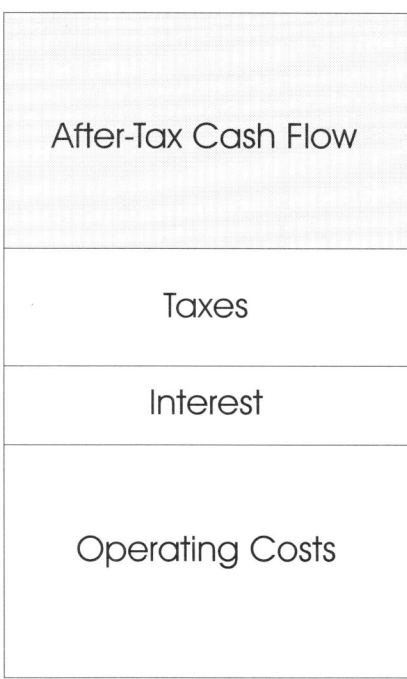

Fig. 5.3 Pictorial representation of after-tax cash flow.

when no borrowed money is involved, and it equals gross income less deductions, not including depreciation. *Before-tax* and *loan cash flow* is used when borrowed money is involved, and it equals gross income less deductions, not including either depreciation or principal or interest on the loan.

Example 5.10

Recall the earlier computer examples for which $P = \$10\,500$, $F = \$500$, and $n = 6$. We expect it to be responsible for reducing operating expenses by \$3 000 per year. Therefore, before-tax and interest cash flow will increase by \$3 000 during each year of the computer's useful life. We plan to depreciate it by the straight-line method and want to know the after-tax cash flows given that taxes are levied at a rate of 50% of taxable income. We will then apply the present worth method as an effectiveness measure to see whether the alternative of purchasing the computer yields at least a 10% after-tax rate of return. The cash flow calculations are given in Table 5.9.

Table 5.9 After-Tax Cash Flow Profile Using Straight-Line Depreciation

End of Year, A	Before-Tax Cash Flow, B	Depre- ciation, C	Taxable Income B−C, D	Tax D×0.50, E	After-Tax Cash Flow B−E, F
0	−$10 500				−$10 500.00
1	3 000	$1 666.67	$1 333.33	$666.67	2 333.33
2	3 000	1 666.67	1 333.33	666.67	2 333.33
3	3 000	1 666.67	1 333.33	666.67	2 333.33
4	3 000	1 666.67	1 333.33	666.67	2 333.33
5	3 000	1 666.67	1 333.33	666.67	2 333.33
6	3 000	1 666.67	1 333.33	666.67	2 333.33
6	500 (salvage)				500.00

Calculating the present worth of the after-tax cash flows, we have

$$PW(10) = -\$10\,500 + \$2\,333.33(P|A\ 10,6) + \$500(P|F\ 10,6)$$
$$= -\$10\,500 + \$2\,333.33(4.3553) + \$500(0.5645)$$
$$= -\$55.38$$

This example shows that the after-tax cash flow profile is nothing more than lumps of money flowing in or out at different points in time over a planning period. Although the present worth effectiveness measure was used in the example, a rate of return, future worth, or other equivalent analysis could have been applied. Since the present worth of the discounted cash flows in our example was negative (−$55.38), we are able to conclude that the alternative as described does not yield a 10% after-tax rate of return.

When we look at the taxable income and after-tax cash flow representations in Figures 5.2 and 5.3, it is natural to wonder about the tax effects of the depreciation method and the effect of interest on borrowed money. We will show that these factors have a substantial effect on taxes and thus on cash flow profiles.

5.11 Effect of Depreciation Method

We have seen how the accelerated depreciation methods provide for a higher depreciation allowance during the early years of an asset's life and a corresponding lower allowance in later years. This will place a lower tax burden on the asset during the early years, followed by a higher burden in later years. In most cases, the total undiscounted business income tax dollars paid will be the same regardless of the depreciation method used. This is always true when the tax rate remains constant and the total depreciation allowance for each depreciation method is the same. Due to the time value of money, however, we may expect the after-tax cash flow profile under accelerated depreciation to be superior to one based on a non-accelerated method such as straight-line depreciation.

Example 5.11

Let us repeat the previous example, now using sum of the years' digits depreciation. The actual depreciation values were calculated earlier (in Example 5.2), and the after-tax results of this method are as given in Table 5.10. Note that the after-tax cash flows for years 1 to 6 result in a decreasing gradient series when we have a uniform series of before-tax cash flows and the sum of the years' digits depreciation is used.

Calculating the present worth of the after-tax cash flows at $i = 10\%$:

$$PW(10) = -\$10\,500 + [\$2\,928.57 - \$238.10(A|G\,10,6)](P|A\,10,6)$$
$$+ \$500(P|F\,10,6)$$
$$= -\$10\,500 + [\$2\,928.57 - \$238.10(2.2236)](4.3553)$$
$$+ \$500(0.5645)$$
$$= \$231.18$$

Comparing Tables 5.9 and 5.10, we can see that the accelerated depreciation method caused after-tax cash flows to be higher in years 1, 2, and 3 and lower in years 4, 5, and 6. Merely changing depreciation methods caused our effectiveness measure to jump from −$55.38 to $231.18 when cash flows were discounted at the *MARR* of 10%.

Table 5.10 After-Tax Cash Flow Profile Using Sum of the Years' Digits Depreciation

End of Year, A	Before-Tax Cash Flow, B	Depre- ciation, C	Taxable Income B−C, D	Tax D×0.50, E	After-Tax Cash Flow B−E, F
0	−$10 500	—	—	—	−$10 500.00
1	3 000	$2 857.14	−$ 142.86	−$ 71.43	2 928.57
2	3 000	2 380.95	619.05	309.53	2 690.47
3	3 000	1 904.76	1 095.24	547.62	2 452.38
4	3 000	1 428.57	1 571.41	785.71	2 214.29
5	3 000	952.38	2 047.62	1 023.81	1 976.19
6	3 000	476.19	2 523.81	1 261.91	1 738.09
6	500 (salvage)				500

Example 5.12

Let us see how double declining balance depreciation affects the cash flow profile for our computer. Note that we previously calculated depreciation for this asset in Example 5.4. The after-tax results are shown in Table 5.11.

Table 5.11 After-Tax Cash Flow Profile Using Double Declining Balance Depreciation

End of Year, A	Before-Tax Cash Flow, B	Depre- ciation, C	Taxable Income B−C, D	Tax D×0.50, E	After-Tax Cash Flow B−E, F
0	−$10 500				−$10 500.00
1	3 000	$3 500.00	−$ 500.00	−$ 250.00	3 250.00
2	3 000	2 333.33	666.67	333.33	2 666.67
3	3 000	1 555.56	1 444.44	722.22	2 277.78
4	3 000	1 037.04	1 962.96	981.48	2 018.52
5	3 000	691.36	2 308.64	1 154.32	1 845.68
6	3 000	460.91	2 539.09	1 269.55	1 730.45
6	500 (salvage)				500.00

The present worth of the after-tax cash flows when discounted at 10% is:

$$PW(10) = -\$10\,500 + \$3\,250(P|F\,10,1) + \$2\,666.67(P|F\,10,2)$$
$$+ \$2\,277.78(P|F\,10,3) + \$2\,018.52(P|F\,10,4)$$
$$+ \$1\,845.68(P|F\,10,5) + \$1\,730.45(P|F\,10,6)$$
$$+ \$500(P|F\,10,6)$$
$$= -\$10\,500 + \$3\,250(0.9091) + \$2\,666.67(0.8265)$$
$$+ \$2\,277.78(0.7513) + \$2\,018.52(0.6830)$$
$$+ \$1\,845.68(0.6209) + \$1\,730.45(0.5645)$$
$$+ \$500(0.5645)$$
$$= \$153.59$$

This accelerated depreciation method also improved upon the straight-line depreciation and, for the specific example used above, double declining balance was slightly worse than the sum of the years' digits. In the first year, depreciation was quite high. In fact, the allowance exceeded before-tax and interest cash flow and yielded a negative taxable income, resulting in a negative tax. This simply implies that the company has other positive taxable income and tax obligations in year 1 that may be offset by the negative taxable income and tax for this alternative. If there were no positive taxable income to offset, the taxable income and tax entries for year 1 would have been zero, and losses could then be carried backward or forward in time to be applied against positive taxable incomes in other years. Carry-back and carry-forward provisions are discussed later in this chapter.

The three previous examples illustrate the important point that the method of depreciation does affect cash flow and, therefore, the economic desirability of a project. Can we make any generalized statements *from these examples* about the preferability of depreciation methods with respect to effects on taxes and after-tax cash flows? No, because these examples are worked out only for a particular set of conditions.

Certain methods of depreciation are superior to others in that they provide a higher present worth of tax savings, assuming that the effective tax rate remains the same from year to year. For example, the sum of the years' digits method is always superior to the straight-line method. The declining balance method is also preferable to straight-line when using a rate of $p = 1.5/n$ for high values of the *MARR* and high values of the salvage to cost basis ratio, $F|P$. The double declining balance method is preferable to the sum of the years' digits for high values of the *MARR* and high values of $F|P$. It is beneficial to switch from double declining balance depreciation to straight-line if the imputed salvage value (book value at the end of the last year of useful life) is not as low as the extimated salvage value. Within the context of Canadian tax laws, however, one must remember that the *maximum* rate allowed is prescribed by government regulations. No matter which method we use, the capital cost allowance is limited by that rate.

5.12 Effect of Interest on Borrowed Money

Investment alternatives may be financed using equity (owners' funds) or debt (borrowed funds). Until now we have implicitly assumed all financing to be through equity, although many companies use a mix of debt and equity for financing plant and equipment. Borrowed funds must be repaid, including both principal and interest. The interest paid each year affects taxable income and, consequently, taxes. Both the principal and interest payments affect after-tax cash flows.

There are four common (and many less common) ways in which money can be repaid. First is the periodic payment of interest over the stipulated repayment period with the entire principal being repaid at the end of that time. The second requires an annual payment that uniformly repays the principal and also covers the annual interest. With this method, the annual payments decrease as the interest on the unrepaid principal decreases. Third is the method requiring a uniform annual payment for the sum of principal plus interest. In each payment the proportion of principal gradually increases as the proportion of interest decreases. The fourth method repays nothing, neither interest nor principal, until the end of a specified period.

Example 5.13

Let us illustrate these four basic plans for repaying principal and interest on borrowed money. Assume that a business borrows $5 000 to be used in financing an investment, and the interest rate on this loan is 8% compounded annually. The stipulated repayment period is six years.

A summary of all relevant components of this example is presented in Table 5.12. In method 1, the interest equals $5\ 000(0.08) = \$400$ per year. Only the interest is paid, and only the principal of $5 000 is owed after each year's payment. Method 2 repays the principal in equal amounts of

Table 5.12 Illustration of Four Common Methods of Principal and Interest Repayment

End of Year	Interest Accrued during Year	Total Money Owed before Yearly Payment	Interest Payment	Principal Payment	Total Payment	Total Money Owed after Yearly Payment
Method 1						
0						$5000
1	$400.00	$5 400.00	$400.00	$0	$400.00	5000
2	400.00	5 400.00	400.00	0	400.00	5000
3	400.00	5 400.00	400.00	0	400.00	5000
4	400.00	5 400.00	400.00	0	400.00	5000
5	400.00	5 400.00	400.00	0	400.00	5000
6	400.00	5 400.00	400.00	5 000.00	5 400.00	0
Method 2						
0						$5 000.00
1	$400.00	$5 400.00	$400.00	$833.33	$1 233.33	4 166.67
2	333.33	4 500.00	333.33	833.33	1 166.67	3 333.33
3	266.67	3 600.00	266.67	833.33	1 100.00	2 500.00
4	200.00	2 700.00	200.00	833.33	1 033.33	1 666.67
5	133.33	1 800.00	133.33	833.33	966.67	833.33
6	66.67	900.00	66.67	833.33	900.00	0
Method 3						
0						$5 000.00
1	$400.00	$5 400.00	$400.00	$681.58	$1 081.58	5 318.42
2	345.47	4 663.90	345.47	736.10	1 081.58	3 582.32
3	286.59	3 868.91	286.59	794.99	1 081.58	2 787.33
4	222.99	3 010.31	222.99	858.59	1 081.58	1 928.74
5	154.30	2 083.04	154.30	927.28	1 081.58	1 001.46
6	80.12	1 081.58	80.12	1 001.46	1 081.58	0
Method 4						
0						$5 000.00
1	$400.00	$5 400.00	$0	$0	$0	5 400.00
2	432.00	5 832.00	0	0	0	5 832.00
3	466.56	6 298.56	0	0	0	6 298.56
4	503.88	6 802.44	0	0	0	6 802.44
5	544.20	7 346.64	0	0	0	7 346.64
6	587.73	7 934.37	2 934.37	5 000	7 934.37	0

$$\frac{\$5\,000}{6} = \$833.33$$

as well as the interest for year t, which is given by

$$[\$5\,000 - \$833.33(t-1)]0.08$$

Clearly, the interest payment, total payment, and total money owed after yearly payments form a decreasing gradient series. Method 3 requires equal annual payments. This annual payment is equal to

$$\$5\,000(A|P\ 8,6) = \$1\,081.58$$

The principal component of this annual payment for year t can be found quickly as

$$\$1\,081.58(P|F\ 8,6-t+1)$$

using the method given in Chapter 3. It is interesting to note that the principal payment is an 8% increasing compound series. Also, the interest payment and the total money owed after yearly payment decreases each year, and the amount by which they decrease is an 8% increasing compound series. In method 4, the interest accrued during each year is added to the principal such that the total amount owed after t years is

$$\$5\,000(1.08)^t$$

When payment is made at the end of year 6, everything over $5 000 is considered interest.

We have seen that the interest on borrowed money is deductible for tax purposes, whereas the principal repayment does not reduce taxable income. In addition, both the interest and principal portions of a payment are real and must be taken into account when calculating cash flows.

Example 5.14

Let us illustrate the effect of borrowed money by assuming that $5 000 of the $10 500 paid for our computer is through debt funding. The loan is to be repaid in equal annual installments (method 3) at 8% over six years. The remaining $5 500 will be equity money. Applicable depreciation and loan activity for this method have previously been calculated in Example 5.13. The resulting after-tax cash flow profile is detailed in Table 5.13.

Table 5.13 After-Tax Cash Flow Profile Using Double Declining Balance Depreciation and Using $5 000 Borrowed Money at 8%

End of Year	Before-Tax and Loan Cash Flow, A	Loan Principal Payment, B	Loan Interest Payment, C	Depreciation Charges, D	Taxable Income B–D–E, E	Tax F×0.50, F	After Tax Cash Flow B–C–D–G, G	H
0	−$10 500	−$5 000	—	—	—	—	−$5 500.00	
1	3 000	681.58	$400.00	$3 500.00	−$900.00	−$450.00	2 368.42	
2	3 000	736.10	345.47	2 333.33	321.20	160.60	1 757.83	
3	3 000	794.99	286.59	1 555.56	1 157.85	578.93	1 339.50	
4	3 000	858.59	222.99	1 037.04	1 739.97	869.99	1 048.44	
5	3 000	927.28	154.30	691.36	2 154.34	1 077.17	841.25	
6	3 000	1 001.46	80.12	460.91	2 458.97	1 229.48	688.94	
6	500	(salvage)					500.00	

$$PW(10) = -\$5\,500 + \$2\,368.42(P|F\,10,1) + \$1\,757.83(P|F\,10,2)$$
$$+ \$1\,339.50(P|F\,10,3) + \$1\,048.44(P|F\,10,4)$$
$$+ \$841.25(P|F\,10,5) + \$688.94(P|F\,10,6)$$
$$+ \$500(P|F\,10,6)$$
$$= -\$5\,500 + \$2\,368.42(0.9091) + \$1\,757.83(0.8265)$$
$$+ \$1\,339.50(0.7513) + \$1\,048.44(0.6830)$$
$$+ \$841.25(0.6209) + \$688.94(0.5645)$$
$$+ \$500(0.5645)$$
$$= \$1021.92$$

The value of our effectiveness measure jumped to $10 21.92 for this particular example. Note that the present worth calculation on cash flows was made using a discount rate equal to our 10% *MARR*. The 8% loan rate was used only in determining the loan repayments. If we actually needed to borrow the $5 000 to implement the project, our estimates indicate that a handsome monetary return will be received on the $5 500 equity investment. If, on the other hand, we had at least the other $5 000 available, borrowing allowed us to invest that money as equity capital in another alternative that will earn a return at least equal to the *MARR*. This, of course, depends on the availability of investments that will yield a rate at least equal to the *MARR*.

We cannot conclude from the above example that borrowing money is always favourable. The desirability of borrowed funds depends on the terms of the loan, including method of repayment, interest, and repayment period. Furthermore, collateral is frequently required, which may be lost (after legal action) if the principal and interest cannot be paid on schedule. In summary, each investment alternative and financing strategy should be compared on its own merits.

We have used the same basic example to illustrate the implications of depreciation and financing stragegies thus far throughout the chapter. The present worth of after-tax cash flows was seen to increase from −$55.38 to $1021.92 simply by going from the commonly used equity financing and straight-line depreciation to the more sophisticated mix of debt and equity financing and double declining balance depreciation. Nothing about the basic incomes or costs relating to the asset were changed.

By now one should have a good idea how to calculate and assess after-tax cash flows under differing depreciation schedules and financing arrangements, and should be able to apply these techniques in the analysis of alternative investments.

Among the other factors affecting income tax paid are carry-back and carry-forward rules, capital gains and losses, treatment of capital assets, recaptured capital cost allowance, incentives, lease-buy considerations, and depletion. Since US regulations differ greatly from the Canadian laws in these areas, they have little bearing on most Canadian engineering projects, and go beyond the scope of this book. In the discussion which follows, US legislation will be considered only in the area of recaptured capital cost allowance.

5.13 Carry-back and Carry-forward Rules

In three of the previous examples we saw that an alternative's taxable income and tax can be negative for a particular year. It was implicitly assumed that those values were used to offset positive tax liabilities from other corporate activities. According to the Canadian Income Tax Act, when there are insufficient profits to be offset, the taxable income (noncapital) losses can be carried back three years or forward seven years in order to reduce positive taxable income. That is, if a loss occurs in 2000 we can reopen books from 1997, or apply the loss against positive taxable income until 2007. We will continue to assume that negative taxable incomes are used to offset positive values for the year in question.

5.14 Capital Gains and Losses

Assets receiving capital treatment are important for tax purposes because they receive preferential tax treatment, being less severely taxed than business income. Many of the common capital gain and loss situations faced in economic analysis will be presented in this chapter. However, any uncertainty should be resolved by professional (legal and accounting) counsel. If a depreciable asset is sold and the proceeds are in excess of the book value but less than original cost, the difference is handled as recapture of capital cost allowance (depreciation for tax purposes) and treated as business income. If the asset is sold for a value in excess of its original cost, this excess is considered a *capital gain*. Three-quarters of the capital gains are then added on to the rest of the company's taxable income. This new amount is then taxed at the normal tax rates, i.e., capital gains are taxed at 75% of the normal rate.

Capital property is defined as depreciable property and any other property in respect of which a gain or loss on disposition would be a capital gain or loss. Under this definition, the question of whether a property is capital property depends on the definition of capital gain and capital loss. For instance, a capital gain or loss does not arise on the disposition of the following types of property:

1. Property, the proceeds of which would be taken into account in computing business income. Inventory of a business or property acquired as part of a trade venture would fall into this category.

2. Eligible capital property, which includes goodwill and other nondeductible intangibles acquired in connection with a business.

3. Canadian and foreign resource properties, which include petroleum, natural gas, and mining rights, oil and gas wells, mining properties (excluding depreciable property), rents and royalties from oil, gas and mineral production, and interests in any of the foregoing.

4. A life insurance policy or an annuity contract.

5. A timber resource property, such as a renewable right or licence to cut or remove timber.

Generally, a capital property is any property other than one listed in categories (1) to (5) above. However, anomalies arise in some circumstances due to the fact that shares of a corporation are normally capital property, while the assets of a corporation may to a large extent be noncapital property. Therefore, the consequences for income tax calculation will vary widely depending upon whether shares or assets are disposed of.

5.15 Tax Treatment of Capital Assets

Eligibility is the key difficulty in complying with Revenue Canada capital gains codes. For example, the courts have used purpose of acquisition and disposal, frequency and extent of sales, length of time the property was owned, nature of the taxpayer's business, the taxpayer's sales efforts, and so forth, to determine whether or not capital treatment is allowable. Any items sold to customers in the ordinary course of trade or business may not be considered capital assets by the seller, and they are therefore subject to ordinary gains or losses. That is, they are included in the seller's income and are taxed at the normal rates.

Before applying capital treatment, the extent of the capital gain or loss must be determined. A capital gain or loss may occur when there is a sale or exchange of a capital asset. The excess of selling price over original cost is a *capital gain*. If the selling price is less than the book value, then a capital loss has occurred. Three-quarters of capital losses, known as "allowable capital losses" may be offset against capital gains. Three-quarters of capital losses incurred in a year may be deducted against three-quarters of capital gains realized in the year, but any deductible excess cannot be applied against other income. Corporations may carry losses back three years and forward an indefinite number of years, but *only* against capital gains realized, until the losses are absorbed.

Example 5.15

Now, suppose our corporation engaged in manufacturing and processing has $170 450 taxable income, of which we found $50 450 to be a net capital gain taxable as a capital gain. Therefore $120 000 would be business income. Computing the federal taxes, we have:

Basic Tax = (38% − 10%) × $120 000 =	$33 600.00
Less:	
Small Business Deduction = 16% × $120 000 =	(19 200.00)
Capital Gains Tax = (38% − 10%) × (0.75 × $50 450) =	10 594.50
Total Federal Tax	$24 994.50

5.16 Recaptured Capital Cost Allowance

Depreciable property and real property used in a trade or business may, under certain conditions, receive capital treatment after disposal. Taxpayers were quick to recognize the monetary advantages of capital gains as compared to business income. In the United States, a favourite loophole was then to declare a low salvage value on a depreciable asset, depreciating it far below the market value. The excess depreciation was used to offset business income, thereby saving taxes at the rate of about 50%. After disposing of the asset, the lower capital gains tax rate was applied to the difference between fair market value and book value. In 1964, the US Congress ruled that gains realized at the time of sale are taxed, to a limited extent, as business income. Similar legislation was passed by the Canadian Parliament in 1972 to remedy the same problem. This is known as *depreciation recapture*. In both countries this legislation effectively closed the loophole of overdepreciating as a means of converting business income into capital gains.

Depreciation recapture can best be explained by considering a depreciable asset depreciated by the straight-line method. If the selling price is below the original cost and above the book value, the gain is considered a recapture of previously charged depreciation and is taxed as business income. If the item actually appreciates and sells for more than the original cost and if book value represents depreciation, then the difference between book value and the original cost is taxed as recaptured business income, while the difference between selling price and original cost is a capital gain. Finally, if the selling price is below the book value, the loss is treated as a terminal loss, provided that the asset class is eliminated completely; that is, there are *no* assets remaining in that particular class.

Example 5.16

In Example 5.1, straight-line depreciation was used to depreciate a computer from $10 500 down to $500 over a six-year period. The computer was estimated to provide an additional before-tax cash flow of $3 000 per year. The asset, we will now assume, was sold for $2 000, resulting in a $1 500 gain. Since this gain is considered to be depreciation recapture, it is taxed at the business income rate (we will again use 50%). That is,

$$\text{Tax on gain} = 0.50(\$2\,000 - \$500) = \$750$$

Table 5.14 illustrates the after-tax cash flow profile for this example when depreciation recapture is involved.

Table 5.14 After-Tax Cash Flow Profile with Depreciation Recapture

End of Year, A	Before-Tax Cash Flow, B	Depreciation, C	Taxable Income B−C, D	Tax D ×0.50, E	After-Tax Cash Flow B − E, F
0	−$10 500				−$10 500.00
1	3 000	$1 666.67	$1 333.33	$666.67	2 333.33
2	3 000	1 666.67	1 333.33	666.67	2 333.33
3	3 000	1 666.67	1 333.33	666.67	2 333.33
4	3 000	1 666.67	1 333.33	666.67	2 333.33
5	3 000	1 666.67	1 333.33	666.67	2 333.33
6	3 000	1 666.67	1 333.33	666.67	2 333.33
6	2 000	(salvage)			2 000.00
6		−1 500.00	1 500.00	750.00	−750.00

Note that three after-tax cash flows are shown for year 6. The first, $2 333.33, is for regular year 6 operations. The second, $2 000, is the salvage value received after six years. The third, −$750, represents the business income tax liability on the $1 500 of excess depreciation recaptured. The present worth of the after-tax cash flows may now be calculated using a *MARR* of 10%.

$$PW(10) = -\$10\,500 + \$2\,333.33(P|A\ 10,6)$$
$$+ \$2\,000(P|F\ 10,6)$$
$$- \$750(P|F\ 10,6)$$
$$= -\$10\,500 + \$2\,333.33(4.3553)$$
$$+ \$2\,000(0.5645) - \$750(0.5645)$$
$$= \$367.98$$

5.17 Income Tax Incentives

A number of measures are provided in federal and provincial income tax legislation that are more in the nature of incentives than direct government assistance.

(a) *Accelerated capital cost allowances.* Through the Canadian capital cost allowance system, accelerated tax depreciation is often used as an incentive to encourage investment in selected assets or industries. This incentive, together with a reduced rate of tax on manufacturing and processing income, has produced substantial stimuli to the manufacturing sector.

(b) *Investment tax credit.* To stimulate new business capital spending in Canada, the federal government established an investment tax credit for expenditures incurred in acquiring certain buildings, machinery or equipment, or on scientific research and development. The investment tax credit amounts to a certain percentage of the cost of qualifying expenditures. It depends upon the region in which the expenditure is incurred and ranges from 7% to 20%. The high credit is designed to promote industrial development in those parts of Canada most adversely affected by economic disparities and is applicable to areas under the Regional Development Incentives Act.

Generally, scientific research expenditures of a current and capital nature incurred in Canada qualify for the credit. The rate of tax credit on scientific research expenditure for a Canadian-controlled private corporation qualifying for the small business tax rate is 35%, regardless of the region in Canada in which the expenditure is incurred.

Other qualifying expenditures consist of capital assets which meet the following criteria. The assets must be:

1. Buildings, machinery, or equipment of the type prescribed by the Income Tax Act.
2. Unused when acquired by the taxpayer.
3. Acquired primarily for use by the taxpayer in Canada in a designated activity including manufacturing or processing of goods for sale or lease, prospecting or exploring for or developing a mineral resource, logging, fishing, farming, and storing of grain.

For tax depreciation purposes, the capital cost of the property acquired will be reduced by any investment tax credit claimed, as will the deductible amount of related scientific expenditures.

The federal and provincial income tax laws contain similar rules regarding the taxation of government assistance granted for the acquisition of capital property. The general rule is that this type of incentive or assistance reduces the capital costs of the acquired property for both capital gains and capital cost allowance purposes. Investment tax credits and most cash grants come under this general rule.

The federal and Ontario acts do, however, exclude from the definition of government assistance the grants and allowances described below, with the result that these amounts do not reduce the cost of the related capital property for tax purposes.

1. An amount authorized to be paid under an appropriation act and on terms and conditions approved by the Treasury Board in respect of scientific research expenditures incurred for the purposes of advancing or sustaining the technological capability of Canadian manufacturing or other industry. Enterprise Development Program grants are examples of this particular exemption. However, these grants do reduce the amount that can be deducted in respect of current and capital expenditures.
2. Depletion allowance (see details in Section 5.19).

Example 5.18

A firm invests $500 000 in an asset expected to be used for four years, and $1 200 000 in another property to be used for 12 years. Both qualify for a 7% investment tax credit. Tax liability for the tax year 2000 is $200 000. Let us determine the tax credit used in year 2000.

Depreciation and Income Tax Considerations

$$\text{Tax credit} = \$1\,700\,000 \times 0.07$$
$$= \$119\,000$$
$$\text{Tax credit used} = \$15\,000 + (\$200\,000 - \$5\,000) \times 0.50$$
$$= \$107\,500$$

Note that the excess tax credit of $\$119\,000 - \$107\,500 = \$11\,500$ can be carried forward to apply against taxes in other years.

Example 5.19

Let us again consider Example 5.10 in which straight-line depreciation is used to depreciate a computer from $10 500 down to $500 over six years. The computer was estimated to provide an additional before-tax cash flow of $3 000 per year. Now, however, we will assume that the investment tax credit of 10% applies.

The tax credit will be as follows:

$$\text{Tax credit} = \$10\,500 \times 0.10$$
$$= \$1\,050$$

The new depreciation base = $10 500 - $1 050 = $9 450.

If the purchase occurs at the beginning of fiscal year 1, the $1050 tax credit applies against the first-year tax liability. This is illustrated in Table 5.17.

The present worth of the after-tax cash flows may now be calculated.

$$PW(10) = -\$10\,500 + \$3\,337.50(P|F10,1)$$
$$+ \$2\,287.50(P|A10,5)(P|F10,1) + \$500(P|F10,6)$$
$$= -\$10\,500 + \$3\,337.50(0.9091)$$
$$+ \$2\,287.50(3.7908)(0.9091) + \$500(0.5645)$$
$$= \$699.59$$

Table 5.17 After-Tax Cash Flow Profile with Investment Tax Credit

End of Year, A	Before-Tax Cash Flow, B	Depre- ciation, C	Taxable Income B−C, D	Tax D ×0.50, E	Investment Tax Credit, F	After-Tax Cash Flow B − E+F, G
0	−$10 500	—	—	—	—	−$10 500.00
1	3 000	$1 575	$1 425	$712.50	$1 050.00	3 337.50
2	3 000	1 575	1 425	712.50	0	2 287.50
3	3 000	1 575	1 425	712.50	0	2 287.50
4	3 000	1 575	1 425	712.50	0	2 287.50
5	3 000	1 575	1 425	712.50	0	2 287.50
6	3 000	1 575	1 425	712.50	0	2 287.50
6	500 (salvage)					500.00

The investment tax credit increased the after-tax cash flow present worth from −$55.38 to $699.59. Obviously, the investment tax credit is a significant factor in an economic analysis. In the problems at the end of this chapter, *the tax credit, if applicable, will be applied at the yearend preceding the commencement of the project (time 0)*.

5.18 Lease-Buy Considerations

An important aspect of a capital budgeting decision is the financing of the investment project and, in particular, the question whether the asset to be acquired should be bought or leased. A decision to buy or to lease can be taken only after there has been an affirmative investment decision to acquire the asset; once that decision has been taken the choice between buying the asset or leasing it becomes a question of financing.

If the asset is to be bought, funds must be obtained either from internal or from external sources. Where the acquisition is financed from internal sources such as equity capital or retained earnings, the suppliers of the acquisition (i.e., the owners of the business) will expect an acceptable rate of return on the money so invested, the acceptable rate being set by that obtainable from alternative investment opportunities. Where the funds for buying the asset are to be obtained from external sources in the form of bonds or other long-term debt, the lenders will expect interest payments over an extended period of time and the repayment of the capital sum at a fixed or determinable future date. In the case of leasing, the cash outflows will be in the form of monthly, quarterly, or yearly payments to be made during the term of the lease.

Since leasing is an alternative form of financing, its cost must be compared with the cost of other sources of finance and the analysis of a buy or lease decision will therefore take the form of a comparison of the cash flows associated with the various ways in which financing is to be provided. If a comparison is made between the cash flows connected with leasing and with buying the asset, the terms of the lease agreement must be considered in order to determine what benefits (such as free maintenance) accrue under the lease agreement and which would have to be paid for by the owner if the asset were bought. On the other hand, if the asset is bought, there is an additional cash inflow that must be taken into account, namely the expected proceeds from the sale of the used asset at the end of its economic life. In analyzing lease-buy decisions, therefore, any value attached to the asset at the end of the period over which the comparison is made must be treated as a benefit under the purchase scheme.

The cost of leasing (i.e., the effective rate of interest paid in terms of the lease agreement) may be calculated by finding the rate that will discount the future lease payments to a present value equal to the cash purchase price of the asset. When this calculation is made the differential benefits of leasing, such as free maintenance and taxation allowances, as compared with other ways of financing the acquisition must be taken into account. The effective rate of interest paid in terms of the lease agreement must be compared with the effective rate of interest when borrowing the cash required for immediate purchase of the asset (taking into account the residual value of the asset at the end of its economic life) and with the effective cost of capital if equity capital funds were used for the acquisition of the asset.

The lowest of the three rates will indicate which method of financing is the most beneficial for the undertaking.

An alternative way of analyzing a lease-buy decision is to make a detailed comparative statement of the cash flows under each of the different methods of financing, taking into account the relevant cash flows, appropriate tax rates, the effect of capital cost allowances, and the expected length of the project life.

It is often more attractive to lease property than to own it. When an asset is leased, a schedule of payments over time is agreed upon. Lease charges, being expenses, apply directly against taxable income during the year in which they occur. Thus, each dollar spent on a leased item has the effect of costing only $0.50 (assuming a 0.50 tax rate). This is not true for dollars spent on a purchased item, because this cost must be written off over the life of the asset. Thus, considering the time value of money, the effective cost per dollar spent on a purchased item is more than $0.50.

Example 5.20

Our computer has a $10 500 first cost, salvage value of $500, and life of six years. The computer may be purchased using equity capital and depreciated using double declining balance depreciation. This case was illustrated in Table 5.11 and had a $153.19 present worth of after-tax cash flows using a cost of capital of 10% after tax.

A lessor offers to buy the computer and lease it to us for $3 037.55 − $100(t − 1) during the tth year. This decreasing gradient is equivalent to 120% of the pretax equivalent uniform annual cost of the computer. The after-tax cash flows for this lease alternative are presented in Table 5.18.

Table 5.18 After-Tax Cash Flow for Lease Alternative

End of Year, A	Reduced Expenses Due to Asset, B	Cost to Lease Asset, C	Taxable Income B−C, D	Tax D×0.50, E	After-Tax Cash Flow B−C−E, F
0	—	—	—	—	—
1	3000	$3037.55	−$37.55	−$18.78	−$18.77
2	3000	2937.55	62.45	31.23	31.22
3	3000	2837.55	162.45	81.23	81.22
4	3000	2737.55	262.45	131.23	131.22
5	3000	2637.55	362.45	181.23	181.22
6	3000	2537.55	462.45	231.23	231.22

The present worth of the after-tax cash flows at a MARR of 10% is as follows:

$$PW(10) = [-\$18.77 + \$50(A|G\,10,6)](P|A\,10,6)$$
$$= [-\$18.77 + \$50(2.2236)](4.3553)$$
$$= \$402.47$$

Thus, leasing appears to be more favourable than purchasing. However, if we are eligible for an investment tax credit, the decision will change. That is, assuming the 10% tax credit is applied to the purchase alternative in year 0, the present worth of the after-tax cash flows will be $153.19 + $10 50.00, or $1 203.19.

In Example 5.14 we calculated the present worth of the after-tax cash flows when we assumed that $5 000 of the $10 500 paid for our minicomputer is through debt funding using an 8% loan rate. The present worth was found to be $1 021.92 without taking any investment credit into consideration.

Some of the non-monetary advantages of leasing may be summarized as follows:

1. The risks of ownership, such as obsolescence, remain with the lessor.

2. By leasing, the company can borrow 100% of the cost of the asset, compared to less than 100% by borrowing from banks or mortgagors.

3. Financing via leasing has less practical effect on a company's subsequent ability to borrow.

4. It is useful for financing small expansions or equipment additions and other piecemeal financing.

5. It avoids restrictive covenants that often accompany long-term debt.

6. It allows the company to write off for tax purposes the total cost of the leased property, including land that is not depreciable for tax purposes.

7. Lease financing is often easier and less costly to obtain than other types of financing.

5.19 Depletion

Depletion is a gradual reduction of minerals, gas and oil, timber, and natural deposits. In a sense, depletion is closely akin to depreciation. The difference is that while a depleting asset loses value by actually being removed and sold, a depreciable asset loses value through wear and tear in the manufacture of goods to be sold and obsolescence. Money recovered through the depletion allowance is likely to be used in the exploration and development of depletable assets, just as depreciation reserves are reinvested for new equipment. As in most tax-related matters, laws relating to lessors, lessees, royalties, and sales are complex and will probably require expert assistance.

5.20 Summary

This chapter has presented the most important elements of Canadian depreciation and income tax law as they pertain to economic analyses. Some clarification of US income tax law has been included as well. It is clear that the depreciation (or depletion) method, useful life, financing, and tax credits can have significant effects on the desirability of making an investment. Many of these factors are law related, and changes are being made frequently. Therefore, in cases of uncertainty regarding depreciation and tax treatment, it is wise to seek competent legal and accounting advice.

Capital gains and losses are often not important in engineering economic analyses. However, it is wise to know when a capital gain or loss has occurred and when to apply capital gains treatment versus depreciation recapture.

Finally, unless there are overriding considerations, lease versus buy alternatives should be analyzed closely. Quite often, a lease alternative will compare with surprising favour against an equivalent purchase alternative.

Problems

1. An electrostatic precipitator is purchased for $130 000; it has an anticipated life of 10 years, and terminal salvage value of $20 000. Determine the book value for the precipitator at the end of the ninth year of its life and the depreciation charge for the tenth year of its life, using:

    (a) Straight-line depreciation.

    (b) Sum of the years' digits depreciation.

    (c) Sinking fund depreciation at $i = 10\%$. (5.4)

2. Rework Problem 1 using the following methods of depreciation:

    (a) Double declining balance switching to straight-line depreciation.

    (b) Declining balance depreciation using the rate that will yield a book value of $20 000 at the end of year 10.

    (c) Double declining balance depreciation. (5.4)

3. A firm purchases machinery for $100 000. It has a life of nine years and terminal salvage value of $10 000 at that time. Determine the depreciation charge for year 6 and the book value at the *beginning* of year 6, using:

    (a) Straight-line depreciation.

    (b) Sum of the years' digits depreciation.

    (c) Double declining balance depreciation.

    (d) Sinking fund depreciation at $i = 10\%$. (5.4)

4. A furnace has a first cost of $50 000 and salvage value after four years of $0. The method of depreciation to be used is 20% declining balance. What will be the depreciation charge each year? (5.4)

5. A compressor is purchased for $4 000 and has an estimated salvage value of $1 000 after a useful life of three years. Interest is 10%.

    (a) Determine capital recovered, return on the capital unrecovered, and capital recovered plus return for each year using straight-line depreciation.

    (b) Determine capital recovered, return on the capital unrecovered, and capital recovered plus return for each year using sum of the years' digits depreciation.

(c) Determine the annual equivalent capital recovered plus return for each of parts (a) and (b).

(d) Determine the annual equivalent before-tax cost of the asset using $(P-F)(A|Pi,n)+Fi$. (5.4)

6. For the compressor in Problem 5, what would be the capital recovered plus return for the second year, using sinking fund depreciation? (5.4)

7. A numerically controlled lathe is purchased by a small machine shop for $8 000. It is to be used for six years, will have salvage value of $400, and is to be depreciated using the straight-line method. Assuming it qualifies for 20% "additional first-year depreciation," prepare a depreciation schedule for each year of its life. (5.6)

8. An automatic control mechanism is estimated to provide 3000 h of service during its life. The mechanism costs $4800 and has salvage value of $300 after 3000 h of use. Its use is projected over a four-year period as follows:

Year	Hours of Use
1	1500
2	800
3	400
4	300

Calculate the depreciation charge for each year using a method similar to the "operating day" method, but based on operating hours. (5.7)

9. A utility trailer costs $1 000 and is rented out by the hour, day, or week. It is expected to depreciate to zero by the time it has been rented out for a total of $5 000 in gross income. If its forecast annual revenues are $1 800, $1 500, $900, and $800, what will be the appropriate annual depreciation charges? (5.7)

10. What is the federal income tax for each of the following corporate taxable incomes? Consider 28% federal tax and small business deduction in applicable.

    (a) $ 75 000
    (b) $150 000
    (c) $250 000

(d) $1 million

Plot a graph of federal income tax versus taxable income. Be sure to label critical values on both scales and note appropriate assumptions. (5.9)

11. A dust collector with a first cost of $30 000 will be depreciated to zero by the straight-line method. The collector will be used for five years, and will yield an annual gross income less operating expenses of $15 000. The effective tax rate is 50%. Determine the after-tax rate of return. (5.10)

12. An automatic soldering machine is purchased for $18 000. Installation cost is $3 000 because of the extreme care and provisions needed. It is estimated that the asset can be sold for $2 500 after a useful life of 10 000 h of operation. However, before selling, it must be removed at a cost of $500 to Company A. Inspection and maintenance costs $6 per operating hour. It requires one-half hour to solder one unit on the machine, and a total of 10 000 units are soldered per year. The after-tax MARR is 15%, and the effective tax rate is 50%. What will be the cost of soldering per unit assuming depreciation is based on hours of service? (5.10)

13. An investment proposal is described by the following tabulation:

	Year		
	1	2	3
Gross income during year	$8 000	$16 000	$20 000
Investment at *beginning* of year	24 000	0	0
Operating cost	2 000	3 000	4 000
Depreciation charge (straight-line)	8 000	8 000	8 000

The company considering this proposal is profitable in its other activities. Determine the taxable income, tax, and after-tax cash flow for time 0 and years 1, 2, and 3. Assume an effective tax rate of 50%. (5.10)

14. An investment opportunity has the following financial outlook over a five-year period.

EOY	Before-Tax Cash Flow	Depreciation Charges	Taxes
0	−$30 000	—	—
1	10 000	$10 000	$ 0
2	10 000	8 000	800
3	10 000	6 000	1 600
4	10 000	4 000	2 400
5	10 000	2 000	3 200

If a firm requires a 12% rate of return after taxes, should it invest? (5.10)

15. A firm purchases a centrifugal separator for $8 000; it is estimated to have a life of five years. Operating and maintenance costs are anticipated to increase by $500 per year, with the cost for the first year estimated to be $500. Salvage value of $500 is anticipated. Using straight-line depreciation, a 50% income tax rate, and a 10% after-tax MARR, determine the annual equivalent cost of the separator. (5.10)

16. A firm may either invest $10 000 in a numerically controlled lathe, that will last for five years and have zero salvage value at that time, or invest $X in a methods design study. Both investment alternatives yield an increase in income of $4 000 per year for six years. With straight-line depreciation used for all investments in capital assets and with a 50% tax rate, for what value of X will the firm be indifferent between the two investment alternatives? Assume an after-tax MARR of 10%. (5.10)

17. An investment proposal has the following estimated net cash flow before taxes.

EOY	0	1	2	3	4
	−$70 000	42 000	14 000	8000	3000+20 000 salvage value

The effective tax rate is 50% and the after-tax MARR is 10%. Assume the company considering this proposal is profitable in

Depreciation and Income Tax Considerations 275

its other activities. Find the after-tax cash flows and the present worth of those cash flows if the sum of the years' digits depreciation is used. (5.11)

18. An earth-moving company is contemplating an investment of $35 000 in equipment that will have a useful life of five years with $5 000 salvage value at that time. Increases in the firm's income from cut-and-fill contracts due to the investment will total $10 600 per year for five years. The company pays taxes at a rate of 50%. Using sum of the years digits' depreciation and an after-tax MARR of 10%, should the firm undertake the investment? (5.11)

19. An automatic copier is being considered that will cost $8 000, have a life of four years, and no prospective salvage value at the end of that time. It is estimated that because of this venture, there will be a yearly gross income of $8 000. The operating cost during the first year will be $2 000, increasing by $100 each year thereafter. The company contemplating the purchase of this asset has an effective tax rate of 50%. Determine the after-tax cash flow for each year. First use straight-line depreciation, and then use declining balance depreciation. (5.11)

20. Two mutually exclusive alternatives, A and B, are available. Alternative A requires an original investment of $80 000, has a life of five years, zero annual operating costs, and salvage value of $20 000. Alternative B requires an original investment of $120 000, has a life of seven years, annual operating costs of $2 000, and salvage value of $36 000. The after-tax MARR is 10%. A 50% tax rate is to be used. Use an annual worth comparison and recommend the least-cost alternative, using a planning period of

   (a) five years.
   (b) seven years.
   (c) thirty-five years.

For tax purposes, as well as predicting salvage values for incomplete life cycles, use 10% declining balance depreciation. If replacements are required, assume that they have identical cash flow profiles. (5.11)

21. Suppose a back-hoe is purchased for $12 000 and has estimated salvage value of $2 000 at the end of five years. Annual revenues and annual operating costs, excluding depreciation, are $5 000 and $1 000, respectively. If $i = 10\%$ and the firm's tax rate is considered 50%,

    (a) Determine the present worth of "taxes paid" for the five-year period (1) if straight-line depreciation is used, and (2) if the sum of the years' digits depreciation method is used.

    (b) Determine the present worth of the depreciation charges for the five-year period (1) if straight-line depreciation is used, and (2) if the sum of the year's digits depreciation method is used.

    Do your answers from parts (a) and (b) suggest that the present worth of taxes is minimized if the present worth of depreciation charges is maximized? (5.11)

22. The ABC Company is considering the purchase of a computer-controlled printing press that will cost $10 000, have a life of six years, and estimated salvage value of zero. The press will develop a gross income of $4 000 per year. The annual cost connected with the press (exclusive of depreciation) will amount to $500 per year. The company will use straight-line depreciation and the effective tax rate is 50%. If the *MARR* is 15% after taxes, determine the present worth of the after-tax cash flows. Now, assuming that the press is depreciated over four years, although it is used for six years as estimated, determine the present worth of the after-tax cash flows. (5.11)

23. A company has the opportunity to invest in a water purification system requiring $100 000 capital. The firm has only $60 000 available and must borrow the additional $40 000 at an interest rate of 10% per year. The system has a life of four years and estimated salvage value at the end of its life of $0. The before-tax and loan cash flow for each of years 1 to 4 is $25 000. Only the loan interest is paid each year, and the entire principal is repaid at the end of year 4. Assuming that depreciation charges are based on the sum of the years' digits method, the tax rate is 50%, and the firm is able to sell the system for $10 000 at the end of year 4 (note depreciation recapture), construct a table showing the following for each of the four years:

    (a) loan principal payment

Depreciation and Income Tax Considerations

(b) loan interest payment
(c) depreciation charges
(d) taxable income
(e) taxes
(f) after-tax cash flow (5.12)

24. Rework Problem 23, assuming that the loan is paid back as follows:

    (a) in equal annual amounts
    (b) the principal in equal annual amounts plus yearly interest (5.12)

25. Complete the partial table below.

EOY	Before-Tax and Loan Cash Flow	Loan Principal Payment	Loan Interest Payment	Depre- ciation Charges	Taxable Income	Taxes (Rate = 0.50)	After-Tax Cash Flow
0	−$100 000	−$40 000					−$60 000
1	20 000		$4 000	$18 000	?	?	?
2	20 000		4 000	17 000	?	?	?
3	?		?	15 000	?	$5 500	?
4	?		?	?	?	3 000	?
5	40 000		4 000	?	?	13 000	?
6	20 000		4 000	?	?	4 000	?
7	?		4 000	7 000	$19 000	?	?
8	?		4 000	5 000	31 000	?	?
9	45 000		4 000	3 000	?	?	?
10	30 000		4 000	2 000	?	?	?
10		40 000					−40 000

Note 1: There is no uniformity to the before-tax and loan cash flows.
Note 2: Depreciation does not follow any well-known pattern, but totals $100 000.
Note 3: The effective income tax rate is 50%. (5.12)

26. A firm purchases a heat exchanger by borrowing the $10 000 purchase price. The loan is to be repaid with six equal annual payments at an annual compound rate of 7%. It is anticipated that the exchanger will be used for nine years and then be sold for $1 000. Annual operating and maintenance expenses are estimated to be $5 000 per year. Assume sum of years' digits

depreciation, a 50% tax rate, and an after-tax *MARR* of 8%. Compute the after-tax, equivalent uniform annual cost for the heat exchanger. (5.12)

27. A firm borrows $10 000, paying back $2 000 per year on the principal, plus 10% on the unpaid balance at the end of the year. The $10 000 is used to purchase a digital frequency meter that lasts five years and has salvage value of $1 000 at that time. The firm uses 20% declining balance depreciation; it requires a 15% after-tax *MARR* in such analyses, and has a 50% tax rate on taxable income. Determine the after-tax equivalent uniform annual cost for the meter. (5.12)

28. An electronic multiaxis wheel balancer was purchased for $10 000 three years ago. At that time, the estimated life was eight years, with estimated salvage value of $1 000. Today, it appears that the balancer will only last two more years with estimated salvage value of $2 000. The sum of years' digits depreciation schedule will *not* be changed. The effective tax rate is 0.50, the capital gains tax rate is 0.3, and the straight-line depreciation method is being used. Operating expenses are $4 000 per year and gross income is $8 000 per year. Assume that total capital gains are always greater than total capital losses for the company. How much will be paid in taxes for *each of the next two years?* Do not neglect the effect of the capital loss. (5.15)

29. A company had the following capital gains and losses and ordinary taxable income in year 2000:

Capital gains	$25 000
Capital losses	4 000
Ordinary taxable income	80 000

If the company's effective tax rate is 25%, how much federal tax was paid that year? (5.15)

30. A firm purchases a crane for $100 000. Originally, it was estimated that the crane would be retained for nine years and sold for $10 000. However, the crane is used for five years and sold for $40 000. Income less operating and maintenance costs for the crane have been $30 000 per year. Using a 50% income tax rate and straight-line depreciation, compute the

after-tax cash flows for the five-year period. Do not neglect depreciation recapture or capital loss treatment as required. Assume that capital gains are taxed at 0.7 times the normal rate. (5.16)

31. A $39 000 investment in an automobile control panel is proposed. It is anticipated that this investment will cause a reduction in net annual operating disbursements of $12 000 per year for eight years. The investment will be depreciated for income tax purposes by the declining balance method, assuming an eight-year life and $3000 salvage value. The effective tax rate is 50%. A 10% investment tax credit applies at the end of year 0. With a required after-tax return of 10%, what is the equivalent present worth cost for the investment? (5.17)

32. A sum of $10 000 is borrowed at 10% compounded annually and paid back with equal payments over a two-year period. The $10 000 is combined with $15 000 of equity funds to purchase a transformer that has a service life of eight years and terminal salvage value of $3 400. Annual operating and maintenance costs are anticipated to be $5 000.

    A 10% investment tax credit applies in year 0, along with an income tax rate of 50%. Straight-line depreciation is used. The after-tax *MARR* is 20%. Determine the present worth of costs for this machine. (5.17)

33. Rework Problem 32 assuming that the transformer qualifies for additional first-year depreciation. (5.17)

34. A $50 000 heat recovery incinerator is expected to cause a reduction in net out-of-pocket costs of $11 000 per year for 10 years. The incinerator will be depreciated for income tax purposes using straight-line depreciation with a life of 10 years and zero salvage value. A 10% investment tax credit is allowed in time 0. If the tax rate is 50%, and the after-tax *MARR* is 10%, determine whether or not the incinerator should be purchased. (5.17)

35. A firm is considering investing $765 000 in a material handling system that will reduce annual operating costs by $150 000 per year over a 15-year planning period. Perform a

before-tax analysis using a *MARR* of 20%. Is the investment justified? Suppose further analysis indicates $540 000 of the investment is for equipment; the remaining $225 000 is for expense items. A 50% tax rate, 10% investment tax credit at time 0, eight-year equipment write-off period, 20% declining balance depreciation method, and 15% after-tax *MARR* are to be used. Is the investment justified on an after-tax basis? (5.17)

36. A tractor originally costs $10 000. Operating costs for year j are equal to $2000 + 200(j-1)$. A 50% tax rate, 10% investment tax credit in time 0, straight-line depreciation, zero salvage value at all time, 10% minimum attractive rate of return after taxes, and indefinite planning period is to be assumed. Determine the optimum replacement interval. (5.17)

37. A firm is contemplating either purchasing or renting a mainframe computer. If purchased, the computer will cost $400 000 and have annual operating and maintenance costs of $75 000. Because of the obsolescence rate on computing equipment, a five-year study period is chosen. At the end of five years, it is anticipated the computer will have a value of $150 000. The same computer can be leased, with *beginning of year* lease charges of $100 000, which includes maintenance. If leased, the firm will have end-of-year operating costs of $50 000. Using straight-line depreciation, a 50% tax rate, and an after-tax *MARR* of 20%, perform an annual worth comparison to determine if the computer should be leased. (5.18)

38. A firm is considering purchasing a communications system for $200 000. The system is estimated to have a life of nine years and salvage value of $20 000. Operating and maintenance costs are estimated to be $25 000 per year. Declining balance (30%) depreciation is used, and a 10% investment tax credit applies in time 0. Alternatively, the system can be leased for $X at the *beginning* of each year for a nine-year period. If leased, the annual operating and maintenance cost to be paid by the firm reduces to $15 000. Using a 50% tax rate, an after-tax *MARR* of 10%, and annual worth comparison, determine the value of X that will yield equivalent annual costs between buying and leasing. (5.18)

Depreciation and Income Tax Considerations

39. A firm is considering purchasing a supercomputer for $500 000. The computer is estimated to have a life of five years and salvage value of $50 000. End-of-year operating and maintenance costs are estimated to be $60 000 per year. Declining balance (30%) depreciation is used. A 10% investment tax credit applies in time 0. Alternatively, the computer can be leased for $X per year for the five-year period with the lease payment due at the *beginning* of the year. If leased, the annual end-of-year operating and maintenance costs are estimated to be $40 000. Using a 50% tax rate, an after-tax *MARR* of 10%, and an annual worth comparison, determine the value of X that will yield equivalent annual costs between buying and leasing. (5.18)

40. A highway construction firm purchased an item of earth-moving equipment three years ago for $50 000. Assume $2 400 yearly depreciation and use salvage value of 34% of first cost at $t = 10$. The firm earns an average annual gross revenue of $45 000 with the equipment, and the average annual operating costs have been and are estimated to be $26 000.

   Now, $(t = 3)$, the firm has the opportunity to sell the equipment for book value and subcontract the work normally done by the equipment over the next seven years. The firm will not have this opportunity after the present time. If the subcontracting is done, the average annual gross revenue remains $45 000, but the subcontractor charges $35 000 per year for these services.

   If the firm's income tax rate is 50% and a 25% rate of return after taxes is desired, determine whether the firm should subcontract or not by the *annual worth method*. (5.18)

41. The XYZ Construction Company has been subcontracting a particular job at a cost of $30 000 annually, payable at the end of the year. Annual revenues associated with this particular job have averaged $40 000. The agreement with the subcontractor has now expired but could be renewed for another five years. The same revenues and subcontract charges are expected to continue.

   The XYZ Company is also considering the purchase of equipment to handle this same job and terminate the subcontracting. The initial cost of the equipment plus installation expenses is $50 000. The economic life of the equipment is established as 10 years with salvage value of $5 000 at this time. However, the company is only interested in a planning period of five years at which time a salvage value of $20 000 is anticipated. With the equipment, annual revenues are expected to remain at $40 000, but average annual operating expenses expected are $20 000.

The company uses the straight-line depreciation method of accounting. If the company's annual tax rate is 50%, a 10% after-tax rate of return is desired, and a five-year planning period is assumed, should the equipment be purchased or the subcontract agreement be renewed? (5.18)

Chapter 6

Economic Analysis of Projects in the Public Sector

6.1 Introduction

Knowing how to evaluate and select projects to be approved, paid for, and operated by the government is at least as important to today's engineer as a similar knowledge about the private sector. Fortunately, analysis methods for public and private projects are very similar, even though there are some basic, significant differences between the two. The methods considered in this chapter are those most frequently used in evaluating government (federal, provincial, or local) projects—*benefit-cost* and *cost-effectiveness* analysis. More emphasis will be spent on benefit-cost methods, which require that benefits and costs be evaluated on a monetary basis. Although cost effectiveness requires a numerical measure of effectiveness, that measure need not be in terms of money.

6.2 The Nature of Public Projects

There are many types of government projects and many agencies involved. Four classes reasonably cover the spectrum of projects entered into by government. They include cultural development, protection, economic services, and natural resources. *Cultural development* is fostered through education, recreation, radio and television services, historic, and similar institutions. *Protection* is achieved through military services, police and fire protection, and the judicial system. *Economic services* include transportation, power generation, and housing loan programs. *Natural resource* projects might entail wildland management, pollution control, and flood control. Although these are obviously incomplete project lists in each class, it is not so obvious that some projects belong in more than one area. For example, flood control is certainly a form of protection for some, as well as being related to natural resources.

Government projects have a number of interesting characteristics that set them apart from projects in the private sector. Many government projects are huge, having first costs of tens of millions of dollars. They tend to have extremely long lives, such as 50 years for a bridge or a dam. The multiple-use concept is common, as in wildland management projects, where economic (e.g., timber), wildlife preservation, and recreation projects (e.g., camping, hiking) are each considered uses of import for the land. The benefits or enjoyment of government projects are often completely out of proportion to the financial support of individuals or groups. Also, there are almost always multiple government agencies that have an interest in a project. Public sector projects are not easily evaluated, since it may be many years before their benefits are realized. Usually, there is not a clear-cut measure of success or failure, such as the rate of return or net present worth criteria that exists in the private sector.

6.3 Objectives in Project Evalution

If large, complex, lengthy, multiple-use projects of interest to several groups are to be evaluated for their desirability, the criteria for the evalution must first be agreed upon. The setting for modern evaluation of public projects dates back to the US River and Harbor Act of 1902, which "required a board of engineers to report on the desirability of river and harbor projects, taking into account the amount of commerce *benifited* and the *cost.*" Even more applicable to today is the criterion specified in the US Flood Control Act of June 22, 1936, which stated "...that the government should improve or participate...if the *benefits to whomsoever they may accrue* are in excess of the *estimated costs*..." Obviously, the idea of benefit-cost analysis has been around for a long time, even though little has been written about it until recent years.

Benefit-cost analysis is a way to assess the desirability of projects over a long period of time and with respect to the wider effects on the organization mounting the project, or, in the case of projects in the public sector, the community at large. One of the greatest difficulties in applying benefit-cost analysis to public sector projects is to quantify the effects in monetary terms. Positive effects are referred to as *benefits,* while negative effects are *disbenefits*, and all positive and negative effects are included in the analysis. In contrast, the "effects" of primary importance in the private sector are those that relate to income being returned to the organization. *Costs* of construction, financing, operation, and maintenance are estimated in much the same way in both the public and private sectors.

6.4 Benefit-Cost Analysis

The notion of benefit-cost analysis is simple in principle. It follows the same systematic approach used in selecting an economic investment alternative, including the following eight steps from Chapter 4.

1. Define the set of feasible, mutually exclusive alternatives to be compared.
2. Define the planning period to be used.
3. Develop the cost-savings and benefit-disbenefit profiles in monetary terms for each alternative.
4. Specify the interest rate to be used.
5. Specify the measure(s) of merit or effectiveness to be used.
6. Compare the alternatives using the measure(s) of merit on effectiveness.
7. Perform supplementary analyses.
8. Select the preferred alternative.

The heart of many engineering problems is in identifying all of the *alternatives* available to achieve a particular goal. Some alternatives can be excluded from further consideration immediately. Each remaining alternative should then be described thoroughly. This involves specifying all of the good and bad effects that a project will have on the public, including direct effects on people, land values, and the environment. In addition, all aspects of project development, operation, maintenance, and eventual salvage must be stated.

It is essential to define the *planning period* over which the best project is to be selected. If the planning period is longer than the life of a nonrenewable public project, there may be several years during which no benefits or costs are considered. If the planning period is shorter than the life of a project, residual values (same concept as salvage values) may be estimated and used.

Costs and savings are then quantified. This is done by determining all the expenditures made and incomes received for the project during the planning period. Disbursements include all first and continuing costs of a project, while income may result from tolls, fees, or other charges to the user public. Residual or salvage values may exist if the project is believed to still be in operation at the end of the planning period, and these are treated as savings or negative costs at that time. Costs, unlike benefits, are quantified for public projects in a way similar to that used for private projects.

Each *benefit* or *disbenefit* during the planning period must also be quantified monetarily. Benefits, or the positive effects of a government investment, refer to desirable consequences for the public, not to positive effects on any governmental body. Disbenefits are the negative effects on the public. (Again, the effects on the public, and not the effects on any governmental organization are considered.) Unfortunately, placing dollar values on benefits received by a diverse public is often not an easy task. Neither is deciding on the interest rate to be used.

A base measure such as annual equivalent benefits and costs or present worths of benefits and costs is then established, and the *measure of merit* is chosen. Benefit-cost analyses frequently use the benefit-cost ratio (B/C) or, to a lesser extent, a measure of benefits less costs $(B-C)$. If

B_{jt} = public benefits associated with project j during year t, $t = 1, 2, \ldots n$

C_{jt} = governmental costs associated with project j during year t, $t = 0, 1, 2, \ldots, n$

and

i = appropriate interest rate

then the B/C criterion may be expressed mathematically, using a present worth base measure such as,

$$B/C_j(i) = \frac{\sum_{t=1}^{n} B_{jt}(1+i)^{-t}}{\sum_{t=0}^{n} C_{jt}(1+i)^{-t}}$$

(6.1)

Note that the *B/C* ratio is an alternative name for the savings/investment ratio treated in Chapter 4. The *B − C* criterion is expressed as

$$(B-C)_j(i) = \sum_{t=0}^{n} \left(B_{jt} - C_{jt}\right)(1+i)^{-t}$$

(6.2)

which is similar to the present worth method described in Chapter 4.

When two or more project alternatives are being compared using a *B/C* ratio, the analysis should be done on an *incremental basis.* That is, first the alternatives should be ordered from lowest to highest cost (present worth, annual equivalent, etc.). Then, the incremental benefits of the second alternative over the first, $\Delta B_{2-1}(i)$, are divided by the incremental costs of the second over the first, $\Delta C_{2-1}(i)$. That is,

$$\Delta(B/C)_{2-1}(i) = \frac{\Delta B_{2-1}(i)}{\Delta C_{2-1}(i)} = \frac{\sum_{t=1}^{n}(B_{2t} - B_{1t})(1+i)^{-t}}{\sum_{t=0}^{n}(C_{2t} - C_{1t})(1+i)^{-t}}$$

(6.3)

Note that if the first alternative is "do nothing," the incremental *B/C* ratio is also the straight *B/C* ratio for the second alternative. As long as $\Delta(B/C)_{2-1}(i)$ exceeds 1.0, Alternative 2 is preferable to Alternative 1. Otherwise, Alternative 1 is preferred to Alternative 2. The better of these is then compared on an incremental basis with the next most costly alternative. These comparisons of successive pairs continue until all alternatives have been exhausted, and only one "best" project remains. The procedure used is very similar to that specified for the rate-of-return method in Chapter 4.

With the *B − C* criterion, an incremental approach may be used following the same rules as for the *B/C* ratio, but preferring Alternative 2 to Alternative 1 as long as the following condition holds:

$$\Delta(B-C)_{2-1}(i) = \Delta B_{2-1}(i) - \Delta C_{2-1}(i)$$
$$= \sum_{t=1}^{n}\left[B_{2t}(i) - B_{1t}(i)\right](1+i)^{-t}$$
$$-\sum_{t=0}^{n}\left[C_{2t}(i) - C_{1t}(i)\right](1+i)^{-t} \geq 0$$

(6.4)

Where benefits and costs are known directly, the value of $(B-C)_j$ for each Alternative j may be calculated and the maximum value selected.

Apparently straightforward, there are a number of potential pitfalls in the evaluation of government projects. That is, economic analyses of public projects can be easily biased unknowingly by a project evaluator. The most significant of these pitfalls are discussed in subsequent sections.

After the comparison of project alternatives, *supplementary analyses* such as risk analysis, sensitivity analysis, and break-even analysis may be performed. These are often useful in determining how critical various inputs can be and how close various alternatives are economically, and they otherwise quantify much other information that would likely remain intangible.

The final step is *selecting the preferred alternative.* As this selection is made, all quantitative and qualitative supporting considerations should be recorded in detail. This is particularly important in the case of public sector projects because of public accountability.

Example 6.1

We are given the task of choosing one of three highway alternatives to replace a winding, old, dangerous road. The length of the current route is 26 km. Alternative A is to overhaul and resurface the old road at a cost of $2 200 000. Resurfacing will then be required at a cost of $2 million at the end of each 10-year period. Annual maintenance for Alternative A will cost $8 000/km. Alternative B is to cut a new road following the terrain; it will be only 22 km long. Its first cost will be $8 800 000, and surface renovation will be required every 10 years at a total cost of $1 800 000. Annual maintenance will be $8 000/km. Alternative C also involves a new highway which, for practical reasons, will be built along a 20.5 km straight line. Its first cost, however, will be $17 300 000, because of the extensive additional excavating necessary along this route. It, too, will require resurfacing every 10 years at a cost of $1 800 000. Annual maintenance will be $15 500/km. This increase over the maintenance cost of Route B is due to required additional roadside bank retention work.

Our task is to select one of these alternatives, considering a planning period of 30 years, with negligible residual value for each of the highways at that time. One of these alternatives is required, since the old road has deteriorated below acceptable standards. We can calculate the annual equivalent first cost and maintenance cost of each alternative using an interest rate of 6%.

Construction and resurfacing cost:

Route A:

[$2 200 000 + $2 000 000 (P|F 6, 10) + $2 000 000(P|F 6, 20)] (A|P 6, 30)
= [$2 200 000 + $2 000 000 (0.5584) + $2 000 000(0.3118)] (0.0726)
= $286 269 per year

Route B:

[$8 800 000 + $1 800 000 (P|F 6, 10) + $1 800 000(P|F 6, 20)] (A|P 6, 30)
= [$8 800 000 + $1 800 000 (0.5584) + $1 800 000(0.3118)] (0.0726)
= $753 115 per year

Route C:

[$17 300 000 + $1 800 000 (P|F 6, 10) + $1 800 000(P|F 6, 20)] (A|P 6, 30)
= [$17 300 000 + $1 800 000 (0.5584) + $1 800 000(0.3118)] (0.0726)
= $1 370 640 per year

Maintenance cost:

Route A: $\left(\dfrac{\$8\,000}{km}/year\right)(26\ km) = \$208\,000$ per year

Route B: $\left(\dfrac{\$8\,000}{km}/year\right)(22\ km) = \$176\,000$ per year

Route C: $\left(\dfrac{\$15\,500}{km}/year\right)(20.0\ km) = \$317\,750$ per year

Clearly, Route A costs less than Route B, which itself costs less than Route C. Can we now conclude that Route A should be selected?

Absolutely not. So far we have analyzed the cost, which is only one side of the problem. Now, we must attempt to quantify the public benefits along each of the routes.

Traffic density along each of the three routes will fluctuate widely from day to day, but will average 4 000 vehicles per day throughout the year. This volume is composed of 350 light commercial trucks, 250 heavy trucks, 80 motorcycles, and the remainder are automobiles. The average cost of operation per kilometre for these vehicles is $0.20, $0.40, $0.05, and $0.12, respectively.

There will be a time savings because of the different distances along each of the routes, as well as the different speeds that each of the routes will sustain. Route A will allow heavy trucks to average 70 km/h, while other traffic can maintain an average speed of 90 km/h. Routes B and C will allow heavy trucks to average 80 km/h, and the rest of the vehicles can average 100 km/h. The cost of time for all commercial traffic is valued at $18 per vehicle per hour, and for noncommercial traffic, $6 per vehicle per hour. Twenty-five percent of the automobiles and all of the trucks are considered commercial.

Finally, there is a significant safety factor that should be included. Along the old winding road, there has been an excessive number of accidents per year. Route A will probably reduce the number of vehicles involved in accidents to 105, and Routes B and C are expected to involve only 75 and 70 vehicles in accidents, respectively, per year. The average cost per vehicle in an accident is estimated to be $4 400, considering actual physical property damages, lost wages because of injury, medical expenses, and other relevant costs.

We now set about to analyze the various benefits in monetary terms. We have considered savings in vehicle operation, time, and accident prevention. The costs incurred by the public for these items are calculated in the following steps:

Operational costs:

Route A:

$$\left(350 \tfrac{\text{light trucks}}{\text{day}}\right)\left(26 \tfrac{\text{km}}{\text{light truck}}\right)\left(0.20 \tfrac{\$}{\text{km}}\right)\left(365 \tfrac{\text{days}}{\text{year}}\right)$$

$$+\left(250 \tfrac{\text{heavy trucks}}{\text{day}}\right)\left(26 \tfrac{\text{km}}{\text{heavy truck}}\right)\left(0.40 \tfrac{\$}{\text{km}}\right)\left(365 \tfrac{\text{days}}{\text{year}}\right)$$

$$+\left(80 \tfrac{\text{motorcycles}}{\text{day}}\right)\left(26 \tfrac{\text{km}}{\text{motorcycle}}\right)\left(0.05 \tfrac{\$}{\text{km}}\right)\left(365 \tfrac{\text{days}}{\text{year}}\right)$$

$$+\left(3320 \tfrac{\text{automobiles}}{\text{day}}\right)\left(26 \tfrac{\text{km}}{\text{automobile}}\right)\left(0.12 \tfrac{\$}{\text{km}}\right)\left(365 \tfrac{\text{days}}{\text{year}}\right)$$

=$5 432 076 per year

Route B:

$$[\,350(\$0.20) + 250(\$0.40) + 80(0.05) + 3\,320(\$0.12)\,](22)(365)\,/\,\text{year}$$
$$=\$4\,596\,372 \text{ per year}$$

Route C:

$$[\,350(\$0.20) + 250(\$0.40) + 80(0.05) + 3\,320(\$0.12)\,](20.5)(365)\,/\,\text{year}$$
$$=\$4\,282\,983 \text{ per year}$$

Time costs:

Route A:

$$\left(350\,\tfrac{\text{light trucks}}{\text{day}}\right)\left(26\,\tfrac{\text{km}}{\text{light truck}}\right)\left(\tfrac{1}{90}\,\tfrac{\text{hour}}{\text{km}}\right)\left(365\,\tfrac{\text{days}}{\text{year}}\right)\left(18\,\tfrac{\$}{\text{hour}}\right)$$
$$+\left(250\,\tfrac{\text{heavy trucks}}{\text{day}}\right)\left(26\,\tfrac{\text{km}}{\text{heavy truck}}\right)\left(\tfrac{1}{70}\,\tfrac{\text{hour}}{\text{km}}\right)\left(365\,\tfrac{\text{days}}{\text{year}}\right)\left(18\,\tfrac{\$}{\text{hour}}\right)$$
$$+\left(80\,\tfrac{\text{motorcycles}}{\text{day}}\right)\left(26\,\tfrac{\text{km}}{\text{motorcycle}}\right)\left(\tfrac{1}{90}\,\tfrac{\text{hour}}{\text{km}}\right)\left(365\,\tfrac{\text{days}}{\text{year}}\right)\left(6\,\tfrac{\$}{\text{hour}}\right)$$
$$+\left(3\,320\,\tfrac{\text{automobiles}}{\text{day}}\right)\left(26\,\tfrac{\text{km}}{\text{heavy truck}}\right)\left(\tfrac{1}{90}\,\tfrac{\text{hour}}{\text{km}}\right)\left(365\,\tfrac{\text{days}}{\text{year}}\right)$$
$$\times\left(0.25\times 18\,\tfrac{\$}{\text{hour}} + 0.75\times 6\,\tfrac{\$}{\text{hour}}\right)$$
$$=\$4\,475\,665 \text{ per year}$$

Route B:

$$\left[\tfrac{350}{100}(\$18)+\tfrac{250}{80}(\$18)+\tfrac{80}{100}(\$6)+\tfrac{3320}{100}(0.25\times\$18+0.75\times\$6)\right](22)(365))\,/\,\text{year}$$
$$=\$3\,395\,486 \text{ per year}$$

Route C:

$$\left[\tfrac{350}{100}(\$18)+\tfrac{250}{80}(\$18)+\tfrac{80}{100}(\$6)+\tfrac{3320}{100}(0.25\times\$18+0.75\times\$6)\right](20.5)(365))\,/\,\text{year}$$
$$=\$3\,163\,975 \text{ per year}$$

Safety costs:

Route A: $\left(105 \dfrac{\text{vehicles}}{\text{year}}\right)\left(4\,400 \dfrac{\$}{\text{vehicle}}\right) = \$462\,000$ per year

Route B: $(75)(4\,400)\left(\dfrac{\$}{\text{year}}\right) = \$330\,000$ per year

Route C: $(70)(4\,400)\left(\dfrac{\$}{\text{year}}\right) = \$308\,000$ per year

We can summarize all relevant government and public costs as in Table 6.1.

Table 6.1 Summary of Annual Equivalent Government and Pubic Costs

Yearly Cost ($ per year)	Route A	Route B	Route C
Government first cost of highway	286 269	753 115	1 370 640
Government cost of highway maintenance	208 000	176 000	317 750
Public operational costs	5 432 076	4 596 372	4 282 983
Public time costs	4 475 665	3 395 486	3 163 975
Public saftey costs	462 000	330 000	308 000
Total government costs	494 269	929 115	1 688 390
Total public costs	10 369 741	8 321 858	7 754 958

We can compare these three alternative routes using benefit-cost criteria. Let our first criterion be the popular benefit-cost ratio. Since one of these alternatives must be selected, we will assume the lowest government cost alternative, Route A, will be selected unless the extra expenditures for Routes B or C prove more worthy. Since we have not defined "benefits" per se, user benefits will be taken as the incremental reduction in user costs from the more expensive to the less expensive alternatives. Since we are looking at incremental benefits, it makes sense to compare these against the respective incremental costs needed to achieve these additional benefits.

The incremental benefits and costs for Route B as compared to Route A for $i = 6\%$ are given as follows:

$$\Delta B_{B-A}(6) = \text{public costs}_A(6) - \text{public costs}_B(6)$$
$$= \$10\,369\,741 - \$8\,321\,858 = \$2\,047\,883 \text{ per year}$$
$$\Delta C_{B-A}(6) = \text{government costs}_B(6) - \text{government costs}_A(6)$$
$$= \$929\,115 - \$494\,269 = \$434\,846 \text{ per year}$$

That is, for an incremental expenditure of $434 846 per year, the government can provide added benefits of $2 047 883 per year for the public. The appropriate benefit-cost ratio is then

$$\Delta(B/C)_{B-A}(6) = \frac{\Delta B_{B-A}(6)}{\Delta C_{B-A}(6)} = \frac{\$2\,047\,883}{\$434\,846} = 4.71$$

This clearly indicates that the additional funds for Route B are worthwhile, and we prefer Route B to Route A.

Using a similar analysis, we now calculate the benefits, costs, and $\Delta(B/C)$ ratio to determine whether or not Route C is preferable to Route B.

$$\Delta B_{C-B}(6) = \$8\,321\,858 - \$7\,754\,958 = \$566\,900 \text{ per year}$$
$$\Delta C_{C-B}(6) = \$1\,688\,390 - \$929\,115 = \$759\,275 \text{ per year}$$
$$\Delta(B/C)_{C-B}(6) = \frac{\$566\,900}{\$759\,275} = 0.75$$

This benefit-cost ratio, being less than 1.00, indicates that the additional expenditure of $759 275 per year to build and maintain Route C would not provide commensurate benefits to the public. In fact, the user savings would be only $566 900 per year. Of the three alternative routes, Route B is preferred.

The next benefit-cost criterion is based on the fact that if $\Delta(B/C) > 1$ then $\Delta(B-C) > 0$. That is, the difference in incremental benefits and costs may be used in place of the incremental benefit-cost ratio.

Example 6.2

Applying this measure to the routes in the previous example results in the following calculations for comparing Routes A and B. We know that

$$\Delta B_{B-A}(6) = \$2\,047\,883 \text{ per year}$$
$$\Delta C_{B-A}(6) = \$434\,846 \text{ per year}$$

This results in

$$\Delta(B-C)_{B-A}(6) = \Delta B_{B-A}(6) - \Delta C_{B-A}(6)$$
$$= \$2\,047\,883 - \$434\,846$$
$$= \$1\,613\,037 \text{ per year}$$

leaving us again to conclude that Route B is preferable to Route A. Similarly,

$$\Delta(B-C)_{C-B}(6) = \Delta B_{C-B}(6) - \Delta C_{C-B}(6)$$
$$= \$566\,900 - \$759\,275$$
$$= -\$192\,375 \text{ per year}$$

which indicates that Route C is not worthy of the additional expenditure required, and that Route B should be constructed.

The benefit-cost criteria are consistent with the methods of alterntive evaluation presented in Chapter 4. For example, suppose that the decision is to be based on the present worth criterion, minimizing the sum of government construction and maintenance costs as well as public user costs.

Example 6.3

The present worth of all costs for each highway alternative is given by the following formula:

$$PW(\text{total}) = PW_{\text{total government costs}}(i) + PW_{\text{total public costs}}(i)$$

Then:

Route A: $PW_A(6) = \$494\,269(P|A\,6,30) + \$10\,369\,741(P|A\,6,30)$
$= \$494\,269(13.7648) + \$10\,369\,741(13.7648)$
$= \$149\,541\,263$

Route B: $PW_B(6) = \$929\,115(P|A\,6,30) + \$8\,321\,858(P|A\,6,30)$
$= \$929\,115(13.7648) + \$8\,321\,858(13.7648)$
$= \$127\,338\,081$

Route C: $PW_C(6) = \$1\,688\,390(P|A\ 6,30) + \$7\,754\,958(P|A\ 6,30)$
$= \$1\,688\,390(13.7648) + \$7\,754\,958(13.7648)$
$= \$129\,986\,091$

Since the present worth of all costs for Route B is smallest, we again see that this alternative is preferred.

The annual equivalent cost is calculated by simply multiplying a constant times the present worth. Thus, the benefit-cost criteria are also comparable to the annual worth measure of merit.

Example 6.4

Calculating the equivalent uniform annual cost using

$$AW_{total}(i) = AW_{total\ government\ costs}(i) + AW_{total\ public\ costs}(i)$$

we have

Route A:
$AW_A(6) = \$494\,269 + \$10\,369\,741 = \$10\,864\,010$ per year

Route B:
$AW_B(6) = \$929\,115 + \$8\,321\,858 = \$9\,250\,973$ per year

Route C:
$AW_C(6) = \$1\,688\,390 + \$7\,754\,958 = \$9\,443\,348$ per year

Again, Route B is preferred.

Note that the differences in equivalent uniform annual costs for Routes A, B, and B, C are the same as the differences in incremental benefits minus incremental costs for Routes A, B, and B, C. That is,

$$AW_A(6) - AW_B(6) = \$10\,864\,010 - \$9\,250\,973$$
$$= \$1\,613\,037 \text{ per year}$$

$$\Delta B_{B-A}(6) - \Delta C_{B-A}(6) = \$2\,047\,883 - \$434\,846$$
$$= \$1\,613\,037 \text{ per year}$$

and

$$AW_B(6) - AW_C(6) = \$9\,250\,973 - \$9\,443\,348$$
$$= -\$192\,375 \text{ per year}$$

$$\Delta B_{C-B}(6) - \Delta C_{C-B}(6) = \$566\,900 - \$759\,275$$
$$= -\$192\,375 \text{ per year}$$

Finally, the benefit-cost criteria can even be related to the internal rate-of-return approach discussed in Chapter 4. That is, the analyst determines the interest rate that causes the annual equivalent incremental benefits less incremental costs to equal zero.

Example 6.5

The internal rate-of-return approach may be used by finding the value of the interest rate i that equates the incremental benefits of Alternative k over Alternative j to their incremental costs. That is,

$$AW_{k-j}(i) = 0 = \Delta B_{k-j}(i) - \Delta C_{k-j}(i)$$

Using this expression to consider Route B versus Route A, we have

$\Delta B_{B-A}(i)$
$= \$10\,369\,741 - \$8\,321\,858 = \$2\,047\,883$ per year

$\Delta C_{B-A}(i)$
$= [\$8\,800\,000 + \$1\,800\,000(P|F\ i,10) + \$1\,800\,000(P|F\ i,20)](A|P\ i,30)$
$\quad + \$176\,000$
$\quad - \{[\$200\,000 + \$2\,000\,000(P|F\ i,10) + \$2\,000\,000(P|F\ i,20)](A|P\ i,30) + \$208\,000\}$
per year

Searching for the value of i yielding

$$AW_{B-A}(i) = \Delta B_{B-A}(i) - \Delta C_{B-A}(i) = 0$$

we find

$$AW_{B-A}(30) = \$103\,827 \text{ per year}$$
$$AW_{B-A}(35) = -\$226\,726 \text{ per year}$$

The interest rate we are seeking is between 30 and 35%, or approximately 32%. In any event, i easily exceeds 6%, which indicates that Route B is preferable to Route A.

Using the rate-of-return approach, we can also compare Routes B and C.

$\Delta B_{C-B}(i)$
$= \$8\,321\,858 - \$7\,754\,958 = \$566\,900$ per year

$\Delta C_{C-B}(i)$
$= [\$17\,300\,000 + \$1\,800\,000(P|F\,i,10) + \$1\,800\,000(P|F\,i,20)](A|P\,i,30)$
$+ \$317\,750$
$- \{[\$8\,800\,000 + \$1\,800\,000(P|F\,i,10) + \$1\,800\,000(P|F\,i,20)](A|P\,i,30) + \$176\,000\}$
per year

The rate of interest causing $AW_{C-B}(i) = \Delta B_{C-B}(i) - \Delta C_{C-B}(i) = 0$ is between 2.5 and 3%, since

$$AW_{C-B}(2.5) = \$19\,020 \text{ per year}$$
$$AW_{C-B}(3) = -\$8\,520 \text{ per year}$$

Since i is less than 6%, we prefer Route B to Route C.

Several different approaches to the same problem have been illustrated and shown to be consistent project evaluators. Typically, where government projects are involved, one of the benefit-cost criteria $(B/C$ or $B-C)$ is used. More often than not, the criterion used is the benefit-cost ratio. This is unfortunate because, just like rate-of-return analyses in the private sector, the benefit-cost ratio is easy to misuse and misinterpret, and it is very sensitive to the classification of problem elements as "benefits" or "costs." These problems will be discussed later.

The present worth, annual worth, and rate-of-return methods are seldom used in government analyses. Even though the mechanics underlying these approaches are perfectly suitable, their underlying philosophy is more attuned to return on investment, discounted net profits, and minimum attractive rate of return, implying measures of return to *investors* based on capital investment *by the same investors*. Government investments, however, do not necessarily result in *any* monetary incomes to the government and are seldom evaluated solely on the basis of monetary return. Government employees and politicians also prefer benefit-cost analysis, because it allows them to point out specific "benefits" derived by the public as a result of the government's expenditures of the public's money.

6.5 Important Considerations in Evaluating Public Projects

There are a number of pitfalls that can affect benefit-cost analysis of government projects. In fact, opportunities for error pervade benefit-cost analyses from the very initial philosophy through to the interpretation of a B/C ratio. It is important to examine the most significant of these, both to help prevent analysts from erring and to help those who may be reviewing a biased evaluation.

The major topics to be considered are as follows:

1. point of view (federal, provincial, local, individual)
2. selection of the interest rate
3. assessment of benefit-cost factors
4. overcounting
5. unequal lives
6. tolls, fees, and user charges
7. multiple-use projects
8. problems with the ratio

6.5.1 Point of View

The stance taken by the engineer in analyzing a public venture can have an extensive effect on the economic "facts." The analyst may take any of several viewpoints, including those of

1. an individual who will benefit or lose
2. a particular governmental organization
3. a local area such as a city or county
4. a regional area such as a province
5. the entire nation

The first of these viewpoints is not particularly interesting from the standpoint of economic analysis. Nonetheless, all too frequently, an isolated road is paved, a remote stretch of water or sewer line is extended under exceptional circumstances, or a seemingly ideal location for a public works facility is suddenly eliminated from consideration. In these cases, the "benefit-cost analysis," its review, and the implementation decision are usually made from the perspective of a small, select group.

The other four viewpoints are, however, of considerable interest to those involved in public works evaluation. Analyzing projects or project components from viewpoint 2, that of a particular government agency, is analogous to making economic comparisons in private enterprise. That is, only the gains and losses to the organization involved are considered. This viewpoint, which seems contrary to benefit-cost optimization to the public as a whole, may be appropriate under certain circumstances.

Example 6.6

> Consider a Department of Public Works construction project for which the water table must be lowered in the immediate area before work can proceed. Any of several water cutoff or dewatering systems may be employed. Water cutoff techniques include driving a sheet pile diaphragm or using a bentonite slurry trench to cut off the flow of water to the construction area. Dewatering methods include deep well turbines, or wellpoints for lowering the water level. It is sometimes appropriate to evaluate these different techniques from an "organization" point of view, since each of the feasible methods provides the same service or outcome—a dry construction site. Therefore, the most economical decision from the department's standpoint is also correct from the viewpoint of the public as a whole, since the benefits or contributions to the project are the same regardless of the method chosen.

The third point of view, that of a city or county, is popular among local government employees and elected officials. Unfortunately, seemingly localized projects often have an impact on a much wider range of the citzenry than those in the immediate vicinity.

Example 6.7

> County officials are to decide whether future refuse service should be county owned and operated or whether a private contractor should be employed. The job requires front-end loader compaction trucks, as well as roll on-off container capability. These trucks would be travelling primarily along rural roads, and from 1 container (e.g., a roadside picnic area) to 50 containers (e.g., a large rurally located industrial plant) must be collected at each stop. Front-end loader containers range from 2 to $8 m^3$ while roll on-off containers are sized from 15 to $45 m^3$. Several trucks and drivers

will be required, along with a base for operations and maintenance.

The cost in dollars per tonne of refuse collected, removed, and disposed of, is given below:

Personnel services	$ 6.08 per tonne
Materials, supplies, utilities	5.13
Maintenance and repair	5.62
Overhead	3.41
Depreciation	3.61
Five percent interest on the half financed by bonds	1.80
Total county cost	$25.65 per tonne
Federal taxes foregone	$ 1.87 per tonne
Provincial taxes foregone	0.62
Property taxes foregone	1.54
Not necessarily paid by county	$ 4.03 per tonne

County cost to provide refuse service will be $25.65/t. However, the county is not required to pay the additional $4.03/t for federal and provincial taxes and local property taxes, as would a private firm.

It is obvious from the example that a "local" decision can affect a much wider public. Suppose, based on $25.65/t, that the county decided to own and operate the needed refuse service. Federal taxes of $1.87/t that would have been paid by a private contractor will not be paid. Since the federal government will still have the same revenue requirements, the difference will be made up by passing on an infinitesimally small burden to the national public as a whole. Although this is easily rationalized at the local level—to spread a portion of the cost of county refuse service over the entire county—the effect would be great if every town, city, and county took this attitude.

An analogous argument follows for provincial income taxes foregone; however, now the burden is being spread over the people of the province. Even though this is much smaller than the national population, the burden per person to make up the lost provincial tax income is still very small. Again, providing refuse service at less cost to the local

Economic Analysis of Projects in the Public Sector

populace at the expense of the province is tempting from a parochial point of view.

If the county were to plan on not having to pay the property tax, the slack would be taken up by increasing the property tax rates in the county. Although this approach increases the burden on county property owners, that burden may be entirely disproportionate when compared to the refuse service each requires.

The local government's choice of tax alternative will depend on whether it adopts a local, regional, or national point of view. Although a good case can be made for considering the nationwide consequences of a decision, experience indicates that the primary concern of public works officials and politicians is their particular constituency.

Perhaps the best advice for evaluators and decision makers in the public realm is to examine multiple viewpoints, since most projects will affect people and groups at numerous levels. Project evaluators should instead present a thorough analysis clearly indicating benefits and costs from a number of perspectives. Similarly, decision makers should require multiple points of view so they can be aware of the kind and degree of repercussions resulting from their actions.

6.5.2 Selection of the Interest Rate

Acceptable interest rates, discount rates, and minimum attractive rates of return must also be determined before public works projects are evaluated. In the route selection examples, the interest rate was taken at 6% with no question of the appropriateness of such a figure. Clearly, however, the interest rate has a significant effect on the net present worth or annual worth of cash flows in the private sector. Similarly, the interest rate has a significant effect on the net present worth of benefits minus costs or the benefit-cost ratio in public sector analyses.

Example 6.8

Let us consider the highway route selection Examples 6.1 to 6.5 again, this time using a minimum attractive rate of return of 2.5%. Since Route B was preferable to Route A at $i = 6\%$, that decision continues to hold. However, Route C was not preferred to Route B using a *MARR* of 6%. In the rate-of-return approach, it was determined that the incremental "rate of return" of Route C over Route B exceeded 2.5% but fell short of 3%. Now, however, using an interest rate of 2.5%, we find that spending the substantial additional funds for Route C is worthwhile.

One may argue with the above example, saying that $i = 2.5\%$ is lower than would ever be used. Such an argument is probably true for private industry. In public works, however, various knowledgeable individuals have explicitly proposed values on i ranging from 0 to 13%, and rather vague support has been seen ranging from 0 to over 20%. Before considering various schools of thought on the appropriate interest rate, it is helpful to know how public activities are financed.

Financing of Government Projects. There are several different ways that governments finance public sector projects. The most obvious way is, of course, through taxation such as income tax, property tax, goods and services tax, and road user tax. Another popular approach is through the issuance of bonds for either specific projects or general use. Other forms of borrowing, such as notes, may be thought of in the same category as bonds. A third type of fund raising includes income generating activities such as a municipally owned power plant, a toll road, or other activity where a charge is made to cover (or partially offset) the cost of the service performed. Although these are the primary sources of government funds, there are a number of ways in which this money may be passed from one government authority to another by way of direct payments, loans, subsidies, and grants.

Federal funds are raised through tax money and federal borrowing. Federal projects may then be financed through direct payment. In this case, no monetary return is expected by the government; however, the "return" is expressed through the benefits gained by the public. Direct payment financing may be total, as in the case of many public works projects, or partial, for example 90%, as in cost sharing with the provinces for provincial highways.

Financing for projects of national interest and impact may also be available through no-interest or low-interest-rate loans from the federal or provincial government. Both are available for long periods of time, say up to 40 years, with terms obviously more favourable than could normally be expected from conventional sources of money. Such loans are available for financing large projects where revenues resulting from the projects are used to pay back the loans plus interest, if any. There are also occasions when principal payment deferment is permitted during the early years of the project.

Other forms of federal financing include subsidies and federal loan insurance. Subsidies are used to encourage projects or services believed to be in the public's best interest, such as in transportation. Loan insurance is used to eliminate the private lending institution's risk, allowing lower interest loans over longer periods than conventionally available. Insured loans for housing are available through the Canada Mortgage and Housing Corporation, a federal government agency.

Considerations in the Selection of the Interest Rate for the Purposes of Benefit-Cost Analysis. Many arguments about the correct philosophy to use in selecting the interest rate have surfaced over time. For practical purposes, most of these philosophies are aligned to a certain degree with one of the following positions:

1. A zero interest rate is appropriate when tax monies are used for financing.
2. The interest value need only reflect society's time preference rate.
3. The interest rate should match that paid by government for borrowed money.
4. The appropriate interest rate is dictated by the opportunity cost of those investments foregone by private investors who pay taxes or purchase bonds.
5. The appropriate interest rate is dictated by the opportunity cost of those investments foregone by government agencies due to budget constraints.

Advocates of a zero interest rate when tax money is used argue that current taxes (e.g., highway user taxes), requiring no principal or interest payment at all, should be considered "free" money, and no interest or discount rate applied. Counter-arguments point out that a zero (or even low) interest rate will allow very marginal projects or marginal "add-on" project enhancements to achieve a B/C ratio greater than 1. This, in turn, takes money away from other projects that are truly deserving. If, in fact, it is not true that more deserving projects are precluded while very marginal projects (say 1 or 2% "return") are being approved, perhaps there is an excess of money available and taxes should be lowered. A final position against the zero interest rate advocates that if government does not invest funds in high-benefit or "profitable" projects, the people should be able to retain and invest their own money to provide benefits more economically.

Many people back the use of an interest rate that matches that paid by government for borrowed money. This seems reasonable in that government bonds are in direct competition with other investment opportunities available in the private sector. One way of obtaining an interest figure is to determine the rate on "safe" long-term government bonds. Another possibility is to use the rate paid on borrowings by the particular government unit in question. That is, if a project will be financed by borrowing, the rate of interest to be paid is used. If non-borrowed funds are to be used, the appropriate rate is the average rate of interest being paid on long-term (over 15 years) borrowing.

There are also good arguments that the cost assigned to government money may be too low. For example, the opportunities foregone by other government agencies or investors in the private sector may have provided a far higher "return" than approved investments with interest rates equivalent to the cost of government money. In addition, using a rate equal to the cost of government borrowings includes no provision for risk. Finally, it is argued that plain good judgement dictates that projects worthy of approval under higher interest rates are more justifiable in cases where benefits and disbenefits received by the public are nonuniform.

The opportunity cost approach takes into account the rate of return that could have been generated in the private sector with the funds that must be raised by for examle, taxes or bonds, for the public project under consideration.

Consider a situation in which private investments would yield an average annual return of i_1%. Also consider the rate of return on consumption, which is measured by the rate that consumers are willing to pay to consume now instead of later. This rate could be 6% to 14% for a house, 10% to 18% for a new or used car, and so on. Let this rate of return on consumption average out to i_2%. Now, if a fraction α of government financing precludes private investments, while $1-\alpha$ precludes consumption, the rate of return foregone to finance public projects is given as i.

$$i = \alpha i_1 + (1-\alpha)i_2 \qquad 0 \leq \alpha \leq 1 \qquad (6.5)$$

The last philosophy also requires an opportunity cost approach in which an artificial interest rate reflects the rates of return foregone on government projects by virtue of having insufficient funds. That interest rate is found by continuing to increase the value of i until the only projects that remain have $B/C > 1$ or $B - C > 0$, which can be afforded with monies available.

What conclusions can be reached from these philosophical arguments? What guidelines are available for evaluators of public works projects? Clearly, no one answer is universally applicable. But, as a general rule, we recommend using a rate that is at *least* as high as the average effective yield on long-term government bonds. That is, projects or enhancements to projects not providing at least this rate of return should not be undertaken. This is not to say, however, just because a project is attractive at the MARR, that it should be implemented. In an environment of limited funds, only the most worthy projects should be undertaken. These projects may be identified by conducting several analyses of prospective projects, continually raising the interest rate (e.g., $i = 6, 8, 10, 12, \ldots \%$) until only the projects remain that may be accomplished with available or legally borrowable funds.

6.5.3 Assessment of Benefit-Cost Factors

The benefit-cost analyst knows, before starting a study, that placing a monetary figure on certain "societal benefits" may be difficult. Actually, there is even a more fundamental problem—choosing what factors to assess. Some insight can be gained by considering the following four types of factors.

1. *"Internal"* effects are those which accrue directly or indirectly to the individual(s) or organization(s) with which the analyst is primarily concerned.

2. *"External" technological (or real)* effects are those which cause changes in the physical opportunities for consumption or production, for example, effects on navigation and water sport recreation due to a new hydroelectric plant.

3. *"External" pecuniary* effects relate to changes in the distribution of incomes through changes in the prices of goods, services, and production factors. For example, if a firm's expansion is sufficiently large to affect industry prices, an increase in its output is most likely to result in lower prices for output from other firms in the industry as a whole. Many authors agree that these effects can safely be ignored.

4. *Secondary effects* involve changes in the demand for and supply of goods, services, resources, and production factors which *arise from* a particular project. As an example, mining on government lands will bring instant population increases

to nearby small towns. Secondary effects include increasing the incomes of various producers such as vehicle repairmen and barbers who provide the vehicle repairmen with haircuts.

Example 6.9

A dam and reservoir are contemplated in an effort to reduce flood damage to homes and crops in the Red River district of southern Manitoba. Annual damage to property varies from year to year, but averages approximately $230 000 per year. The dam and reservoir contemplated should virtually eliminate damage to the area in question. No other benefits (e.g., irrigation, power generation, recreation) will be provided.

In performing a benefit-cost analysis of the flood control project, the engineer notes that the primary benefit to the public will be the $230 000 per year damage prevented. However, the engineer argues that this will also cut back on money paid to contractors and mechanics for home and car repair, to health care units, for insurance premiums, and so forth. In other words, the building of a dam, it could be argued, will provide disbenefits to those who would normally receive part of their livelihood from helping flood-damaged families. Should the engineer include in the evaluation only the direct benefits to the flood damage victims, or should the other effects be included as well? That depends on how the analyst chooses to handle secondary effects, as indicated in the following discussion.

The disbenefits to those who would lose income if the dam and reservoir were built are considered *secondary effects.* That is, there would be a decrease in the demand for and supply of post-flood restoration goods and services. It is argued that only the *incremental* incomes or incremental profits (losses or lost profits in this case) should be considered when secondary effects are involved.

Another argument calls upon the "ripple" effect of the economy. That is, every secondary effect disbursement by one person or organization is a receipt to another person or organization. Each receipt then contributes to another disbursement, and so on. If the ripple effect were tracked or followed for the secondary effects of a particular alternative, the sum of receipts less disbursements would equal zero, and there would be no economic evaluation.

With either philosophy, the secondary effects of the example's flood control dam and reservoir would be small, if not negligible. This is intuitively reasonable, because the dam's main benefits represent the measure of the direct usefulness of the dam serving its intended purpose, whereas the diseconomies described are, in fact, secondary and diffuse.

Example 6.10

> Now, reconsider the previous flood control dam and reservoir of Example 6.9. Suppose the reservoir would cause a loss of agricultural land for grazing and crops. Should this loss be considered in the benefit-cost analysis? Yes, because it is an external real effect causing changes in the physical opportunities for consumption or production.

External technological effects are often well defined and, in practice, are usually included in benefit-cost analyses. External pecuniary effects, however, may not be vividly apparent, as illustrated in the following example.

Example 6.11

> A large irrigation project is being considered in the heart of the Okanagan Valley. The irrigation will significantly affect the quantity and quality of the fruit grown there. This additional supply of fruit will, however, depress the price of fruit, lowering the profitability of other fruit growers. Also, the same effect will be felt throughout, say, the fruit processing industry, and these firms will likely have to reduce prices. At the same time, producers of complementary goods (items that are used in conjunction with fruit products) will note increased demands, and may increase prices because of an insufficient supply.
>
> Which of these effects would we include in an evaluation of the irrigation project? Of the factors mentioned above, none would be included in the analysis. Each of the effects described relates to changes in the distribution of incomes through changes in the prices of goods, services, and production factors. As such, they are considered external pecuniary effects, which are not "real" benefits or disbenefits and, hence, are not included.

In the above examples, internal and external technological effects were considered to be factors in an analysis, while external pecuniary and secondary effects were not included. There is, of course, no complete agreement on either of these classifications or whether or not they should be evaluated. For the practitioner, good practical judgement is probably the best asset in deciding what factors to include.

As a guide, all identifiable effects of a project should be delineated. Some effects clearly will provide direct benefits or disbenefits to the public that should be counted. There may also be some factors that obviously should not be included. The third group—the controversial factors, if any—should be studied in depth and either included for good reason or be used implicitly as supporting information. The reason for including or not considering these controversial factors should be stated in writing and become a part of the evaluation for the record.

6.5.4 Overcounting

A common disfunction of trying to consider a large variety of effects in a benefit-cost analysis is to overcount, or unknowingly count some factors twice.

Example 6.12

Many years ago, a department of health calculated the following loss from *preventable disability*, hoping to increase interest in health problems.

Individual income loss

Number of wage earners affected	350 000
Average days per year lost due to preventable accidents	4
Average hourly wage	$10.00
Total individual loss per year	$14 000 000

Industry loss:

Loss to industry in disorganization, idle overhead, and lessened production is 2 1/2 times the wage loss, or

Total industry loss per year	$35 000 000

These figures were totally inadequate because of a classic (but not overly apparent) case of double counting. If we assume correctly that the loss to individuals is their foregone earnings, then the loss to industry would be *its* foregone earnings, or its lost profits, and not the entire value of lessened production. As the figures stand above, lost wages were counted once from the viewpoint of the employee, and are again included as part of industry's losses

Example 6.13

Reconsider Example 6.11 involving the irrigation system. The increased quantity of fruit will require that additional employees be hired to work in the canning factory, removing a number of persons from the welfare rolls. The amount of their new wages, equal to the sum of their old welfare payments plus some increase, represents an increase in real output and constitutes a legitimate national benefit of the project. To then add the reduction in welfare payments from the taxpayers to the unemployed would be to double count welfare payments, once from the standpoint of the recipient, and once from the taxpayer's viewpoint.

6.5.5 Unequal Lives

When comparing one-shot public works projects having unequal lives, the planning period will commonly coincide with the longest-lived alternative. This selection of a planning period was recommended in Chapter 4. When some projects are expected to have a long life while others have a shorter life, changes in the discount rate could change the attractiveness of some projects with respect to the alternatives.

Example 6.14

Two projects each have first costs of $200 000, with annual operating costs of $30 000. The life of Project A is 15 years, while that of Project B is 30 years, and benefits accrued to the public are estimated at $60 000 per year and $52 808 per year, respectively. Each is a one-shot project; hence, Project A will have no benefits or costs after year 15. A 30-year planning period will be used. If the MARR is set at 5%, which project is more attractive?

Using the $B-C$ measure of merit and a present worth base, we have

$(B-C)_A(5) = \$60\,000(P|A\ 5,15) - \$30\,000(P|A\ 5,15) - \$200\,000$

$= \$60\,000(10.380) - \$30\,000(10.380) - \$200\,000$

$= \$111\,400$

$$(B-C)_B(5) = \$52\,808(P|A\ 5{,}30) - \$30\,000(P|A\ 5{,}30) - \$200\,000$$
$$= \$52\,808(15.372) - \$30\,000(15.372) - \$200\,000$$
$$= \$150\,605$$

Over a 30-year planning period, and using an interest rate of 5%, Project B is clearly better. Now suppose that insufficient money is available to implement all of the many projects, B included, that a government agency would like to do. A decision is made to consider all projects using an interest rate that reflects the opportunity cost of investments foregone by government agencies. That is, the interest rate is increased until only those projects that can be afforded remain. Assuming the interest rate is up to 12%, now which project is more desirable?

$$(B-C)_A(12) = \$60\,000(P|A\ 12{,}15) - \$30\,000(P|A\ 12{,}15)$$
$$- \$200\,000$$
$$= \$60\,000(6.811) - \$30\,000(6.811) - \$200\,000$$
$$= \$4\,330$$

$$(B-C)_B(12) = \$52\,808(P|A\ 12{,}30) - \$30\,000(P|A\ 12{,}30)$$
$$- \$200\,000$$
$$= \$52\,808(8.055) - \$30\,000(8.055) - \$200\,000$$
$$= -\$16\,282$$

At the higher interest rate, Project A is more favourable because of the increased emphasis on early year net benefits as opposed to the heavily discounted net benefits during the latter years of the planning period.

When a planning period shorter than some project durations is selected, a residual value must be estimated for those projects. The residual value is handled in the same way as a salvage value.

Example 6.15

Reconsidering the previous example and selecting a 15-year planning period, let us now determine the more favourable alternative, assuming Project B has a residual value of 40% of first cost. Let $i = 12\%$.

$$(B-C)_A(12) = \$60\,000(P|A\ 12,15) - \$30\,000(P|A\ 12,15)$$
$$- \$200\,000$$
$$= \$60\,000(6.811) - \$30\,000(6.811) - \$200\,000$$
$$= \$4\,330$$

$$(B-C)_B(12) = \$52\,808(P|A\ 12,15) + 0.4(\$200\,000)(P|F\ 12,15)$$
$$- \$30\,000(P|A\ 12,15) - \$200\,000$$
$$= \$52\,808(6.811) + \$80\,000(0.1827)$$
$$- \$30\,000(6.811) - \$200\,000$$
$$= -\$30\,039$$

6.5.6 Tolls, Fees, and User Charges

Tolls, fees, and user charges have an interesting effect on the fiscal aspects of public projects. If a toll, fee, or user charge is regarded as a payment or partial payment for benefits derived, it can be argued that net benefits received are reduced by the amount of the payment. Similarly, the amount of the payment decreases the cost of the project to the government. Thus, the B/C ratio will change, but the $B-C$ measure of merit will remain constant so long as total user benefits remain constant.

Example 6.16

Suppose 10 000 people per year attend a public facility that has an equivalent uniform annual cost of $20 000. The people, on the average, receive recreational benefits in the amount of $3 each. The B/C ratio would be

$$B/C = \frac{\$3(10\,000)}{\$20\,000} = 1.5$$

and the $B-C$ measure of merit is

$$B-C = \$3(10\,000) - \$20\,000 = \$10\,000 \text{ per year}$$

Based on either criterion, the public facility appears worthwhile. Now suppose that a fee of $1.5 per season is charged. The net benefits are now $3 − $1.5, or $1.5 per person, and the government cost is reduced by $15 000 per year. Thus, the B/C and B−C measures are as follows:

$$B/C = \frac{\$30\,000 - \$15\,000}{\$20\,000 - \$15\,000} = 3$$

$$B - C = \$30\,000 - \$15\,000 - (\$20\,000 - \$15\,000) = \$10\,000 \text{ per year}$$

Note that in the example B/C changed while B − C did not. This phenomenon will be discussed in Section 6.7, "Problems with the B/C Ratio."

It might be concluded that tolls, fees, and user charges are irrelevant, at least with respect to the B − C measure of merit. This, however, *is not* true if the number of users or degree of use is linked to the fee charged, as it almost always will be.

Example 6.17

As an extreme, suppose that the 10 000 users of the public facility in the previous example receive different levels of benefits, but they average out to $3 per person. The actual breakout is that 8 000 persons perceive $1.45 worth of enjoyment, 1 000 persons perceive $3 worth, and 1 000 persons expect to derive $15.40 in recreational benefits. With a user fee of $1.50 per person, only 2 000 will patronize the facility. Thus, the B/C and B−C measure would be:

$$B/C = \frac{1\,000(\$3) + 1\,000(\$15.40) - 2\,000(\$1.50)}{\$20\,000 - 2\,000(\$1.50)}$$

$$= \frac{\$15\,400}{\$17\,000} = 0.91$$

$$B - C = [1\,000(\$3) + 1\,000(\$15.40) - 2\,000(\$1.50)]$$
$$- [\$20\,000 - 2\,000(\$1.50)]$$
$$= -\$1\,600$$

In this case, the reaction of demand to a fee would cause the costs to exceed realized benefits. Thus, when tolls, fees, and user charges are expected, their effect on user demand, and hence total user benefits, must be determined and accounted for.

6.6 Multiple-Use Projects

Multiple-use projects receive a great deal of attention, both pro and con. Multiple uses, and hence multiple benefits, are often available at slight incremental costs over single-use projects. Of course, the incremental capital and net operating costs required for an additional use must provide at least a like worth of benefits.

Example 6.18

A dam and reservoir for irrigation will provide present worth benefits of $25 million over the next 50 years. The present worth cost of construction, operation, and maintenance of the irrigation facility will be $14 500 000. A single-purpose flood control dam providing present worth benefits of $6 million would cost a present worth of $9 million. Suitable design modifications can be made to the irrigation dam and reservoir to provide the flood control benefits too, at a total package present worth cost of $18 500 000. Funds permitting, what should be done?

First, it must be understood that the benefits and costs discounted to the present were discounted at the *MARR* deemed suitable by the decision maker. Assuming this to be true, it is clear that the irrigation project is worthwhile, providing a benefit-cost ratio of

$$B/C_{\text{irrigation}} = \frac{\$25\,000\,000}{\$14\,500\,000} = 1.72$$

As a single-purpose facility, a flood control dam would not provide benefits commensurate with its costs, yielding a *B/C* ratio of

$$B/C_{\text{flood control}} = \frac{\$6\,000\,000}{\$9\,000\,000} = 0.67$$

Note that $B/C_{\text{irrigation}}$ and $B/C_{\text{flood control}}$ are also the incremental *B/C* ratios of irrigation and flood control over doing nothing. As a multiple-use facility, however, the flood control benefits may be provided at a sufficiently low incremental cost to be justifiable. The incremental *B/C* ratio is

$$\Delta B/\Delta C_{\substack{\text{irrigation plus} \\ \text{flood control}}} = \frac{\$6\,000\,000}{\$18\,500\,000 - \$14\,500\,000} = 1.5$$

Thus, a multiple-use facility should be built.

The example illustrates how multiple uses can draw on each other, providing benefits economically that could never have been provided using a single-purpose facility (e.g., the B/C flood control ratio of 0.67). Multiple-purpose projects also have their problems. For example, it is frequently desirable to "allocate" the costs of a project to its various uses.

Example 6.19

> A city's refuse is used to fire a power generation facility owned by a municipality. Not only is electrical power supplied to a segment of the city, but burning of the refuse after processing has virtually eliminated the need for an expensive landfill operation. Since this project is self-supporting, construction, operation, and maintenance costs must be allocated between the disposal and power benefits provided in order to determine the user charge for refuse disposal and electrical energy. Arguments for cost allocation range from (1) no costs should be allocated to refuse disposal, because the refuse is being used in place of fuel and, in fact, a credit should be issued, to (2) refuse disposal should receive sufficient cost allocation to raise rates above those for conventional disposal to include the aesthetic benefits of no unsightly public landfill.[1]

Another problem with multiple-purpose projects is that they arouse opposition by various interest groups on the basis of politics, environment, and the like. Yet another problem has to do with coordinating project finances, in which some aspects may be federally financed with no payback required, other aspects may be self-supporting, and yet others may require provincial funds. Finally, there may be conflicts between the various purposes involved. That is, an enhancement of one use may detract from another use. Nevertheless, multiple-use projects are here to stay and will be encountered by engineers in the public sector.

6.7 Problems with the B/C Ratio

There are two frequent problems with the B/C ratio that require an explanation and warning. Either can give misleading results that may cause an otherwise perfect analysis to point toward the wrong project.

---

[1] These extremes in arguments have actually been used by public officials of one major city.

First, it is sometimes difficult to decide whether an item is a benefit to the public or a cost savings to the government. Similarly, there is often uncertainty between disbenefits and costs.

Example 6.20

A project provides annual equivalent benefits of $100 000, disbenefits of $60 000 per year, and annual costs of $5 000. What is the benefit-cost ratio?

Let us first calculate the B/C ratio for the problem as stated.

$$B/C = \frac{\$100\,000 - \$60\,000}{\$5\,000} = 8$$

Such a high ratio leads one to believe that the project is outstanding.

Now another analyst notes that the government will reimburse those incurring damages from the project in an amount equivalent to $60 000 per year. Thus, the analyst concludes that the public disbenefits have been compensated for by the government and calculates a B/C ratio of

$$B/C = \frac{\$100\,000}{\$5\,000 + \$60\,000} = 1.54$$

which is considerably lower.

Example 6.20 shows that a wide range of B/C ratios reasonably may be obtained on a single project simply by interpreting certain elements of the problem differently. The resolution of this problem is not difficult. The analyst should simply calculate the net benefits less the net costs.

Example 6.21

Let us reconsider Example 6.20 and calculate the annual net benefits less net costs.

The first analyst would have calculated

$$B - C = (\$100\,000 - \$60\,000) - (\$5\,000) = \$35\,000$$

and the second analyst would have calculated

$$B - C = (\$100\,000) - (\$5\,000 + \$60\,000) = \$35\,000$$

Calculating $B - C$ eliminates the inherent bias in the B/C ratio and does not require an incremental approach between alternatives where benefits and costs are known directly for each alternative. That is, if mutually exclusive alternatives are involved, over the same period, the one having the highest $B - C$ value should be selected. Unfortunately, the B/C ratio is by far the more popular criterion of the two.

The other B/C problem was illustrated in the dam and reservoir irrigation example. That is, when the B/C ratio is used, it should be based on *incremental benefits* and *incremental costs*.[2] Simply to calculate the B/C ratio of each alternative and take the one with the largest ratio is incorrect and will frequently lead to errors in project selection.

Example 6.22

In the dam and reservoir of Example 6.18 we calculated the B/C irrigation ratio to be 1.72. To compare this value against a total project $B/C_{\text{irrigation + flood control}}$ ratio of

$$\frac{\$25\,000\,000 + \$6\,000\,000}{\$18\,500\,000} = 1.68$$

would cause us to select irrigation only, in error.

A related error is to require the incremental B/C ratio to be above that for the previous incremental B/C ratio. In the dam and reservoir example, had the incremental B/C ratio, $\Delta B / \Delta C_{\text{irrigation + flood control}} = 1.5$ been compared against the $B/C_{\text{irrigation}} = 1.72$ ratio, again an incorrect conclusion would have resulted. As long as the incremental B/C ratio exceeds 1, the incremental benefits justify the incremental costs. In this regard, the B/C ratio criterion is closely akin to the rate-of-return criterion discussed in Chapter 4.

6.8 Cost-Effectiveness Analysis

To this point it has been assumed that public project effects were measurable, either directly or indirectly, in monetary terms. There are circumstances, however, when project outputs are not measurable

---

[2] Actually, the B/C irrigation ratio in the referenced example is a ratio of incremental benefits to incremental costs where the comparison of pairs is between irrigation and doing nothing.

monetarily and must be expressed in physical units appropriate to the project. In these cases, *cost-effectiveness analysis* has proven to be a useful technique for deciding among projects or systems for the accomplishment of certain goals. Although cost-effectiveness is most often associated with the economic evaluation of complex defence and space systems, it has also proven useful in the social and economic sectors. In fact, it has roots dating back to Arthur M. Wellington's *The Economic Theory of Railway Location* in 1887.

Cost-effectiveness analyses require that common goals or purposes must be identifiable and attainable; that there be alternative means of meeting the goals; and that there must be perceptible constraints for bounding the problem.

Common goals are required in order to have a basis for comparison. For example, it would not make sense to compare a submarine with a communication network. Obviously, alternative methods of accomplishing the goals must be available in order to have a comparison. Finally, reasonable bounds for constraining the problem by time, cost, and/or effectiveness are necessary to limit and better define the alternatives to be considered.

6.8.1 The Standardized Approach

Kazanowski[3] presents 10 standardized steps that constitute a correct approach to cost-effectiveness analyses. They are, in their usual order, presented below.

1. *Define* the *goals*, purpose, missions, etc., that are to be met. Cost-effectiveness analysis will identify the best alternative way of meeting these goals.

2. *State* the *requirements* necesssary for attainment of the goals. That is, state any requirements which are essential if the goals are to be attained.

3. *Develop alternatives* for achieving the goals. There must be at least two alternative ways of meeting or exceeding the goals.

---

[3] Kazanowski, A. D., "A Standardized Approach to Cost-Effectiveness Evaluations." In J. Morley English, ed., Cost Effectiveness. New York: John Wiley, 1968.

4. *Establish evaluation measures* which relate capabilities of alternatives to requirements. Typical measures are performance, availability, reliability, maintainability, etc.

5. *Select* the fixed-effectiveness or fixed-cost *approach*. The fixed-effectiveness criterion is minimum alternative cost to achieve the specified goals or effectiveness levels. Alternatives failing to achieve these levels may either be eliminated or assessed penalty costs. The fixed-cost criterion is the amount of effectiveness achieved at a given cost. "Cost" is usually taken to mean a present worth or annual equivalent of "life cycle cost" which includes research and development, engineering, construction, operation, maintenance, salvage, and other costs incurred throughout the life cycle of the altnernative.

6. *Determine capabilities* of the alternatives in terms of the evaluation measures.

7. *Express* the alternatives and their *capabilities* in a suitable manner.

8. *Analyze* the various *alternatives* based upon the effectiveness criteria and cost considerations. Often, some alternatives are clearly dominated by others and should be removed from consideration.

9. *Conduct a sensitivity analysis* to see if minor changes in assumptions or conditions cause significant changes in alternative preferences.

10. *Document all* considerations, analyses, and decisions from the above nine steps.

Clearly, one cannot become a cost-effectiveness expert by reading these ten steps. They do, however, give an indication of what is involved in a cost-effectiveness study.

Example 6.23

Three different propulsion systems are under consideration. The life-cycle cost is not to exceed $2.4 million. A single effectiveness measure is decided on, that being reliability. Reliability would be defined as "the probability that the propulsion system will perform without failure under given conditions for a given period of time."

Four contractors submit candidate systems, which are evaluated as follows:

Propulsion System	Life-Cycle Cost (Millions)	Reliability
1	2.4	0.99
2	2.4	0.98
3	2.0	0.98
4	2.0	0.97

Since there is only one effectiveness measure, these results may be expressed graphically as shown in Figure 6.1.

If we look at these systems in pairs, comparing just 1 and 2 is equivalent to a fixed-cost comparison in which we prefer system 1 because of its higher reliability. Similar reasoning leads us to prefer system 3 over system 4. If we were to compare only systems 2 and 3, this would be a fixed-effectiveness comparison, in which case we prefer system 3 due to its lower cost. Clearly, system 1 dominates 2, 3 dominates 2, and 3 dominates 4. We are left with making the decision between systems 1 and 3. This choice will depend on whether or not an additional percent reliability justifies the expenditure of an additional $400 000.

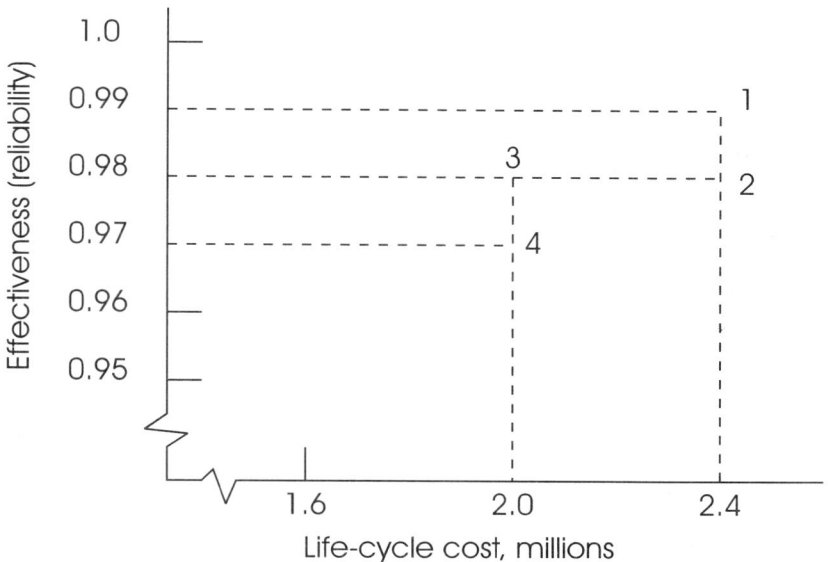

Fig. 6.1 Effectiveness (reliability versus life-cycle cost).

In this example, it is tempting to go the next step and say that based on the ratios of reliability to cost 0.99/$2.4 million and 0.98/$2.0 million, system 3 should be selected. Unfortunately, apparent as such a decision may seem, the correct decision may depend on many other considerations, not the least of which are the payload (perhaps human life) and the consequences of system failure (perhaps total destruction with no payload recovery).

6.9 Summary

Benefit-cost and cost-effectiveness analyses are accepted facts of life in the public sector. It is especially important to engineers to be familiar with these assessment methods, since each year large amounts are being spent on public projects such as road works, railroad networks, bridges, airports, harbour installations, water conservation and irrigation systems, environmental protection systems, communication networks, hospitals, schools and other public buildings, defence installations, research projects, etc., employing large numbers of engineers.

Problems

1. Identify the benefits, both monetary and intangible, that would accrue to the public for the following projects: (a) a museum, (b) a fence-enclosed walkway over a busy four-lane road connecting two city subdivisions, one having a playground and park, (c) a city sanitary system, and (d) a visiting troupe of performers to appear in a downtown city mall for several days. (6.3)

2. Two four-lane roads intersect, and traffic is controlled by a standard green, yellow, red stoplight. From each of the four directions, a left turn is permitted from the inner lane; however, this impedes the flow of traffic while the person who wants to turn left waits until a safe turn may be accomplished. The light operates on a cycle allowing 30 s of green-yellow, followed by 30 s of red light for each direction. Approximately an eighth of the 8 000 vehicles using the intersection daily are held up for one full minute cycle of the light, and made to perform one extra stop-start operation, solely because of the left-turn bottleneck. These delays are only during the 260 working days per year. A stop-start costs 0.5¢ per vehicle. A widening of the

intersection to accommodate a left-turn lane plus a new lighting system to provide for a left-turn signal will cost $48 000. The cost of time for commercial traffic is three times that of private traffic, and private traffic accounts for 90% of the vehicles. How much must commercial traffic time be worth to justify the intersection changes if the interest rate is 7% and a lifetime of 10 years is expected? (6.4)

3. Three alternatives are available, A, B, and C. Their respective annual benefits, disbenefits, costs, and savings are as follows:

	A	B	C
Benefits	$200 000	$300 000	$400 000
Disbenefits	37 000	69 000	102 000
Costs	150 000	234 000	312 000
Savings	15 000	31 000	42 000

(a) Calculate the B/C ratios for each project. Can you tell from these calculations which should be selected?
(b) Determine which should be selected using the incremental B/C ratio.
(c) Calculate B−C for each alternative. (6.4)

4. The following costs and benefits have been listed for a dam and reservoir that will be used for both flood control and electrical power generation.

Investment

Dam, including access roads, clearing and foundation treatment	$31 330 000
Generation equipment and transmission apparatus	16 176 000
Land	2 200 000
Highway relocation	2 770 000
Miscellaneous	180 000

Operating and maintenance costs

Two percent of investment during the first year and increasing by 5% of itself each subsequent year.

Annual benefits
- Flood losses prevented $ 835 000
- Property value enhancement   149 000
- Power value   3 190 000

For a planning period of 50 years and $i = 5\%$, what is the B/C ratio? (6.4)

5. A city is trying to decide among coal, fuel oil 3, low-sulphur fuel oil, and natural gas to power their electrical generators. Fuel forecasts indicate the following needs for the upcoming year and the gradient increase during each subsequent year:

Fuel	First Year	Gradient Each Subsequent Year
Coal	460 000 t	23 000 t
Fuel oil 3	52 800 000 L	2 600 000 L
Low-sulphur fuel oil	54 000 000 L	2 700 000 L
Natural gas	$110\,000 \times 10^6$ L	$5\,500 \times 10^6$ L

The cost of the fuel, transportation, and various pollution effects have been estimated as follows:

Fuel	Cost	Transportation and Storage	Health	Crops	Uncleanliness
Coal	$4.85/t	$2.25/t	$0.70/t	$1.75/t	$1.05/t
Fuel oil 3	1.80/L	0.20/L	0.10/L	0.25/L	0.15/L
Low-sulphur fuel oil	2.25/L	0.20/L	0.03/L	0.075/L	0.045/L
Natural gas	0.00048/L	0.00002/L	Negligible	Negligible	Negligible

Calculate the annual equivalent benefits and costs of these fuels considering $i = 7\%$ and a life of 30 years. Use an incremental B/C analysis to determine the best fuel to use. (6.4)

6. A highway is to be built connecting Baldhill with Broken Arrow. Route A follows the old road and costs $4 million initially and $210 000 per year thereafter. A new Route, B, will cost

$7 million initially and $180 000 per year thereafter. Route C is simply an enhanced version of Route B with wider lanes, shoulders, and so on. It will cost $9 million at first, plus $240 000 per year to maintain. Relevant annual user costs considering time, operation, and safety are $1 million for A, $700 000 for B, and $500 000 for C. Using a MARR of 7%, a 15-year study period, and a residual (salvage) value of 50% of first cost, which should be constructed? Use a B/C analysis. Which route is preferred if $i = 0$? (6.4)

7. Solve problem 6 using the $B - C$ criterion. (6.4)

8. A municipal zoo is to be enlarged, and the initial cost of the enlargement for physical facilities will be $225 000. Animals for the addition will cost another $44 000. Maintenance, food, and animal care will run $31 000 per year. The zoo is expected to be in operation for an indefinite period; however, a study period of 20 years is to be assumed, with a residual (salvage) value of 50% for all physical facilities. Interest is 7%. An estimated 240 000 persons will visit the zoo each year, and they will receive, on the average, an additional $0.35 per person in enjoyment when the new area is complete. Should the new area be built? Use a $B - C$ measure of merit. (6.4)

9. A proposed expressway is under study and is found to have an unfortunately high annual cost of $2 200 000 as compared to benefits of only $1 300 000. However, with only inconsequential changes in design, the highway may be eligible for incorporation into the interprovincial highway system, in which case, 90% of the cost would be paid by the federal government. If so, it is argued that the expressway would cost only $220 000 per year, yielding a handsome B/C ratio of 5.91. Is this reasoning sound? Discuss the reasoning from different points of view. (6.5)

10. Seven projects are available as summarized in the table below. The evaluator is restricted to selecting only the projects that government can afford on an available budget of $75 000 for first cost. Operating and maintenance costs are no worry. All have a favourable B/C ratio at the cost of money, $i = 6\%$. Raise the interest rate until only those projects remain that continue to have a $B/C > 1$ and that government can afford. Use a 10-year planning period. Which projects are selected? What

is the opportunity cost of those investments foregone by government? (6.5)

Project	First Cost	Operating and Maintenance	Residual Value after 10 Years	Benefits per Year
A	$33 000	$3 000	$16 000	$14 400
B	27 000	1 500	8 000	7 630
C	41 000	2 500	12 000	10 800
D	38 000	1 600	14 500	8 460
E	30 000	4 200	16 000	8 600
F	34 000	600	8 000	5 400
G	25 000	5 100	3 000	8 980

11. Many benefits resulting from public sector projects may be argued to be at the expense of someone else, thus counterbalancing the supposed direct benefits. In Example 6.1, operating costs saved include the cost of fuel, oil, tires, wear and tear, and the like. This is money that will *not* be spent at service stations, tire stores, and new car dealers. The safety costs saved include fees that otherwise would go to lawyers, hospitals, doctors, auto repair shops, and so forth. Do you think the fact that the direct savings represent lost revenues to others should nullify these types of benefits in a benefit-cost analysis? Why or why not? (6.5)

12. Three projects, each having a first cost of $1 million and annual operating costs of $100 000, are proposed. The lives of Projects A, B, and C are 20, 30, and 40 years, respectively, after which the project will be over, providing no benefits and requiring no costs. The annual benefits provided over the lives of Projects A, B, and C will be $268 000, $250 000, and $243 000, respectively.

Only one project may be implemented. Use a present worth basis and a $B-C$ measure of merit. First, use a *MARR* of 5%, and a study period of 40 years. Projects A and B will be void during the last 20 and 10 years of the study period, respectively. Second, continually increase the interest rate, again using a study period of 40 years, until only one project remains with a favourable B – C value. Do the two methods point to the same project? Why or why not? (6.5)

13. Rework problem 12 using a value of $i = 12\%$, a planning period of 20 years, and residual values of 20% and 40% of first cost for projects B and C, respectively. (6.5)

14. A recreation area suitable for camping, picnicking, hiking, and water sports is to be developed at an annual equivalent cost of $87 600, including initial cost of construction (e.g., clearing, water, restrooms, road, etc.), operation, upkeep, and security. An average of 60 families will camp each night throughout the year. In addition, another 100 persons will be admitted for free day use of the recreational facilities. The perceived benefits provided to overnight and day users will vary, due to the subjective nature of people. However, the average family camping at the area is estimated to be willing to pay $5 for the privilege, and the average day user $1. What is the B/C ratio of this recreational area? B – C? Should it be built? (6.5)

15. In Problem 14, assume that 50% of the potential camping families perceive $10.50 in benefits, 25% perceive $15 in benefits, and 25% expect to receive $14 worth of recreation. If a charge of $12 is imposed on campers only, and day users are still permitted free, recalculate B/C and B – C. Should the facility be built? (6.5)

16. Reconsider problem 8 and assume that half of the people will derive $0.50 of enjoyment per person, while the other half anticipate only $0.20 in benefits. With this information available, and assuming that an incremental $0.25 entrance fee per person will be charged separately for attendance to the addition, do you recommend building the addition? (6.5)

17. A community must develop a new power generation capability. One alternative is a coal-burning power plant having a first cost of $18 million, a life of 40 years, with no residual value, and operating and maintenance costs of $5 570 000 per year, including delivered fuel. If this alternative is selected, rates per kilowatt hour will be set so as to make the plant just self-supporting. Another alternative is to use a dam and reservoir costing $70 million, with a life of 40 years, and having an annual operating and maintenance cost of

$2 million. Electrical energy benefits would be no greater than for the coal-fired plant, and thus the energy charges would be the same per kilowatt hour. However, if the dam and reservoir are used, for another $1 500 000 first cost and $175 000 per year, a fine boating and recreation area can be developed as an enhancement to the reservoir. Such a facility would have a first cost of $6 million if constructed by itself. It would provide recreational benefits estimated at $900 000 per year for 40 years to the general public. The community has several alternatives from which to choose, including (A) a coal-fired power plant, (B) a coal-fired power plant plus a separate recreational area, (C) a dam and reservoir for power generation only, and (D) a dam and reservoir for power generation and recreation. Use a $B-C$ analysis to decide among these alternatives, assuming that the cost of money is 8%. One of these alternatives must be chosen because of the need for new power generation capability. (6.6)

18. Comment on the following analysis if you think it is incorrect. Benefits of $1.75 each are now received by 9 000 persons and benefits are perceived to be $3 each by another 9 000 users. The annual cost of the recreational facility is $36 000. Commissioners have argued that an entrance fee of $2 should be charged to make the facility self-supporting. They argue that $2 is below the average benefit received per person. They do point out, though, that 9 000 persons perceive only a $1.75 recreational value, and hence a $0.25 disbenefit per person should be noted. The B/C ratio is then

$$B/C = \frac{\$3(9\,000) + \$1.75(9\,000) - \$0.25(9\,000)}{36\,000 - \$2(18\,000)} = \infty$$

and the user fee of $2 should definitely be implemented immediately. What do you think? (6.5)

19. Analysts recognize that with a $2 entrance fee as in Problem 18, only 9 000 persons will utilize the facility, and the benefits derived by 9 000 more persons at $1.75 each will be lost. They argue that the *true* benefits are those received only by 9 000 persons at $3 each, less the $2 fee, less a disbenefit of $1.75(9 000) for the persons who chose not to pay $2 to enter

and lost the previous recreational enjoyment from the facility. As such, the B/C ratio is as follows:

$$B/C = \frac{\$3(9\,000) + \$2(9\,000) - \$1.75(9\,000)}{36\,000 - \$2(9\,000)} = -0.38$$

Do you agree with this analysis? Why or why not? (6.5)

20. A government has the following estimates:

Annual benefits	$250 000
Annual disbenefits	200 000
Annual costs	150 000
Annual savings	145 000

(a) Calculate the B/C ratio.
(b) Mistakenly treating disbenefits as costs and savings as benefits, calculate the B/C ratio.
(c) Calculate B − C.
(d) Which do you prefer, B − C or B/C? Why? (6.7)

21. Six subsystem designs have been proposed for a critical part of a communications network. Each will be "burned in" to eliminate the infant mortality problem. Hence, reliability of the subsystem will have a constant hazard rate. That is, the failure density function will be exponential and reliability may be expressed as

$$R(t) = e^{-\lambda t}$$

where t is the mission time. Mission time for the subsystem will be one year, as a thorough annual preventative maintenance and recalibration procedure is standard, returning the subsystem to "new" condition. The unit will be on call 24 hours a day, seven days a week. The design data are as follows:

System	Annual Equivalent Initial Cost	Annual Maintenance Cost	λ
A	$ 9 000	$1 700	$1.144\,688\,6 \times 10^{-5}$
B	9 800	1 400	$1.098\,901\,1 \times 10^{-5}$
C	10 200	1 800	$1.175\,824 \times 10^{-5}$
D	11 000	1 300	$1.070\,000 \times 10^{-5}$
E	11 000	1 500	$1.081\,250 \times 10^{-5}$
F	13 100	900	$0.998\,000 \times 10^{-5}$

Using cost-effectiveness analysis, plot reliability versus annual cost. Which subsystems can be eliminated from consideration? How might you decide among the rest? (6.8)

22. Four designs are under consideration for a space shuttle designed to carry a particular type of payload. It is estimated that each will have equal lives; however, the first cost, operating cost, and renovation cost, as well as the salvage value, will differ, as will the size of the payload that may be carried. These characteristics are summarized below for shuttles having lives of three years, and fired once per year at the end of the year.

Shuttle Design	First Cost	Operating Cost/Firing	Renovation Cost/Firing	Salvage Value	Payload in Units
1	$18 000 000	$5 000 000	$1 400 000	$2 600 000	176
2	22 000 000	4 400 000	2 400 000	3 100 000	152
3	24 000 000	5 200 000	1 900 000	3 000 000	134
4	30 000 000	3 600 000	1 200 000	5 000 000	168

Based on $i = 7\%$, and using cost-effectiveness analysis, which designs may be eliminated and which should be considered further? (6.8)

Chapter 7

Break-Even, Sensitivity, and Risk Analyses

7.1 Introduction

In the previous chapters we have assumed that all of the parameter values of the economic models were known with certainty. In particular, correct estimates of the values of the length of the planning period, the minimum attractive rate of return, and each of the individual cash flows were assumed to be available. In this chapter, we consider the consequences of having to estimate parameters for which values are not known.

The discussion will concentrate on answering a number of "what if..." questions concerning the effects of different parameter values on the measure of economic effectiveness of interest. For example, when we are completely *uncertain* of the possible values a parameter can take on, we will be interested in determining the set of values for which an investment alternative is justified economically and the set of values for which an alternative is not justified; this process is called *break-even analysis*. Alternatively, when we are reasonably sure of the possible values a parameter can take on, but *uncertain* of their chances of occurrence, we will be interested in the sensitivity of the measure of merit to various parameter values; this process is referred to as *sensitivity analysis*. Finally, when probabilities can be assigned to the occurrence of the various values of the parameters, we can make probability statements concerning the values of the measures of merit for the various alternatives; this process is referred to as *risk analysis*.

The conditions that lead to break-even and sensitivity analyses are described in the economic analysis literature as conditions under *uncertainty*, since one is completely uncertain of the chances of a parameter taking on a given value. When the conditions are such that probabilities can be assigned to the various parameter values, the decision environment that results is said to be a decision under *risk*; hence, the term risk analysis is used.

In Chapter 8, we will present *prescriptive* or *normative* models under uncertainty and risk conditions, models that allow us to *prescribe* the best or optimal course of action using a number of different decision criteria. In this chapter, we present *descriptive* models under uncertainty and risk conditions. Instead of attempting to *prescribe* the action to be taken, we attempt to *describe* the behaviour of the measure of merit under conditions of uncertainty and risk. Thus, we view break-even, sensitivity, and risk analyses as descriptive processes, not normative processes. Our objective in this chapter will be to gain insight into the behaviour of the measure of merit under conditions of uncertainty and risk; the decision on how to respond, given that the information from the sensitivity analysis depends on the decision process employed by the decision maker. Hence, the decision models from Chapter 8 may be used to assist the decision maker in making the final selection.

7.2 Linear Break-Even Analysis

Although we did not label the process as such, in Chapter 3 we performed a number of break-even analyses; there we referred to the process as *equivalence*. In particular, in presenting the concept of equivalence, a situation was posed and the reader was asked to determine the value of a particular parameter in order for two cash flow profiles to be equivalent. When the cash flow profiles for two alternatives are equivalent, a break-even situation can be said to exist between the two alternatives. Therefore, another way of stating the problem could have been: "Determine the value of X that will yield a break-even situation between the two alternatives." In this case, X denotes the parameter whose value is to be determined.

Another application of break-even analysis occurs in the use of the rate-of-return method. Specifically, the rate of return can be interpreted as the break-even value of the reinvestment rate, since such a reinvestment rate will yield a future worth of zero for either an individual alternative or the differences between two alternatives.

Break-even analysis is certainly not an unfamiliar concept. Furthermore, the information obtained from a break-even analysis can be of considerable aid to anyone faced with an investment alternative where some parameter's value is uncertain to any degree.

The *break-even point* is the point on the *break-even chart* (see Figure 7.1) where the performance curves intersect. For example, in Figure 7.1,

Break-Even, Sensitivity, and Risk Analyses

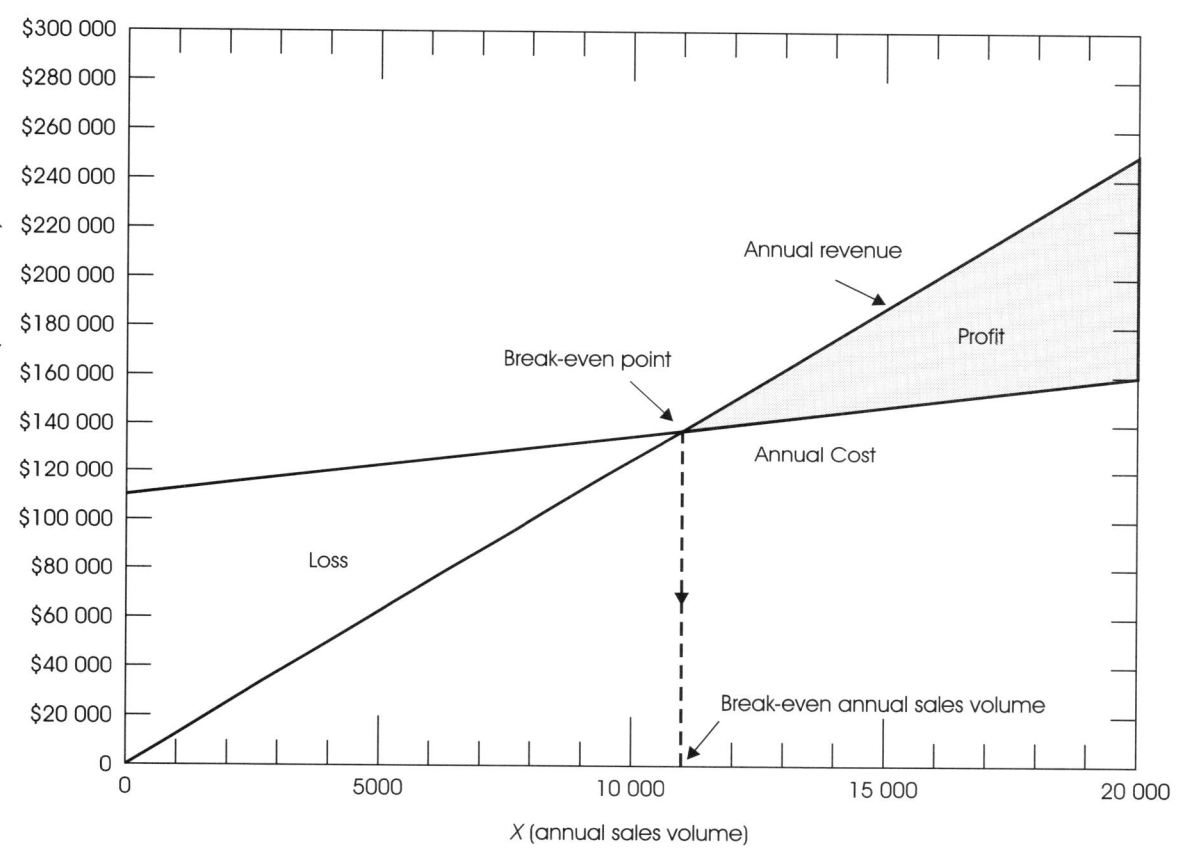

Fig. 7.1 Break-even chart.

the break-even point is the point of intersection of the revenue and cost curves. At this point the annual revenue equals the annual cost, and the profit is zero. The objective of the break-even analysis is to determine the value of a given parameter (e.g., the break-even annual sales volume in Figure 7.1) at the break-even point. The operation represented by the curves on Figure 7.1 is profitable only if the annual sales volume exceeds the break-even value. *Point of indifference* is another term that is sometimes used to denote the break-even point.

Example 7.1

Suppose a firm is considering manufacturing a new product and the following data have been provided:

Sales price	$12.50 per unit
Equipment cost	$200 000
Overhead cost	$50 000 per year
Operating and maintenance cost	$25 per operating hour
Production time	0.1 hours per unit
Planning period	5 years
Minimum attractive rate of return	15%

Assuming zero salvage value for all equipment at the end of five years, and letting X denote the annual sales for the product, the annual worth for the investment alternative can be determined as follows:

$$AW(15\%) = -\$200\,000(A|P\ 15,5) - \$50\,000$$
$$-0.100(\$25)X + \$12.50X$$
$$= -\$109\,660 + \$10.00X$$

Solving for X when $AW(15\%) = 0$ yields *break-even* sales value of 10 966 units per year. If it is felt that annual sales of at least 10 966 units can be achieved each year, then the alternative appears to be worthwhile economically. Even though one does not know with certainty how many units of the new product will be sold annually, it is felt that the information provided by the break-even analysis will assist management in deciding whether or not to undertake the new venture.

A graphical representation of the example is given in Figure 7.1. The chart is referred to as a *break-even* chart, since one can determine graphically the *break-even point* by observing the value of X when annual revenue equals annual cost.

Example 7.2

Consider a contractor who experiences a seasonal pattern of activity for compressors. The company currently owns eight compressors and suspects that this number will not be adequate to meet the demand. The contractor realizes that situations will arise where more than eight compressors will be required, and is considering purchasing an additional compressor for use during heavy demand periods.

A local equipment rental firm will rent compressors at a cost of $25 per day. Compressors can be purchased for $3 000. The difference in operating and maintenance costs between owned and rented compressors is estimated to be $1 500 per year.

Letting X denote the number of days a year that more than eight compressors are required, the following break-even analysis is performed. A planning period of five years, zero salvage values, and 10% minimum attractive rate of return are assumed.

Annual worth (purchasing compressor)

$$AW_1(10\%) = \$3\,000(A|P\ 10,5) - \$1\,500$$
$$= -\$2\,291.40$$

Annual worth (renting compressor)

$$AW_2(10\%) = -\$25X$$

Setting the annual worths equal for the two alternatives yields *break-even value* of $X = 91.656$ days per year.

Hence, if the contractor anticipates that demand will exist for an additional compressor more than 91 days per year over the next five years, then an additional compressor should be purchased.

Table 7.1 Forecasts of Demand for Compressors

X Number of Compressors Demanded on a Given Day	$f(X)$ Number of Days per Year that Demand equals X
≤ 8	140
9	20
10	30
11	30
12	30
13	10
≥ 14	0

N Number of Compressors Owned by the Contractor	Number of Compressor Rental Days if N are Owned	Difference in Number of Compressor Rental Days if N versus $N+1$ Are Owned
8	20(1)+30(2)+30(3)+30(4)+10(5)=340	—
9	30(1)+30(2)+30(3)+10(4) = 220	120
10	30(1)+30(2)+10(3) = 120	100
11	30(1)+10(2) = 50	70
12	10(1) = 10	40
13	0	10

Extending the previous example problem to the question of how many compressors should be purchased, it is interesting to note that compressors should continue to be purchased as long as each one reduces the number of compressor rental days by at least 91 days per year over the five-year period. To illustrate, if the contractor provides a forecast of daily demand for compressors as given in Table 7.1, it is seen that two additional compressors should be purchased.

To show that two more compressors should be purchased, note that if only one additional compressor is purchased, the annual worth will be

$$AW(10\%) = -\$3\,000(A|P\ 10,5) - \$1\,500 - \$25(220)$$
$$= -\$7\,791.40$$

Purchasing two more compressors yields an annual worth of

$$AW(10\%) = -\$6\,000(A|P\ 10,5) - \$3\,000 - \$25(120)$$
$$= -\$7\,582.80$$

Purchasing three more compressors yields an annual worth of

$$AW(10\%) = -\$9\,000(A|P\ 10,5) - \$4\,500 - \$25(50)$$
$$= -\$8\,124.20$$

Hence, two additional compressors should be purchased in order to provide a total of 10 compressors to meet annual demand.

Example 7.3

As a third illustration of break-even analysis consider a manufacturing company that has decided to replace a number of obsolete milling machines by a numerically controlled machining centre for a specific manufacturing application. Two different machines are being considered for purchase and installation:
1. Machine A would cost $500 000, and Machine B $750 000. If Machine A is purchased, the estimated labour cost per unit of production would be $25, and materials would cost $15 per unit.
2. The labour cost would be reduced to $20 per unit of production on Machine B, with material costs remaining at $15. The expected useful life of each machine is five years, with estimated salvage value of $100 000 for each. Determine the range of production for which Machine A is economically

superior to Machine B, based on a minimum attractive rate of return of 15%.

Some of the cost items listed above are identical for the two alternatives: the first $500 000 of the initial machine cost, the first $20 of the unit labour cost, $15 unit material cost, and $100 000 salvage value. These items have no influence on the result of the analysis; therefore, in the calculation only incremental cost items will be considered.

The incremental total production cost for Machine A in a five-year period is

$$\Delta TC_A = (\$5)(P|A\ 15,5)X = \$16.761X$$

where X is the number of units of production. For Machine B the incremental cost is

$$\Delta TC_B = \$250\ 000$$

At break-even volume of production, denoted by X^*, the two incremental cost values are equal; thus,

$$\$16.761X^* = \$250\ 000$$

and $X^* = 14\ 916$ units per year is the break-even point, or point of indifference. The result indicates that Machine A is more economical than Machine B if the yearly quantity of production is below 14 916 units. At a production quantity above 14 916 units per year, Machine B would be superior. For example, at a production of 6 000 units per year: $\Delta TC_A = \$16.761(6\ 000) = \$100\ 566$, and $\Delta TC_B = \$250\ 000$. Thus, purchasing Machine A instead of Machine B would result in a total saving during the five-year operating period of $\Delta TC_B - \Delta TC_A = \$250\ 000 - \$100\ 566 = \$149\ 434$.

7.3 Nonlinear Break-Even Analysis

Performance functions representing the measure of economic effectiveness of the alternatives in terms of the parameters can be linear or nonlinear. For example, consider the cost of machining an aluminum pump casing. The pump casing can be machined on any one of the following machines owned by the company: (1) vertical milling machine, (2) numerically controlled milling machine, (3) special purpose automatic machine. The cost of machining, consisting of fixed and variable costs, is different for each machine. The set-up cost is the only fixed cost if the work is done on the vertical milling machine. The sum of the

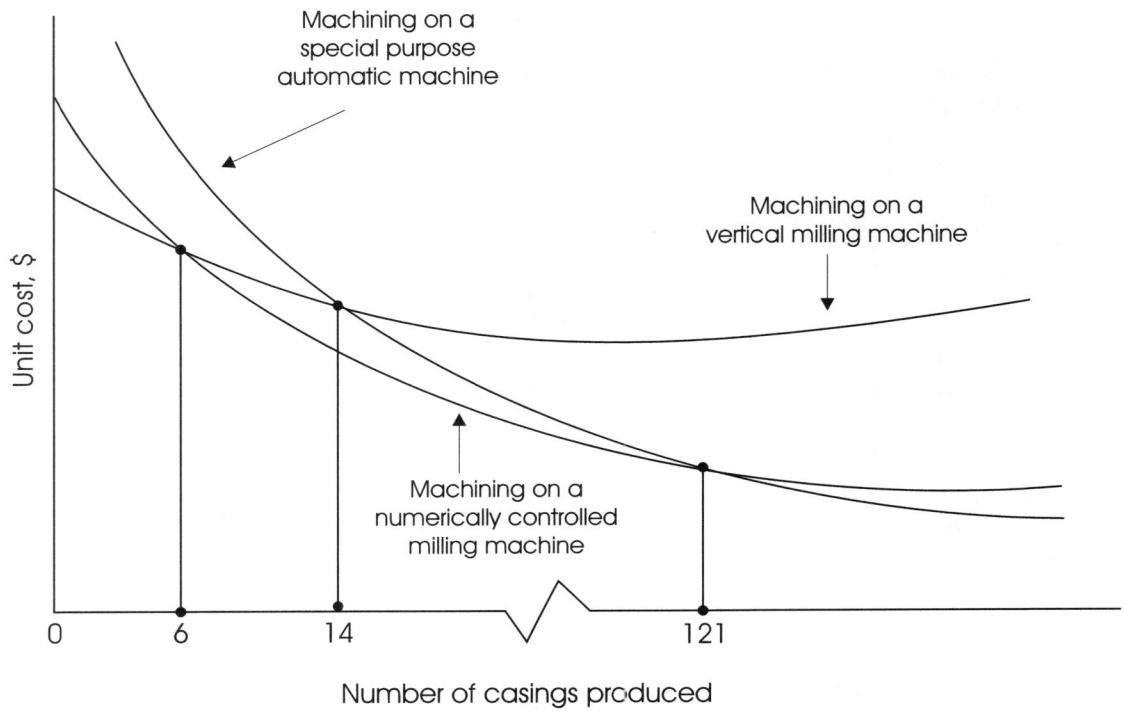

Fig. 7.2 The cost of machining aluminum pump casings.

costs of part programming, control tape preparation, and set-up are the fixed costs if the numerically controlled machine is used. For the automatic machine, the fixed cost includes the cost of special fixtures and cams, in addition to the cost of machine adjustment and set-up.

The variable costs are also different for the three machines. The unit variable cost is high for the vertical milling machine, lower for the numerically controlled machine, and lower still for the automatic machine. The total unit cost of producing a pump casing is a nonlinear function of the total number of casings produced. There are three such unit-cost curves, one for each of the machines considered, and they are depicted in Figure 7.2. The break-even points, where the curves intersect, are at production quantities of 6, 14, and 121 casings. It can be concluded that, in order to keep production cost as low as possible, the vertical milling machine should be used for orders of less than 6 casings. For lots of 6 to 121 casings, the work should be done on the numerically controlled machine, and for batches of more than 121 casings, the automatic machine should be used.

Break-even analysis is frequently used to determine the profitable range of a production process. This is done by comparing the process with a no-action alternative having zero fixed and variable costs, and thus zero profit. Assuming that the profit function of the production process to be analyzed is $P(X)$, where X is the quantity of production, the break-even point is obtained by solving the equation

$$P(X^*) = 0$$

where X^* is the break-even production quantity. If the profit function is nonlinear, more than one break-even point is generally obtained.

Example 7.4

The total annual revenue of a manufacturing plant producing identical small steam turbines of a certain size and type for a specific application is

$$TR(X) = \$(300X - 0.01X^2)$$

where X is the number of turbines manufactured yearly. The total annual cost of production is

$$TC(X) = \$(180\,000 + 100X + 0.01X^2)$$

where \$180 000 are the fixed cost, and $\$(100X + 0.01X^2)$ is the variable cost. Determine the break-even production volume.

The net profit of the operation (before tax) is

$$P(X) = TR(X) - TC(X)$$
$$= \$(200X - 0.02X^2 - 180\,000)$$

At the break-even quantity of production the profit is zero, so that

$$\$[200X^* - 0.02(X^*)^2 - 180\,000] = 0$$

Thus, there are two break-even points of production: $X^* = 1\,000$ or $X = 9\,000$ turbines per year. The operation is profitable if the production quantity is between 1 000 and 9 000 turbines per year. For yearly production below 1 000 or above 9 000 turbines the operation results in a loss. Note that the break-even point with the higher production quantity is normally termed, *profit limit point.*

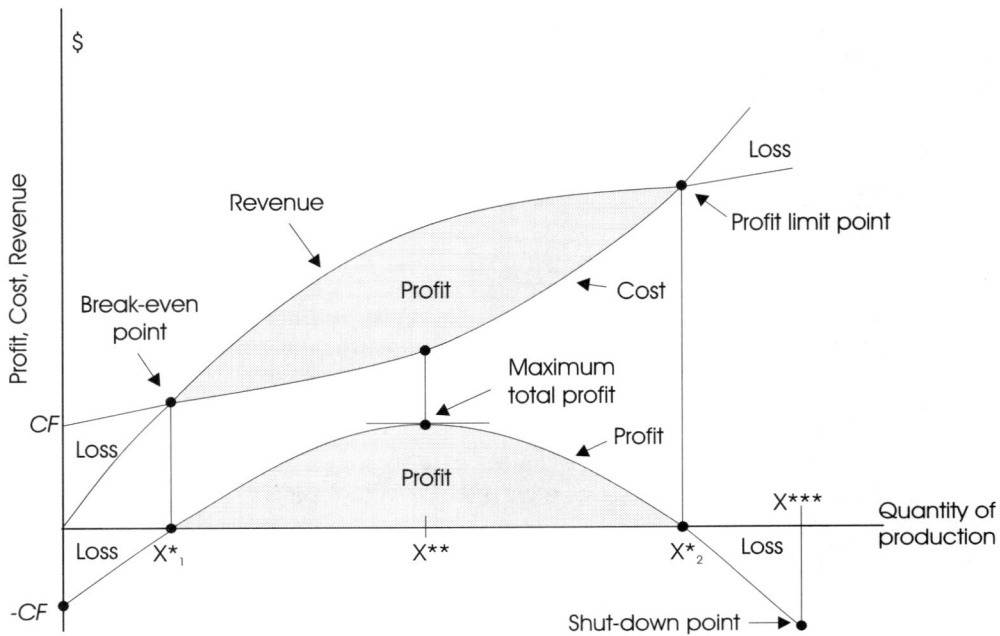

Fig. 7.3 Graphical representation of a non-linear profit function.

A typical nonlinear profit function in terms of the quantity of production is depicted in Figure 7.3. The total profit is maximum at a quantity of production of X^{**}, where the derivative of the profit function $\dfrac{dP(x)}{dx}$ is zero. The derivative of the profit function $\dfrac{dP(x)}{dx}$ is termed the *marginal profit* (or *incremental profit*). It represents the increase (or decrease) in total profit corresponding to a unit increase in production. For linear profit functions the marginal profit is constant; for nonlinear profit functions, it is a variable function of the quantity of production.

Example 7.5

The profit as a function of the quantity of production for the manufacturing operation described in Example 7.4 is

$$P(X) = \$(200X - 0.02X^2 - 180\,000)$$

where X is the number of turbines manufactured yearly. Determine the maximum profit of this operation.

The marginal profit is

$$\frac{dP(X)}{dX} = \$(200 - 0.04X)/\text{unit}$$

and $\frac{dP(X)}{dX}$ is zero at $X^{**} = 5\,000$ turbines per year.

The maximum profit is, therefore,

$$\begin{aligned} P_{max} &= \$P(5\,000) \\ &= \$\left[\,200(5\,000) - 0.02(5\,000)^2 - 180\,000\,\right] \\ &= \$320\,000 \end{aligned}$$

Marginal profit can be positive, zero, or negative. Positive marginal profit indicates that the total profit will increase if production is increased. For instance, in Example 7.5 at a production of 1 000 turbines per year the marginal profit is $\$[\,200 - 0.04(1\,000)\,] = \$160$ per unit, indicating that one unit increase in production at this production level would result in a profit increase of \$160. If the marginal profit is positive, it is advantageous to increase production. On the other hand, if the marginal profit is negative, only a reduction of production will result in an increase of profit (or reduction in loss).

The *average profit*, profit per unit of product, or *unit profit* at a given level of output is the total profit divided by the total production

$$p(X) = \frac{P(X)}{X}$$

The average profit is generally variable with respect to the production quantity, and at the production quantity where the average profit is maximum, the derivative $\frac{dP(X)}{dX}$, termed the *marginal average profit*, is zero.

The profit terminology that we have just discussed can be extended to revenue and cost values as well. The *marginal revenue* is defined as the derivative of the total revenue function $\dfrac{dTR(X)}{dX}$. The total revenue is maximum at the quantity of production where the marginal revenue is zero. The *average revenue* (or *revenue per unit of product*) is the total revenue divided by the total quantity of production:

$$r(X) = \dfrac{TR(X)}{X}$$

and the *marginal average revenue* is the derivative of the average revenue function $\dfrac{dr(X)}{dX}$. The average revenue reaches its maximum value at a production quantity where the marginal average revenue is zero.

The *marginal cost* is the derivative of the total cost function, and the total cost of production is minimum at a quantity of production where the marginal cost is zero. *Unit cost* is also frequently used together with its derivative, the *marginal unit cost.*

Example 7.6

The yearly revenue of a manufacturing company producing small steam turbines is

$$TR(X) = \$(300X - 0.03X^2)$$

Determine the maximum total revenue, and the corresponding revenue per unit of product.

The marginal revenue is

$$\dfrac{dTR(X)}{dX} = \$(300 - 0.06X)/\text{unit}$$

The marginal revenue is zero if $X = 5\,000$ units per year. The maximum revenue is, therefore

$$\begin{aligned} TR_{max} &= \$TR(5\,000) \\ &= \$[300(5\,000) - 0.03(5\,000)^2] \\ &= \$750\,000 \end{aligned}$$

The average revenue is,

$$r(X) = \$\frac{TR(X)}{X}/\text{unit}$$
$$= \$(300 - 0.03X)/\text{unit}$$

and

$$r(5\,000) = [300 - 0.03(5\,000)]/\text{unit}$$
$$= \$150/\text{unit}$$

Another important point on the profit curve is the *shut-down point*. At the shut-down point the yearly loss is equal to the yearly fixed cost. When reaching this point, it is more economical to shut down the plant than to continue production. When production is shut down the yearly loss is limited to the value of the yearly fixed cost.

Example 7.7

The profit of a manufacturing plant producing small steam turbines is,

$$P(X) = \$(200X - 0.02X^2 - 180\,000)$$

Determine the shut-down point.

At the shut-down point the yearly loss is equal to the fixed cost of production; therefore,

$$-P(X) = -\$(200X - 0.02X^2 - 180\,000) = \$180\,000$$

Solving this equation, we obtain $X = 0$ and 10 000 units. If production exceeds 10 000 units, the yearly loss is larger than the fixed cost of production and, therefore, it is more economical to shut down the plant than to continue production. For example, at a production quantity of 11 000 turbines per year the total yearly profit is

$$P(11\,000) = \$[200(11\,000) - 0.02(11\,000)^2 - 180\,000]$$
$$= -\$400\,000$$

That is, there is a $400 000 loss per year. By shutting down the plant, the yearly loss is only $180 000. Alternatively, one can argue that production should be reduced to make the operation profitable.

7.4 Sensitivity Analysis

Break-even analysis is normally used when an accurate estimate of a parameter cannot be provided, but intelligent judgements can be made as to whether or not the parameter's value is less than or greater than some break-even value. Sensitivity analysis is used to analyze the effects of making errors in estimating these parameter values.

Although the analysis techniques employed in break-even and sensitivity analysis are quite similar, there are some subtle differences in the objectives of each. Because of the similarities in the two, however, it is not uncommon to see the terms used interchangeably.

Example 7.8

To illustrate what we mean by sensitivity analysis, consider the investment alternative depicted in Figure 7.4. The alternative is to be compared against the do-nothing alternative, which has zero present worth. Suppose that errors are made in estimating the size of the initial investment ($10 000), the magnitude of the annual receipt ($3 000), the duration of the project (five years), the minimum attractive rate of return (12%), and/or the form of the series of receipts (uniform versus nonuniform). Then, the economic desirability of the alternative might be affected. It is anticipated that the future states (possible values) for each parameter will be contained

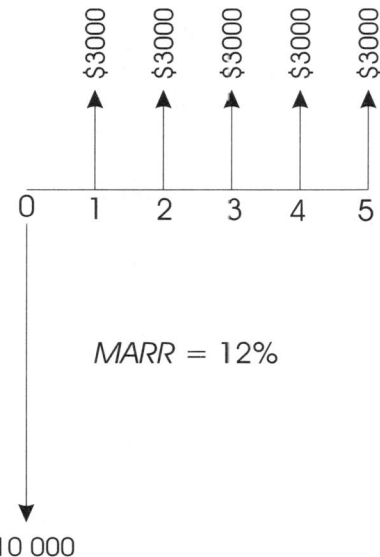

Fig. 7.4 Cash flow diagram for investment alternative.

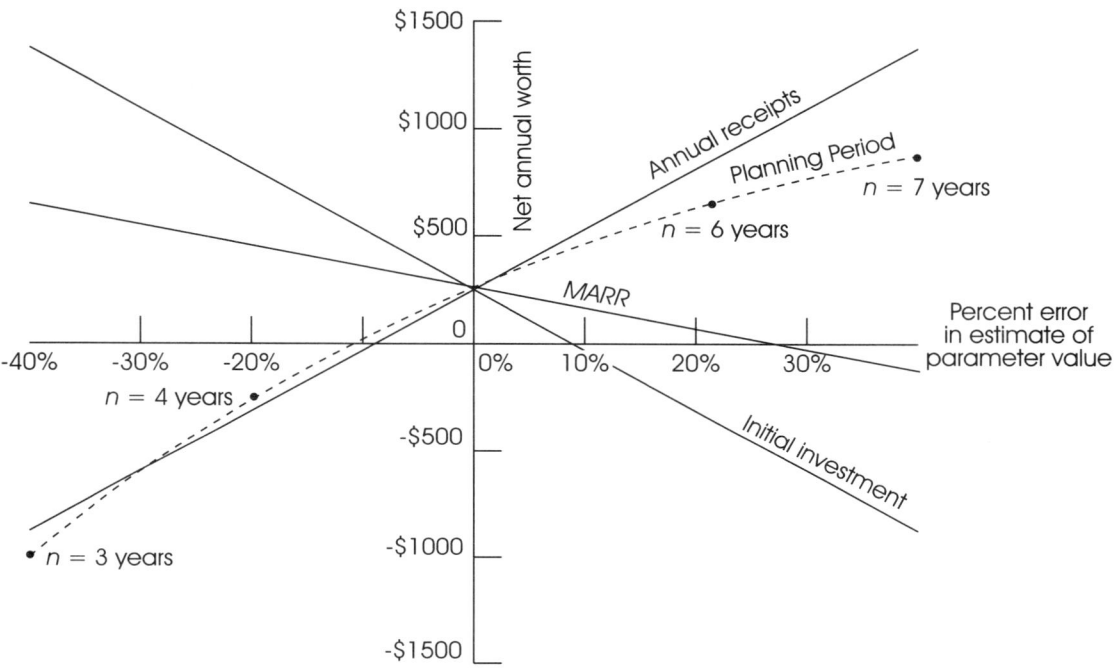

Fig. 7.5 Deterministic sensitivity analysis.

within an interval having a range from −40 to +40% of the initial estimate. Hence in the uncertain environment an infinite number of future states are to be considered in the contnuum from −40 to +40% of the initial estimate.

If it is assumed that all estimates are correct except the estimate of annual receipts, the annual worth for the alternative can be given as

$$AW(12\%) = -\$10\,000(A|P\ 12,5) + \$3\,000(1+X)$$

where X denotes the percent error in estimating the value for annual receipts. Plotting annual worth as a function of the percent error in estimating the value of annual receipts yields the straight line having positive slope shown in Figure 7.5. This line intersects the horizontal axis at −7.47%, indicating that the annual worth of the project becomes zero only if the annual receipts drop by as much as 7.47% to $3\,000(1−0.0747) = \$2\,776$. Performing similar analyses for the initial investment required, the duration of the investment (planning period) and the minimum attractive rate of return yields the results given in Figure 7.5.

As shown in Figure 7.5, the net annual worth for the investment is affected differently by errors in estimating the values of the various parameters. The net annual worth remains positive provided that the error in estimating the *MARR* does not exceed 27.07%, the error in estimating the initial investment does not exceed 8.14%, and the error in estimating the annual receipt is not less than −7.47%. Therefore, the net annual worth is relatively insensitive to changes in the minimum attractive rate of return; in fact, as long as the *MARR* is less than approximately $12(1+0.2707)\% = 15.25\%$, representing a 27.07% error in estimating the nominal 12% *MARR* value, the investment will be recommended. A *break-even* situation exists if either the annual receipts decrease by approximately 7.47% to $2 776 per year or the initial investment increases by approximately 8.14% to $10 814. If the project life is four years or less, the net annual worth is negative; therefore, the investment will not be recommended.

The analysis depicted in Figure 7.5 examines the sensitivity of individual parameters one at a time. In practice, estimation errors can occur for more than one parameter. In such a situation, instead of a sensitivity curve, a sensitivity surface is needed.

Example 7.9

Consider an investment alternative involving the modernization of a warehousing operation in which automated storage and retrieval equipment is to be installed in a new warehouse facility. The building has a projected life of 30 years, and the equipment has a projected life of 15 years. A minimum attractive rate of return of 15% is to be used in the economic analysis. It is anticipated that the new warehouse system will require 70 fewer employees than the present system. Each employee costs approximately $18 000 per year. The building is estimated to cost $2 500 000, and the equipment is estimated to cost $3 500 000. Annual operating and maintenance costs for the building and equipment are estimated to be $150 000 per year more than the current operation. Existing equipment and buildings not included in the new warehouse have terminal salvage values totalling $600 000. Investing in the new warehouse will negate the need to replace existing equipment in the future; the present worth savings in replacement cost is estimated to be $200 000. The estimated cash flow profile for the investment alternative is given in Table 7.2.

A 30-year planning period is to be employed in the analysis. Since equipment life is estimated to be 15 years, it is assumed that identical replacement equipment will be purchased after 15 years. Furthermore, since constant worth dollar estimates are being used to contend with inflationary effects, it is assumed that the replacement equipment will have cash flows that are identical to those that occur during the first 15 years.

Table 7.2. Estimated Cash Flows for Warehouse System

EOY	Building	Equipment	Labour Savings	Operating and Maintenance	Salvage*	Total
0	-$2 500 000	-$3 500 000			$800 000	$5 200 000
1-15			$1 260 000	-$150 000		1 110 000
15		-3 500 000				-3 500 000
16-30			1 260 000	150 000		1 110 000
30	0	0				0

*Includes present worth of savings in equipment replacement.

The architectural and engineering estimate of $2 500 000 for the building is believed to be quite accurate as are the estimates of $150 000 for annual operating and maintenance costs, for terminal salvage values totalling $600 000, and for savings of $200 000 in replacement costs. However, it is felt that the equipment savings estimate of $3 500 000 for equipment and labour savings estimate of 70 employees are subject to error. A sensitivity analysis for these two parameters is to be performed on a before-tax basis.

Letting x denote the percent error in the estimate of equipment cost and y denote the percent error in estimating the annual labour savings, it can be seen that the warehouse modernization will be justified economically if

$$PW = -\$2\,500\,000 - \$3\,500\,000(1+x)[1+(P|F\ 15,15)]$$
$$- \$150\,000(P|A\ 15,30) + \$18\,000(70)(1+y)(P|A\ 15,30)$$
$$+ \$800\,000 \geq 0$$

or if

$$PW = \$1\,658\,110 - \$3\,930\,150x + \$8\,273\,160y \geq 0$$

Solving for y gives,

$$y \geq -0.2004 + 0.47505x$$

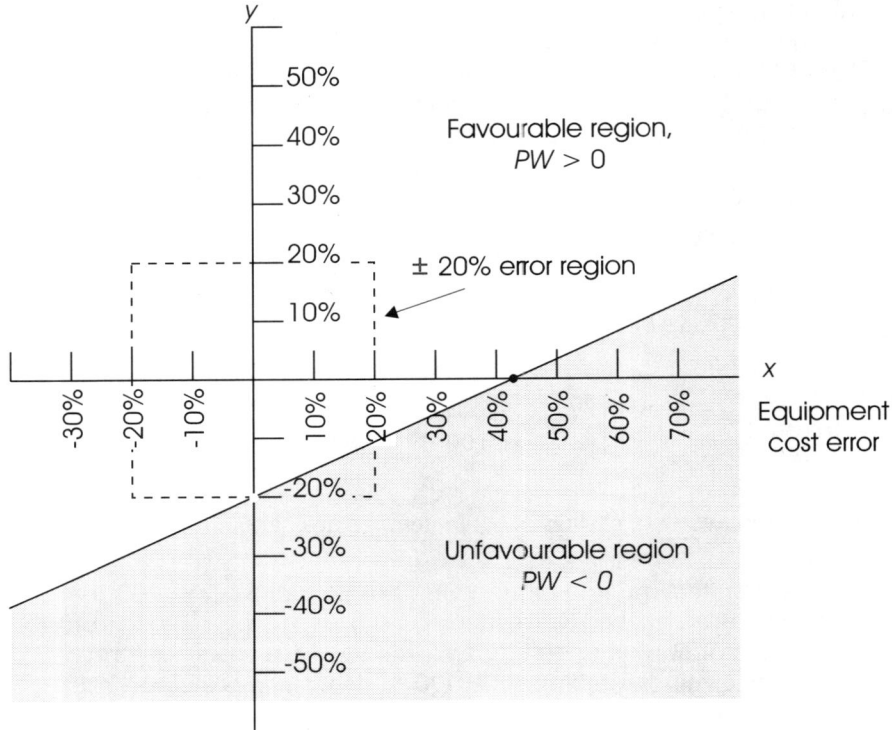

Fig. 7.6 Multiparameter sensitivity analysis.

Plotting the equation, as shown in Figure 7.6, indicates the favourable region ($PW > 0$) lies above the break-even line and the unfavourable region ($PW < 0$) lies below the break-even line.

If no errors are made in estimating the equipment cost (i.e., $x = 0$), then up to a 20.04% reduction in annual labour savings can be tolerated. Likewise, if no errors are made in estimating the annual labour savings (i.e., $y = 0$), then up to a 42.185% increase in equipment cost could occur and the warehouse modernization would continue to be justified economically.

Since very little of the ±20% estimation error zone results in a negative present worth, it appears that the recommendation to modernize the warehouse is insensitive to errors in estimating either equipment cost, labour savings, or both. However, the decision is more sensitive to errors in estimating labour savings than in estimating equipment costs.

This particular example illustrates the difficulties in distinguishing between break-even and sensitivity analyses. However, no matter what you call it, the analysis can be quite beneficial in gaining added understanding of the possible outcomes associated with a new venture involving a large capital investment.

Example 7.10

To illustrate the use of sensitivity analysis in comparing investment alternatives, recall the three mutually exclusive alternatives described in Chapter 4 having cash flow profiles as given in Table 7.3. The comparison performed in Chapter 4 assumed a minimum attractive rate of return of 10%. Suppose some uncertainty exists concerning the appropriate MARR to use in the analysis. As depicted in Figure 7.7, Alternative 3 is preferred if the MARR is less than approximately 17%; Alternative 1 is preferred if the MARR is greater than 17%; and Alternative 2 is never preferred.

Table 7.3
Cash Flow Profiles for Three Mutually Exclusive Investment Alternatives

t	A_{1t}	A_{2t}	A_{3t}
0	$ 0	− $10 000	− $15 000
1	− 12 000	− 9 000	− 9 000
2	− 12 000	− 9 000	− 8 000
3	− 12 000	− 9 000	− 7 000
4	− 12 000	− 9 000	− 6 000
5	− 12 000	− 9 000	− 5 000

In Chapter 4, when the internal rate-of-return method was used to compare the alternatives, the differences in Alternatives 1 and 2 were analyzed, and a rate of return (break-even value) of 15.02% was obtained; next, Alternatives 2 and 3 were compared and a break-even value of 19.48% was obtained. The combination of Alternatives 1 and 3 was not considered when the rate-of-return method was used.

Example 7.10 illustrates two points made earlier: the rates of return obtained from the internal rate-of-return method are actually break-even values for the minimum attractive rate of return for the combinations of alternatives considered; and sensitivity analysis and break-even analysis are closely related. The example will be treated again in Chapter 8.

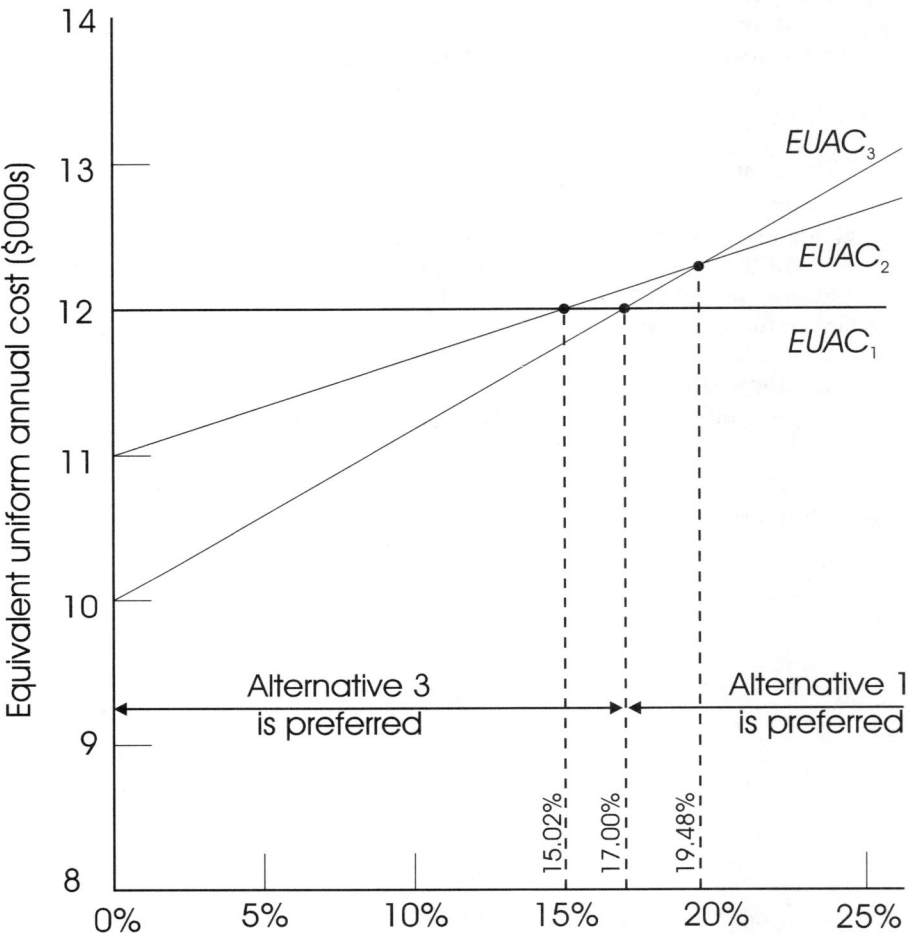

Fig. 7.7 Sensitivity of equivalent uniform annual cost to the *MARR*.

The final approach we consider in performing sensitivity analyses assumes the possible number of values for the parameters can be reasonably represented by three values for each: pessimistic; most likely; and optimistic estimates will be provided. Given the three estimates of the value of each parameter, the value of the measure of merit is determined for each combination of estimates.

Example 7.11

To illustrate the suggested approach, consider once more the example problem depicted in Figure 7.4 involving an initial investment of $10 000 followed by annual receipts of $3 000 for five years. Realizing that any or all of the parameter estimates could be in error, pessimistic, most likely, and optimistic estimates are provided in Table 7.4 for the parameters. As can be seen, there are 3^m combinations to consider, where m is the number of parameters to be included in the sensitivity analysis. Thus, in this case, 81 combinations are to be considered. Four representative combinations are given below.

Table 7.4 Pessimistic, Most Likely, and Optimistic Estimates

	Pessimistic*	Most Likely	Optimistic
Investment	− $12 000	− $10 000	− $8 000
Annual receipts	$ 2 500	$ 3 000	$4 000
Discount rate	20%	15%	12%
Planning period	4 years	5 years	6 years

*Here we interpret the pessimistic value as a value which will lower the present worth; hence, a high discount rate is viewed as a pessimistic value.

$$PW = -\$12\,000 + \$4\,000(P|A\ 15,4) = -\$580$$
$$PW = -\$8\,000 + \$2\,500(P|A\ 20,6) = \$313.75$$
$$PW = -\$10\,000 + \$3\,000(P|A\ 15,4) = -\$1\,435$$
$$PW = -\$12\,000 + \$3\,000(P|A\ 12,5) = -\$1185.60$$

The range of present worth can be obtained by evaluating the totally pessimistic case and the totally optimistic case.

$$PW(\text{pesimistic}) = -\$12\,000 + \$2\,500(P|A\ 20,4) = -\$5\,528.25$$
$$PW(\text{optimistic}) = -\$8\,000 + \$4\,000(P|A\ 12,6) = \$8445.60$$

A tabulation of the present worths for the 81 possible combinations of pessimistic (P), most likely (M), and the optimistic (O) estimates is provided in Table 7.5. The legend (PMOM) indicates that a pessimistic estimate for the investment required, a most likely estimate for annual receipts, an optimistic estimate for the discount rate, and a most likely estimate for the planning period were used to determine the present worth of −$1185.60. Plotting the results in the form of a frequency histogram yields the frequencies provided in Table 7.6.

Table 7.5 Present Worth Values for Combination of Estimates

PPPP	−$5 528.25	MPPP	−$3 528.25	OPPP	−$1 528.28
PPPM	− 4 523.50	MPPM	− 2 523.50	OPPM	− 523.50
PPPO	− 3 686.25	MPPO	− 1 686.25	OPPO	313.75
PPMP	− 4 862.50	MPMP	− 2 862.50	OPMP	− 862.50
PPMM	− 3 619.50	MPMM	− 1 619.50	OPMM	380.80
PPMO	− 2 538.75	MPMO	− 538.75	OPMO	1 461.25
PPOP	− 4 406.75	MPOP	− 2 406.75	OPOP	− 406.75
PPOM	− 2 988.00	MPOM	− 988.00	OPOM	1 012.00
PPOO	− 1 721.50	MPOO	278.50	OPOO	2 278.50
PMPP	− 4 233.90	MMPP	− 2 233.90	OMPP	− 233.90
PMPM	− 3 028.20	MMPM	−1 028.20	OMPM	971.80
PMPO	− 2 023.50	MMPO	− 23.50	OMPO	1 976.50
PMMP	− 3 435.00	MMMP	−1 435.00	OMMP	565.00
PMMM	− 1 943.40	MMMM	56.50	OMMM	2 056.60
PMMO	− 646.50	MMMO	−1 353.50	OMMO	3 353.50
PMOP	− 2 888.10	MMOP	− 888.10	OMOP	1 111.90
PMOM	− 1 185.60	MMOM	814.40	OMOM	2 814.40
PMOO	334.20	MMOO	2 334.20	OMOO	4 434.20
POPP	− 1 645.20	MOPP	354.80	OOPP	2 354.80
POPM	− 37.60	MOPM	1 962.40	OOPM	3 962.40
POPO	1 302.00	MOPO	3 302.00	OOPO	5 302.00
POMP	− 580.00	MOMP	1 420.00	OOMP	3 420.00
POMM	1 408.80	MOMM	3 408.80	OOMM	5 408.80
POMO	3 138.00	MOMO	5 138.00	OOMO	7 138.00
POOP	149.20	MOOP	2 149.20	OOOP	4 149.00
POOM	2 419.20	MOOM	4 419.20	OOOM	6 419.20
POOO	4 445.60	MOOO	6 445.60	OOOO	8 445.60

Table 7.6 Frequency Tabulation of Present Worths

Present Worth	Frequency
−$6 000 to − $5 000.01	1
− 5 000 to − 4 000.01	4
− 4 000 to − 3 000.01	5
− 3 000 to − 2 000.01	8
− 2 000 to − 1 000.01	9
− 1 000 to − 0.01	11
0 to 999.99	10
1 000 to 1 999.99	9
2 000 to 2 999.99	7
3 000 to 3 999.99	6
4 000 to 4 999.99	4
5 000 to 5 999.99	3
6 000 to 6 999.99	2
7 000 to 7 999.99	1
8 000 to 8 999.99	1

Unfortunately, it is difficult to draw any significant conclusions from the data given in Tables 7.5 and 7.6 concerning the parameter that has the greatest effect on present worth. Instead, the use of the three estimates for each parameter provides us with a feel for the possible values of present worth and their relative frequencies. However, we should not place too much weight on the relative frequencies, since we have not verified that each combination of estimates is equally likely to occur. The assignment of probabilities to combinations of parameter estimates lies in the domain of risk analysis and will be treated next.

7.5 Risk Analysis

Risk analysis will be defined as the process of developing probability distributions for some measure of merit for an investment proposal. Typically, probability distributions are developed for either present worth, annual worth, or the rate of return for an individual investment alternative. Consequently, probability distributions are required for random variables such as the cash flows, planning period and discount rate. The probability distributions are then aggregated analytically or through simulation to obtain the desired probability distribution for the measure of merit.

The cash flow occurring in a given year is often a function of a number of other variables such as selling prices, size of the market, share of the market, market growth rate, investment required, inflation rate, tax rates, operating costs, fixed costs, and salvage values of all assets. The values of a number of these random variables can be correlated with each other, as well as autocorrelated.[1] Consequently, an analytical development of the probability distribution for the measure of merit is not easily achieved in most real-world situations. Thus, simulation is widely used in performing risk analyses. Risk analysis has gained acceptance in a large number of industries. One industry survey indicated that over 50% of the companies responding to the survey were using risk analysis for operational and/or strategic planning.

By incorporating the concept of utility theory, it is possible to extend risk analysis and a normative or prescriptive model. To be more specific, by combining a manager's utility function and the probability distribution for, say, present worth, it is possible to specify the alternative that maximizes expected utility. Our discussion will concentrate on the contributions of risk analysis as a descriptive technique, not a normative technique.

7.5.1 Distributions

The risk analysis procedure was developed to take into consideration the imprecision in estimating the values of the inputs required for economic evaluations. The imprecision is represented in the form of a probability distribution.

Probability distributions for the random variables are usually developed on the basis of subjective probabilities. Typically, the further an event is into the future, the less precise is our estimate of the value of the outcome. Hence, by letting the variance reflect our degree of precision, we would expect the variance of the probability distributions to increase with time.

---

[1] The term autocorrelated means correlated with itself over time.

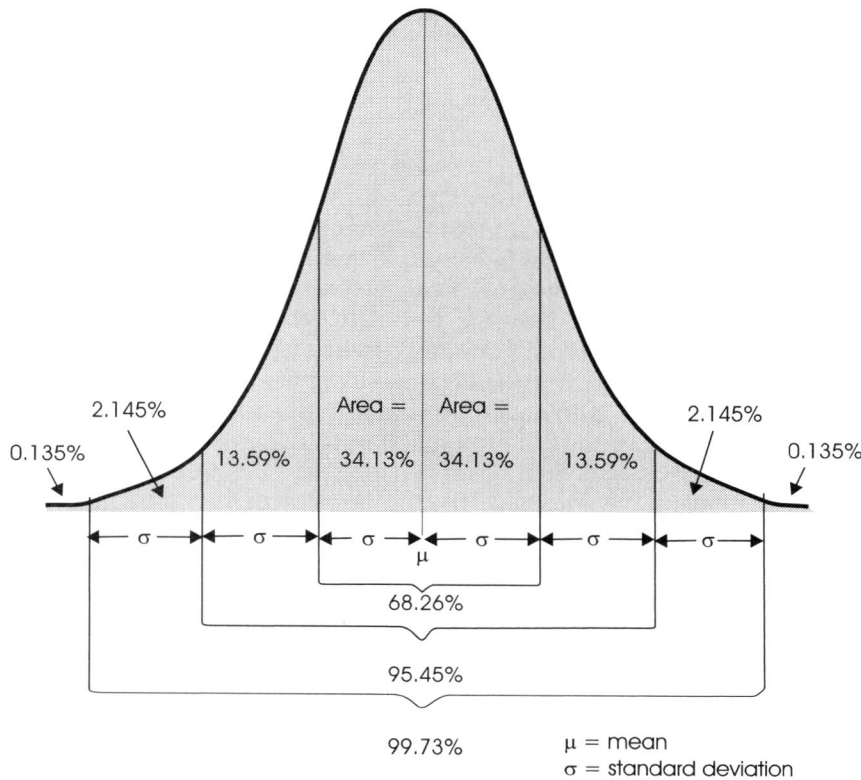

Fig. 7.8 Normal distribution

Among the theoretical probability distributions commonly used in risk analysis are the normal distribution and the beta distribution. The normal distribution is depicted in Figure 7.8. In some situations, the subjective probability distribution cannot be represented accurately using a well-known theoretical distribution. Instead, one must estimate directly the probability distribution for the random variable. One approach that can be used to estimate the subjective probability distribution is to provide optimistic, pessimistic, and most likely estimates for the random variable. The optimistic and pessimistic values should be ones that you do not anticipate will be exceeded with a significant probability (e.g., 1% chance). Given the practical limits on the range of values anticipated for the random variable, an estimate is provided of the chance that the most likely estimate will not be exceeded. Next, a smooth curve is passed through the three points obtained; the resulting cumulative distribution function for the random variable is divided into an appropriate number of intervals and the individual probabilities estimated.

Example 7.12

To illustrate the process, suppose we wish to develop the probability distribution for, say, the anticipated salvage value of a machine in five years. We estimate that the salvage value will range from $0 to $3 000, with the most likely value being $1 250. We estimate that there is a 40% chance of salvage value being less than $1 250. Since we believe the extreme values of $0 and $3 000 are not likely to occur, we use an S-shaped curve to represent the cumulative distribution function, as depicted in Figure 7.9(a). Using intervals of $500, the cumulative distribution function is transformed into the probabilty distribution function shown in Figure 7.9(b). By letting the midpoints of the intervals represent the probabilities associated with the intervals, the probability density function is given as in Figure 7.9(c). Depending on the use to be made of the probabilities obtained, any of the three representations of the probability distribution could be used.

The process described above is quite subjective, since a number of different curves could be used to describe the cumulative distribution function. However, the probability distribution itself is subjectively based. What is sought is a probability distribution that best describes one's beliefs about the outcomes of the random variable. If the probability distribution obtained above does not reflect your beliefs, the process should be repeated until a satisfactory probability distribution is obtained.

The critics of risk analysis cite the degree of subjectivity involved in developing probability distributions. However, it is argued by those who favour the technique that the only alternative to using subjectively based probabilities is to use the traditional single estimate approach, which implies that conditions of certainty exist. Interestingly, the final distribution obtained for the measure of merit is often quite insensitive to deviations in the shapes of the distributions for the parameters.

7.5.2 Risk Aggregation

Given the essential factors and their associated probability distributions, we are in a position to aggregate the distributions and obtain the probability distribution for the measure of merit. Three measures of merit have been mentioned: present worth, annual worth, and rate of return. In practice, a combination of the rate of return and either the

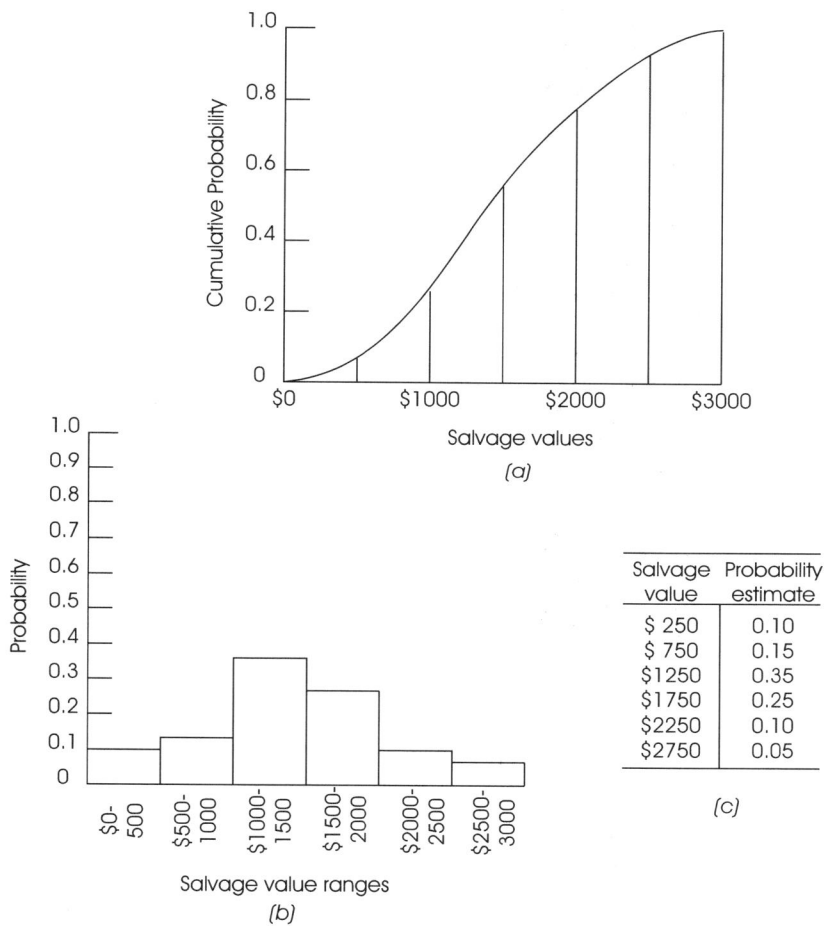

Fig. 7.9 Developing the subjective probability distribution.

present worth or annual worth measures of merit are often used. Each method of comparing investment alternatives has its limitations. For example, if the rate of return is being determined, the external rate-of-return method is recommended, since multiple rates of return can arise when using the internal rate-of-return method.

Risk aggregation is achieved in basically two ways: analytically and by using simulation. Analytic approaches can be used in a number of simple cases. For more complex situations involving a large number of variables, simulation is used.

Analytic Approaches. As an illustration of the use of analytic approaches in developing the probability distribution for present worth, consider the following present worth relation:

$$PW = \sum_{j=0}^{n} C_j (1+i)^{-j}$$

(7.1)

Suppose the cash flows, C_j, are *random variables* with expected values $E(C_j)$ and variances $V(C_j)$. Since the *expected value* of a sum of random variables is given by the sum of the expected values of the random variables, the expected present worth is given by

$$E(PW) = \sum_{j=0}^{n} E\left[C_j (1+i)^{-j}\right]$$

(7.2)

Furthermore, the expected value of the product of a constant and a random variable is given by the product of the constant and the expected value of the random variable,

$$E\left[C_j (1+i)^{-j}\right] = E(C_j)(1+i)^{-j}$$

(7.3)

Substituting Equation 7.3 into Equation 7.2 yields

$$E(PW) = \sum_{j=0}^{n} E(C_j)(1+i)^{-j}$$

Hence, we see that the expected present worth of a series of cash flows is found by summing the present worths of the expected values of the individual cash flows.

To determine the variance of the present worth, we first recall that if X_1, X_2, X_3 and Y are random variables related as follows,

$$Y = X_1 + X_2 + X_3$$

(7.4)

then the variance of Y is given by

Break-Even, Sensitivity, and Risk Analyses

$$\boxed{\mathrm{Var}(Y) = E(Y^2) - [E(Y)]^2} \tag{7.5}$$

or

$$\boxed{\mathrm{Var}(Y) = E\left[(X_1 + X_2 + X_3)^2\right] - [E(X_1 + X_2 + X_3)]^2}$$

Expanding and collecting terms yields

$$\begin{aligned}
\mathrm{Var}(Y) &= E\left(X_1^2 + X_2^2 + X_3^2 + 2X_1X_2 + 2X_1X_3 + 2X_2X_3\right) \\
&\quad - [E(X_1) + E(X_2) + E(X_3)]^2 \\
&= E\left(X_1^2\right) + E\left(X_2^2\right) + E\left(X_3^2\right) + 2E(X_1X_2) \\
&\quad + 2E(X_1X_3) + 2E(X_2X_3) - [E(X_1)]^2 \\
&\quad - [E(X_2)]^2 - [E(X_3)]^2 - 2E(X_1)E(X_2) \\
&\quad - 2E(X_1)E(X_3) - 2E(X_2)E(X_3) \\
&= E\left(X_1^2\right) - [E(X_1)]^2 + E\left(X_2^2\right) - [E(X_2)]^2 \\
&\quad + E\left(X_3^2\right) - [E(X_3)]^2 \\
&\quad + 2[E(X_1X_2) - E(X_1)E(X_2)] \\
&\quad + 2[E(X_1X_3) - E(X_1)E(X_3)] \\
&\quad + 2[E(X_2X_3) - E(X_2)E(X_3)]
\end{aligned} \tag{7.6}$$

Since $E\left(X_k^2\right) - [E(X_k)]^2$ defines the variance of the random variable X_k and $E(X_pX_k) - E(X_p)E(X_k)$ defines the covariance of the random variables X_p and X_k, we see that

$$\boxed{\begin{aligned}\text{Var}(Y) &= \text{Var}(X_1)+\text{Var}(X_2)+\text{Var}(X_3)+2\text{Cov}(X_1 X_2)\\&\quad +2\text{Cov}(X_1 X_3)+2\text{Cov}(X_2 X_3)\end{aligned}}$$
(7.7)

where $\text{Cov}(X_p X_k)$ denotes the covariance of X_p and X_k.

Generalizing to the sum of $(n + 1)$ random variables, if

$$\boxed{Y = \sum_{j=0}^{n} X_j}$$
(7.8)

then the expected value and variance of Y are given by

$$\boxed{E(Y) = \sum_{j=0}^{n} E(X_j)}$$

$$\boxed{\text{Var}(Y) = \sum_{j=0}^{n} \text{Var}(X_j) + 2\sum_{j=0}^{n-1}\sum_{k=j+1}^{n}\text{Cov}(X_j X_k)}$$
(7.9)

Additionally, if a_1 and a_2 are constants and X_1, X_2, and Y are random variables related by

$$\boxed{Y = a_1 X_1 + a_2 X_2}$$
(7.10)

recall that

$$\boxed{\text{Var}(Y) = a_1^2 \text{Var}(X_1) + a_2^2 \text{Var}(X_2) + 2 a_1 a_2 \text{Cov}(X_1 X_2)}$$
(7.11)

Combining the relationships given in Equations 7.9 and 7.11, the variance of present worth is found to be

Break-Even, Sensitivity, and Risk Analyses

$$\text{Var}(PW) = \sum_{j=0}^{n} \text{Var}(C_j)(1+i)^{-2j}$$
$$+ 2\sum_{j=0}^{n-1} \sum_{k=j+1}^{n} \text{Cov}(C_j C_k)(i+1)^{-(j+k)} \tag{7.12}$$

Hillier[2] argues that it is probably unrealistic to expect investment analysts to develop accurate estimates for covariances. Hence, the net cash flow in any year should be divided into those components of cash flow that are reasonably independent from year to year and those that are correlated over time. Specifically, it is assumed that

$$C_j = X_j + Y_{j1} + Y_{j2} + \cdots + Y_{jm} \tag{7.13}$$

where the X_j values are mutually independent over j but, for a given value of h, $Y_{0h}, Y_{1h}, \ldots, Y_{nh}$ are *perfectly* correlated.

When two random variables X and Y are perfectly correlated, one can be expressed as a linear function of the other. Hence, if

$$Y = a + bX \tag{7.14}$$

where a and b are constants, then the covariance of X and Y is given by

$$\text{Cov}(XY) = E(XY) - E(X)E(Y) \tag{7.15}$$

However, from Equation 7.14, we note that

$$E(Y) = a + bE(X) \tag{7.16}$$

and

---

[2] Hillier, F. S., "The Derivation of Probabilistic Information for the Evaluation of Risky Investments," *Management Science*, 9 (3), April 1963.

$$\text{Var}(Y) = b^2 \text{Var}(X) \tag{7.17}$$

Substituting Equations 7.14 and 7.16 into Equation 7.15 gives

$$\begin{aligned} \text{Cov}(XY) &= E(aX + bX^2) - E(X)[a + bE(X)] \\ &= aE(X) + bE(X^2) - aE(X) - b[E(X)]^2 \\ &= b\text{Var}(X) \end{aligned} \tag{7.18}$$

Recalling that the *standard deviation* of a random variable is defined as the square root of the variance of the random variable, we see that the Equation 7.18 can be expressed as

$$\text{Cov}(XY) = b\,SD(X)SD(X) \tag{7.19}$$

where $SD(X)$ denotes the standard deviation of X. However, from Equation 7.17 we see that $SD(Y)$ equals $bSD(X)$. Hence,

$$\text{Cov}(XY) = SD(X)SD(Y) \tag{7.20}$$

where X and Y are perfectly correlated.

The model suggested by Hillier yields the following expressions for the expected present worth and variance of present worth:

$$E(PW) = \sum_{j=0}^{n} E(X_j)(1+i)^{-j} + \sum_{j=0}^{n}\sum_{h=1}^{m} E(Y_{jh})(1+i)^{-j} \tag{7.21}$$

and

$$\text{Var}(PW) = \sum_{j=0}^{n} \text{Var}(X_j)(1+i)^{-2j} + \sum_{h=1}^{m} \left[\sum_{j=0}^{n} SD(Y_{jh})(1+i)^{-j} \right]^2$$

(7.22)

where $SD(Y_{jh})$ denotes the standard deviation of the random variable Y_{jh}.

A very important theorem from probability theory, the *central limit theorem*, is usually invoked at this point. The central limit theorem establishes under very general conditions that the sum of independently distributed, random variables tends to be distributed normally as the number of terms in the summation approaches infinity. Hence, it is argued that present worth, as defined by Equation 7.1, is normally distributed with mean and variance, as given by Equations 7.21 and 7.22. Of course, this assumes that the C_j are statistically independent. However, the view is usually taken that the normal distribution is a reasonable approximation to the distribution of present worth.

Given that present worth is assumed to be normally distributed, one can compute for each investment alternative the probability of achieving a given desired level. For example, one is usually interested in knowing the probability that present worth is less than zero. Some analysts interpret this to be a *measure of the risk* associated with an investment alternative.

In performing an analysis of the risk associated with an investment alternative, we have treated the simplest situation; only the cash flows were considered to be random variables and the measure of effectiveness employed was present worth, not rate of return. When either the discount rate or the planning period are treated as random variables and when the probability distribution for rate of return is desired, simulation approaches are usually employed.

Example 7.13

As an illustration of the analytic approach in developing the probability distribution for *PW*, consider the following.

A flight school operator is considering the alternatives of purchasing a utility category training aircraft versus purchasing an acrobatic version of the same aircraft. Having been in the flight training business for a number of years, the operator has reason to believe that income and expense from a utility category aircraft will be nearly independent from year to year. The pertinent data are summarized in Table 7.7.

Table 7.7 Estimated Net Cash Flows for a Utility Model Aircraft

Year	Source	Expected Value	Range	Standard Deviation
0	Purchase	−$11 000	$9 500 – 12 500	$500
1	Income-expense	2 200	2 050 – 2 350	50
2	Income-expense	2 200	1 900 – 2 500	100
3	Income-expense	2 200	1 900 – 2 500	100
4	Income-expense	2 000	1 700 – 2 300	100
5	Income-expense	1 000	700 – 1 300	100
5	Salvage	6 000	4 800 – 7 200	400

Investment in the acrobatic aircraft is a more risky but promising investment. The acrobatic aircraft is also expected to have a life of five years. The operator feels that maintenance costs will be nearly independent year to year with this aircraft. However, since the flight school has never offered an acrobatic course before, there is some uncertainty regarding the demand for time in such an aircraft. It is felt that the net cash flow from the sale of flight time for each of the five years will be perfectly correlated. The pertinent data are summarized in Table 7.8. An interest rate of 10% is to be used in the analysis.

From Equations 7.21 and 7.22, we obtain the following:

$$E(PW_1) = \$190 \qquad E(PW_2) = \$355$$
$$\text{Var}(PW_1) = 334\,741 \qquad \text{Var}(PW_2) = 11\,631\,930$$

Assuming normally distributed PW, the probability of an equivalent present worth less than zero is found to be

$$Pr(PW_1 < 0 | i = 10\%) = 0.37$$

for the utility model aircraft and

$$Pr(PW_2 < 0 | i = 10\%) = 0.46$$

for the acrobatic model aircraft. Here we are faced with a choice between the alternative having the greatest expected value and the alternative having the smallest variance. The choice will depend on the owner's attitudes toward risk. If the decision were yours to make, which model aircraft would you choose? Why?

Table 7.8 Estimated Net Cash Flows for an Acrobatic Model Aircraft

Year	Source	Expected Value	Range	Standard Deviation
0	Purchase	−$14 000	$13 100 – 14 900	$ 300
1	Expense	− 10 000	9 100 – 10 900	300
2	Expense	− 10 000	9 100 – 10 900	300
3	Expense	− 11 000	10 100 – 11 900	300
4	Expense	− 12 000	10 800 – 13 200	400
5	Expense	− 12 000	10 800 – 13 200	400
1	Income	12 500	9 500 – 15 500	1000
2	Income	13 500	10 500 – 16 500	1000
3	Income	13 500	10 800 – 16 200	900
4	Income	14 500	12 100 – 16 900	800
5	Income	13 500	11 400 – 15 600	700
5	Salvage	7 500	6 000 – 9 000	500

7.6 Computer Simulation

Even though analytic approaches can be used in some situations to perform the risk aggregation step in risk analysis, simulation approaches are typically required to cope with the complexities of a real-world situation. Simulation, in the general sense, may be thought of as performing experiments on a model. Basically, simulation is an "if…, then…" device (i.e., *if* a certain input is specified, *then* the output can be determined). Some of the major reasons for using simulation in risk analysis are:

1. Analytic solutions are impossible to obtain without great difficulty.

2. Simulation is useful in selling a system modification to management.
3. Simulation can be used as a verification of analytical solutions.
4. Simulation is very versatile.
5. Less background in mathematical analysis and probability theory is generally required.

Some of the major disadvantages of simulation are:

1. Simulations can be quite expensive.
2. Simulations introduce a source of randomness not present in analytic solutions.
3. Simulations do not reproduce the input distributions exactly (especially the tails of the distribution).
4. Validation is easily overlooked in using simulation.
5. Simulation is so easily applied it is often used when analytic solutions can be easily obtained at considerably less cost.

Example 7.14

To illustrate the simulation approach in risk analysis, consider the following situation. An individual is planning on purchasing a used computer for $1 000 and performing certain billing and accounting functions for several retail businesses in the neighbourhood. It is anticipated that the business will last only four years because of the growth of competition. If income in the third year exceeds expenses by more than $400, operations will continue the fourth year. However, if income is less than or equal to $400 more than expenses in the third year, the computer and software will probably be sold at the end of the third year. For simplicity, a probability of 0.70 is assigned to the possibility of selling out at the end of the third year, given that income is not greater than $400 above expenses in the third year. Let

E_j = expense for year j

I_j = income for year j

n = life of the investment

S_n = salvage value based on an n-year life.

Assuming a zero discount rate for simplicity of calculations, the present worth of the investment is given as

$$PW = -\$1\,000 + \sum_{j=1}^{n}\left(I_j - E_j\right) + S_n$$

The probability distributions assumed to hold for this example are provided in Table 7.9. In practice, a computer would be used to perform the simulation. However, to illustrate the technique, we will manually perform 10 simulations of the investment. A table of two-digit random numbers is given in Table 7.10. Using the worksheet given in Table 7.11, 10 simulations of the investment yielded an average present worth of $805 for the investment. Of course, 10 simulations is not an adequate number of trials to draw strong conclusions concerning the investment. However, the example does illustrate the simulation approach.

To illustrate the approach taken, the first random number selected will provide the simulated value for expenses in the first year. A random number of 90 was obtained from row 1, column 1 in Table 7.10. Consulting Table 7.9, it is seen that a random number of 90 represents an expense of $400; hence, $400 is entered appropriately on the worksheet given in Table 7.11. The second random number is selected from row 2, column 1 of Table 7.10 to generate the income for year 1. A random number of 78 is obtained and, from Table 7.9, a simulated income of $600 is obtained for the first year. Continuing through the third year, it is found that income exceeds expenses by $500; hence the business will continue through the fourth year. A random number of 97 is drawn to generate the salvage value of $400 for the investment.

Note that in the second trial, it was decided that the business should be discontinued after three years. Furthermore, income was never less than expenses in any years; this illustrates that insufficient observations (trials) were obtained since, in year 2, expense can exceed income with probability

$$Pr\left(E_2 = 600 \text{ and } I_2 = 500\right) = 0.10(0.20) = 0.02$$

and in year 4 expense can exceed income with probability

$$Pr\left(E_4 = 700 \text{ and } I_4 = 600\right) + Pr\left(E_4 = 800 \text{ and } I_4 = 600\right) = 0.125$$

As an alternative to using the discrete probability distributions given in Table 7.9, we might employ continuous distributions such as the normal, gamma, or beta distributions to represent income and expenses.

Table 7.9 Data for the Risk Analysis Example Problem

E_1	$p(E_1)$	RN	I_1	$p(I_1)$	RN
$200	0.25	00–24	$ 400	0.50	00–49
300	0.50	25–74	600	0.50	50–99
400	0.25	75–99			

E_2	$p(E_2)$	RN	I_2	$p(I_2)$	RN
$300	0.10	00–09	$ 500	0.20	00–19
400	0.40	10–49	750	0.40	20–59
500	0.40	50–89	1000	0.40	60–99
600	0.10	90–99			

E_3	$p(E_3)$	RN	I_3	$p(I_3)$	RN
$400	0.20	00–19	$ 800	0.30	00–29
500	0.30	20–49	1000	0.50	30–79
600	0.30	50–79	1200	0.20	80–99
700	0.20	80–99			

E_4	$p(E_4)$	RN	I_4	$p(I_4)$	RN
$500	0.25	00–24	$ 600	0.25	00–24
600	0.25	25–49	800	0.25	25–49
700	0.25	50–74	1000	0.25	50–74
800	0.25	75–99	1200	0.25	75–99

N	$p(N|I_3 - E_3 \leq \$400)$	RN
3	0.70	00–69
4	0.30	70–99

N	$p(N|I_3 - E_3 > \$400)$
4	1.00

S_3	$p(S_3)$	RN	S_4	$p(S_4)$	RN
$400	0.50	00–49	$ 300	0.60	00–59
500	0.50	50–99	400	0.40	60–99

Table 7.10 Two-Digit Random Numbers

90	43	78	83	82	99	54	02
78	31	58	98	68	09	87	80
51	81	42	35	21	42	03	62
93	97	15	95	07	56	60	39
27	37	12	63	31	35	66	93
79	39	44	22	83	96	51	00
89	61	73	29	43	84	91	34
29	38	30	84	90	18	00	10
97	64	33	29	17	48	26	04
07	64	15	02	44	32	92	99
82	13	50	83	35	39	50	51
59	83	21	30	86	90	16	09
04	46	19	63	60	53	33	97
96	54	91	43	44	40	09	02
31	27	71	18	03	65	53	62
03	45	70	42	22	16	67	13
08	35	45	92	79	97	46	02
37	60	80	55	05	35	75	57
90	43	63	17	56	21	69	09
22	07	69	85	38	74	02	58
05	33	79	00	69	29	67	08
48	97	91	14	53	00	03	42
94	68	64	58	97	32	27	80
15	39	85	87	82	38	52	16
09	37	81	73	37	01	66	84

Table 7.11 Simulation of the Sample Investment Problem

Trial	Year	RN(E)	E	RN(I)	I	I−E	RN(N)	RN(S)	S	PW
1	1	90	$400	78	$ 600	$200				
	2	51	500	93	1000	500				
	3	27	500	79	1000	500	—			
	4	89	800	29	800	0		97	$400	$600
2	1	07	200	82	600	400				
	2	59	500	04	500	0				
	3	96	700	31	1000	300	03	08	400	100
	4	—	—	—	—	—				
3	1	37	300	90	600	300				
	2	22	400	05	500	100				
	3	48	500	94	1200	700	—			
	4	15	500	09	600	100		43	300	500
4	1	31	300	81	600	300				
	2	97	600	37	750	150				
	3	39	500	61	1000	500	—			
	4	38	600	64	1000	400		64	400	750
5	1	13	200	83	600	400				
	2	46	400	54	750	350				
	3	27	500	45	1000	500	—			
	4	35	600	60	1000	400		43	300	950
6	1	07	200	33	400	200				
	2	97	600	68	1000	400				
	3	39	500	37	1000	500	—			
	4	78	800	58	1000	200		42	300	600
7	1	15	200	12	400	200				
	2	44	400	73	1000	600				
	3	30	500	33	1000	500	—			
	4	15	500	50	1000	500		21	300	1100
8	1	19	200	91	600	400				
	2	71	500	70	1000	500				
	3	45	500	80	1200	700	—			
	4	63	700	69	1000	300		79	400	1300
9	1	91	400	64	600	200				
	2	85	500	81	1000	500				
	3	83	700	98	1200	500	—			
	4	35	600	95	1200	600		63	400	1200
10	1	22	200	29	400	200				
	2	84	500	29	750	250				
	3	02	400	83	1200	800	—			
	4	30	600	63	1000	400		43	300	950
										8050

As an alternative to using the discrete probability distributions given in Table 7.9, we might employ continuous distributions such as the normal, gamma, or beta distributions to represent income and expenses. If such distributions are to be used, appropriate techniques for generating simulated values of the random variables are required.

Risk analysis offers a number of important advantages over traditional deterministic approaches. Klausner[3] summarizes some of the most significant advantages as follows:

1. Uncertainty Made Explicit: The uncertainty which an estimator feels about his estimate of an element value is brought out into the open and incorporated into the investment analysis. The analysis technique permits maximum information utilization by providing a vehicle for the inclusion of "less likely" estimates in the analysis.

2. More Comprehensive Analysis. This technique permits a determination of the effect of simultaneous variation of all the element values on the outcome of an investment. This approximates the "real world" conditions under which an actual investment's outcome will be determined. The Probabilistic Cash Flow Simulation generates an overall indication of potential variation in outcome and project risk. This indicator, in the form of a probability distribution, accounts statistically for element interaction.

3. Variability of Outcome Measured. One of the most significant advantages of this analysis technique is that it gives a measure of the dispersion around the investment outcome based on the expected cash flow. This dispersion, or variability, is an important consideration in the comparison of alternative investments. Other things being equal, lower variability for the same return is usually desirable. The probability distribution associated with each investment's outcome gives a clear picture of this important evaluation consideration.

4. Promotes More Reasoned Estimating Procedures. By requiring that element values be given as probability distributions

---

[3] R. F. Klasuner, "The Evaluation of Risk in Marine Capital Investments," The Engineering Economist, 14 (4), Summer 1969.

rather than as single values, more reasoned consideration is given to the estimating procedure. Judgment is applied to the individual element values rather than to the investment's outcome which is jointly determined by all of the elements. Thinking through the uncertainties in a project and recognizing what is known and unknown will go far toward ensuring the best investment decision. Understanding and dealing effectively with uncertainty and risk is the key to rational decision making.

Risk analysis is a technique that has been used by a number of firms to improve the decision-making process in a risk environment. When applied properly, risk analysis can enhance significantly the manager's understanding of the risks associated with an investment alternative. The major premise underlying risk analysis is the belief that a manager can make better decisions when he or she has a fuller understanding of the implications of the decision.

7.7 Summary

Descriptive approaches were presented in this chapter for coping with risk and uncertainty conditions in performing economic analyses. In particular, break-even, sensitivity, and risk analyses were discussed. In Chapter 8, we employ normative or prescriptive approaches for dealing with risk and uncertainty.

The extent to which supplementary analyses are justified depends largely on the magnitude of the initial investment. In particular, if an analysis has been performed using the techniques described in Chapters 4, 5, and 6 and one of the alternatives is not clearly the superior alternative, a risk analysis will probably yield additional insight concerning the decision to be made. If either small amounts of money are involved or if the choice is not close using the techniques described in Chapters 4, 5, and 6 under assumed certainty, then supplementary analyses are not justified. In those cases where supplementary analysis is justified and a sensitivity analysis is not sufficient to allow a decision, a risk analysis is performed.

Problems

1. A plastic extrusion plant manufactures a particular product at a variable cost of $0.10 per unit, including material cost. The fixed costs associated with manufacturing the product equal $30 000 per year. Determine the break-even value for annual sales if the selling price per unit is (a) $0.60, (b) $0.40, and (c) $0.20. (7.2)

2. A consulting engineer is considering two pumps to meet a demand of 56 000 L/min at 3.7 m total dynamic head. The specific gravity of the liquid being pumped is 1.50. Pump A operates at 70% efficiency and costs $8 000; Pump B operates at 75% efficiency and costs $12 000. Power costs $0.0374/kW·hr. Continuous pumping for 365 days a year is required (i.e., 24 hours a day). Using a MARR of 10% and assuming equal salvage values for both pumps, how many years of service are required for Pump B to be justified economically? (*Note:* Dynamic head times litres per minute times specific gravity divided by 6116.2 equals the number of kilowatts required.) (7.2)

3. Two 100 kW motors are under consideration by the Mighty Machinery Company. Motor Q costs $2 000 and operates at 90% efficiency. Motor R costs $1 500 and is 88% efficient. Annual operating and maintenance costs are estimated to be 15% of the initial purchase price. Power costs 2.4 ¢/kW·hr. How many hours of full-load operation are necessary each year in order to justify the purchase of Motor Q? Use a 15-year planning period; assume that salvage values will equal 20% of the initial purchase price; and let the MARR be 10%. (7.2)

4. Owners of a motel chain are considering locating a new motel in Barrie, Ontario. The complete cost of building a 150-unit motel (excluding furnishings) is $2 million; the firm estimates that the furnishings in the motel must be replaced at a cost of $750 000 every five years. Annual operating and maintenance cost for the facility is estimated to be $50 000. The average rate for a unit is anticipated to be $18 per day. A 15-year planning period is used by the firm in evaluating new ventures of this type; terminal salvage value of 20% of the original building cost is anticipated; furnishings are estimated to have no salvage value at the end of each five-year replacement

interval; land cost is not be included. Determine the break-even value for the daily occupancy percentage based on a *MARR* of (a) 0%, (b) 10%, (c) 20%, and (d) 30%. (Assume that the motel will operate 365 days a year.) (7.2)

5. A manufacturing plant in Quebec has been contracting snow removal at a cost of $200 a day. The past three years have produced heavy snowfalls, resulting in the cost of snow removal being of concern to the plant manager. The plant engineer has found that a snow-removal machine can be purchased for $35 000; it is estimated to have a useful life of 10 years, and zero salvage value at that time. Annual costs for operating and maintaining the equipment are estimated to be $8 000. Determine the break-even value for the number of days per year that snow removal is required in order to justify the equipment, based on a *MARR* of (a) 0%, (b) 10%, and (c) 30%. (7.2)

6. A machine can be purchased at $t = 0$ for $10 000. The estimated life is five years, with estimated salvage value of zero at that time. The average annual operating and maintenance expenses are expected to be $4 000. If *MARR* = 20%, what must the average annual revenues be in order to be indifferent between (a) purchasing the machine, or (b) "do nothing"? (7.2)

7. A business firm is contemplating the installation of an improved material handling system between the packaging department and the finished goods warehouse. Two designs are being considered. The first consists of a driverless tractor system involving three tractors on the loop, with four trailers pulled by each tractor. The second design consists of a pallet conveyor installed between packaging and the warehouse. The driverless tractor system will have an initial equipment cost of $185 000 and annual operating and maintenance costs of $7 800. The pallet conveyor has an initial cost of $220 000 and annual operating and maintenance costs of $500. The firm is not sure what planning period to use in the analysis; however, salvage value estimates given in the following table have been developed for various planning periods. Using a *MARR* of 20%, determine the break-even value for *n*, the planning period. (7.2)

Salvage Value Estimates

n	Driverless Tractor	Pallet Conveyor
4	$108 000	$140 000
5	90 000	125 000
6	74 000	110 000
7	60 000	95 000
8	48 000	80 000
9	38 000	70 000
10	30 000	60 000
11	23 000	50 000
12	17 000	40 000
13	12 000	30 000
14	8 000	20 000
15	5 000	20 000

8. The motor on a gas-fired furnace in a small foundry is to be replaced. Three different 15 kW electric motors are being considered. Motor X sells for $500 and has an efficiency rating of 90%; Motor Y sells for $400 and has a rating of 85%; and Motor Z sells for $300 and is rated to be 80% efficient. The cost of electricity is $0.03/kW · hr. An eight-year planning period is used, and zero salvage values are assumed for all three motors. A *MARR* of 15% is to be used. Assume that the motor selected will be loaded to capacity. Determine the range of values for annual usage of the motor (in hours) that will lead to the preference of each motor. (7.2)

9. An investment of $10 000 is to be made into a variable rate GIC. The interest rate to be paid each year is uncertain; however, it is estimated that it is twice as likely to be 7% as it is to be 6%, and it is equally likely to be either 6% or 8%. Determine the probability distribution for the amount in the fund after 3 years, assuming the interest rate is not autocorrelated. (7.5)

10. Two condensers are being considered by the Ajax Company. A copper condenser can be purchased for $1 000; annual operating and maintenance costs are estimated to be $100.

Alternatively, a ferrous condenser can be purchased for $900; since the Ajax Company has not had previous experience with ferrous condensers, they are not sure what annual operating and maintenance cost estimate is appropriate. A five-year planning period is to be used, salvage values are estimated to be 10% of the original purchase price, and a MARR of 20% is to be used. Determine the break-even value for the annual operating and maintenance cost for the ferrous condenser. (7.2)

11. A firm has decided to manufacture widgets. There are two production processes available for consideration. Process A involves the purchase of a $22 000 machine that will last for 10 years and have salvage value of $2 000 at that time. Annual operating and maintenance costs amount to $3 000 per year for the machine. In addition, using Process A requires additional costs amounting to $0.15 per widget produced.

    The second process, Process B, requires an investment of $12 000 in a machine that will last for 10 years and have salvage value of $2 000 at that time. Annual operating and maintenance costs amount to $2 000 for the machine. There are additional costs of $0.20 per widget produced when Process B is used.

    With an 8% interest rate, for what annual production volume is Process A preferred? (7.2)

12. Two manufacturing methods are being considered. Method A has a fixed cost of $500 and a variable cost of $10. Method B has a fixed cost of $200 and a variable cost of $50. For what production volume would one prefer (a) Method A and (b) Method B? (7.2)

13. The ABC Company is faced with three proposed methods for making one of their products. Method A involves the purchase of a machine for $5 000. It will have a seven-year life, with a zero salvage value at that time. Using Method A involves additional costs of $0.20 per unit of product produced per year. Method B involves the purchase of a machine for $10 000. It will have $2 000 salvage value when disposed of in seven years. Using Method B involves additional costs of $0.15 per unit of product produced per year. Method C involves buying a machine for $8 000. It will have a $2 000 salvage value when

disposed of in seven years. Additional costs of $0.25 per unit of product per year arise when Method C is used. An 8% interest rate is used by the ABC Company in evaluating investment alternatives. For what range of annual production volume values is each method preferred? (7.2)

14. In Problem 4 suppose the following pessimistic, most likely, and optimistic estimates are given for building cost, furnishings cost, annual operating and maintenance costs, and the average rate per occupied unit.

	Pesimistic	Most Likely	Optimistic
Building cost	$3 000 000	$2 000 000	$1 500 000
Furnishings cost	1 250 000	750 000	500 000
Annual operating and maintenance costs	60 000	50 000	40 000
Average rate	13 per day	15 per day	18 per day

Determine the pessimistic and optimistic limits on the break-even value for the daily occupancy percentage based on a *MARR* of 20%. Assume the motel will operate 365 days a year. (7.4)

15. In Problem 7, suppose a 10-year planning period is specified. Perform a sensitivity analysis comparable to that given in Figure 7.5 to determine the effect of errors in estimating the *differences* in initial investment, the *differences* in annual operating and maintenance costs, and the *differences* in salvage values for the two investment alternatives. (7.4)

16. A warehouse modernization plan requires an investment of $3 million in equipment; at the end of the 15-year planning period, it is anticipated that the equipment will have salvage value of $600 000. Annual savings in operating and maintenance costs due to the modernization are anticipated to total $1 400 000 per year. A *MARR* of 15% is used by the firm. Perform a sensitivity analysis to determine the effects on the economic feasibility of the plan due to errors in estimating the initial investment required, and the annual savings. (7.4)

17. Given an initial investment of $10 000, annual receipts of $3 000, and an uncertain life for the investment, determine the probability of the investment being profitable. Use a 15% MARR. Let the probability distribution for the life of the investment be given as follows: (7.5)

n	p(n)
1	0.10
2	0.15
3	0.20
4	0.25
5	0.15
6	0.10
7	0.05

18. In Problem 17, suppose the MARR is not known with certainty, and the following probability distribution is anticipated to hold.

i	p(i)
0.10	0.20
0.15	0.60
0.20	0.20

What is the probability of the investment being profitable? (7.5)

19. In Problem 17, suppose the magnitude of the annual receipts (R) is subject to random variation. Assume that each annual receipt will be identical in value, and the annual receipt has the following probability distribution:

R	p(R)
$2000	0.20
3000	0.50
4000	0.30

What is the probability of the investment being profitable? (7.5)

20. In Problem 17, suppose the minimum attractive rate of return is distributed as given in Problem 18, and suppose the annual receipts are distributed as given in problem 19. Determine the probability of the investment being profitable. (7.5)

21. In Problem 20, suppose the initial investment is equally likely to be either $9 000 or $10 000. Determine the probability of the investment being profitable. (7.5)

22. Suppose $n = 4$, $i = 0\%$, a $10 000 investment is made, and the receipt in year j, $j = 1, ..., 4$, is statistically independent and distributed as in problem 19. Determine the probability distribution for present worth. (7.5)

23. Consider an investment alternative having a six-year planning period and expected values and variances for statistically independent cash flows as given below:

j	$E(C_j)$	$V(C_j)$
0	−$25 000	625×10^4
1	4 000	16×10^4
2	5 000	25×10^4
3	6 000	36×10^4
4	7 000	49×10^4
5	8 000	64×10^4
6	9 000	81×10^4

Using a discount rate of 20%, determine the expected values and variances for both present worth and annual worth. Based on the central limit theorem, compute the probability of a positive present worth; compute the probability of a positive annual worth. (7.4)

24. Solve Problem 23 using a discount rate of (a) 0%, (b) 30%. (7.5)

25. Solve Problem 23 when the C_j, $j = 1, ..., 6$ are perfectly correlated. (7.5)

26. Two investment alternatives are being considered. Alternative A requires an initial investment of $13 000 in equipment;

annual operating and maintenance costs are anticipated to be normally distributed, with a mean of $5 000 and a standard deviation of $500; the terminal salvage value at the end of the eight-year planning period is anticipated to be normally distributed with a mean of $2 000 and standard deviation of $800. Alternative B requires end-of-year annual expenditures over the planning period, with the annual expenditure being normally distributed with a mean of $7 500 and standard deviation of $750. Using a MARR of 15%, what is the probability that Alternative A is the most economic alternative? (7.5)

27. In Problem 26, suppose the MARR were 10% with probability 0.25, 15% with probability 0.50, and 20% with probability 0.25. What is the probability that Alternative A is the most economic alternative? (7.5)

28. Recall Example 7.10. Continue the simulation for 10 additional trials and compute the cumulative average present worth. (7.5)

29. In Example 7.10, suppose the individual had used a minimum attractive rate of return of 10% in the calculations. Would the first 10 simulation trials have yielded a positive average present worth? (7.5)

30. In Example 7.10, determine the ERR value for each trial, given a MARR of 10%. (7.5)

31. Two investment alternatives are under consideration. The data for the alternatives are given below. It is assumed the cash flows are not autocorrelated.

EOY	E[CF(A)]	SD[CF(A)]	E[CF(B)]	SD[CF(B)]
0	−10 000	1000	−15 000	1500
1	4 000	400	8 000	800
2	5 000	500	8 000	800
3	6 000	600	8 000	800
4	7 000	700	8 000	800
5	8 000	800	8 000	800
6	9 000	900	8 000	800

(a) For each alternative, determine the mean and standard deviation for present worth and annual worth using a MARR of 10%.

(b) For each alternative, based on the central limit theorem, compute the probability of a positive present worth using a MARR of 10%.

(c) Develop the mean and standard deviation for the *incremental* present worth using a MARR of 10%.

(d) Based on the central limit theorem, compute the probability of a positive incremental present worth using a MARR of 10%. (7.5)

32. Company W is considering investing $10 000 in a machine. The machine will last n years, at which time it will be sold for L. Maintenance costs for this machine are estimated to increase by $200 a year over its life. The maintenance cost for the first year is estimated to be $1 000. The company has a 10% MARR. Based on the probability distributions given below for n and L, what is the expected equivalent uniform annual cost for the machine? (7.5)

n	L	$p(n)$
6	$3 000	0.2
8	2 000	0.4
10	1 000	0.4

33. A company manufactures one single product. The marginal revenue function for the company is

$$\frac{dTR(X)}{dX} = \$(100 - 0.02X) \text{ per unit}$$

where X is the number of products sold. The total cost is

$$TC(X) = \$(2.10^{-4} X^2 + 10\,000)$$

Determine the quantity of production for
(a) minimum unit cost,
(b) maximum profit,
(c) break-even. (7.3)

34. For problem 33, determine
    (a) marginal cost at the profit limit point,
    (b) marginal profit at the shut-down point,
    (c) marginal unit cost at the point of maximum total profit,
    (d) marginal unit profit at the break-even point,
    (e) maximum unit profit. (7.3)

35. A production process is characterized by the following empirical relations:

$$\text{Average revenue: } \frac{TR(X)}{X} = \$(a - bX) \text{ per unit of product}$$

$$\text{Average variable cost: } \frac{CV(X)}{X} = \$(cX^2 + dX^3) \text{ per unit of product}$$

$$\text{Fixed cost} = K$$

where X is the number of units of product produced yearly, and a, b, c, d, and K are constant values. Determine:
   (a) the range of profitable production,
   (b) the production rate at maximum total profit,
   (c) the total profit at the production rate where the unit cost is minimum,
   (d) the marginal unit cost at the shut-down point,
   (e) the marginal unit profit at the profit limit point. (7.3)

Chapter 8
Decision Models

8.1 Introduction

This chapter presents basic models for displaying decision situations involving risk and uncertainty and suggests criteria for choosing among alternatives. These models will be *prescriptive* or *normative* in that they prescribe a decision action to be taken. The risk-analysis discussion and models presented in Chapter 7 described the behaviour of the measure of effectiveness when uncertainty and risk prevailed. This chapter is concerned with specifying the preferred choice from a set of feasible alternatives and represents the final step in the eight-step procedure for comparing alternatives presented in Chapter 4.

The first section of this chapter discusses a matrix model and suggests criteria for choosing among alternatives. The second section presents a network or tree model for decision situations involving a sequence of decisions over a time period where the outcomes of alternatives are uncertain. The final section of the chapter offers an introductory discussion on methodology for choosing among alternatives when multiple measures of effectiveness are involved. The matrix model for decisions made under conditions of risk and uncertainty is now introduced by an example concerning material handling.

Example 8.1

A new product group is to be added by the ABC Company with initial production scheduled two years hence. The new product group, which will consist of a single product in varying sizes and finishes, requires that a new and separate facility be added to the existing plant. The company is presently completing plans for the new facility, and plant construction will begin in the near future. Excess floor space is planned in order to increase production if demand increases; this increase is expected, with demand reaching stability after five years of production.

The manufacturing system consists, essentially, of several machining centres that operate on a job-shop basis instead of on a production-line basis. The manufacturing sequence is fixed except for slight variations, depending on product size and finish required. Thus, a subproblem of the total manufacturing system is to determine the material handling requirements. After considerable study and discussion, it has been decided that industrial lift trucks provide the flexibility of handling required in the job-shop environment. Now, questions arise as to the type of lift truck, the number required, the attachments needed, whether they should be leased or purchased, and, if leased, which of several leasing plans should be chosen, and so forth. The answers to these and other relevant questions depend primarily on the demand for the new product during the first five-year production period. This demand is uncertain, and ABC Company management can influence demand through advertising only to a limited degree.

Management could make a decision on material handling requirements based on the demand forecast for the first year of production, then evaluate the situation at the end of the first year, make a new decision for the next year, re-evaluate, and so on, in sequential fashion. However, let us assume that it has been decided to choose a material handling system now for the first three-year production period and then re-evaluate at the end of that period.

Given the above managerial decision, the analysts investigating the material handling requirements make the judgement that demand for the product group during the three-year planning period will be such that four, five, or six lift trucks are required. They further reason that the feasible alternatives are:

A_1, A_2, A_3 — lease four, five, or six trucks, respectively.

A_4, A_5, A_6 — purchase four, five, or six trucks, respectively.

These lease plans do not include a maintenance contract, since the ABC Company presently employs skilled maintenance personnel. Furthermore, the analysts have not considered any lease plan with an "option to purchase" clause in order to avoid "contractual sales" income tax complexities.

The analysts select before-tax, equivalent uniform annual cost as the single measure of effectiveness for a given combination of A_j versus number of trucks required. It is necessary to make gross estimates of these annual costs, but essentially, for leasing, there is a fixed monthly charge per truck plus a judgemental penalty cost associated with production delays when the number of trucks required exceeds the number of trucks available. On the other hand, the company can obtain a more favourable leasing charge as the number of trucks leased increases. For the purchase alternatives, similar fixed and variable annual operating and maintenance costs, penalty costs, and quantity discounts are involved. After much study, somewhat gross estimations, and a liberal application of judgement, the analysts present the model of Table 8.1 to management for a final decision.

On the basis of the single, equivalent uniform annual cost measure of effectiveness, any *purchase* alternative is preferable to any *lease* alternative. For example, the purchase alternative A_4 has a lower equivalent annual cost than any lease alternative if four, five, or six trucks are required. Thus, Alternative A_4 dominates the A_1, A_2, and A_3 alternatives, and they may therefore be deleted from the decision matrix. (It is noted that this might not have been the case if after-tax annual cost values had been calculated. Furthermore, a consideration of intangible factors by management could result in a lease alternative being chosen over a purchase alternative even though annual costs are higher.) The purchase alternatives A_5 and A_6 also dominate the three lease alternatives, but no one purchase alternative dominates any other purchase alternative. That is, no single purchase alternative has lower annual costs than any other purchase alternative when viewed across all possible number of trucks required. Dominance will be discussed in greater detail subsequently.

Table 8.1 Decision Model for the Lift Truck Example

Number of Trucks Required	4	5	6
Alternatives	Equivalent Uniform Annual Costs		
A_1 – Lease four	$18 000	$20 000	$24 000
A_2 – Lease five	20 000	20 000	22 000
A_3 – Lease six	21 000	21 000	21 000
A_4 – Purchase four	12 500	14 500	18 500
A_5 – Purchase five	14 000	14 000	16 000
A_6 – Purchase six	15 000	15 000	15 000

A solution to the lift truck problem will not be presented now; the example has been cited to illustrate the matrix format of Table 8.1 and to suggest the model as an effective portrayal of many decision situations.

The elements in the decision model of Table 8.1 are now summarized as:

States = the number of lift trucks required in the three-year planning period.

Feasible alternatives = the number of lift trucks available, whether leased or purchased.

Outcomes = the results of particular combinations of state versus alternatives. In this case, the general outcomes are that the number of trucks required is less than, equal to, or greater than the number of trucks available.

Value of the outcomes = the measure of effectiveness for a particular combination of state versus alternatives. In this case, equivalent uniform annual costs.

A matrix decision model with general symbolism is given in Figure 8.1, and further discussed in the next section.

A_j \ S_k	P_1	P_2	...	...	P_k	...	...	P_m
	S_1	S_2	...	...	S_k	...	...	S_m
A_1	$V(\theta_{11})$	$V(\theta_{12})$	...	...	$V(\theta_{1k})$	...	...	$V(\theta_{1m})$
A_2	$V(\theta_{21})$	$V(\theta_{22})$	...	...	$V(\theta_{2k})$	...	...	$V(\theta_{2m})$
⋮	⋮	⋮			⋮			⋮
⋮	⋮	⋮			⋮			⋮
A_j	$V(\theta_{j1})$	$V(\theta_{j2})$	...	...	$V(\theta_{jk})$	...	...	$V(\theta_{jm})$
⋮	⋮	⋮			⋮			⋮
⋮	⋮	⋮			⋮			⋮
A_n	$V(\theta_{n1})$	$V(\theta_{n2})$	...	...	$V(\theta_{nk})$	...	...	$V(\theta_{nm})$

Fig. 8.1 A matrix decision model with general symbolism

8.2 The Matrix Decision Model

The symbolism employed is defined as follows:

A_j = an *alternative* or *strategy* under the decision maker's control, where $j = 1, 2, ..., n$.

S_k = a *state* or *possible future* that can occur given that alternative A_j is chosen, where $k = 1, 2, ..., m$.

θ_{jk} = the *outcome* of choosing alternative A_j and having state S_k occur.

$V(\theta_{jk})$ = the *value* of outcome θ_{jk} which may be in terms of dollars, time, distance, or utility.

p_k = the *probability* that state S_k will occur.[1]

The term p_k was not previously introduced in the lift truck decision problem but will subsequently serve as a means by which decisions may be categorized into decisions under assumed certainty, decisions under risk, and decisions under uncertainty. This categorization has essentially been made in Chapter 7, but will be repeated in order to present decision rules for the normative models of this chapter.

The matrix model, in general, describes a set of alternatives available where a single alternative is to be selected at the present time. It is implied that the outcomes for a given alternative do not necessitate decisions at future times. If a sequence of decisions were involved, the decision tree model of a later section would probably be more appropriate.

It is also assumed in the matrix model that the alternatives are mutually exclusive, feasible, and represent the total set of alternatives the decision maker may wish to consider. Recall from the discussion in Chapter 4 that mutually exclusive alternatives cannot occur together, and thus a single alternative will be chosen from the feasible set. The question now arises, "What constitutes feasibility?" The answer to the question lies in the original statement of objectives that the alternative is to accomplish. For example, if one were considering the replacement of a production machine because of increasing maintenance costs and inadequate capacity, an objective might be that the new ma-

---

[1] It is assumed that the probability of a particular state occurring does not depend on the alternative chosen by the decision maker.

chine must be capable of producing 1 000 units of Product A per day. Those machines incapable of satisfying this constraint of 1 000 units per day are infeasible and rejected from further consideration. Those machines capable of meeting or exceeding the requirement of 1 000 units per day are feasible, and the "best" replacement machine is then selected from the feasible set according to a more discriminating criterion, such as minimum estimated equivalent annual cost or maximum estimated annual profit.

Determining the "total set" of feasible alternatives is also a difficulty deserving additional comment. Rarely does the analyst have the time or even the insight to define all the possible alternatives available in a given decision situation. How to execute a search procedure and when to terminate the search are interesting and important questions, but will not be pursued here. Suffice it to say that experience plays a major role in the search procedure, and when to terminate the search is primarily a matter of judgement in the practical world of business decisions. In deciding that the cost of finding an additional alternative will outweigh the benefits of that other alternative or that the benefits justify the cost, the analyst is actually making a subjective judgement.

Experience and judgement also play an important role in defining the relevant states for a given decision situation and the set of feasible alternatives involved. To convey further the meaning of a "state," other terms in the literature are *possible futures, states of nature, external conditions,* and *future events.*

In the discussion of risk analysis in Chapter 7, states were values for the random variable of purchase price, annual operating expenses, annual revenues, service lives, and salvage values. However, in other decision situations, the states may be defined more grossly; for instance, the market demand for Product A will be low, medium, or high. Furthermore, the number of states may be finite or infinite, but for the examples in this text, the number of states will be assumed to be finite and few in number. This seems a realistic assumption for most business and engineering decisions because, although the states may actually be a very large finite or even infinite number, the decision maker wishes to limit the scope of the problem and uses a sample of state values over the range of possible values. Using the industrial lift truck example presented earlier, knowledge of the machining processes involved and a recorded history of machining times can provide the basis for estimating values for the states in terms of the number of trucks required to service different levels of production output. This same experience can also provide the range of state values to consider. In any case, it will be assumed here that once the states are defined, they are collectively exhaustive (i.e., the total set of possible states). Furthermore:

1. The occurrence of one state precludes the occurrence of the others (i.e., they are mutually exclusive).
2. The occurrence of a state is not influenced by the alternative selected by the decision maker.
3. The occurrence of a state is not known with certainty by the decision maker.

For each alternative-versus-state combination, there is an outcome and a value associated with the outcome. In the previous industrial lift truck example, a given combination of number of trucks available versus number of trucks required resulted in the trucks available being short, over, or equal to the trucks required. For each outcome, it was assumed that an annual equivalent dollar cost could be calculated and was the value of interest. Other value measurements may, of course, be appropriate in other decision situations. The matrix values may be one of two general kinds: objective or subjective. Objective values represent physical quantities such as dollars, hours, litres, kilograms, and so forth. With the exception of the section on multiple goals in this chapter, the subsequent examples are concerned with dollars or objective values.[2] Subjective values, on the other hand, are numbers that represent the decision maker's relative preferences; for example, a score of 90 on an examination is preferred to a score of 60. The values of 90 and 60 are subjective values assigned on an arbitrary scale of 0 to 100. An example of a subjective value matrix will be given in the multiple-goals section of this chapter.

The problems of measurement and value determination are often very substantial in real-world situations. Among the problems of measurement are determining the probability of occurrence of a given state. By looking back at the matrix model in Figure 8.1, we can begin to classify decisions on the basis of how much information is available regarding probability of occurrence.

---

[2] A linear utility function for the decision maker is also assumed. That is, minimizing costs or maximizing profits are both the same as maximizing the decision maker's utility.

8.3 Decisions under Assumed Certainty

It is reasonable to assume in many decision situations that only one state is relevant and then treat the decision as if the state were certain to occur. This kind of case is termed a *decision under assumed certainty*. For example, it may be a matter of company policy to depreciate new equipment purchases fully in five years. If, in an economic analysis of several new equipment candidates, the measure of effectiveness is equivalent annual cost, the determination of these annual costs would be based on a five-year life for each candidate. The outcome of a five-year life and the associated annual operating costs for each candidate are assumed to be certain (i.e., will occur with probability of 1.0). The actual functional life of an equipment candidate might be 4, 5, 6, 7, 8,... years. However, a judgement is made by the decision maker to suppress uncertainty because of convenience or practicality and assume only the single state of a five-year life. All of the analysis in Chapters 3 to 6 was based on such judgements; single-value estimates for operating costs, service life, and the like were examples of assumed certainty.

In terms of the matrix decision model, a decision under assumed certainty would appear as follows:

	S
A_1	$V(\theta_1)$
A_2	$V(\theta_2)$
⋮	⋮
A_j	$V(\theta_j)$
⋮	⋮
A_{n-1}	$V(\theta_{n-1})$
A_n	$V(\theta_n)$

If, in the above model, the values for each alternative were profits (or gains), the principle for choice might be "choose the alternative that maximizes profit." Conversely, if the values were costs (or losses), one might "choose the alternative that minimizes costs."

Example 8.2

Suppose that an investor is considering a $10 000 purchase of government securities. There are several types of such securities (Government of Canada Bonds, provincial bonds, etc.), and they are available on the open market. Assuming that the various securities have common maturity dates, the investor judges that each is financially secure and payoffs are certain. A logical criterion for choice among the securities would be the effective annual yield, or internal rate of return, on the $10 000 purchase. (It is assumed that there is no reason for allocating the $10 000 to different securities.) Suppose the investor considers five government securities and, for a maturity date of five years hence, has calculated the effective annual yield on each security to be

A_1	14.0%
A_2	14.5%
A_3	15.2%
A_4	16.1%
A_5	17.0%

Since the investor would logically want to maximize the yield on the investment, A_5 would be chosen. Recall that equal risk and maturity dates were assumed for each security.

Example 8.3

The cash flow profiles for the three mutually exclusive investment alternatives given in Table 4.8 are repeated in Table 8.2. Considering the minimum attractive rate of return of 10% as the single state that will occur with certainty and the present worth of each alternative as the value of interest, the matrix model of this decision appears in Table 8.3. If the principle for choice is "select the alternative that maximizes the present worth," A_3 would be chosen as before in Chapter. 4.

Table 8.2 Cash Flow Profiles for Three Mutually Exclusive Investment Alternatives (MARR = 10%)

t	A_1	A_2	A_3
0	$ 0	−$10 000	−$15 000
1	4 000	7 000	7 000
2	4 000	7 000	8 000
3	4 000	7 000	9 000
4	4 000	7 000	10 000
5	4 000	7 000	11 000
Present worth	$15 163.20	$16 535.60	$18 397.33

Table 8.3 Present Worths of Three Mutually Exclusive Investment Alternatives (MARR = 10%)

A_j	S_1
A_1	$15 163.20
A_2	16 535.60
A_3	18 397.33

8.4 Decisions under Risk

A decision situation is called a *decision under risk* when the decision maker elects to consider several states, and the probabilities of their occurrence are explicitly stated. In some decision problems, the probability values may be objectively known from historical records or objectively determined from analytical calculations. For example, if an unbiased die is rolled once, the six possible states are that one, two, three, four, five, or six dots will occur on the top face exposed. The objective probability that each state will occur prior to the roll is 1/6. The sum of the individual probabilities over all states equals the value of 1.0. This property of $\sum p_j = 1.0$ agrees with an earlier statement that, once the set of possible states is decided on, the states are treated as if they constituted a set of mutually exclusive and collectively exhaustive events. Indeed, the property of $\sum p_j = 1.0$ is a necessary restriction if probability theory is to be used as a guide to decision making.

The decision maker may not have past records available to arrive at objective probability values. This was the case in the industrial truck example cited earlier. However, if the decision maker feels that experience and judgement are sufficient to assign probability values subjectively to the occurrence of each state, the decision is still treated as one under risk. The restriction of $\sum p_k = 1.0$ still applies in this case. Principles of choice for selecting an alternative in a decision under risk, which commonly appear in the literature on decision theory, are illustrated in Example 8.4.

Example 8.4

The three mutually exclusive investment alternatives presented in Table 8.2 will again be used to illustrate guidelines for a decision under risk. The single state previously defined was a minimum attractive rate of return equal to 10%. For the purpose of this example, the two additional states of $MARR = 20\%$ and $MARR = 30\%$ will be assumed. The realism of considering these states as probabilistic in nature may be questionable, but an argument supporting this claim can be posed. In periods of an unstable economy, the analyst may perceive the MARR as uncertain and be willing to define different values with associated probabilities. The matrix model of Table 8.4 assumes that this is the case, and cell values are present worths for a five-year planning period.

The states of $S_1 = 10\%$, $S_2 = 20\%$, and $S_3 = 30\%$ may have been defined as optimistic, most likely, and pessimistic estimates, as discussed in Chapter 7. A MARR that lowers the present worth of the investment alternative is considered a pessimistic MARR. The chance of occurrence for the states may have been reasoned as follows: S_3 is twice as likely to occur as S_2, which, in turn, is three times as likely to occur as S_1. This reasoning can be stated as the equation $p_1 + 3p_1 + 6p_1 = 1.0$, and $p_1 = 0.10$, $p_2 = 0.30$, and $p_3 = 0.60$ determined.

8.4.1 Dominance

The *dominance* principle is described as follows. Given two alternatives, if one would always be preferred no matter which state occurs, this preferred alternative is said to dominate the other, and the dominated alternative can be deleted from further consideration. Applying this principle will not necessarily solve the decision problem by selecting a unique alternative, but it may reduce the number of alternatives for further consideration.

The dominance principle, then, is a way of deciding which alternatives not to select. If the values are *costs*, a more formal statement of the dominance relation is: *If there exists a pair of alternatives A_j and A_l, such that $V(\theta_{jk}) \leq V(\theta_{lk})$ for all k, A_j is said to dominate A_k. Alternative A_l may then be discarded from the decision problem.* If the values of a decision matrix are *gains*, the condition for A_j to dominate A_l is that $V(\theta_{jk}) \geq V(\theta_{lk})$ for all k.

In Table 8.4, no alternative has a present worth value greater than the present worth value of another alternative for all states. Thus, there is no dominant alternative for this example, but recall that there were dominant alternatives in Table 8.1.

Table 8.4 Matrix Model with States of MARR = 10%, 20%, and 30%

	$p_1 = 0.10$	$p_2 = 0.30$	$p_3 = 0.60$
MARR	S_1 10%	S_2 20%	S_3 30%
A_j			
A_1	$15 163	$11 962	$9 742
A_2	$16 536	$10 934	$7 049
A_3	$18 397	$10 840	$5 679

8.4.2 Expectation-Variance Principle

If x is a discrete random variable that is defined for a finite number of values and p(x) denotes the probability of a particular value occurring, then the expected or mean value of the random variable is defined as

$$E(x) = \sum_{\text{all } x} xp(x) \tag{8.1}$$

The variance of the random variable is defined as Var(x), where

$$\text{Var}(x) = \sum_{\text{all } x} [x - E(x)]^2 p(x) \tag{8.2}$$

Decision Models

For the matrix model of Figure 8.1, the random variable is $V(\theta_{jk})$, and the expected value of an alternative A_j is

$$E(A_j) = \sum_{\text{all } k} V(\theta_{jk}) p_k \qquad (8.3)$$

where p_k denotes the probability of state k occurring.

The expected values for each of the alternatives of Table 8.4 are calculated as

$$E(A_1) = \$15\,163(0.10) + \$11\,962(0.30) + \$9\,742(0.60)$$
$$= \$10\,950.10$$
$$E(A_2) = \$16\,563(0.10) + \$10\,934(0.30) + \$7\,049(0.60)$$
$$= \$9\,163.20$$
$$E(A_3) = \$18\,397(0.10) + \$10\,840(0.30) + \$5\,679(0.60)$$
$$= \$8\,499.10$$

If the principle of maximizing (minimizing) the expected gain (loss) is followed, alternative A_1 would be chosen, since the expected values are positive present worths. A corollary of the expected-value criterion is that, in the event of a tie in expected value for two or more alternatives, the alternative having minimum variance should be chosen. The argument for this secondary criterion is, essentially, that a larger variance is indicative of greater uncertainty, and since most decision makers have an aversion to uncertainty, the alternative with smallest variance should be chosen.

8.4.3 Most Probable Future Principle

If, in a decision under risk, one state has a probability of occurrence considerably greater than any other, the *most probable future principle* is to consider this state as certain and all other states as having a zero chance of occurrence. The decision is thereby reduced to a decision under assumed certainty. Then an alternative is chosen that maximizes (or minimizes) the measure of effectiveness being used. For Table 8.4, S_3 has a probability of 0.60 assigned and is therefore the most probable state. Assuming certainty for S_3, or $p_3 = 1.0$, the reduced matrix is,

	S_3
A_1	$9 742
A_2	$7 049
A_3	$5 679

To obtain the maximum present worth, A_1 is selected as the preferred alternative.

Another example of the most probable future principle is the case of a person commuting to work. Each morning that the person leaves for work there is a nonzero probability that an accident could occur and serious injury result. However, the most probable future is that the commuter will reach work safely. The commuter considers this future certain, or else he (she) would probably not go to work!

As a final remark on this principle of choice, it seems appealing only if one state has a significantly higher probability of occurring than any other and the values in the total matrix do not differ significantly. One would not wish to suppress the consideration of a state that, albeit with low probability, would inflict a severe loss on the decision maker. An example of this is the matrix below, where the positive values are profits.

	$p_1 = 0.95$	$p_2 = 0.05$
A_1	$10	$40
A_2	$50	−$5000

Following the most probable future principle, A_2 would be chosen. However, the suppression of S_2 does not alter the fact that S_2 could occur, by virtue of $p_2 = 0.05$; in that event, the decision maker would suffer a loss of $5000.

8.4.4 Aspiration-Level Principle

In most real-world decisions, the complexity of the decision prevents the discovery and selection of an alternative that will yield the single "best" result, and decision makers set their goals in terms of outcomes that are good enough. That is, decision makers set aspiration levels and then evaluate alternatives against them. An interpretation of this philosophy in terms of a decision under risk is to select an alternative that maximizes the probability of achieving the desired aspiration level. Typical aspiration-level objectives might be to choose the alternative that maximizes the probability of either (1) a 25% internal rate of return, (2) annual costs less than $8 000, or (3) annual profits greater than $9 500. In terms of the example of Table 8.4, let the objective be to choose whichever alternative maximizes the probability that the present worth value will be equal to or greater than $8 000. This objective can be symbolized as $P(PW \geq \$8\,000)$. From Table 8.4, it can be seen that, for A_1, the present worth will be greater than $8 000 when $S_1, S_2,$ or S_3 occurs. Thus, a present worth greater than $8 000 is certain if A_1 is chosen. Because the states are mutually exclusive, the additive-probability law permits the sum of the individual state probabilities to be the probability of the joint event, "$PW \geq \$8000$" for A_1. The calculations for each alternative are:

For A_1, $P(PW \geq \$8\,000) = P(S_1) + P(S_2) + P(S_3)$
$$= 0.10 + 0.30 + 0.60 = 1.0$$

For A_2, $P(PW \geq \$8\,000) = P(S_1) + P(S_2)$
$$= 0.10 + 0.30 = 0.40$$

For A_3, $P(PW \geq \$8\,000) = P(S_1) + P(S_2)$
$$= 0.10 + 0.30 = 0.40$$

and A_1 maximizes $P(PW \geq \$8\,000)$.

A variety of aspiration level objectives are, or course, possible for even a *given* decision under risk. The relevant aspiration-level objective is determined by the decision maker's preference.

Do not interpret any of the principles for choice above or those to follow in the next section as the way an alternative *should* be selected. Instead, the principles for choice presented result from observing how people *seem* to decide on alternatives; therefore, they are offered only as possible guidelines to the decision maker.

8.5 Decisions under Uncertainty

A decision situation where several states are possible and sufficient information is not available to assign probability values to their occurrence is termed a *decision under uncertainty*. It can be argued that any decision maker, if able to define the states, should have sufficient knowledge to make at least gross subjective probability estimates for these states. On the other hand, in situations such as (1) installing new safety devices on machinery where the states are the number of injuries expected per year, or (2) the introduction of a new product where the states are units demanded per year, one may feel quite inadequate to assign probabilities to these states. This may be the case even though the analyst is not totally lacking in knowledge concerning the number of injuries expected per year or the units of new product demanded per year. The principles for choice subsequently presented are based on the premise that probabilities cannot be assigned.

8.5.1 The Laplace Principle

The philosophy of the *Laplace Principle,* named after the early nineteenth century mathematician, is simply that if one cannot assign probabilities to the states, the states should be considered as equally probable. Then, consider the decision as one under risk. Applying this principle to the previous example of Table 8.4, the probability value assigned to each of the three states is $\frac{1}{3}$. Then,

$$E(A_1) = \$15\,163(\tfrac{1}{3}) + \$11\,962(\tfrac{1}{3}) + \$9\,742(\tfrac{1}{3})$$
$$= \$12\,289$$
$$E(A_2) = \$16\,536(\tfrac{1}{3}) + \$10\,934(\tfrac{1}{3}) + \$7\,049(\tfrac{1}{3})$$
$$= \$11\,506$$
$$E(A_3) = \$18\,397(\tfrac{1}{3}) + \$10\,840(\tfrac{1}{3}) + \$5\,679(\tfrac{1}{3})$$
$$= \$11\,638.67$$

and A_1 would be chosen to maximize expected present worth.

8.5.2 Maximin and Minimax Principles

The *maximin* and *minimax* principles for choice represent a single philosophy, depending on whether the matrix values are *gains* or *losses,* respectively. These principles hold considerable appeal for the conservative or pessimistic decision maker. In applying the principles, if the matrix values are *gains,* the minimum gain associated with each alternative (when viewed over all states) is determined, and the maximum value in the set of minimum values designates the alternative to be chosen. In other words, determine the worst value possible for *each* alternative and then choose the best value from the set of worst values. Applying the maximin principle to the example problem of Table 8.4 gives,

	Minimum Present Worth Value	Maximum of These
A_1	$9742	$9742
A_2	7049	
A_3	5679	

Thus, alternative A_1 is selected as the alternative that will maximize the minimum present worth value that could occur. More formally stated, the maximin principle is to

Select the alternative, j, associated with the $\max_j \min_k V(\theta_{jk})$

In the case of a decision dealing with *losses* or *costs*, the minimax philosophy is also one of extreme pessimism. In applying the minimax principle to a matrix of costs, the maximum cost associated with each alternative (when viewed over all states) is determined, and the minimum value in the set of maximum values designates the alternative to be chosen. Again, this is choosing the best (or minimum) cost from the set of worst (or maximum) costs. An example to illustrate the application of the minimax principle is given in Table 8.5, where the cell values are equivalent uniform annual costs for the alternatives. Thus, from Table 8.5, alternative A_1 would be chosen. Formally stated, the minimax principle is to

Select the alternative, j, associated with the $\min_j \max_k V(\theta_{jk})$

8.5.3 Maximax and Minimin Principles

As illustrated in the previous section, the maximin principle for gains and the minimax principle for losses represent a philosophy of extreme pessimism on behalf of the decision maker. For the person who is an extreme optimist in a given decision situation, an *optimistic* rule for choice among alternatives that involve *gains* is the *maximax principle*. That is, the decision maker desires to select the alternative that affords the opportunity to obtain the largest value given in the matrix. A formal statement for the maximax principle is to,

Select the alternative, j, associated with the $\max_j \max_k V(\theta_{jk})$

Then, if the matrix cell values were losses, an optimistic philosophy of choice is to select the alternative that affords the opportunity to obtain the minimum loss value given in the matrix. A formal statement for the *minimin principle* involving *loss* values is

Select the alternative, j, associated with the $\min_j \min_k V(\theta_{jk})$

Applying the maximax principle to the example of Table 8.4, alternative A_3 is selected, and applying the minimin principle to the example of Table 8.5, alternative A_4 is selected.

Table 8.5 Matrix of Equivalent Uniform Annual Costs

States Alternatives	S_1	S_2	S_3	S_4	Maximum Cost	Minimum of These
A_1	$7 000	$10 000	$ 6 500	$ 5 000	$10 000	$10 000
A_2	9 000	4 000	9 000	15 000	15 000	
A_3	4 000	6 000	12 000	8 500	12 000	
A_4	2 000	8 000	3 000	11 000	11 000	

8.5.4 Hurwicz Principle

The *Hurwicz principle* considers that a decision maker's view may, in the case of gains, fall between the extreme pessimism of the maximin principle and the extreme optimism of the maximax principle and offers a method by which various levels of optimism/pessimism may be incorporated into the decision. The Hurwicz principle defines an index of optimism, α on a scale from 0 to 1. A value of $\alpha = 0$ thus indicates zero optimism or extreme pessimism and, conversely, a value of $\alpha = 1$ indicates extreme optimism.

Assuming that a decision maker is able to reflect a degree of optimism by assigning a specific value to α and again emphasizing that the decision is in terms of *gains*, the maximum gain value for each alternative is multiplied by α, and the minimum gain value for each alternative is multiplied by $(1-\alpha)$. The linear sum of these two products for each alternative j is called the Hurwicz value, H_j, and the alternative that maximizes this value is selected. That is,

Select an index of optimism, α, such that $0 \leq \alpha \leq 1$. For each alternative j, compute

$$H_j = \alpha \left[\max_k V(\theta_{jk}) \right] + (1-\alpha) \left[\min_k V(\theta_{jk}) \right]$$

and select the alternative that maximizes this quantity.

Note that if $\alpha = 0$ (extreme pessimism), only the minimum gain for each alternative will be examined, and the maximum of these values would then be selected—the maximin principle of choice. If $\alpha = 1$ (extreme optimism), the maximax principle would be executed.

Suppose the decision maker concerned with the example problem of Table 8.4 was a middle-of-the-road type person and assigns $\alpha = 0.5$. Then,

H_1 for $A_1 = (0.5)(\$15\,163) + (0.5)(\$9\,742) = \$12\,452.50$

H_2 for $A_2 = (0.5)(\$16\,536) + (0.5)(\$7\,049) = \$11\,792.50$

H_3 for $A_3 = (0.5)(\$18\,397) + (0.5)(\$5\,679) = \$12\,038.00$

Choosing the maximum of these values is to select alternative A_1.

Shortcomings of the Hurwicz principle include (1) ignoring intermediate values for each alternative, (2) the inability to select a particular alternative when two or more alternatives have the same Hurwicz value, and (3) the practical difficulty of assigning a specific value to α. This latter objection can be circumvented by the graphic solution discussed below.

It will be noted from the expression for calculating the Hurwicz value that the equation is linear in α, which ranges in value from 0 to 0.1. Any alternative thus yields a linear function that can be plotted and a preferred alternative determined for various ranges of α. An example to illustrate the procedure is given in Table 8.6, where the cell values are gains.

Table 8.6 Data for the Hurwicz Example

	S_1	S_2	S_3
A_1	$ 2	$10	$30
A_2	18	18	18
A_3	24	12	10

Applying the principle for the data of Table 8.6, the Hurwicz values are,

H_1 for $A_1 = \alpha(\$30) + (1-\alpha)(\$2) = 28\alpha + 2$

H_2 for $A_2 = \alpha(\$18) + (1-\alpha)(\$18) = 18$

H_3 for $A_3 = \alpha(\$24) + (1-\alpha)(\$10) = 14\alpha + 10$

These equations are plotted in Figure 8.2.

From Figure 8.2 it can be seen that if a decision maker's index of optimism is less than α_1, alternative A_2 would be preferred; if it is greater than α_1, alternative A_1 would be preferred. A specific value for α_1 can be determined graphically or analytically by, in this case, equating the Hurwicz value functions for any pair of alternatives (since there is a single intersection for the three linear functions). For example,

$H_1 = H_2$
$28\alpha_1 + 2 = 18$ and $\alpha_1 = 4/7$

If the decision problem involves costs or losses, a reinterpretation of the Hurwicz principle is necessary:

Select an index of optimism, α, such that $0 \leq \alpha \leq 1$. For each alternative A_j, compute

$$H_j = \alpha \left[\min_k V(\theta_{jk}) \right] + (1-\alpha) \left[\max_k V(\theta_{jk}) \right]$$

and select the alternative that minimizes this quantity.

Note that if $\alpha = 0$, only the maximum costs would be examined for each alternative and the minimum of these would be selected—the

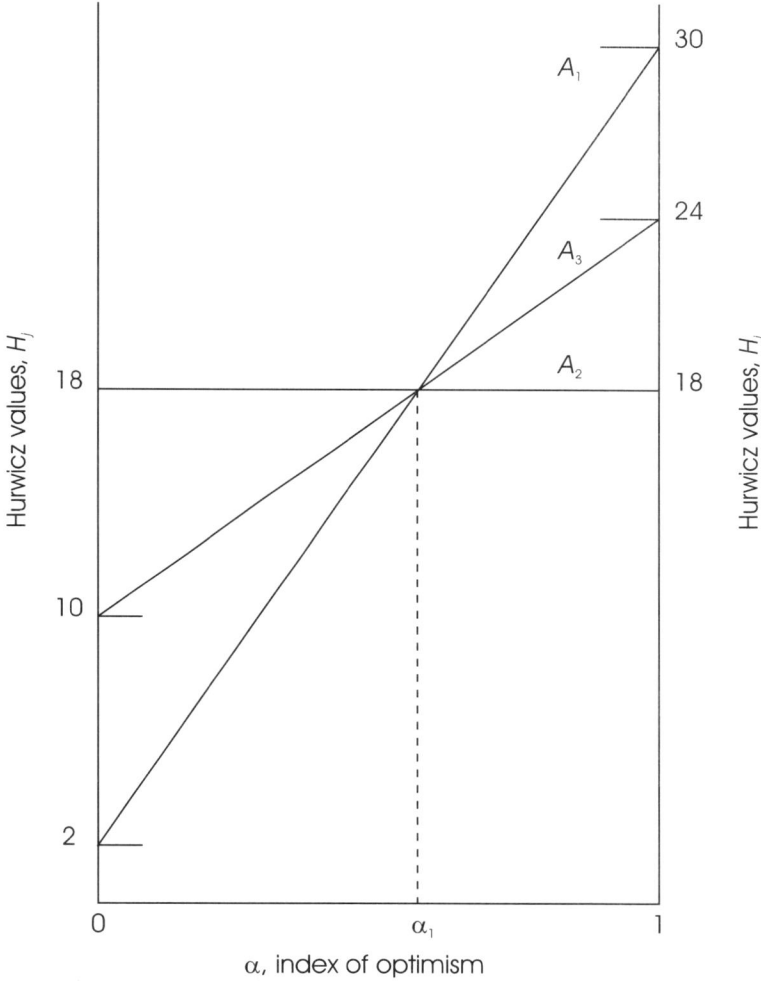

Fig. 8.2 Graphical interpretation of the Hurwicz principle

pessimistic minimax criterion. If $\alpha = 1$, the optimistic minimin principle would be executed.

8.5.5 Savage Principle (Minimax Regret)

This principle, proposed by L.J. Savage, introduces and defines a quantity termed *regret*. A matrix consisting of regret values is first developed. Then the maximum regret value for each alternative A_j is determined, and the alternative associated with the minimum regret value is chosen from the set of maximum regret values.

If the original decision matrix values are *gains,* the procedure for determining the regret matrix is as follows:

1. For a given state S_l, search the matrix column values over all alternatives and determine the largest gain. Assign this gain a zero regret value.
2. For all other gain values *under* the given state S_l, subtract these from the largest gain value from Step 1. The difference is interpreted as units of regret for a particular alternative A_j, given that state S_l occurs.
3. Repeat Steps 1 and 2 for each state S_k, $k \neq l$, until the regret matrix is completed.

Once the regret matrix is developed, examine each alternative over all states and select the maximum regret value. Then *choose the alternative that minimizes the maximum regret.* This principle of choice is conservative and is very similar to the maximin principle. The Savage principle deals in units of regret, not gain values in the original matrix; therefore, if both the maximin principle and the minimax regret principle are applied to a common decision problem, different alternatives may be selected.

Applying the Savage principle to the present-worth example of Table 8.4, the regret matrix given in Table 8.7 is obtained.

Table 8.7 Regret Matrix for the Minimax Regret Example

	S_1	S_2	S_3
A_1	$3 234	$ 0	$ 0
A_2	1861	1 028	2 693
A_3	0	1 122	4 063

The maximum regret values are $3 234, $2 693, and $4 063 for alternatives A_1, A_2, and A_3 respectively. Thus, the minimum of these is $2 693, and alternative A_2 would be preferred.

In a decision problem involving costs, the procedure for creating the regret matrix is the same as the problem involving gains.

1. For a given state S_l, search the matrix column values over all alternatives and determine the smallest cost. Assign this cost a zero regret value.

2. For all other cost values *under* the given state S_l, subtract the smallest cost value determined in Step 1 from these to determine regret values.

3. Repeat Steps 1 and 2 for each state S_k, $k \neq l$, until the regret matrix is completed.

After the regret matrix is obtained, the procedural steps for selecting an alternative are the same as before.

8.6 Sequential Decisions

In the previous sections, the decision situations happened once. We now wish to consider situations that may require multiple decisions in sequential fashion. The reader can no doubt recall from personal experience or imagine real-world decisions that are sequential. For example, medical decisions are typically sequential. Based on the results of a routine blood analysis, a physician may require additional medical tests. The physician may then perform exploratory surgery and, based on these findings, recommend additional surgery for the patient. Following surgery, the recuperative procedure may also be conditional and involve further sequential decisions on behalf of the physician.

Sequential decisions in the business and industrial world are also common. For example, the number and type of material-handling units to purchase may depend on the forecast of product demand in each of the next five years. In the field of quality control, sampling is concerned with sequential decisions. For example, if a sample of five units is taken from a production machine, subsequent actions taken in regard to the production machine might very well depend on whether 0, 1, 2, 3, 4, or 5 defective units were found in the sample of five units.

In this section, sequential decisions under risk will be discussed, the technique of *decision-tree* representation will be used, and the logic of Bayes's theorem will be applied in the solution of the decision-tree problem.

8.6.1 Decision Trees

Magee[3] is generally credited with representing sequential decisions in decision trees and advocating the expectation criterion for solving the problem. Hespos and Strassman[4] developed stochastic decision trees incorporating risk-analysis methods and simulation techniques into the basic decision-tree methodology originated by Magee. Before presenting the graphic technique and a solution procedure, it is necessary to define certain symbolism.

△	=	a decision point, which will be labelled D_i for the ith decision point in the tree
△—	=	a branch emanating from a decision point; represents an alternative that can be chosen at this point
○	=	a fork (or node) in the tree where chance events influence the outcomes of an alternative choice
○—	=	a branch representing a probabilistic outcome for a given alternative. The notation $(\theta; p)$ represents an outcome θ, having an associated value $V(\theta; p)$, which occurs with probability p. It is assumed that branches emanating from a fork in the tree represent mutually exclusive and collectively exhaustive outcomes such that their probabilities sum to 1.0
V	=	a value associated with a particular outcome (branch)

In Figure 8.3, a symbolic decision tree for a simple sequential decision problem is presented. The sequential decision problem depicted involves two decision points, D_1 and D_2. If alternative A_1 is chosen, it can have two outcomes, θ_1 and θ_2. If θ_1 occurs, then a decision, D_2, is

---

[3] John F. Magee, "Decision Trees for Decision Making," *Harvard Business Review*, 42 (4), July-August 1964, and "How to Use Decision Trees in Capital Investment," *Harvard Business Review*, 42 (5), Sept-Oct, 1964.

[4] Richard F. Hespos and Paul A. Strassman, "Stochastic Decision Trees for the Analysis of Investment Decisions," *Management Science*, 11 (10), August 1965.

Decision Models 405

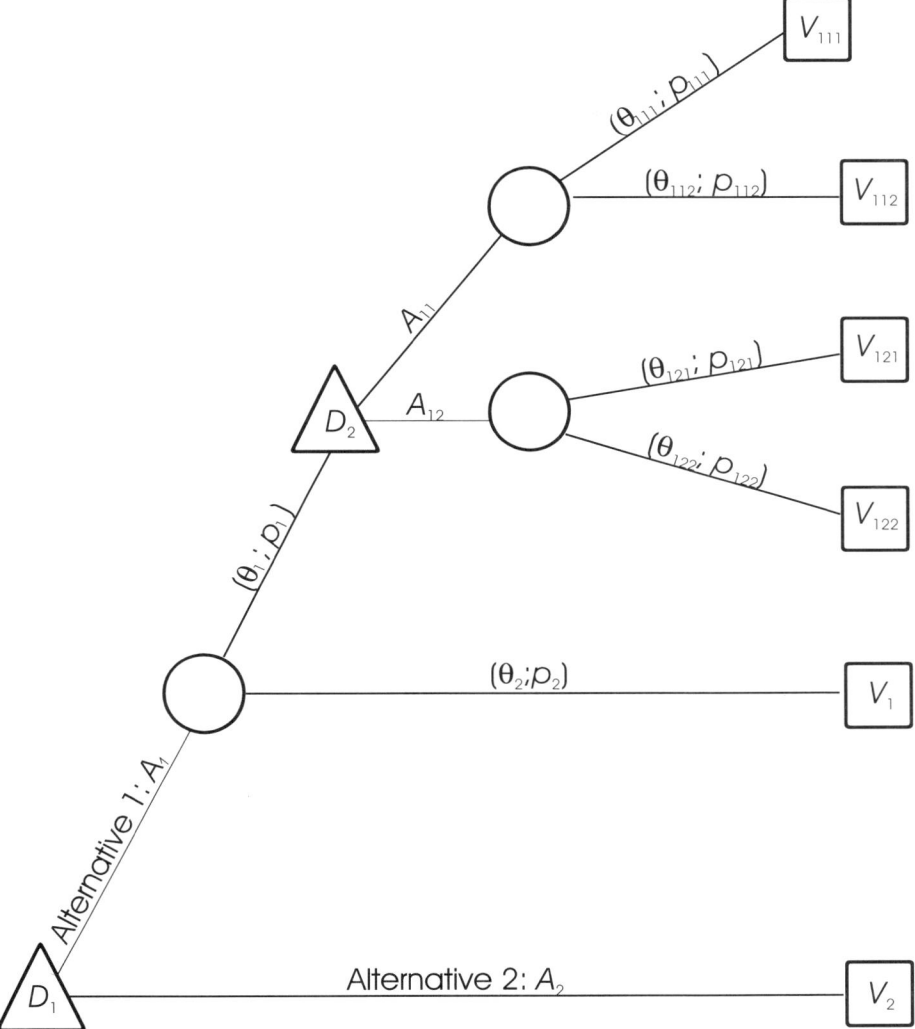

Fig. 8.3 A symbolic decision tree.

required. If alternative A_{11} is chosen, there can be the two outcomes θ_{111} and θ_{112}. If alternative A_{12} is chosen at D_2, there can be the two outcomes θ_{121} and θ_{122}. For each branch in the three there is an associated value. For example, if alternative A_1 is chosen *and* outcome θ_1 occurs *and* alternative A_{11} is chosen *and* outcome θ_{111} occurs, then value V_{111} results with the conditional probability p_{111}. The first decision to be made is at decision point D_1, and if alternative A_1 is chosen, then a second decision is required at decision point D_2. The ultimate question to answer is which alternative to choose at D_1.

The principle of *maximizing expected gain* or *minimizing expected loss* is adopted as the principle of choice at each decision point. The solution procedure advocated is to reach the best decision at the decision point (or points) most distant from the base (the first decision) of the tree. Then, replace this most distant decision point with the best expected value and work backward through decision points until the best decision is made at the initial decision point, D_1. To illustrate the solution procedure, consider the decision tree of Figure 8.3 and assume all the values are *gains*. The procedure follows.

1. At decision point D_2, calculate the expected gains for A_{11} and A_{12}. That is,

$$E(A_{11}) = p_{111}V_{111} + p_{112}V_{112}$$

and

$$E(A_{12}) = p_{121}V_{121} + p_{122}V_{122}$$

If $E(A_{11}) > E(A_{12})$, then alternative A_{11} is chosen as best at D_2. Assume this is the case.

2. Replace D_2 with $E(A_{11})$, and the new reduced decision tree becomes as shown in Figure 8.4.

3. Calculate the expected gains for alternatives A_1 and A_2. That is,

$$E(A_1) = p_1 E(A_{11}) + p_2 V_1$$

and

$$E(A_2) = V_2 \text{ (a certain event with } p = 1.0)$$

If $E(A_1) > E(A_2)$, alternative A_1 is chosen as the best at decision point D_1. If $E(A_2) > E(A_1)$, then A_2 would, of course, be chosen, and if $E(A_2) = E(A_1)$, then one would be indifferent between the choice of A_1 and A_2. Given that $E(A_1) > E(A_2)$ from Step 3, the optimal sequence of decisions is to choose A_1 at D_1 and then choose A_{11} later on if outcome θ_1 does occur.

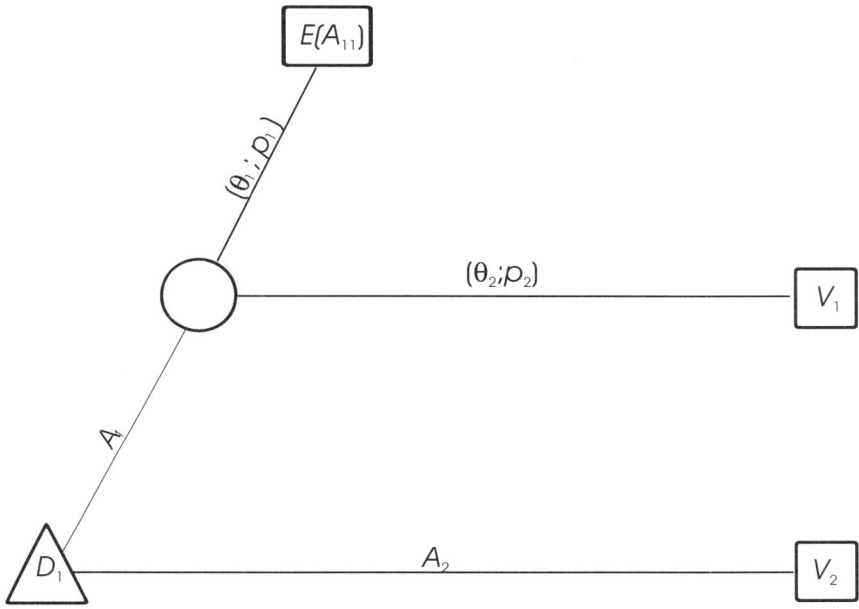

Fig. 8.4 The reduced decision tree for Figure 8.3.

Example 8.5

The Jax Tool and Engineering Company is a medium-size, job-order machine shop and has been in business for 10 years. After surviving two national economic recessions with shrewd and industrious management, the company has built a solid reputation for dependability and high-quality work. During the past two years, sales have sharply increased, and the company is operating near capacity. A market test has recently been made for a new product—electroplating tanks—and the response has been very favourable. However, production floor space to establish the new product line is limited in the present building. The people in management have three alternatives: (1) do nothing, (2) rearrange an area of existing floor space, or (3) build an addition to the present plant. Also, there is the consideration of competition from one other job-order machine shop in the local area. If the electroplating tank venture is successful and the market demand cannot be fully met by the Jax Company, there is a good chance that competition will occur.

Management assigns representatives from production engineering and marketing to make a detailed study of the alternatives. Because management wants a report as soon as possible, it is necessary to make rather gross estimates on many items in the analysis. The analysts decide to use a five-year planning period, a minimum attractive rate of return of 30%, and an incremental present worth value for each alternative relative to the do-nothing alternative. The results of the analysis now follow.

Rearrangement does not require any new construction, but machinery must be moved, some new machinery purchased, and new storage areas created. Production would be limited to 10 tanks per week on a two-shift operation. If there is competition in the first year from the other local firm, the Jax Company would probably take no further action, and a present worth value of $50 000 is estimated for this outcome. On the other hand, if there is no competition in the first year, the Jax Company will not be able to meet the expected demand. Subjective probabilities of 0.4 and 0.6 are assigned to the outcomes of no competition in the first year and competition in the first year, respectively. If no competition occurs, Jax management would then face the decision of either expanding the product line by building an addition or not expanding.

Expansion at this point would be expensive; a building addition plus a layout of the existing plant would be involved. However, no competition is expected if the expansion is made now and demand can be met with increased capacity. A present value of $30 000 is estimated for this outcome. If there is no expansion, two outcomes are possible. One possibility is that the other local machine shop can still enter the competition. However, a one-year advantage in the market will have been gained, and the present worth of this outcome is estimated to be $10 000. It is very unlikely that competition will occur after a two-year period, and the probability of such an event is assumed to be zero. A subjective probability value of 0.7 is assigned to the outcome of competition in the second year. The other possible outcome is, of course, no competition in the second year, with a probability of 0.3 and estimated present worth of $50 000.

The "build addition initially" alternative has higher first and recurring costs than the "new layout" alternative, but production capacity is estimated at 25 tanks per week, which should satisfy average annual demand over the next five years. With adequate production capacity initially, the chance of competition is low and about 20%. The present worth estimate in the event of no competition is $90 000. If there is competition, then the reduced sales and high annual fixed costs for the new building would result in a present worth of −$100 000. A decision-tree representation of this example is given in Figure 8.5 (values are in thousands of dollars).

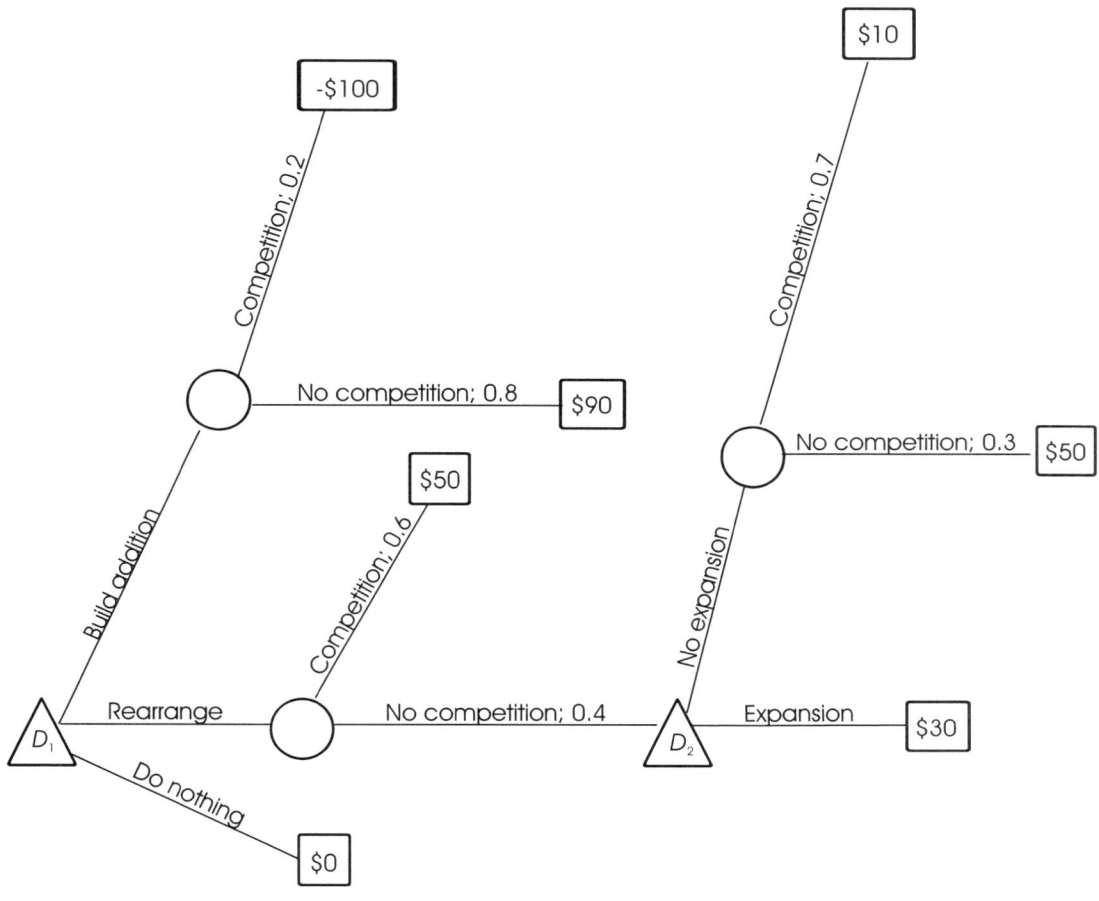

Fig. 8.5 The Jax Company example—original tree.

Using the principle of choice of maximizing expected present worth, a best decision is sought at D_2, the decision point most distant from the base of the tree, D_1. The expected values at D_2 are

$$E(\text{expansion}) = \$30\,000$$

and

$$E(\text{no expansion}) = \$10\,000(0.7) + \$50\,000(0.3) = \$22\,000$$

Thus, the alternative of expansion is chosen to maximize expected present worth, and this *best* value replaces D_2 in the tree, as shown in Figure 8.6.

From Figure 8.6, the expected values at D_1 are calculated as:

$$E(\text{build addition}) = -\$100\,000(0.20) + \$90\,000(0.80)$$
$$= \$52\,000$$
$$E(\text{rearrange}) = \$50\,000(0.60) + \$30\,000(0.40)$$
$$= \$42\,000$$
$$E(\text{do nothing}) = \$0$$

Thus, the alternative of building an addition should be chosen in order to maximize expected present worth. It is interesting to note from Figure 8.6 that if this alternative is indeed chosen, the Jax Company is exposed to the possibility of a −$100000 present

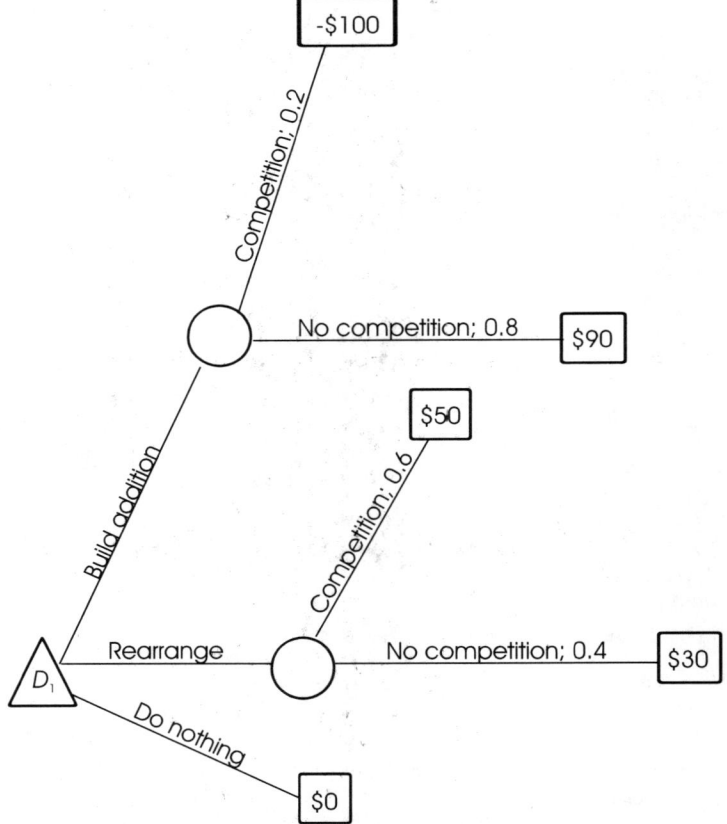

Fig. 8.6 The Jax Company example—reduced tree.

worth. This event is "smoothed" by the expectation principle, on which most decision-tree analysis is based. The $-\$100\,000$ figure is not necessarily a dollar loss, since a $MARR = 30\%$ was assumed in the calculation of the values. However, management may wish to choose the "rearrange" alternative instead, because of the large negative figure; if so, they would be exercising a maximin philosophy.

Example 8.6

For further illustration of the decision-tree approach to a problem and exposure to fundamental laws of probability, consider an urn containing five white (W) balls and seven black (B) balls. Three consecutive, independent, random (without bias in the selection) draws are made from the urn without replacing any ball drawn. The probability of selecting a white or black ball on the first draw is unconditional; that is, it does not depend on a previous draw (event). The probability of selecting a white or black ball on the second draw is, however, conditional on the colour of ball selected on the first draw, and the third draw is conditional on both the first and second draws. In this example, we are interested in all the possible sequences of colours for the three draws and their probabilities of occurrence. A decision-tree representation of this problem is helpful and given in Figure 8.7.

All the eight possible sequences of colours are indicated in Figure 8.7. The objective probabilities of occurrence for each sequence will be determined by a relative-frequency argument. That is, for a general event A, the probability of A, $P(A)$, is calculated as the ratio

$$P(A) = \frac{\text{number of ways favourable for the event}}{\text{total number of ways possible}}$$

In the urn example, the probability of the unconditional event "a white ball on the first draw," W_1, is given by

$$P(W_1) = \frac{5 \text{ white balls (favourable number)}}{12 \text{ balls (total number)}}$$

Similarly, the probability of the unconditional event "a black ball on the first draw," B_1, is

$$P(B_1) = \frac{7}{12}$$

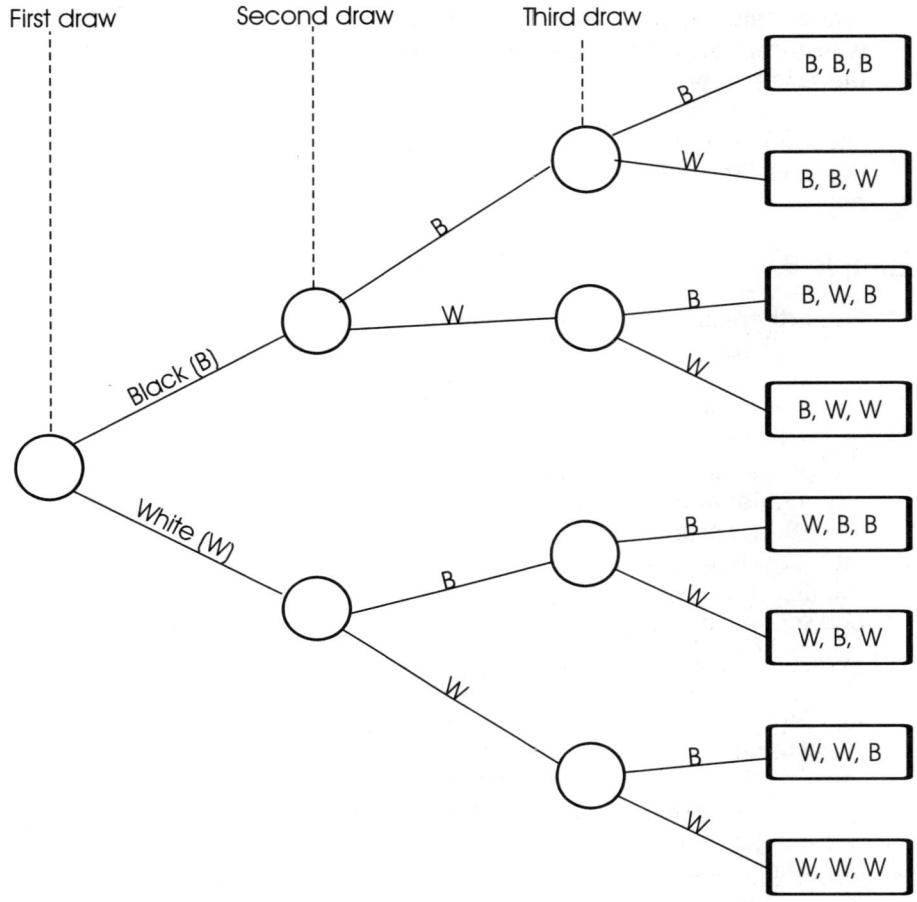

Fig. 8.7 Tree representation of urn example.

For the second draw, the probability of the event "a white ball on the second draw," W_2, is conditional on whether the first draw results in a white or black ball [i.e., $P(W_2|W_1)$ or $P(W_2|B_1)$]. The probability that a white ball is selected on the second draw given that a white ball is selected on the first draw is reasoned to be

$$P(W_2|W_1) = \frac{4 \text{ (white balls remaining after first draw)}}{11 \text{(total balls remaining after first draw)}}$$

Similarly,

$$P(W_2|B_1) = \frac{5}{11}$$

The probability of the event "a black ball on the second draw," B_2, is also conditional on whether the first draw results in a white or black ball, and

$$P(B_2|W_1) = \frac{7}{11}$$

$$P(B_2|B_1) = \frac{6}{11}$$

From the above results, note that

$$P(W_2|W_1) + P(B_2|W_1) = 1.0$$

and

$$P(W_2|B_1) + P(B_2|B_1) = 1.0$$

If a white (or black) ball is in fact drawn first, then the second draw can only yield two possible outcomes (collectively exhaustive): either a white or black ball. Since these two outcomes are mutually exclusive (cannot occur together), the individual probabilities of occurrence are additive and sum to 1.0 or certainty. In general, for two mutually exclusive events A and B, the additive law of probability states

$$P(\text{either } A \text{ or } B) = P(A \cup B) = P(A) + P(B) \tag{8.4}$$

where the symbol ($\cup$) stands for the union of the events. Thus, in this example,

$$P(W_2|W_1 \cup B_2|W_1) = P(W_2|W_1) + P(B_2|W_1)$$

and

$$P(W_2|W_1 \cup B_2|B_1) = P(W_2|B_1) + P(B_2|B_1)$$

The additive law is easily expanded for multiple, mutually exclusive events.

For the third draw, if the ball selected on each of the first and second draws is white, then the probability of selecting a white ball on the third draw is denoted $P(W_3|W_1,W_2)$ and reasoned, by the relative-frequency logic, to be

$$P(W_3|W_1,W_2) = \frac{3 \text{ (white balls remaining after first and second draw)}}{10 \text{ (total balls remaining after first and second draw)}}$$

The symbolism $(W_3|W_1,W_2)$ is interpreted to mean the event of drawing a white ball on the third draw given that a white ball was selected on the first draw *and* a white ball was selected on the second draw.

Similarly,

$$P(B_3|W_1,W_2) = \frac{7}{10}$$

$$P(W_3|W_1,B_2) = \frac{4}{10}$$

$$P(B_3|W_1,B_2) = \frac{6}{10}$$

$$P(W_3|B_1,W_2) = \frac{4}{10}$$

$$P(B_3|B_1,W_2) = \frac{6}{10}$$

$$P(W_3|B_1,B_2) = \frac{5}{10}$$

$$P(B_3|B_1,B_2) = \frac{5}{10}$$

A summary of all these events and their probabilities is shown in Figure 8.8.

Up to this point the probabilities of each individual outcome (event) of the example have been determined. However, the probability of each sequence (branch of the tree) occurring is yet to be determined (see Figure 8.7 to recall the possible sequences of colours). For instance, you may want to know the probability of the joint event of a white ball being selected with each of the three draws or $P(W_1,W_2,W_3)$, which may also be written with set notation as $P(W_1 \cap W_2 \cap W_3)$ to mean the joint occurrence of these three events (conditional events in this particular case). This probability may be calculated as

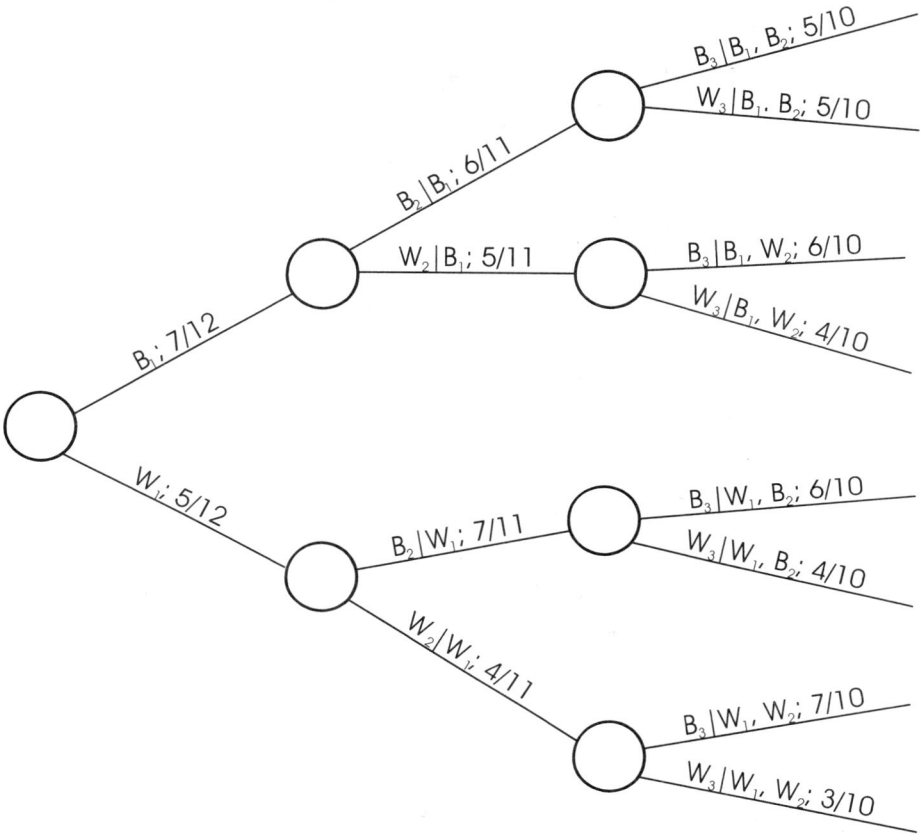

Fig. 8.8 Probability tree for urn example.

$$P(W_1, W_2, W_3) = P(W_1) \cdot P(W_2|W_1) \cdot P(W_3|W_1, W_2)$$
$$= \left(\frac{5}{12}\right)\left(\frac{4}{11}\right)\left(\frac{3}{10}\right) = \frac{60}{1320}$$

A full justification of the above multiplicative calculation requires a probability theory development that is outside the scope of this textbook. However, the calculation of the joint occurrence of three conditional events is a direct extension of the conditional probability theorem presented later in this chapter. For the other possible sequences of this example,

$$P(W_1, W_2, B_3) = P(W_1) \cdot P(W_2|W_1) \cdot P(B_3|W_1, W_2)$$
$$= \left(\frac{5}{12}\right)\left(\frac{4}{11}\right)\left(\frac{7}{10}\right) = \frac{140}{1320}$$

$$P(W_1, B_2, W_3) = P(W_1) \cdot P(B_2|W_1) \cdot P(W_3|W_1, B_2)$$
$$= \left(\frac{5}{12}\right)\left(\frac{7}{11}\right)\left(\frac{4}{10}\right) = \frac{140}{1320}$$

$$P(W_1, B_2, B_3) = P(W_1) \cdot P(B_2|W_1) \cdot P(B_3|W_1, B_2)$$
$$= \left(\frac{5}{12}\right)\left(\frac{7}{11}\right)\left(\frac{6}{10}\right) = \frac{210}{1320}$$

$$P(B_1, B_2, B_3) = P(B_1) \cdot P(B_2|B_1) \cdot P(B_3|B_1, B_2)$$
$$= \left(\frac{7}{12}\right)\left(\frac{6}{11}\right)\left(\frac{5}{10}\right) = \frac{210}{1320}$$

$$P(B_1, B_2, W_3) = P(B_1) \cdot P(B_2|B_1) \cdot P(W_3|B_1, B_2)$$
$$= \left(\frac{7}{12}\right)\left(\frac{6}{11}\right)\left(\frac{5}{10}\right) = \frac{210}{1320}$$

$$P(B_1, W_2, B_3) = P(B_1) \cdot P(W_2|B_1) \cdot P(B_3|B_1, W_2)$$
$$= \left(\frac{7}{12}\right)\left(\frac{5}{11}\right)\left(\frac{6}{10}\right) = \frac{210}{1320}$$

$$P(B_1, W_2, W_3) = P(B_1) \cdot P(W_2|B_1) \cdot P(W_3|B_1, W_2)$$
$$= \left(\frac{7}{12}\right)\left(\frac{5}{11}\right)\left(\frac{4}{10}\right) = \frac{140}{1320}$$

The sum of the probabilities for each of these eight possible sequences is $\frac{1320}{1320} = 1.0$, and the complete probability tree is shown in Figure 8.9.

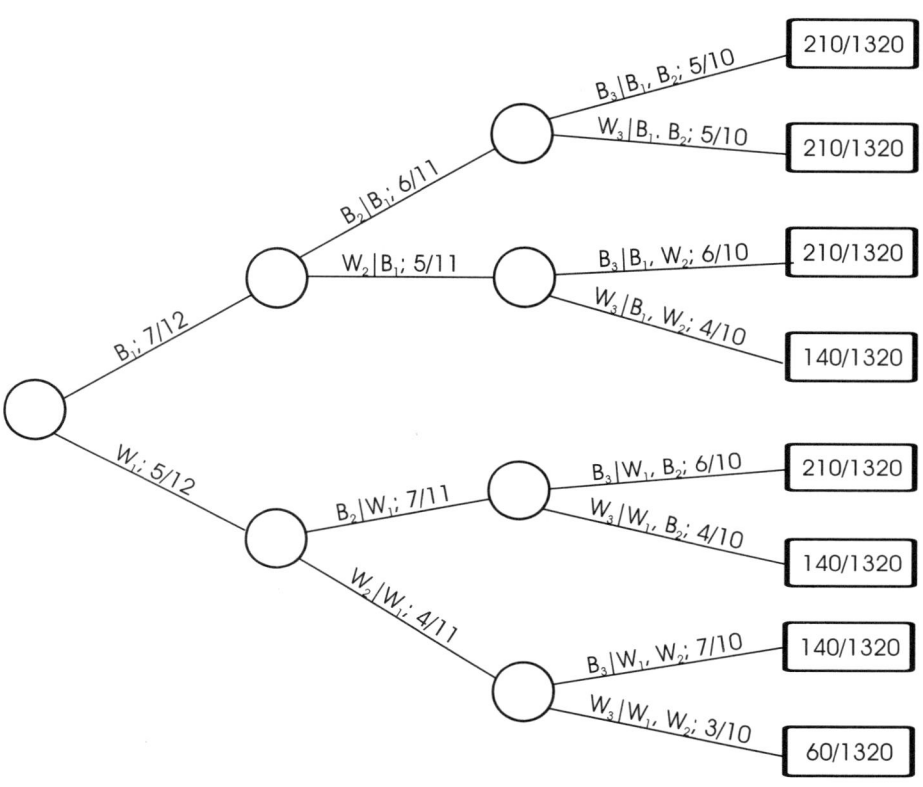

Fig. 8.9 Complete probability tree for urn example.

8.6.2 The Conditional Probability Theorem

The probability of the joint occurrence of two dependent events, A and B, is given by

$$P(A \text{ and } B) = P(A \cap B) = P(A) \cdot P(B|A) \qquad (8.6)$$

where $P(A)$ is an unconditional probability and $P(B|A)$ is the conditional probability of event B given that event A has occurred or will occur. If event B is not conditional on event A, then events A and B are *independent* events and Equation 8.6 may be written as

$$P(A \cap B) = P(A) \cdot P(B) \qquad (8.7)$$

An extension of Equation 8.6 for three dependent events is given by

$$P(A \cap B \cap C) = P(A) \cdot P(B|A) \cdot P(C|A,B) \qquad (8.8)$$

which was the relationship used in the previous urn example.

For the purpose of the development to follow, Equation 8.6 is now written in its complete form as

$$P(A \cap B) = P(A) \cdot P(B|A) = P(A|B) \cdot P(B) \qquad (8.9)$$

from which it follows that

$$P(B|A) = \frac{P(A|B) \cdot P(B)}{P(A)} \qquad (8.10)$$

or

$$P(A|B) = \frac{P(B|A) \cdot P(A)}{P(B)} \qquad (8.11)$$

Example 8.7

Either Equation 8.10 or 8.11 is a form of Bayes's theorem, which may provide a logical guide to sequential decision making under risk. To illustrate this claim, let us consider another urn example where one urn, U_1, contains two white and eight black balls; the other urn, U_2, contains five white and five black balls. Suppose the urn contents are known to you, but the urns themselves are hidden from your view. Now, another person selects a single ball from one of the urns. Without knowing the colour of the ball drawn, which urn would you guess was selected? Then, if the person reported a black ball had been drawn, which urn would you guess had been selected?

Without benefit of the sample information concerning the colour of the ball drawn, most persons would feel that it was equally likely that U_1 or U_2 had been chosen. That is, since $P(U_1) = P(U_2) = \frac{1}{2}$, guess either urn. However, after the sample information of a black ball is received, intuition suggests that U_1 was the urn selected because the probability of a black ball in U_1 is greater than the probability of a black ball in U_2. We now wish to formalize this example and show that Bayes's theorem can be used to support the intuitive guess of U_1 after receiving the report.

Before receiving the report of a black ball, the prior belief that U_1 or U_2 was selected is subjectively stated as

$P(U_1)$ = the *prior* probability of selecting U_1
$$= \tfrac{1}{2}$$

and

$P(U_2)$ = the *prior* probability of selecting U_2
$$= \tfrac{1}{2}$$

Note that

W = the event of a white ball being drawn

and

B = the event of a black ball being drawn

Also, before receiving the sample information, the likelihood of a particular report (W or B), given that U_1 or U_2 was selected, can be determined, based on knowledge of the contents of each urn. That is,

$P(W|U_1)$ = the conditional probability of a white ball being drawn given that U_1 was selected for the draw
$$= \frac{2}{10}$$

Similarly,

$$P(B|U_1) = \frac{8}{10}$$

$$P(W|U_2) = \frac{5}{10}$$

$$P(B|U_2) = \frac{5}{10}$$

In the literature on sequential decision making, these conditional probabilities, either objectively known as above or subjectively assigned, are termed *likelihood statements.*

Ultimately, it is desired to assess the probability of whether U_1 or U_2 was chosen based on the sample information received (the report of a black ball in this case). Thus, the *posterior probabilities* of $P(U_1|B)$, $P(U_2|B)$, $P(U_1|W)$, and $P(U_2|W)$ need to be determined.

In the calculation of these posterior probabilities, it is convenient to calculate the probability of event W and event B before the ball is drawn. It is reasoned that event W could occur if either U_1 or U_2 were chosen. The same is true for the event B. Thus, B could occur if (1) U_1 were selected and B occurred, or (2) U_2 were selected and B occurred. That is,

$$P(B) = P(U_1 \text{ and } B \text{ or } U_2 \text{ and } B)$$
$$= P(U_1 \cap B \cup U_2 \cap B)$$
$$= P(U_1 \cap B) + P(U_2 \cap B)$$

Then, using Equation 8.9,

$$P(B) = P(U_1)P(B|U_1) + P(U_2)P(B|U_2)$$

and, from the previous data,

$$P(B) = \left(\frac{1}{2}\right)\left(\frac{8}{10}\right) + \left(\frac{1}{2}\right)\left(\frac{5}{10}\right) = \frac{13}{20}$$

Similarly,

$$P(W) = P(U_1)P(W|U_1) + P(U_2)P(W|U_2)$$
$$= \left(\frac{1}{2}\right)\left(\frac{2}{10}\right) + \left(\frac{1}{2}\right)\left(\frac{5}{10}\right) = \frac{7}{20}$$

With these results, the posterior probabilities may be calculated from Bayes's theorem as

$$P(U_1|B) = \frac{P(B|U_1)P(U_1)}{P(B)} = \frac{(8/10)(1/2)}{13/20} = \frac{8}{13}$$

$$P(U_2|B) = \frac{P(B|U_2)P(U_2)}{P(B)} = \frac{(5/10)(1/2)}{13/20} = \frac{5}{13}$$

$$P(U_1|W) = \frac{P(W|U_1)P(U_1)}{P(W)} = \frac{(2/10)(1/2)}{7/20} = \frac{2}{7}$$

Decision Models

$$P(U_2|W) = \frac{P(W|U_2)P(U_2)}{P(W)} = \frac{(5/10)(1/2)}{7/20} = \frac{5}{7}$$

It is noted from the results of $P(U_1|B) = \frac{8}{13}$ and $P(U_2|B) = \frac{5}{13}$ that the event of B should therefore increase the belief that U_1 was selected. That is, the prior probability of U_1 was $\frac{1}{2}$ and the posterior probability, given the event B, of U_1 is $\frac{8}{13}$. If, after receiving the information of B, the guess has been U_1, then the above calculations would support the intuitive guess. If a second sample of a ball is taken from an urn, then the posterior probabilities $P(U_1|B)$ and $P(U_2|B)$ become the prior probabilities $P(U_1)$ and $P(U_2)$ for the second trial.

8.6.3 The Value of Perfect Information

In order to use the logic of Bayes's theorem in a more practical way to aid decision making, it is of interest to evaluate the value of the sample information in a monetary sense. Recalling the previous example, suppose the person who drew the ball from the urn had offered to pay a $10 prize if the correct urn were guessed. However, the person will charge for the sample information of whether a white or black ball is drawn. The question then is how much should be paid for the *sample information*. Before answering this particular question, a rationale for determining the *value of perfect information* is considered (i.e., the knowledge of exactly which urn was selected for the draw).

The prior probabilities of $P(U_1) = P(U_2) = \frac{1}{2}$ are now recalled and written with the more suggestive notation of $PR(U_1) = PR(U_2) = \frac{1}{2}$. Without any sample information, one should be indifferent as to which urn is guessed, and the *prior expected profit* (PEP) would be $5, as determined from the matrix of Table 8.8.

Table 8.8 Matrix for the Urn Example

	$PR(U_1) = \frac{1}{2}$	$PR(U_2) = \frac{1}{2}$
Actual Urn Selected	U_1	U_2
Urn Guessed		
U_1	$10	$ 0
U_2	$ 0	$10

The prior expected profit is $5, as determined from either of the two calculations below.

$$E(\text{profit}|\text{guess } U_1) = \left(\frac{1}{2}\right)(\$10) + \left(\frac{1}{2}\right)(\$0) = \$5$$

$$E(\text{profit}|\text{guess } U_2) = \left(\frac{1}{2}\right)(\$0) + \left(\frac{1}{2}\right)(\$10) = \$5$$

If these expected profits were different, the alternative with the larger expected value would be chosen, and the prior expected profit is the larger value.

Now, given that the person drawing the ball states that U_1 was selected and that the person is perfectly reliable, U_1 would obviously be guessed and a $10 payoff received. The same is true, of course, for the report of U_2. The *expected profit given perfect information (EP|PI)* is thus $10 in this example. It is now reasoned that the maximum amount one should pay for perfect information is $10 − $5 (expected profit without any information) = $5. We can formalize the *expected value of perfect information (EVPI)* as the difference between the *expected profit given perfect information (EP|PI)* and the *prior expected profit (PEP)*, or

$$EVPI = EP|PI - PEP \tag{8.12}$$

8.6.4 The Value of Imperfect Information

In most real-world decision situations, information is incomplete and not perfectly reliable, as in the urn example just discussed. The report that a black ball had been drawn gave additional information to the decision maker, but still did not provide the absolute answer as to which urn was selected. In order to define the *expected value of sample information (EVSI)*, the posterior probability (with new notation) results are repeated below.

For the event B, the posterior probabilities were

$$PO(U_1|B) = \frac{8}{13}$$

and

$$PO(U_2|B) = \frac{5}{13}$$

Thus, since $\frac{8}{13} > \frac{5}{13}$, the best guess is that U_1 was selected. The actual urn chosen could have been U_1 or U_2, and the expected profit given a guess of U_1 is

$$E(\text{profit}|\text{guess } U_1) = (\text{expected profit}|\text{guess } U_1 \text{ and is } U_1)PO(U_1)$$
$$+ (\text{expected profit}|\text{guess } U_1 \text{ and is } U_2)PO(U_2)$$
$$= (\$10)\left(\frac{8}{13}\right) + (\$0)\left(\frac{5}{13}\right)$$
$$= \frac{\$80}{13}$$

For the event W, the posterior probabilities were

$$PO(U_1|W) = \frac{2}{7}$$

$$PO(U_2|W) = \frac{5}{7}$$

Thus, if the report had been W, and since $\frac{5}{7} > \frac{2}{7}$, the best guess would be U_2 with

$$E(\text{profit}|\text{guess } U_2) = (\text{expected profit}|\text{guess } U_2 \text{ and is } U_1)PO(U_1)$$
$$+ (\text{expected profit}|\text{guess } U_2 \text{ and is } U_2)PO(U_2)$$
$$= (\$0)\left(\frac{2}{7}\right) + (\$10)\left(\frac{5}{7}\right)$$
$$= \frac{\$50}{7}$$

However, *before* the sample is taken and the report is given, both events B and W are possible with $P(B) = \frac{13}{20}$ and $P(W) = \frac{7}{20}$. The *expected profit given sample information* ($EP|SI$) is

$$EP|SI = (\text{expected profit}|B)P(B) + (\text{expected profit}|W)P(W)$$
$$= \left(\frac{\$80}{13}\right)\left(\frac{13}{20}\right) + \left(\frac{\$50}{7}\right)\left(\frac{7}{20}\right)$$
$$= \$6.50$$

In the above calculations, both expected profit values (expected profit$|B$) and (expected profit$|W$) assume that the best guess would, in fact, be made for the reports of B and W, respectively.

Finally, it is reasoned that the *expected value of sample information* is the difference between the *expected value given sample information* and the *prior expected profit*, or

$$EVSI = EP|SI - PEP$$
$$= \$6.50 - \$5$$
$$= \$1.50 \tag{8.13}$$

which is the maximum amount one should pay for the sample information of this example. This vaue is, of course, less than the $5 value for perfect information.

A decision-tree representation of this example is given in Figure 8.10 for a sample information charge of $1.50.

From Figure 8.10, the appropriate calculations at D_2 are

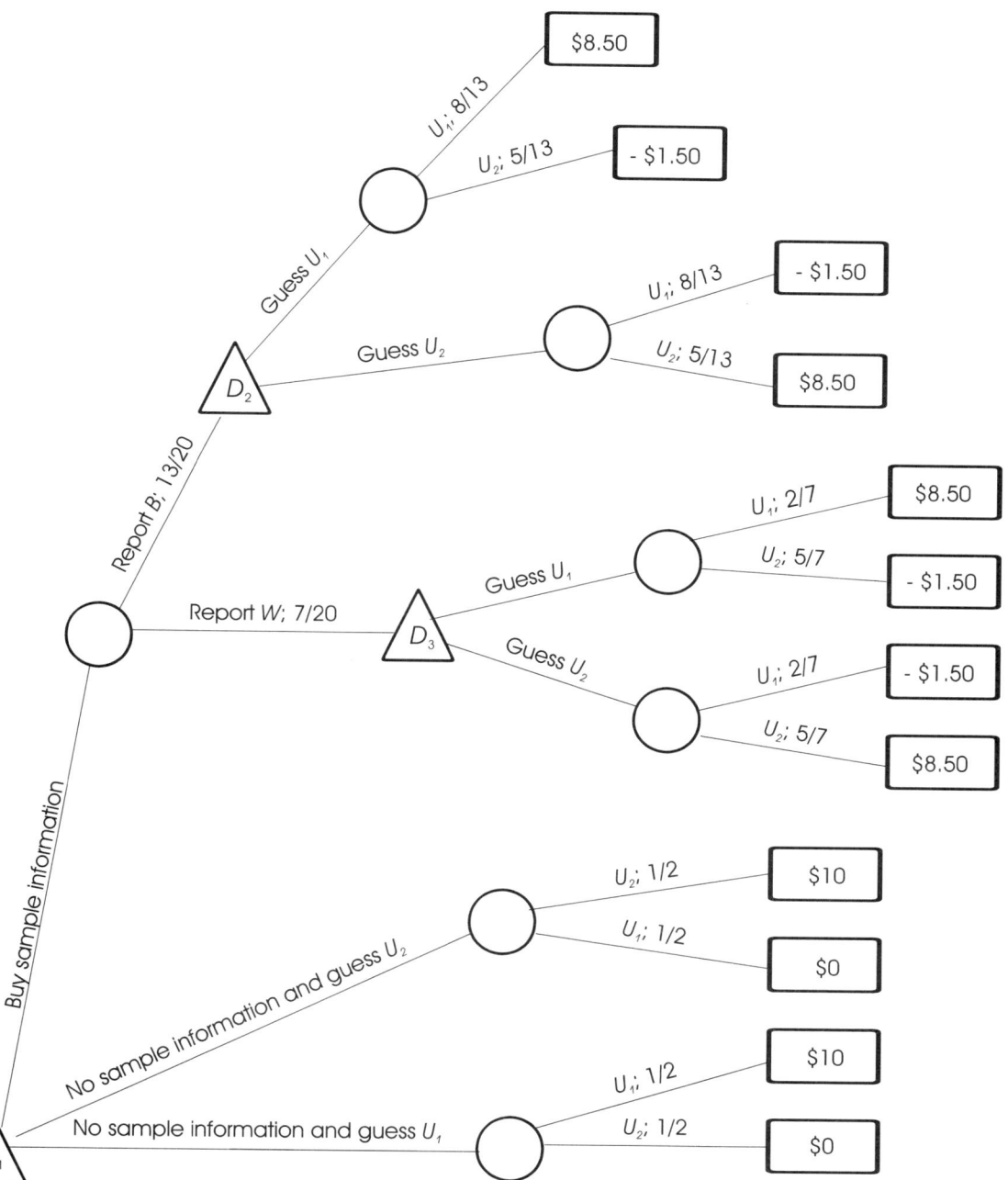

Fig. 8.10 Decision tree for urn example.

$$E(\text{guess } U_1) = (\$8.50)\left(\frac{8}{13}\right) + (-\$1.50)\left(\frac{5}{13}\right)$$

$$= \frac{\$60.5}{13}$$

$$E(\text{guess } U_2) = (-\$1.50)\left(\frac{8}{13}\right) + (\$8.50)\left(\frac{5}{13}\right)$$

$$= \frac{\$30.5}{13}$$

Since $\frac{\$60.5}{13} > \frac{\$30.5}{13}$, the best guess given B would be U_1.

The appropriate calculations at D_3 are

$$E(\text{guess } U_1) = (\$8.50)\left(\frac{2}{7}\right) + (-\$1.50)\left(\frac{5}{7}\right)$$

$$= \frac{\$6.5}{7}$$

$$E(\text{guess } U_2) = (-\$1.50)\left(\frac{2}{7}\right) + (\$8.50)\left(\frac{5}{7}\right)$$

$$= \frac{\$39.5}{7}$$

Since $\frac{\$39.5}{7} > \frac{\$6.5}{7}$, the best guess given W would be U_2.

Replacing D_2 and D_3 with these best expected values yields the reduced decision tree of Figure 8.11. From Figure 8.11, the appropriate calculations at D_1 are

$$E(\text{buy sample information}) = \left(\frac{\$60.5}{13}\right)\left(\frac{13}{20}\right) + \left(\frac{\$39.5}{7}\right)\left(\frac{7}{20}\right)$$

$$= \$5$$

Decision Models

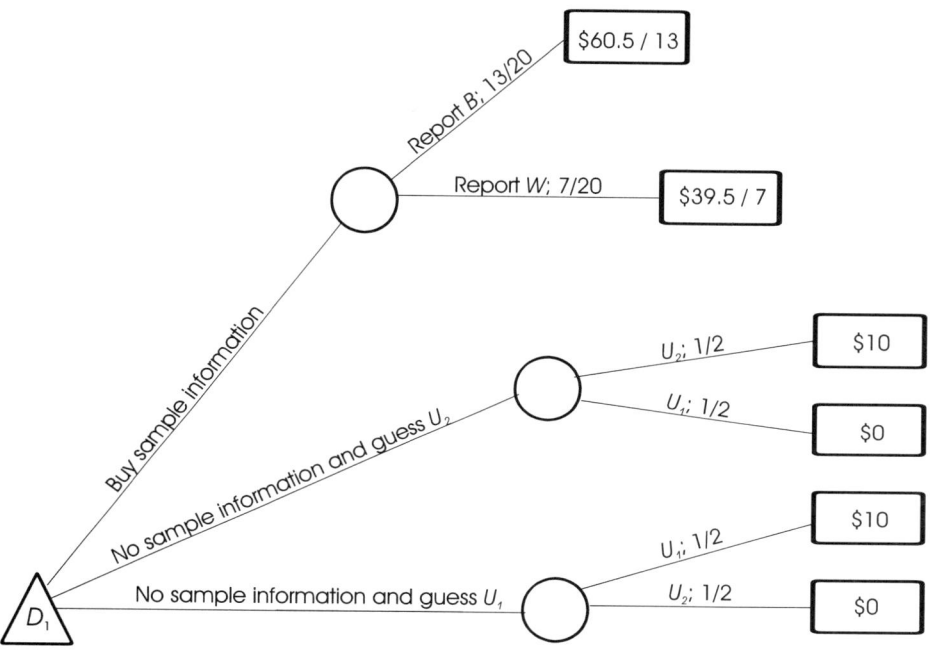

Fig. 8.11 Reduced decision tree for urn example.

$$E(\text{no sample information and guess } U_2)$$
$$= E(\text{no sample information and guess } U_1)$$
$$= \$5$$

Thus, the three alternatives at D_1 are equally desirable given that the charge for sample information is $1.50. If the charge is less than this amount, the decision maker should choose the "buy sample information" alternative in order to maximize expected profits.

For a similar sequential decision problem involving only costs, the basic relationships to determine expected value of information become

$$EVPI = \text{prior expected cost}$$
$$\quad - \text{expected cost given perfect information}$$
$$= PEC - EC|PI \tag{8.14}$$

$EVSI$ = prior expected cost
 − expected cost given sample information
= $PEC - EC|SI$ (8.15)

Example 8.8[5]

The editor of the ABC Book Company is considering a manuscript for an historical novel. The editor estimates the book can be marketed for $5, with the company receiving $4 and the author $1. Since other publishers are interested in the manuscript, the author wants a commitment from the ABC Book Company in the very near future. ABC has already invested $300 in a preliminary review of the manuscript and a quick market survey. The editor considers this manuscript competitive with other recently published historical novels and the author, although relatively new, has other published works that have received good reviews.

The editor identifies the immediate alternatives: (1) reject the manuscript, (2) accept the manuscript, or (3) obtain an additional review of the manuscript by an expert. An additional review will cost $400, and the expert's opinion will not perfectly predict the historical novel's success in the market. The editor reasons that at best the expert can only rate the manuscript good, fair, or poor.

If the manuscript is accepted for publication, the editor speculates on the possible market outcomes; there could be (1) low market demand, or (2) the novel could be a best-seller. Additionally, there is the question of movie rights. There may be no interest in filming the novel, but, on the other hand, a best-seller could yield $30 000 for the movie rights. Even if there is low market demand, a film company might pay $4 000 for the rights to make a low-budget, grade B film.

The ABC Book Company has available historical data on 60 recent manuscripts of similar books, given in Table 8.9.

---

[5] This example was presented at the Sixth Triennial Symposium, Engineering Economy Division, ASEE, June 19-20, 1971, in a paper entitled "Introduction to Decision Theory" by Barnard E. Smith. The article appears in the publication, *Decision and Risk Analysis: Powerful New Tools for Management*, The Engineering Economist, Stevens Institute of Technology, Hoboken, NJ, and is presented here by the permission of the publisher.

Decision Models

Table 8.9 Sixty Manuscript Decisions/Outcomes

Decision/Outcome When Marketed	No Outside Expert Review	Outside Reviewer's Evaluation			Total
		Good	Fair	Poor	
Not published	5	0	14	12	31
Low market demand	2	0	4	3	9
Best-seller	1	2	4	0	7
Best-seller + movie	1	4	3	0	8
Low market + B movie	1	0	3	1	5
Total	10	6	28	16	60

From available sales figures, the editor can estimate reasonably well the revenues given a particular outcome for a published book. The editor considers the present worth of after-tax revenues over the effective life of a book using a 10% expected rate of return. For a best-seller, a present worth of $70 000 is estimated, and a low-demand publication yields an estimated $5 000 present worth. Offers for movie rights, if any, are usually made one year after the manuscript is under contract, and these after-tax revenues should be added to the book sales (assume that $4 000 for a grade B film and $30 000 for a "best-seller film" are appropriate figures.) It costs $15 000 to publish a book.

If it is assumed the editor has a linear utility function for dollars and uses a "maximize present worth" principle for choice among alternatives, what decision should be made at the present time?

The editor faces these immediate alternatives:
A_1—reject the manuscript without any additional review by an expert
A_2—accept the manuscript without obtaining an expert's review
A_3—obtain an expert's review

The value of alternative A_1 is the $300 cost, which has already been spent on a preliminary review [i.e., $V(A_1) = \$300$]. Indeed, the $300 for a preliminary review is a sunk cost and applies whether alternative A_1, A_2, or A_3 is chosen at the present time. This $300 cost can therefore be dropped from consideration and $V(A_1) = 0$ with this adjustment.

For alternative A_2 there could be four possible market outcomes, defined as

θ_1—low market demand

θ_2—the best-seller

θ_3—best-seller plus movie rights

θ_4—low market demand plus grade B movie

Thus, the value associated with each of these outcomes is determined as

$$V(\theta_1|A_2) = \text{revenues} - \text{publication costs}$$
$$= \$5\,000 - \$15\,000 = -\$10\,000$$
$$V(\theta_2|A_2) = \$70\,000 - \$15\,000 = \$55\,000$$
$$V(\theta_3|A_2) = \$70\,000 + \$30\,000 - \$15\,000 = \$85\,000$$
$$V(\theta_4|A_2) = \$5\,000 + \$4\,000 - \$15\,000 = -\$6\,000$$

For alternative A_3, the expert's evaluations could be either

Z_1—Good

Z_2—Fair

Z_3—Poor

For *each* of these evaluations, the editor must make one of two possible decisions, defined as

X_1—reject the manuscript

X_2—accept the manuscript

Given the decision of X_1 for *each* of the three possible evaluations (Z_j), the cost to the ABC Book Company would be the same — the additional cost of an expert's review. That is,

$$V(X_1|A_3, Z_j) = -\$400, \quad \text{for } j = 1, 2, 3$$

Then, for the decision of X_2 associated with each of the reviewer's evaluations, there are the four possible market outcomes, which were defined previously. The value for any *given* market outcome *and* decision X_2 is the same for all Z_j evaluations. This value is equal to the values for alternative A_2 minus $400 for the expert's review. That is,

$$V(\theta_1|A_3, Z, X_2) = -\$10\,000 - \$400 = -\$10\,400$$
$$V(\theta_2|A_3, Z, X_2) = \$55\,000 - \$400 = \$54\,600$$
$$V(\theta_3|A_3, Z, X_2) = \$85\,000 - \$400 = \$84\,600$$
$$V(\theta_4|A_3, Z, X_2) = -\$6\,000 - \$400 = -\$6\,400$$

A decision-tree representation of this example, including probability values for the various outcomes, is given by Figure 8.12, where the values are in thousands of dollars. The probability values for the various outcomes of the decision tree given in Figure 8.12 are computed using the data of Table 8.9. First, consider alternative A_2 (accept for publication without further review). It is noted from Table 8.9 that out of the ten manuscripts that were not reviewed by an outside expert, five were published. Two of the five had low market demand, one was a best-seller, one was a best-seller plus movie rights, and another had low market demand plus the rights sold for a grade B movie. Thus, by a relative frequency logic, the following probabilities can be determined:

$$P(\theta_1|A_2) = \frac{2}{5}$$
$$P(\theta_2|A_2) = \frac{1}{5}$$
$$P(\theta_3|A_2) = \frac{1}{5}$$
$$P(\theta_4|A_2) = \frac{1}{5}$$

For the 50 manuscripts that were reviewed by an expert, 6 received a good evaluation, 28 received a fair evaluation, and 16 received a poor evaluation. Thus,

$$P(Z_1) = \frac{6}{50}$$
$$P(Z_2) = \frac{28}{50}$$
$$P(Z_3) = \frac{16}{50}$$

Similarly, the following conditional probabilities may be determined from the data of Table 8.9:

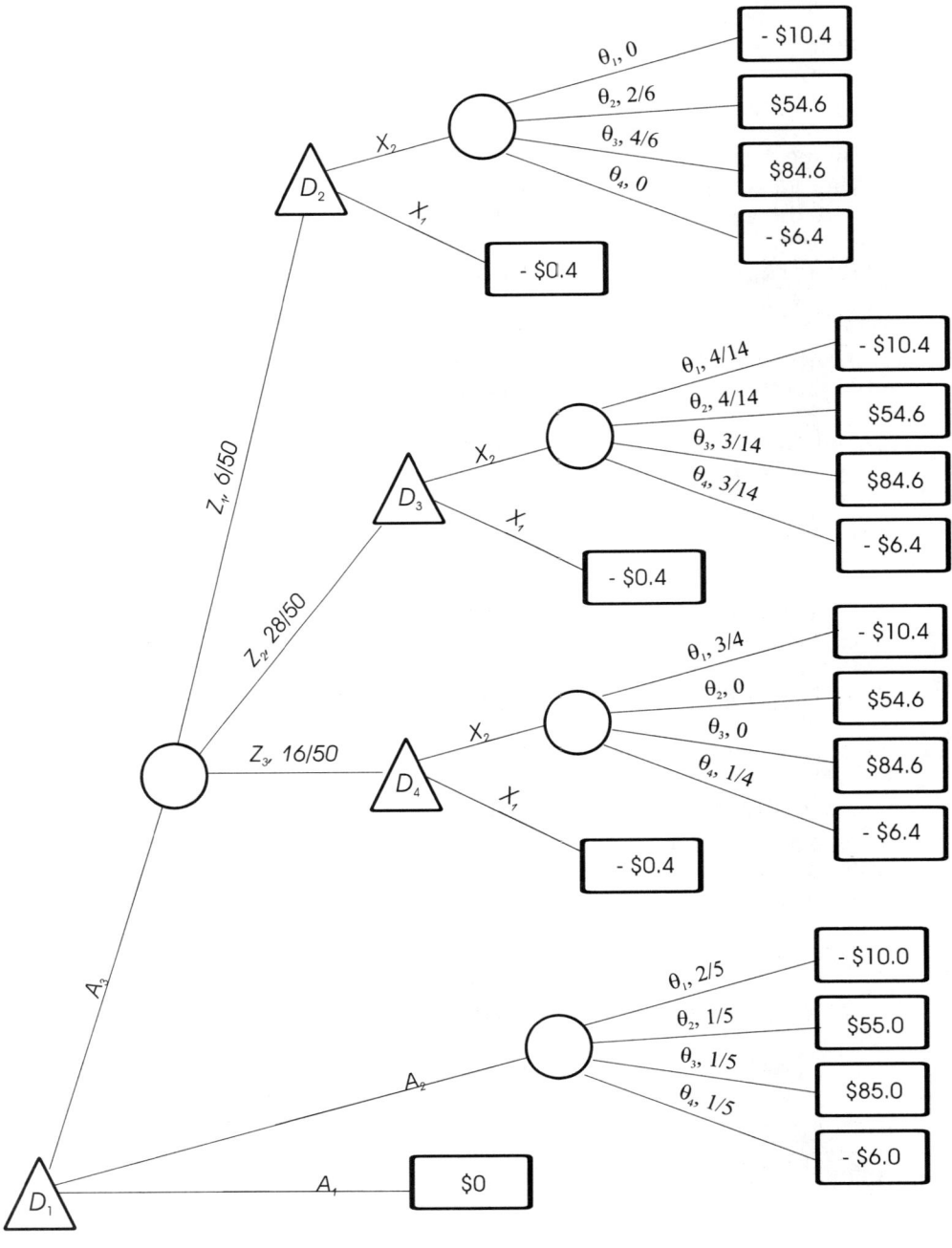

Fig. 8.12 Decision tree for ABC Book Company example (value in thousands of dollars).

$$P(\theta_1|Z_1)=0, \quad P(\theta_1|Z_2)=\frac{4}{14}, \quad P(\theta_1|Z_3)=\frac{3}{4}$$

$$P(\theta_2|Z_1)=\frac{2}{6}, \quad P(\theta_2|Z_2)=\frac{4}{14}, \quad P(\theta_2|Z_3)=0$$

$$P(\theta_3|Z_1)=\frac{4}{6}, \quad P(\theta_3|Z_2)=\frac{3}{14}, \quad P(\theta_3|Z_3)=0$$

$$P(\theta_4|Z_1)=0, \quad P(\theta_4|Z_2)=\frac{3}{14}, \quad P(\theta_4|Z_3)=\frac{1}{4}$$

Although these probability values are estimates, they are likely to be better than purely subjective evaluations made without any historical data as a basis.

The solution to the decision tree of Figure 8.12 follows. The appropriate calculations for D_2 are

$$E(X_1) = -\$0.4 \text{ in thousands} = -\$400$$

$$E(X_2) = (-\$10.4)(0) + (\$54.6)\left(\frac{2}{6}\right) + (\$84.6)\left(\frac{4}{6}\right) + (-\$6.4)(0)$$
$$= \$74.6 \text{ in thousands} = \$74\,600$$

and the best decision at D_2 is therefore X_2 (accept the manuscript). The calculations for D_3 are

$$E(X_1) = -\$400$$

$$E(X_2) = (-\$10.4)\left(\frac{4}{14}\right) + (\$54.6)\left(\frac{4}{14}\right) + (\$84.6)\left(\frac{3}{14}\right) + (-\$6.4)\left(\frac{3}{14}\right)$$
$$= \$29.386 \text{ in thousands} = \$29\,386$$

and the best decision at D_3 is therefore X_2 (accept the manuscript). The calculations for D_4 are

$$E(X_1) = -\$400$$

$$E(X_2) = (-\$10.4)\left(\frac{3}{4}\right) + (\$54.6)(0) + (\$84.6)(0) + (-\$6.4)\left(\frac{1}{4}\right)$$
$$= -\$9.4 \text{ in thousands} = -\$9\,400$$

and the best decision at D_4 is therefore X_1 (reject the manuscript).

The reduced decision tree is given by Figure 8.13, and the appropriate calculations for D_1 are

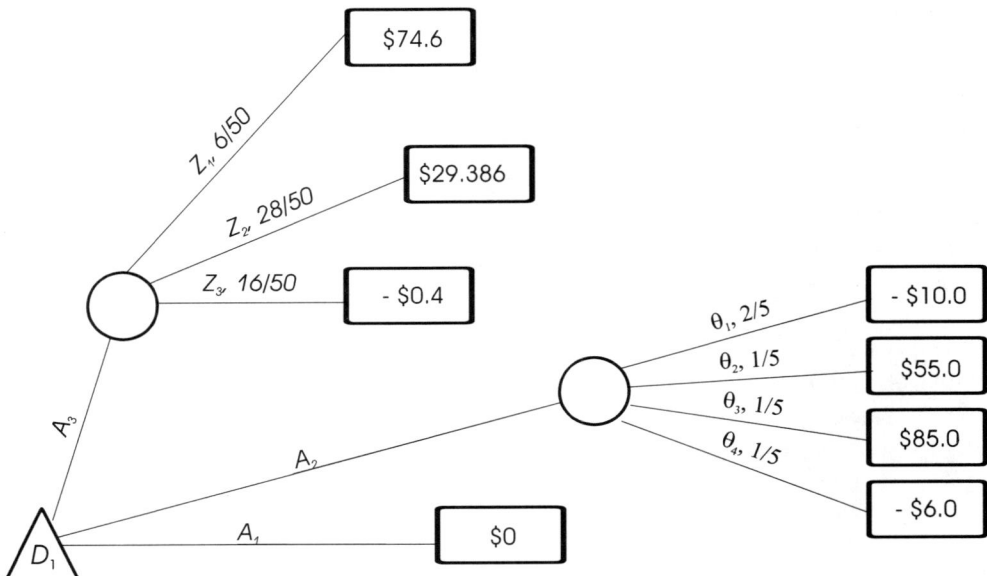

Fig. 8.13 Reduced decision tree for ABC Book Company example (value in thousands of dollars).

$$E(A_3) = (\$74\,600)\left(\frac{6}{50}\right) + (\$29\,386)\left(\frac{28}{50}\right) + (-\$400)\left(\frac{16}{50}\right)$$
$$= \$25\,280$$

$$E(A_2) = (-\$10\,000)\left(\frac{2}{5}\right) + (\$55\,000)\left(\frac{1}{5}\right) + (\$85\,000)\left(\frac{1}{5}\right) + (-\$6\,000)\left(\frac{1}{5}\right)$$
$$= \$22\,800$$

$$E(A_1) = \$0$$

and the best decision at D_1 in order to maximize the expected present worth is alternative A_3 (obtain an expert's opinion). Then, if the evaluation is either good or fair, accept the manuscript for publication; but if the evaluation is poor, reject the manuscript. The expected value of sample information in this example is the difference between $E(A_3)$ and $E(A_2)$ plus the $400 paid to the expert, or $2 880.

$$EVSI = EV|SI - PEP + \$400$$
$$= \$25\,280 - \$22\,800 + \$400$$
$$= \$2\,880$$

Clearly the expert's opinion is well worth the $400 fee in this example.

8.6.5 Sequential Decisions—Summary Comments

The decision-tree representation of a sequential-decision situation is an effective device for visualization. By developing the tree, serious attention is given to explicitly identifying and describing alternatives explicitly, to assessing the outcomes that might occur as a result of the alternative choice, and to making any necessary future decisions. Even if the probabilities for outcomes cannot be adequately assessed or values for the "branches" accurately estimated, considerable insight into the decision situation may be achieved and decision making aided. On the other hand, whether the expectation principle of choice among alternatives is an appropriate criterion or not is a matter of personal preference, and any real-world sequential-decision situation may readily become too large for effective description or convenient solution. As the number of alternatives, the number of outcomes per alternative, and the number of decision points increase, so do the uncertainty of data and the burden of calculation. Furthermore, the expense in time and money to obtain the information required may become prohibitive when compared with the benefits expected from the analysis. An evaluation of such a trade-off is largely, if not primarily, a matter of judgement.

8.7 Multiple Objectives

When managers are unable to describe goals in terms of one objective, they often seek multiple objectives, and desire proposals to achieve these in some combination. Establishing the relevant objectives is an important problem for a given decision maker, and the problem is further complicated by conflicting objectives and the difficulty (if not impossibility) of obtaining a measurement of these objectives.

The literature on the subject of multiple objectives and the theory of measurement is very extensive, encompassing the general areas of psychology, statistics, mathematics, and management. Most of the issues discussed in the literature are beyond the scope of this book. The discussion that follows here is necessarily very limited. For the subsequent discussion, it will be assumed that objectives can be precisely stated and are relevant to the decision.

8.7.1 Classifying Objectives According to Importance

Assessing the relative importance of multiple objectives, measuring the outcomes of alternatives in terms of these objectives, obtaining a common measure for multiple outcomes having different measures, and combining all these into a single measure of merit form the heart of the multiple-objective problem and the source of much difficulty and controversy. We cannot offer a solution to the problem, but can only discuss approaches to it, citing cautions at appropriate points.

When objectives can be sorted into classes such that each objective in the class is equivalent, then this is nothing more than a classification system or *nominal* system of measurement. However, when objectives can be sorted into classes that can then be ordered, this is termed an *ordinal* system of measurement.

To illustrate a nominal classification of objectives, consider a case where objectives are first categorized on a "must" or "desirable" basis. For example, a university library must serve the faculty and student population, but perhaps it is only desirable to serve residents of the local community. Once objectives are classed on a "must" versus "desirable" basis, the desirable objectives may be considered only subjectively thereafter, whereas the must objectives are deserving of greater attention and, if possible, quantification. Some of the approaches taken to quantifying objectives will now be treated; in the discussion, the term *goal* is used synonymously with *objective*, and the symbolism adopted to denote the *k*th goal or objective of interest is G_k.

8.7.2 Ranking

If a decision maker is capable of stating preferences among objectives, objectives may be ranked in terms of relative importance. A technique to assist the decision maker in making consistent preference statements is the method of paired comparisons. In order to illustrate the method, assume that four goals (objectives) are relevant: G_1, G_2, G_3, and G_4. The method of paired comparisons enumerates all possible pairs $\left[\dfrac{n(n-1)}{2}\right]$, makes preference statements, and deduces a ranking for each goal. In this example, the possible pairs are:

G_1 versus G_2	G_2 versus G_3	G_3 versus G_4
G_1 versus G_3	G_2 versus G_4	
G_1 versus G_4		

If the symbol $>$ represents "preferred to" and the symbol $<$ represents "not preferred to," suppose the results of the decision about the pairs were:

$G_1 > G_2$	$G_2 < G_3$	$G_3 > G_4$
$G_1 < G_3$	$G_2 > G_4$	
$G_1 > G_4$		

Rewriting so that all comparisons are of the "preferred to" variety yields:

$G_3 > G_1$	$G_1 > G_2$	$G_2 > G_4$
$G_3 > G_2$	$G_1 > G_4$	
$G_3 > G_4$		

It can then be concluded that G_3 is preferred to all others, G_1 is preferred to two others, G_2 is preferred to one other, and G_4 is preferred to zero others, and the following ranks should be assigned: $G_3 = I, G_1 = II, G_2 = III,$ and $G_4 = IV$.

If, in the above procedure, the preference statements had been $G_1 > G_2$, $G_2 > G_4$, and $G_1 < G_4$, the decision maker should have reconsidered the inconsistent judgements. These results are "intransitive" in a mathematical sense.

The ranking method of classifying objectives according to relative importance is certainly useful; indeed, ranking was implicitly used in the comparison of alternative investments under conditions of assumed certainty in the earlier chapters of this textbook. There are, however, obvious deficiencies in the ranking method. For instance, the ranks do not indicate the *extent* to which one objective is preferred over another. Furthermore, because rankings are an *ordinal scale of measurement,* the mathematical operations of addition, subtraction, multiplication, division, and so forth, cannot be performed. Therefore, a higher-order scale of measurement (interval or ratio scale) is generally required to determine choice among alternatives under conditions of risk and uncertainty. Unfortunately, accepted methods for measuring objectives on a ratio scale do not currently exist.

8.7.3 Weighting Objectives

Before proceeding with a discussion of certain methods to assign relative weights to multiple objectives, a distinction between objectives and outcomes should be made. This is perhaps most easily done by means of an example where a machine-replacement situation is of interest, and the following two objectives are defined:

G_1: minimize the equivalent annual cost over a five-year planning period

G_2: minimize the number of production personnel transferred to other jobs within the firm

Let

θ_{jk} = the outcome resulting from taking the jth alternative course of action in terms of the kth objective or goal

$V(\theta_{jk})$ = the value of the outcome resulting from selecting the jth alternative in terms of the kth objective

Suppose, in this example, that

A_1 = the alternative of keeping the present machining centre

A_2 = the alternative of replacing the present centre with a new machining centre

The outcomes for each of these alternatives in terms of the two objectives, G_1 and G_2 are:

θ_{11} = an annual cost of $80 000

θ_{12} = no persons transferred

θ_{21} = an annual cost of $60 000

θ_{22} = two persons transferred

In order to compare the two alternatives in a quantitative fashion, values must be assigned to each outcome; the values must be weighted in some way to reflect the relative importance of the two objectives, and some type of aggregate number must be determined for each alternative to enable a comparison and final choice. For example, the end result (without regard for validity at this point) might appear as

Decision Models

$$V(A_1) = w_1 V(\theta_{11}) + w_2 V(\theta_{12})$$
$$V(A_2) = w_1 V(\theta_{21}) + w_2 V(\theta_{22})$$

where $V(A_2) > V(A_1)$.

The w_k (or w_1, w_2) values are weighting factors identifying the relative importance of the objectives G_k (or G_1, G_2) and hence are applied to the appropriate outcomes (θ_{jk}) of the jth alternative. Determining the weighting factors is thus one issue on quantifying multiple objectives. The values $V(\theta_{jk})$ may be in different dimensions and, if so, may have to be transformed to another common dimension. Determining the transformation function is another issue in quantifying multiple objectives. Finally, the values for each alternative, $V(A_j)$, were determined by a linear function. Determining the mathematical functional form of the model is yet another issue in quantifying multiple objectives. Again, all these issues will not be resolved in this chapter, but some approaches that have been taken will be presented.

One method for assigning relative weights to multiple objectives in order to indicate the extent by which one objective is preferred to another is simply to assign weights by judgement. That is, in the case of four objectives (G_1, G_2, G_3, G_4), the decision maker may reason that G_3 is twice as important as G_1, three times as important as G_2, and five times as important as G_4. Assigning a weighting value of 1.00 to G_3, it can be reasoned that the set of weighting values is

For G_3: $w_3 = 1.000$
For G_1: $w_1 = 0.500$
For G_2: $w_2 = 0.333$
For G_4: $w_4 = 0.200$

These weights may be transformed to a scale from 0 to 1.0, or "normalized," by dividing each weight by the sum of all weights; that is,

$$w'_3 = \frac{1.000}{2.033} = 0.4918$$

$$w'_1 = \frac{0.500}{2.033} = 0.2459$$

$$w_2' = \frac{0.333}{2.033} = 0.1639$$

$$w_4' = \frac{0.200}{2.033} = 0.0984$$

Another method[6] of assigning weights to multiple objectives by judgement depends upon the validity of the following two assumptions.

For multiple objectives $G_1, \ldots, G_m$:

1. It must be possible for the decision maker to think about and judge the value of any combination of objectives. That is, it must be possible to consider not only the importance or weight value for, say, G_1, but also the weighting value for sums of objectives.

2. Values are assumed to be additive. Given the individual weighting values for, say, G_1 and G_2, it is assumed that the weighting value for both objectives is the sum of their individual weighting values.

The general procedure for the method now follows.

1. Rank the objectives in order of importance, where G_1 indicates the most important, G_2 the next most important, and so forth, and G_m is the least important.

2. Assign the weighting value of 1.00 to G_1 (i.e., $w_1 = 1.00$) and weighting values to the other objectives to reflect their importance relative to G_1. These two steps actually complete the judgement process, and the following steps serve to refine the initial judgements and aid consistency.

3. Compare G_1 to the linear combination of all other objectives. That is, compare G_1 versus $(G_2 + G_3 + \cdots + G_m)$.
   (a) If $G_1 > (G_2 + G_3 + \cdots + G_m)$, adjust (if necessary) the value of w_1 such that $w_1 > (w_2 + w_3 + \cdots + w_m)$. Attempt, in all adjustments

---

[6] Adapted from C, West Churchman et al., *Introduction to Operations Research* (New York: John Wiley, 1957), with commentary from William T. Morris, *The Analysis of Management Decisions*, rev. ed. (Toronto: Irwin, 1964).

of the procedural steps, to keep the relative values within the objective *group* invariant. Proceed to Step 4.

(b) If $G_1 = (G_2 + G_3 + \cdots + G_m)$, adjust w_1 so that $w_1 = (w_2 + w_3 + \cdots + w_m)$ and proceed to Step 4.

(c) If $G_1 < (G_2 + G_3 + \cdots + G_m)$, adjust (if necessary) the value of w_1 so that $w_1 < (w_2 + w_3 + \cdots + w_m)$.

Now compare G_1 versus $(G_2 + G_3 + \cdots + G_{m-1})$,

[i] If $G_1 > (G_2 + G_3 + \cdots + G_{m-1})$, adjust (if necessary) the values so that $w_1 > (w_2 + w_3 + \cdots + w_{m-1})$ and proceed to Step 4.

[ii] If $G_1 = (G_2 + G_3 + \cdots + G_{m-1})$, adjust (if necessary) the values such that $w_1 = (w_2 + w_3 + \cdots + w_{m-1})$ and proceed to Step 4.

[iii] If $G_1 < (G_2 + G_3 + \cdots + G_{m-1})$, adjust (if necessary) the values such that $w_1 < (w_2 + w_3 + \cdots + w_{m-1})$. Then, compare G_1 versus $(G_2 + G_3 + \cdots + G_{m-2})$, and so forth, either until G_1 is preferred or equal to the rest, then proceed to Step 4, or until the comparison of G_1 versus $(G_2 + G_3)$ is completed, then proceed to Step 4.

4. Compare G_2 versus $(G_3 + G_4 + \cdots + G_m)$ and proceed as in Step 3.

5. Continue until the comparison of G_{m-2} versus $(G_{m-1} + G_m)$ is completed.

6. If desired, convert each w_k into a normalized value, dividing w_k by the sum $\sum_{k=1}^{m} w_k$.

Example 8.9

Assume that it is desired to determine weighting values for four objectives where these have been ranked in order of importance such that G_1 is most important and G_4 is the least important. Step 1 of the procedure is thus accomplished.

Step 2. The objectives are tentatively assigned the weights $w_1 = 1.00, w_2 = 0.80, w_3 = 0.50,$ and $w_4 = 0.20$.

Step 3. Assume that G_1 is preferred to the linear combination of the other objectives. That is, $G_1 > (G_2 + G_3 + G_4)$. Since

$w_1 < (w_2 + w_3 + w_4)$, it is necessary to adjust w_1 to be greater than the sum of 1.50, e.g., $w_1 = 1.75$. Then proceed to Step 4.

Step 4. Comparing G_2 versus $(G_3 + G_4)$, assume that $G_2 < (G_3 + G_4)$, and the Step 3 procedure is repeated. In this case, the weighting values do not agree with the assumption $[w_2 = 0.80 \geq (0.50 + 0.20)]$, therefore, w_2 is adjusted to be 0.65. [The adjustment to decrease w_2 does not violate the previous preference of $G_1 > (G_2 + G_3 + G_4)$ with the new value of $w_1 = 1.75$ to reflect this preference.] Since, in this example of only four objectives, the comparison of G_{m-2} versus $(G_{m-1} + G_m)$ or G_2 versus $(G_3 + G_4)$ has been completed at this point, Step 5 of the procedure has been done.

If it is desired to normalize the results, new weighting values are calculated as

$$w'_1 = \frac{1.75}{1.75 + 0.65 + 0.50 + 0.20} = 0.5645$$

$$w'_2 = \frac{0.90}{3.10} = 0.2097$$

$$w'_3 = \frac{0.50}{3.10} = 0.1613$$

$$w'_4 = \frac{0.20}{3.10} = 0.0645$$

Total 1.0000

For a small number of objectives, this method is relatively easy to use, but becomes cumbersome as the number of objectives increases. However, it is believed that many real-world problems reduce to a few primary objectives. It should also be emphasized again that the method still depends on the decision maker's judgement in assigning the relative weights.

8.7.4 Determining the Value of Multiple Objectives

In the previous section on weighting objectives, an example of a machine replacement situation was hypothesized to introduce the discussion. This example is now recalled, where the objectives were:

G_1: minimize the equivalent annual cost over a five-year planning period

G_2: minimize the number of production personnel transferred to other jobs within the firm

For the two alternatives of A_1 (keep present machining centre) and A_2 (replace with new machining centre), the outcomes were

A_1: $80 000 annual cost ($\theta_{11}$); no persons transferred (θ_{12})

A_2: $60 000 annual cost ($\theta_{21}$); two persons transferred (θ_{22})

Let it be assumed that, by some method, the objectives have been weighted as $w_1 = 0.60$ and $w_2 = 0.40$.

One method to assign values to the outcomes of the two alternatives is to make a second series of judgements as to the degree to which each outcome succeeds in meeting the objective. Suppose these judgements are made in the form of numbers on an arbitrary scale from zero to one where the higher scale value, the closer to meeting the desired objective. Assume the results of such judgements are as given below.

Outcome	Value
θ_{11}	0.50
θ_{12}	1.00
θ_{21}	0.85
θ_{22}	0.60

Then, assuming that a linear model is appropriate, the values for each alternative are calculated as

$$V(A_1) = w_1 V(\theta_{11}) + w_2 V(\theta_{12})$$
$$= (0.60)(0.50) + (0.40)(1.00) = 0.70$$
$$V(A_2) = w_1 V(\theta_{21}) + w_2 V(\theta_{22})$$
$$= (0.60)(0.85) + (0.40)(0.60) = 0.75$$

Thus, since $V(A_2) > V(A_1)$ alternative A_2 would be chosen, and the present machining centre would be replaced.

Clearly, these results are sensitive to the weighting values assigned by judgement. Furthermore, the magnitude of the difference in values between $V(A_1)$ and $V(A_2)$ is a function of the arbitrary scale (from zero to one) chosen. The linear model assumed is also questionable.

All these points are criticisms of this method of quantifying multiple objectives. In the literature on the theory of utility functions, it is argued that if a decision maker's utility function can be established for each objective, then values of outcomes (e.g., $80 000 annual cost) can be converted to a utility value (e.g., $80 000 annual cost equals 0.75) from the decision maker's utility curve. Such a utility value could be determined for each objective, and our model for the alternatives above would become

$$U(A_1) = w_1 U(\theta_{11}) + w_2 U(\theta_{12})$$
$$U(A_2) = w_1 U(\theta_{21}) + w_2 U(\theta_{22})$$

Then, the decision maker should choose to maximize the weighted utility value. Whether such utility functions for each objective can be readily obtained for a single decision maker is still questionable in our opinion, and even if this could be accomplished, there is yet another question of whether interaction between multiple objectives would be totally missed. As stated in Chapter 2, the study of value measurement, and utility theory in particular, is interesting, but questions remain on the practical application and implementation of utility theory to business and engineering problems. In any event, a treatment of utility theory is outside the scope of this textbook.

8.8 Summary

In this chapter, some prescriptive or normative models for decisions under risk and uncertainty have been presented. For decisions under risk and uncertainty where the matrix model is appropriate, criteria that are commonly cited in the literature for choosing among mutually exclusive alternatives were discussed. None of these criteria were cited as the best or optimal basis for selecting an alternative; they merely represent observations on how decision makers seem to decide.

The decision-tree model for a sequence of decisions where the outcomes of alternatives are chance events was also presented, and the expectation-variance criterion was suggested as a guideline for choice among alternatives. Some fundamental laws of probability were reviewed, and Bayes's theorem of conditional probability was illustrated as a guideline for revising one's opinion in sequential decisions under risk when only imperfect information is available to the decision maker.

The last section of the chapter was an introductory presentation of some approaches to quantifying multiple objectives. The discussion offered and examples used were restricted to decisions under assumed certainty. However, the methods illustrated are easily extended to decisions under risk and uncertainty and the matrix model used for analysis.

Problems

1. Assume the following decision matrix where the cell values are units of gain.

	S_1	S_2	S_3	S_4
A_1	$4	$2	$0	$1
A_2	−4	10	3	7
A_3	4	8	2	3
A_4	2	4	9	5

   If a person's index of optimism is estimated as $\alpha = 0.25$, predict the person's choice of alternative by applying the various principles of choice for a decision under uncertainty. (8.4.1, 8.5)

2. Analyze the cost matrix below by the use of the Hurwicz principle. (8.5.4)

	S_1	S_2	S_3
A_1	$5	$5	$5
A_2	9	8	0
A_3	4	6	3

3. Apply the various decision-under-certainty principles of choice to the matrix below. The values in the matrix are cost units. (8.4.1, 8.5)

	S_1	S_2	S_3	S_4
A_1	$13	$13	$5	$9
A_2	8	8	8	8
A_3	0	21	5	5
A_4	9	17	5	5
A_5	5	7	7	5

4. A particular production department supervisor is known to be very conservative in business matters. For the decision model below, where values are equivalent annual profits (in thousands of dollars) for a $MARR = 15\%$ and a five-year

planning period, predict the supervisor's choice of an alternative by each of the principles for a decision under certainty. (8.4.1, 8.5)

	S_1	S_2	S_3	S_4
A_1	$16	$16	$ 8	$12
A_2	9	20	10	6
A_3	15	10	8	9
A_4	3	14	6	11
A_5	8	16	9	12

5. Assume the following decision matrix.

	p_1 S_1	p_2 S_2	p_3 S_3
A_1	$ 20	$100	$1 200
A_2	190	190	190
A_3	500	120	100

(a) Treat the decision as one under uncertainty and the values in the matrix as costs.
   (1) If α = an index of optimism, for what value of α is one indifferent between A_2 and A_3? (8.5.4)
   (2) Which alternative would be chosen by the Savage principle? (8.5.5)
(b) Treat the decision as one under risk where $p_1 = 0.20$, $p_2 = 0.70$, and $p_3 = 0.10$, and the values in the matrix are profits. Which alternative would be chosen by the expectation-variance principle? (8.4.2)

6. Given the decision matrix shown below (with *cost* elements), determine the preferred alternative using the following principles of choice. (8.5)

(a) minimax
(b) minimin
(c) minimax regret

(d) expected cost, if each future state is expected to occur with equal probability

(e) Hurwicz (with $\alpha = 0.40$)

	S_1	S_2	S_3
A_1	$120	$50	$10
A_2	60	60	60
A_3	70	50	60
A_4	20	20	150

7. Given the decision matrix shown below, determine the recommended alternative under the following principles of choice: expectation, most probable future, maximax, maximin, and minimax regret. Entries in the matrix are profits. (8.4, 8.5)

	$p_1 = 0.1$ S_1	$p_2 = 0.3$ S_2	$p_3 = 0.4$ S_3	$p_4 = 0.2$ S_4
A_1	−$50	$100	$200	$400
A_2	−20	50	500	100
A_3	100	−100	50	100
A_4	200	−50	50	200

8. Shown below is a matrix of costs for three investment alternatives under three future states. Determine the preferred alternative using the following decision rules. (8.5.2, 8.5.3, 8.5.5)

(a) minimax

(b) minimin

(c) minimax regret

	S_1	S_2	S_3
A_1	$300	$200	$100
A_2	150	180	200
A_3	210	110	175

9. An analysis yields a decision under risk given in the matrix below, where the matrix values are annual profits in thousands of dollars.

$p_k =$	0.1	0.1	0.4	0.2	0.1	0.1
$S_k =$	S_1	S_2	S_3	S_4	S_5	S_6
A_1	$12	$5	−$8	−$3	$6	$9
A_2	7	0	1	5	20	7
A_3	3	3	7	9	−5	5
A_4	0	12	15	2	8	−200
A_5	−10	22	9	0	4	12

(a) Which alternative should be chosen in order to minimize the probability of a loss? (8.4.4)

(b) Which alternative should be chosen in order to maximize the probability of an annual profit of at least $9 000? (8.4.4)

(c) Which alternative should be chosen in order to maximize the probability that annual profits will be between $3 000 and $10 000? (8.4.4)

(d) Would the most probable future principle be a reasonable one to follow in selecting an alternative? (8.4.3)

10. An electronics firm has recently received a government contract to produce a certain quantity of expensive electronic guidance systems. The contracted number of units can be produced in two years, and the contract terminates then. The firm won the contract with cost estimates based on using present manufacturing equipment. If the firm had more specialized equipment (with almost zero salvage value immediately after purchase), the unit cost of production would be reduced, and profit per unit would thus be increased. However, this would be true only if the contract were for at least four years instead of two years. The firm feels there is a 50% chance for an additional two-year contract, a 30% chance for an additional four-year contract, and a 20% chance of no contract renewal.

Using only present equipment, total profit on this job for a two-year, four-year, and six-year contract is estimated as

$40 000, $80 000, and $120 000, respectively. If the specialized equipment is purchased, total profit on this job for two-year, four-year, and six-year contracts is estimated as −$100 000, $90 000, and $180 000. If an expectation principle is used, should the specialized equipment be purchased? Does the most probable future criterion seem reasonable in this decision? (8.4.1, 8.4.2, 8.4.3)

11. A highway construction firm is considering the purchase of a used mobile crane. Two such cranes are available, A_1 and A_2. If either is purchased, the firm will use a six-year depreciation schedule. The cranes differ slightly in capacity, age, and mechanical condition, but both are presently in operating condition and have the capacity to do the job expected. The firm expects that a major overhaul of each will eventually be required, but when this will be necessary is uncertain. Estimates of the operating expenses (excluding labour) for each crane are given below:

	A_1		A_2	
Year	Expense Schedule 1	Expense Schedule 2	Expense Schedule 1	Expense Schedule 2
1	$ 5 000	$ 5 000	$14 000	$ 6 000
2	5 000	11 000	6 000	6 000
3	5 000	5 000	6 000	14 000
4	11 000	5 000	6 000	6 000
5	5 000	5 000	6 000	6 000
6	5 000	5 000	6 000	6 000

Other data for the two used cranes are:

	A_1	A_2
First cost	$30 000	$20 000
Salvage value at the end of six years	12 000	10 000

If the firm uses a before-tax $MARR = 20\%$ and a present worth method of comparing alternatives, which crane would be purchased if the Laplace principle were used? (8.4.1, 8.5.1)

12. Suppose a coin is biased such that, when the coin is tossed, the probability of a head (H) showing is 0.60. Now assume that the coin is tossed three consecutive times.

   (a) Sketch a probability tree for the three tosses and determine the probability values for each of the eight possible outcomes: HHH, HHT, and so on. For the three tosses, the following states are now defined:
   S_1 = the event of three heads (3H)
   S_2 = the event of two heads, one tail—occurring in any order (2H, T)
   S_3 = the event of two tails, one head—occurring in any order (2T, H)
   S_4 = the event of three tails (3T) (8.6.1, 8.6.2)

   (b) Create a decision matrix where four alternatives are to guess the events defined above. Now assume that before tossing the coin three times, a person offers a $4 prize if the (3H) event is guessed correctly, a $3 prize if the (2H, T) event is guessed correctly, a $4 if the (2T, H) event is guessed correctly, and a $5 prize if the (3T) event is guessed correctly. However, there is a $1 charge to play the game. If the game is played, would the expectation principle and the most probable future principle select the same alternative? (8.2, 8.4.2, 8.4.3)

13.[7] A vehicle manufacturer plans to develop and operate a public bus system for a community. The manufacturer's purpose is to demonstrate the profitability of the bus system and then sell the system to the community. The manufacturer's most serious concern is whether the Public Utilities Commission will authorize a competing bus system. It is reasoned that there is a small chance that such an event will happen; therefore, the manufacturer subjectively assigns a probability of 0.2 to the event of competition.

A detailed analysis by the manufacturer results in an estimate of 50 vehicles needed in the event of no competition, and 25 vehicles needed if competition results. Furthermore, it is reasoned that operating either a 50-vehicle or a 25-vehicle

---

[7] Problem 13 taken from deNeufville and Stafford, *Systems Analysis for Engineers and Managers*, New York: McGraw-Hill, 1958, by permission of the publisher.

system will not affect the probability of the Public Utilities Commission authorizing or not authorizing competition.

If the manufacturer initially develops a 50-vehicle system and there is no competition, it is estimated the system will yield a $250 000 profit. In the event of competition, a $120 000 loss is projected.

If a 25-vehicle system is initially developed and competition occurs, the manufacturer plans to sell the system as quickly as possible and estimates a $25 000 profit in this instance. On the other hand, if there is no competition, the 25-vehicle system will be inadequate. Poor service would therefore result, and the value of the demonstration would be reduced, thereby adversely affecting the sales price. Thus, in the event of no competition, the manufacturer reasons that another decision must be faced: whether to expand from 25 vehicles to 50 vehicles or not to expand. If there is no expansion, the manufacturer would sell soon to avoid the bad publicity for expected poor service, estimating a $35 000 profit in this case. If, on the other hand, the system is expanded and the total system is sold as soon as possible, two outcomes are predicted. One outcome is that poor service could result during the expansion and a net $10 000 loss would result (a probability of 0.10 is subjectively assigned to this event). The second outcome is that poor service would not result, and a net $140 000 profit is expected in this instance. Create a decision-tree model of this situation and solve by use of the expectation principle. (8.6.1, 8.6.2, 8.4.2)

14. Company Able purchases two models, Model R and Model G, of Product Tau from three suppliers, A, B, and C. These suppliers have produced, and are expected to continue to produce, according to the following table:

Supplier	Proportion of Total Product Tau	Proportion of Product Tau by Model	
		R	G
A	0.60	0.20	0.80
B	0.10	0.55	0.45
C	0.30	0.60	0.40

Decision Models

The Quality Control Department of Company Able has recently determined that many type R models have been defective and intends to send a representative to each of the suppliers to observe their manufacturing procedures.

Using the logic of Bayes's theorem, which supplier is the most likely, next most likely, and least likely defect producer? (8.6.1, 8.6.2)

15. Let it be supposed that, out of your sight, an experimenter rolls a green, red, and white die. The experimenter covers two of these. Data concerning the dice are that (1) the green die has five surfaces marked H, one surface marked T, (2) the red die has two surfaces marked H, four surfaces marked T, and (3) the white die has three surfaces marked H, three surfaces marked T. Also

   A = the event that the green die is uncovered
   B = the event that the red die is uncovered
   C = the event that the white die is uncovered
   h = the event that a surface marked H is showing on the uppermost surface of the uncovered die
   t = the event that a surface marked T is showing

Now, the experimenter reports that the letter H is showing on the uppermost surface of the uncovered die.

(a) Using the logic of Bayes's theorem, what is the best guess concerning the colour of the uncovered die? (8.6.1, 8.6.2)

(b) Assume that the experimenter, after the report but before the guess, offers to pay $6 if the green die is guessed correctly, $15 if the red die is guessed correctly, and $15 if the white die is guessed correctly. However, there is a $3 charge to play the game. Given that the game is played, model this situation as a decision-under-risk matrix and determine the best guess by the expectation principle. (8.2, 8.4.2)

16.[8] After receiving somewhat unreliable information concerning enemy troop movement along a supply route, a military commander of an artillery battalion faces the following combat

---

[8] Problem 16 taken from Agee, Marvin et al., *Quantitative Analysis for Management Decision*, Englewood Cliffs, NJ, Prentice-Hall, 1976, by permission of the publisher.

decision (assume the matrix values are loss units, arbitrarily chosen).

	$p_1 = 0.7$ S_1 = Troop Movement	$p_2 = 0.3$ S_2 = No Troop Movement
A_1 = bombard supply route	0	4
A_2 = not bombard supply route	12	0

The commander can choose between A_1 and A_2 without additional information, or he can send out a reconnaissance plane for additional information. If the plane is sent out, he reasons the mutually exclusive and collectively exhaustive outcomes of the flight are:

Outcome O_1: the plane is shot down, with a loss of 3 units.

Outcome O_2: the plane returns safely with negative information about troop movement, with a loss of 0.2 units.

Outcome O_3: the plane returns safely with a report of suspicious activity, with a loss of 0.2 units.

The commander assigns the following subjective conditional probabilities to these outcomes:

$P(O_1|S_1) = 0.4$ $P(O_1|S_2) = 0.2$

$P(O_2|S_1) = 0.1$ $P(O_2|S_2) = 0.8$

$P(O_3|S_1) = 0.5$ $P(O_3|S_2) = 0.0$

(a) Determine the posterior probabilities. (8.6.2)

(b) Create a decision tree for this problem. (8.6.1)

(c) Using the minimizing-expected-loss principle of choice, what action should the commander take? (8.4.2)

(d) What is the expected value of sample information in this problem? (8.6.3, 8.6.4)

Decision Models

17. A manufacturer is considering the possibility of introducing a new product and the advisability of a test marketing prior to making the final decision. The alternatives are

   $A_1:$ = market the product
   $A_2:$ = do not market the product

For simplicity, only three possible futures are considered; they are shown below together with the prior probabilities associated with them.

		Profit	Prior Probability
$S_1:$	the product captures 10% of the market	$10 000 000	0.70
$S_2:$	the product captures 3% of the market	1 000 000	0.10
$S_3:$	the product captures less than 1% of the market	− 5 000 000	0.20

If the test marketing is made, three possible results are considered.

   $Z_1:$ test sales of more than 10% of the market
   $Z_2:$ test sales of 5 to 10% of the market
   $Z_3:$ test sales of less than 5% of the market

The conditional probabilities of the test results are

	Z_1	Z_2	Z_3
$P(Z|S_1)$	0.6	0.3	0.1
$P(Z|S_2)$	0.3	0.6	0.1
$P(Z|S_3)$	0.1	0.1	0.8

(a) Determine the prior expected profit. (8.6.1, 8.6.2, 8.6.3)

(b) Determine *EVPI*. (8.6.3)

(c) Determine *EVSI*. (8.6.4)

(d) If the test marketing costs $250 000, what would be the expected net gain from the sample information? (8.6.4)

18. A textile firm in a rural town owns two trucks that transport raw materials from and finished goods to a nearby city. One of the trucks (tractor and trailer) is eight years old, and annual maintenance expenses are increasing significantly. Management is considering the purchase of a new truck but, because of a mild national economic recession, textile sales have declined during the past two years, and management is reluctant to make a large capital expenditure at present. Although in the midst of a recession, inflation is also occurring; if the decision to purchase is delayed for a year, the cost of a new truck will probably increase 10% or more.

Adopting a two-year planning period, the following estimates and judgements are made.

Present Truck

The present market value is approximately $9 000. If kept one more year, operating (excluding labour and depreciation) and maintenance expenses are estimated as either $6 000, $10 000, or $12 000 with subjective probabilities of 0.25, 0.50, and 0.25, respectively. The salvage value at the end of the year is about $8 000. If kept one more year, a new decision to keep an additional year or buy a new truck will be made.

If the truck is kept two additional years, operating and maintenance expenses (excluding depreciation) for the second year are estimated to be either $7 000, $10 000, or $15 000 with subjective probabilities of 0.10, 0.70, and 0.20, respectively. The salvage value at the end of the second year is about $7 500.

New Truck

A new truck can be purchased now for $30 000. Average annual operating, maintenance, and depreciation expenses for each of the first two years are estimated to be $15 000 with reasonable certainty.

If the purchase of the truck is delayed for one year, an increase in purchase price is virtually certain. It is judged there is a 50-50 chance that the purchase price will be $32 000 or $34 000. If it is $32 000, total operating, maintenance, and depreciation expenses will be $16 000; if it is $34 000, total expenses will be $17 500.

For simplicity, assume MARR $= 0\%$ and (a) create a decision-tree model for this situation, and then (b) determine the best present decision by the expectation principle. (8.6.1, 8.4.2)

19. The Jax Tool and Engineering Company is experiencing considerable problems with in-process material handling and storage of finished goods. A new layout of machinery is possible, but this alternative is subjectively judged by management to be prohibitively expensive, and the alternative is discarded. They wish, therefore, to consider a new material-handling system and identify the following objectives:

  G_1: minimize annual costs over a 10-year planning period
  G_2: minimize the disruption of production during the installation of the new system
  G_3: install material-handling equipment that has flexibility to meet different handling requirements
  G_4: the material-handling equipment should have a high index of repairability
  G_5: material-handling personnel should require little training in order to operate the equipment

Management ranks these goals in the following order of importance: $G_1 =$ I, $G_3 =$ II, $G_2 =$ III, $G_5 =$ IV, and $G_4 =$ V. Using the Churchman, Ackoff, and Arnoff method of weighting objectives (see p. 440, assign weights to these five objectives. (8.7.4)

20. Suppose a decision may be modelled as

	$p_1 = 0.6$ $S_1 =$ Contract Renewed	$p_2 = 0.4$ $S_2 =$ Contract Not Renewed
$A_1 =$ use present equipment	$V(\theta_{11})$	$V(\theta_{12})$
$A_2 =$ purchase new equipment	$V(\theta_{21})$	$V(\theta_{22})$

Assume that analysis produces the following descriptions of the outcomes:

θ_{11} = no new capital required, 20 people required for at least four years, average annual profit units are 85

θ_{12} = no new capital required, 20 people required for two years and then laid off, average annual profit units are 67

θ_{21} = two units of new capital required, 8 people required for at least four years, average annual profit units are 150

θ_{22} = two units of new capital required, 8 people required for two years and then laid off, average annual profit units are 400

Presume the relevant values for objectives are

V (profit units) $= +0.003$ (profit units) $+ 1.0$

V (capital units required) $= e^{-u}$, where u = the number of capital units required

V (work-years required) $= -0.0145$ (work-years required) $+ 1.0$

Furthermore, the goals of "capital units required" and "annual profit units" are considered equally important, but the labour consideration is only half as important as either of the other two.

On the basis of this information, which alternative would you recommend? (8.2, 8.7, 8.7.2)

Chapter 9
Accounting Principles

9.1 Introduction

As already mentioned, the engineer should have some understanding of basic accounting practice and cost accounting techniques in order to obtain data from the firm's accounting system. If accounting is classified into general accounting and cost accounting, then cost accounting is judged the most important to the engineer as a source of data for making cost estimates pertinent to engineering projects. Cost accounting will therefore receive the greater emphasis in this text; in either case, the treatment of accounting is cursory and directed toward fundamental accounting concepts instead of comprehensive accounting detail.

In virtually all businesses, general accounting information is summarized in at least two basic financial reports:

1. A *balance sheet*, or statement of financial conditions provides a summary listing of the assets, liabilities, and owners' equity (shareholder's equity) accounts of the firm as of a particular date.

2. An *income statement* (*profit and loss statement* or *statement of earnings*) shows the revenue and expenses incurred by the firm during a stated period of time—a month, quarter, or year.

9.2 Balance Sheet

The records of financial transactions and the variety of internal reports that are put into the accounting system provide information on sales and other revenues and the expenses incurred in obtaining the revenues. Revenues and expenses for a specified period are then summarized on the profit and loss statement. The net profit or loss resulting for this period is then transferred to the owners' equity section of the balance sheet. Thus, although the two basic financial reports provide different financial pictures of the firm, they are directly related. Before illustrating this fact, discussion of certain terminology is necessary.

The items listed on a balance sheet are usually classified into three main groups: assets, liabilities, and owners' equity items. Subgroups may also be identified, such as current and fixed assets, and current and long-term liabilities. *Assets* are properties owned by the firm, and *liabilities* are debts owed by the firm against these assets. The difference between assets and liabilities is the owners' equity, or shareholders' equity, which is the investment made by the owners or shareholders of the business plus any accumulated profits left in the business by the owners or shareholders. A fundamental accounting equation is thus defined:

$$\text{assets} - \text{liabilities} = \text{owners' equity}$$

Rewriting, we have

$$\text{assets} = \text{liabilities} + \text{owners' equity}$$

and the usual format of a balance sheet follows the equation in this form.

Current assets include cash and other assets that can be readily converted into cash; an arbitrary period of one year is usually assumed as a criterion for conversion. Similarly, *current liabilities* are the debts that are due and payable within one year from the date of the balance sheet in question. *Fixed assets,* then, are the properties owned by the firm that are not readily converted into cash within a one-year period, and *long-term liabilities* are debts due and payable after one year from the date of the balance sheet. Typical current-asset items are cash, accounts receivable, notes receivable, raw material inventory, work in progress, finished goods inventory, and prepaid expenses. Fixed-asset items are land, buildings, equipment, furniture, and fixtures. Items that are typically listed under current liabilities are accounts payable, notes payable, interest payable, taxes payable, prepaid income, and dividends payable. Long-term liabilities may also be notes and bonds payable, mortgages payable, and so forth. Owners' equity items appearing on a balance sheet are less standard and, to a degree, depend on whether the business is a sole proprietorship, a partnership, or a corporation. The size of the corporation is also an influencing factor on item designation. However, items such as capital stock, retained earnings, capital surplus or earned surplus always appear under owners' equity. An example of a balance sheet for a hypothetical firm is exhibited in Table 9.1.

Table 9.1 Sample Balance Sheet

Jax Tool and Engineering Company, Inc.
Balance Sheet
December 31, 2000

Assets

Current assets			
Cash		$ 25 000	
Accounts receivable		115 000	
Raw materials		8 500	
Work in progress		7 000	
Finished goods inventory		15 500	
Total current assets			$171 000
Fixed assets			
Land		30 000	
Building	$200 000		
Less: Accumulated depreciation	50 000	150 000	
Equipment	750 000		
Less Accumulated depreciation	150 000	600 000	
Office equipment		10 000	
Total fixed assets			790 000
Total assets			$961 000

Liabilities and Owners' Equity

Current liabilities			
Accounts payable		$ 32 000	
Taxes payable		15 000	
Total current liabilities			$ 47 000
Long-term liabilities			
Mortgage loan payable		130 000	
Equipment loan payable		350 000	
Total long-term liabilities			480 000
Total liabilities			527 000
Owners' Equity			
Common stock		325 000	
Retained earnings		109 000	
Total equity			434 000
Total liabilities and equity			$961 000

The sample balance sheet in Table 9.1 is balanced (i.e., assets = liabilities + owners' equity), and it gives a statement of the financial condition of the organization as of a specific date—the close of an accounting period. Although depreciation is covered in Chapter 5, note the inclusion of accumulated depreciation in the fixed-assets portion of the balance sheet. For example, the building originally cost $200 000, and depreciation expenses have been charged annually so that the total depreciation charges as of the date of the balance sheet have been $50 000, which is the amount entered as the accumulated depreciation for the building. The first cost of the depreciable asset (the building, in this case) minus the accumulated depreciation equals the *book value*. A similar explanation applies for the fixed asset of equipment. In this balance sheet, the equipment account is an aggregate for all equipment owned by the company instead of an individual listing of the equipment, which could be the case, depending on accounting practice.

9.3 Income Statement

The second basic financial report compiled by the accounting system is the *income statement (profit and loss statement* or *statement of earnings)*. For the current accounting period, the income statement provides management with (1) a summary of the revenues received, (2) a summary of the expenses incurred to obtain the revenues, and (3) the profit or loss resulting from business operations. The format of the income statement varies, and the revenue and expense items depend on the type of business involved. Thus, the income statement given in Table 9.2 is illustrative only and still concerns our hypothetical Jax Tool and Engineering Company. Let us further assume that the period of time covered is one year, which has ended as of the date of the balance sheet given in Table 9.1.

9.4 Interpretation of Financial Statements

Recall that one purpose of the accounting system is to interpret the financial data of an organization. Among the most commonly used techniques for interpretation is the calculation of accounting ratios and other accounting quantities derived either from balance sheet or income statement information. From balance sheet data *working capital, current ratio,* the *acid-test ratio,* the *equity ratio,* the *debt to equity ratio,* and other less frequently used accounting ratios can be calculated.

Table 9.2 Sample Income Statement

Jax Tool and Engineering Company, Inc.
Income Statement
Year Ended December 31, 2000

Sales		$1 200 000
Less cost of goods sold		
Direct labour	$420 000	
Direct materials	302 000	
Indirect labour	112 000	
Depreciation	98 000	
Repairs and maintenance	41 500	
Utilities	11 500	985 000
Gross profit		$ 215 000
Less other expenses		
Administration	$ 76 000	
Marketing	49 000	
Interest payments	35 000	160 000
Net profit before tax		$ 55 000
Less income taxes		26 000
Net profit		$ 29 000

Working capital, or the capital used to meet the daily commitments of the organization, is the excess of current assets over current liabilities, so that

$$\text{working capital} = \text{current assets} - \text{current liabilities}$$

The working capital is an important reflection of the financial position of the company. From Table 9.1, the working capital of the Jax Tool and Engineering Company can be determined:

$$\$171\,000 - \$47\,000 = \$124\,000$$

The company's working-capital condition can also be expressed as a ratio. This ratio, called the *current ratio*, is defined as

$$\text{current ratio} = \frac{\text{current assets}}{\text{current liabilities}}$$

From Table 9.1, the current ratio for the Jax Company is

$$\frac{\$171\,000}{\$47\,000} = 3.638$$

which implies that current assets would cover short-term debts 3.638 times.

The current ratio assumes that the current assets of inventories are convertible to cash within one year. A more conservative ratio of the liquidity of the company is the *acid-test* or *quick asset ratio*, which is defined as

$$\text{acid test ratio} = \frac{\text{current assets} - \text{inventories} - \text{prepaid expenses}}{\text{current liabilities}}$$

The acid-test ratio for the Jax Company can be calculated as follows: (In this case, inventories include the cost of raw materials, work in progress and finished goods, and there are no prepaid expenses.)

$$\frac{\$171\,000 - \$31\,000}{\$47\,000} = 2.979$$

A ratio that measures the financial strength of the firm is the *equity ratio*:

$$\text{equity ratio} = \frac{\text{owners' equity}}{\text{total assets}}$$

From Table 9.1, the equity ratio is

$$\frac{\$434\,000}{\$961\,000} = 0.451$$

which means that 45.1% of the Jax Tool and Engineering Company is owned by the shareholders. (The $109 000 retained earnings are also owned by the shareholders.)

The *debt to equity ratio* expresses the relative magnitude of debt to equity capital. It is defined as

$$\text{debt to equity ratio} = \frac{\text{long-term liabilities}}{\text{owners' equity}}$$

From Table 9.1, the debt to equity ratio is

$$\frac{\$480\,000}{\$434\,000} = 1.106$$

Therefore, for every dollar of equity the shareholders have in their company, the creditors have put in $1.106.

In addition to these four ratios that can be computed using only balance sheet data, a number of common accounting ratios are calculated using data from both the balance sheet and the income statement. The most important of these are the *operating ratio,* the *income ratio,* the *"gross" rate of return on investment,* the *inventory turnover ratio,* and the *collection period.* The operating ratio is defined as follows:

$$\text{operating ratio} = \frac{\text{total revenues}}{\text{total expenses}}$$

Using Table 9.2, the operating ratio is

$$\frac{\$1\,200\,000}{\$1\,145\,000} = 1.048$$

where the total expense figure excludes the income taxes paid. (Income taxes may or may not be included in calculating the ratio.) An operating ratio greater than 1.0 indicates that a net profit (before income taxes in the example used) is being made and is, therefore, desirable. Such operating ratios are perhaps more meaningful when they are computed for different product lines or different plants within a multiplant corporation. They can then be used by management to assess the effectiveness of individual product lines or plants.

The *income ratio* is defined as

$$\text{income ratio} = \frac{\text{net profit}}{\text{total revenue}} \times 100\%$$

and, using the after-tax data from Table 9.2, is calculated to be

$$\frac{\$29\,000}{\$1\,200\,000} \times 100\% = 2.417\%$$

This figure for the Jax Company is rather low for a manufacturing firm and is more characteristic of large retail organizations. The ratio indicates the net after-tax profit margin on gross revenues; management would be particularly interested in trends (rising, decreasing, or static) of the ratio.

The net after-tax profit for a given year can also be used to calculate a *"gross" rate of return on investment*. That is,

$$\text{gross rate of return} = \frac{\text{net profit}}{\text{total investment}} \times 100\%$$

Different versions of this ratio arise primarily because of widely varying definitions of total investment. The total investment of the Jax Company is defined as total assets, or the sum of all fixed assets at their original cost minus accumulated depreciation plus all current assets (see Table 9.1). Thus, the gross rate of return for the accounting period covered by the income statement is

$$R/R = \frac{\$29\,000}{\$961\,000} \times 100\% = 3.02\%$$

This is a very low rate of return on investment and not typical of successful manufacturing firms.

Although this ratio is commonly used in the business world, none of its many versions gives a completely accurate financial picture, since the ratio does not take into account the time value of money. This ratio also has little bearing on the discussion in this text, most of our study being based on the time value of money.

The *inventory turnover ratio* shows how many times the organization has sold its complete inventory during the past accounting period, thus measuring the rate at which inventory items are moving. It is defined as

$$\text{inventory turnover ratio} = \frac{\text{cost of goods sold}}{\text{inventory}}$$

where the inventory includes the cost of all finished goods the company has produced during the accounting period. Similarly, inventory turnover ratios can be calculated for inventories of raw material and work in progress. For the Jax Tool and Engineering Company the finished goods inventory turnover ratio is

$$\frac{\$985\,000}{\$15\,500} = 63.55 \text{ per year.}$$

A small inventory turnover ratio indicates an accumulation of slow moving goods.

The *collection period* indicates the promptness with which receivables are collected. It is defined as

$$\text{collection period} = \frac{\text{accounts receivable}}{\text{average daily sales}}$$

For the Jax Tool and Engineering Company, the collection period is

$$\frac{\$115\,000}{(\$1\,200\,000/365 \text{ days})} = 34.98 \text{ days}$$

indicating that accounts are paid in 34.98 days on the average.

All the data necessary for calculating these accounting ratios can be found easily in the consolidated financial statements of a public company. The following example illustrates the method of determining ratios from this source.

Example 9.1

Determine the accounting ratios for Excor Industries and Subsidiaries on the basis of its Second Quarterly Report in 2000. The Consolidated Financial Statements of the company are reproduced in Table 9.3.

The total current assets of the company are $1 942 805 000, and the total current liabilities are $924 437 000; therefore, the working capital is

$$\text{working capital} = \text{current assets} - \text{current liabilities}$$
$$= \$1\,942\,805\,000 - \$924\,437\,000$$
$$= \$1\,018\,368\,000$$

The current ratio is

Table 9.3 Consolidated Financial Statements — June 30, 2000 (In Thousands of Dollars) of Excor Industries and Subsidiaries

Statement of Earnings
Second Quarter 2000

Net sales.	$720 953
Other income	7 259
	$728 212
Cost of sales and operating expenses.	$487 547
Selling, general and administrative expenses.	72 349
Research and development	12 945
Exploration	6 079
Interest, net of amounts capitalized.	41 506
Currency translation adjustments.	8 882
	$629 308
Earnings before income and mining taxes	$ 98 904
Income and mining taxes	52 782
Net earnings	$ 46 122
Dividends on preferred shares.	6 592
Net earnings aplicable to common shares	$ 39 530
Net earnings per common share	$0.53
Common shares outstanding at end of period.	75 526 290

Balance Sheet as at June 30, 2000

Cash and securities	$ 71 824	Notes payable	$374 986
Accounts receivable	558 232	Accounts payable	436 811
Inventories	1 287 310	Current taxes payable	112 658
Prepaid expenses	25 439	Total current liabilities	$ 924 437
Total current assets	$1 942 805	Long-term debt	1 015 160
		Deferred taxes	453 000
Property, plant and equipment—net	$2 517 057	Other liabilities	70 231
		Preferred shares	346 948
Cost in excess of net assets acquired	28 145	Common shares	116 016
Other assets	94 332	Retained earnings and capital surplus	1 656 547
	$4 582 339		$4 582 339

Accounting Principles

$$\text{current ratio} = \frac{\text{current assets}}{\text{current liabilities}}$$

$$= \frac{\$1\,982\,805\,000}{\$924\,437\,000} = 2.10$$

The inventories are $1 287 310 000, the prepaid expenses $25 439 000; therefore,

$$\text{acid test ratio} = \frac{\text{current assets} - \text{inventories} - \text{prepaid expenses}}{\text{current liabilities}}$$

$$= \frac{\$1\,942\,805\,000 - \$1\,287\,310\,000 - \$25\,439\,000}{\$924\,437\,000}$$

$$= 0.682$$

The total assets are $4 582 339 000, and the shareholders' equity (the sum of preferred shares, common shares, and retained earnings and capital surplus) is $2 119 511 000; therefore,

$$\text{equity ratio} = \frac{\text{owners' equity}}{\text{total assets}}$$

$$= \frac{\$2\,119\,511\,000}{\$4\,582\,339\,000} = 0.463$$

The long-term liabilities are $1 015 160 000; thus,

$$\text{debt to equity ratio} = \frac{\text{long-term liabilities}}{\text{owners' equity}}$$

$$= \frac{\$1\,015\,160\,000}{\$2\,119\,511\,000} = 0.479$$

From the Statement of Earnings, the total revenues are $728 212 000 for the quarter (the sum of net sales and other income) and the total expenses are $629 308 000 (including cost of sales and operating expenses; selling, general, and administrative expenses; research and development; exploration; and interest and currency translation adjustments); therefore,

$$\text{operating ratio} = \frac{\text{total revenues}}{\text{total expenses}}$$

$$= \frac{\$728\,212\,000}{\$629\,308\,000} = 1.157$$

The net profit for the quarter is $46 122 000; accordingly,

$$\text{income ratio} = \frac{\text{net profit}}{\text{total revenue}} \times 100\%$$

$$= \frac{\$46\,122\,000}{\$728\,212\,000} \times 100\% = 6.33\%$$

The total investment is $4 582 339 000; therefore,

$$\text{gross rate of return} = \frac{\text{net profit}}{\text{total investment}} \times 100\%$$

$$= \frac{\$46\,122\,000}{\$4\,582\,339\,000} \times 100\% = 1.007\%$$

The inventory is $1 287 310 000, the cost of goods sold is

$$\$487\,547\,000 + \$72\,349\,000 = \$559\,896\,000$$

(cost of sales and operating expenses plus selling, general and administrative expenses), and the inventory turnover is calculated as:

$$\text{inventory turnover} = \frac{\text{cost of goods sold}}{\text{inventory}}$$

$$= \frac{\$559\,896\,000}{\$1\,287\,310\,000}$$

$$= 0.435 \text{ per quarter}$$

The accounts receivable are $558 232 000, and the average daily sales are

$$\frac{\$720\,953\,000}{91\,\text{days}} = \$7\,922\,560/\text{day}$$

Therefore,

$$\text{collection period} = \frac{\text{accounts receivable}}{\text{average daily sales}}$$

$$= \frac{\$558\,232\,000}{\$7\,922\,560/\text{day}}$$

$$= 70.46 \text{ days}$$

9.5 Cost Accounting

The balance sheet and the income statement in Section 9.4 are considerably removed both in time and in detail from decisions at the usual engineering project level. More important to the engineer as a source of cost information is the cost accounting system within a particular firm. The firm may be involved in manufacturing or providing services, and if it is involved in manufacturing, production may be on a job-shop or process basis. There are some fundamental differences in cost accounting procedures for determining manufacturing costs versus determining the cost of providing a service; also, there are differences in accounting procedures if manufacturing is on a job-shop or process basis. In order to concentrate on basic principles instead of details, the cost accounting system assumed will be that of a job-shop manufacturing firm. Thus, the emphasis will be on determining the per-order costs for a job order.

The total cost of producing any job order consists of direct material, direct labour, and overhead costs. An additional item of cost could be special tooling or equipment purchases strictly for the job order in question. In order to simplify the presentation, this definition of total cost does not break down the overhead cost into factory overhead, general overhead, and marketing expenses. Materials for a given job order may include purchased parts and in-house fabricated parts, and the cost for direct materials is determined primarily from purchase invoices. Questions of scrap allowances and averaging material costs, which fluctuate over time, present problems in obtaining accurate direct material costs, but determining such costs is reasonably straightforward. Direct labour time spent on a job order is normally recorded by operators on labour time cards, and direct labour cost is determined by applying the appropriate labour cost rates. The labour rates, as determined by the accounting system, will normally include the cost of employee fringe benefits in addition to the basic hourly rate. Although accurately determining the direct labour cost for a given job is a major accounting problem, it is more readily determined than the overhead cost.

Overhead costs cannot be allocated as direct charges to any single job order and must, therefore, be prorated among all the job orders on some arbitrary basis. Common methods for distribution are:

1. The rate per direct labour hour.
2. A percentage of direct labour cost.
3. A percentage of prime cost (direct material plus direct labour cost).

If a single overhead rate for the entire manufacturing firm is to be used, this would, of course, be an average overhead rate for the entire factory, assuming the expense of providing and using an hour of factory facilities is about the same throughout the factory. To illustrate the three methods listed, it is assumed that a company experienced the following costs in the previous year.

Total direct labour hours	$ 48 000
Total direct labour cost	$480 000
Total direct material cost	$600 000
Total overhead cost	$360 000

Then, the *overhead rate per direct labour hour* would be

$$\text{overhead rate} = \frac{\text{overhead costs}}{\text{direct labour hours}}$$

$$= \frac{\$360\,000}{48\,000 \text{ h}}$$

$$= \$7.50 \text{ per direct labour hour}$$

If a particular job order requires 40 h of direct labour with an average rate of $12.50 per hour and $850 of direct materials, then the total cost of the job would be computed as

Direct material cost	=	$ 850
Direct labour cost	= 40 h × $12.50/h =	500
Overhead cost	= 40 h × $7.50/h =	300
	Total cost	$1650

Determining the overhead rate as a *percentage of direct labour cost* for this company will yield

$$\text{overhead rate \%} = \frac{\text{overhead costs}}{\text{direct labour cost}} \times 100\%$$

$$= \frac{\$360\,000}{\$480\,000} \times 100\%$$

$$= 75\% \text{ of direct labour cost}$$

For the same job above, the total cost would be computed as

Direct material cost	=	$ 850
Direct labour cost	=	500
Overhead cost	= 75% × $500 =	375
	Total cost	$1725

Determining the overhead rate as a *percentage of prime cost* for this company will yield

$$\text{overhead rate \%} = \frac{\text{overhead costs}}{\text{direct labour cost} + \text{direct material cost}} \times 100\%$$

$$= \frac{\$360\,000}{\$1\,080\,000} \times 100\%$$

$$= 33\tfrac{1}{3}\%$$

For the same job above, the total cost would be computed as

$$\begin{aligned}\text{Direct material cost} &= & \$\ 850 \\ \text{Direct labour cost} &= & 500 \\ \text{Overhead cost} &= 33\tfrac{1}{3}\% \times \$1350 = & 450 \\ & \text{Total cost} & \$1800\end{aligned}$$

Determining the overhead cost for a job order by the "rate per direct labour hour" method will yield the same result as the "percentage of direct labour cost" method, provided that the rate per direct labour hour used on the job is equal to the factory average rate per labour hour. The "percentage of prime cost" method will normally yield a different assignment of overhead to a job order than the other two methods. The choice among these three methods is arbitrary; indeed, other methods are used by cost accountants in distributing overhead costs to a given job order. The rate per direct labour hour method is perhaps most commonly used.

Whatever method is chosen from the above for distributing overhead costs to job orders in a current year, the rates or percentages are based on the previous year's cost figures. Thus, overhead rates may change from year to year within a particular firm.

Since an average overhead rate for the entire factory may very well be too gross an estimate when actual overhead costs differ among departments within the factory, cost accounting may determine individual overhead rates for departments or cost centres. Furthermore, the hourly rates for direct labour may vary among these cost centres. A further refinement can be made by determining overhead rates for individual machines within cost centres. Then, as particular job orders progress through departments and/or machines, the direct labour time (or machine time) spent on the job order in the various cost centres is recorded, the appropriate labour or machine rates and overhead rates are applied, and the total cost for the job is calculated.

The following example illustrates one of the variety of methods used to calculate the overhead rate for a given cost centre and the total cost of a job.

Example 9.2

The following information has been accumulated for the Deetco Company's two departments during the past year. (See table below.)

The Deetco Company distributes depreciation overhead based on (1) the first cost of equipment in each department, (2) zero salvage value of the equipment in 10 years, and (3) a constant annual (or straight-line) rate of depreciation. All overhead other than depreciation is first distributed to each department according to the number of employees in each department, and an overhead rate per direct labour hour is computed.

What selling price should the company quote on Job Order D if raw material costs are estimated as $900, estimated direct labour hours required in Departments A and B are 30 hours and 100 hours, respectively, and profit is to be calculated as 25% of selling price? The summary of costs incurred by the two departments of Deetco Company during the last financial year is given below.

	Department A	Department B	Total
Direct material cost	$720 000	$240 000	$960 000
Direct labour cost	$260 000	$140 000	$400 000
Direct labour hours	25 200	16 200	41 400
Number of employees	14	9	23
First cost of equipment	$250 000	$200 000	$450 000
Annual depreciation	$ 25 000	$ 20 000	$ 45 000
Other factory overhead			$150 000
General overhead			$350 000

For Department A, the total overhead allocated is determined as

Annual depreciation	=	$ 25 000
Other factory overhead	= (14/23)($150 000) =	91 304
General overhead	= (14/23)($350 000) =	213 043
	Total overhead costs	$329 347

Thus, the overhead rate for Department A per direct labour hour is

$$\frac{\$329\,347}{25\,200\text{ h}} = \$13.07 \text{ per direct labour hour}$$

For Department B, the total overhead allocated is determined as

Annual depreciation		=	$ 20 000
Other factory overhead	= (9/23)($150 000) =		58 696
General overhead	= (9/23)($350 000) =		136 957
	Total overhead costs		$215 653

Thus, the overhead rate for Department B per direct labour hour is

$$\frac{\$215\,653}{16\,200\text{ h}} = \$13.31 \text{ per direct labour hour}$$

Then the estimated total cost for Job Order D is computed as

Direct material cost		=	$ 900.00
Direct labour cost for Dept. A	$= \dfrac{\$260\,000}{25\,200/\text{h}} \times 30\text{ h}$	=	309.52
Overhead cost for Dept. A	$= (\$13.07/\text{h})(30\text{ h})$	=	392.10
Direct labour cost for Dept. B	$= \dfrac{\$140\,000}{16\,200/\text{h}} \times 100\text{ h}$	=	864.20
Overhead cost for Dept. B	$= (\$13.31/\text{h})(100\text{ h})$	=	1331.00
	Total cost	=	$3796.82

If x = the selling price of Job Order D, then

$$x = \text{total cost} + \text{profit}$$
$$= \$3\,796.82 + 0.25x$$

and

$$x = \frac{\$3\,796.82}{0.75} = \$5\,062.43$$

Problems

1. K.Z. Moley purchased a small engine repair shop and opened on July 1, 2000. At the date of opening, he had invested $8 000 of equity funds with the following breakdown: $4 000 in equipment, $3 000 in inventory items, and $1 000 in operating cash.

   (a) Prepare a balance sheet for the business as of July 1, 2000. (9.2) The data below summarize the gross sales and expenses for the business during the first three-month period.

Gross sales	$13 500
Purchases	7 000
Salaries	3 000
Advertising expense	250
Rent expense	600
Expense for utilities	400

   For the purchases, $6 000 was paid with cash and $1 000 is still owed. At the close of the three-month period, the end-of-period inventory is worth $2 600.

   (b) Prepare an income statement for the three-month period covered. (9.3)

   (c) If, at the end of the three-month period on October 1, 2000, the owners' equity account is $9 850, determine the amount of the cash account in order to balance the accounting equation as of October 1, 2000. (9.2)

2. A successful building contractor purchased a farm to operate on a part-time basis and raise beef cattle. The purchase price of the farm was $75 000. The contractor paid $25 000 cash and financed the remainder over 10 years at a 9% interest rate with a mortgage, payable to a local bank. Soon after the purchase of the farm, the contractor purchased cattle for $8 000, paid $3 000 in cash, and gave a promissory note to the seller for the balance. The note carried an 8% annual interest rate and was to be paid off within five years. During the first full year, the farm operation resulted in the following revenues and expenses.

Calves sold	$6 500
Hay sold	200
Labour expenses	500
Expenses for machinery rented	2 000
Veterinarian fees	175
Fertilizer purchased	1 200
Property taxes	450
Expenses for repairs	375
Internal expenses	4 900
Expenses for miscellaneous supplies	125

Prepare an income statement for the farming operation and determine the net profit (loss) before income taxes. (9.3)

3. An income statement and balance sheet for the WAC Company covering a calendar year ending December 31 are as below:

Income Statement

Gross income from sales		$247 000
Less: cost of goods sold		138 800
Net income from sales		$108 200
Operating exenses		
Rent	$ 9 700	
Salaries	30 200	
Depreciation	5 800	
Advertising	4 300	
Insurance	1 500	$51 500
Net profit before income taxes		56 700
Less: income taxes		23 973
Net profit after income taxes		$32 727

Balance Sheet

Assets			Liabilities	
Cash		$ 94 227	Notes payable	$ 25 000
Accounts receivable		8 000	Accounts payable	6 000
Raw materials inventory		10 000	Declared dividends	20 000
Work-in-progress inventory		15 000	Total liabilities	$ 51 000
Finished goods inventory		18 500		
Land		30 000	*Shareholders' Equity*	
Building	$80 000		Capital stock	$20 000
Less: accumulated depreciation	$ 8 000	72 000	Retained earnings	32 727
Equipment	$40 000		Total net worth	$232 727
Less: accumulated depreciation	$ 4 000	36 000	Total liabilities and shareholders' equity	
Total assets		$238 72		$283 727

If land, building, and equipment are fixed assets, and notes payable is a long-term liability, compute (a) the current ratio, (b) the acid-test ratio, (c) the equity ratio, (d) the operating ratio, (e) the income ratio for net profit after taxes, (f) the debt to equity ratio, (g) the gross rate of return, (h) the inventory turnover ratio, and (i) the collection period for the WAC Company. (9.4)

4. An order for 500 units of Part D-142 is received by the J.T. Kling Engineering Company, a small machine shop. The finished dimensions of the rectangular part are 19 mm x 17 mm x 45 mm (neglecting tolerances). The raw material for this part is 20 mm x 20 mm x 1 000 mm steel rectangular bar stock, with each unit costing $12 and yielding 20 part blanks. The basic manufacturing sequence, with standard machining times per part, and machine overhead rates (per machining hour) is given below.

Operating	Standard Time per Part	Machine Overhead Rate
Cutoff on power saw	1 min	$0.30/h
Mill two sides	4 min	$0.50/h
Drill three 4-mm-diameter holes	2 min	$0.40/h
Surface grind one side	2 min	$0.35/h
Package	0.25 min	—

The direct labour time per part is the same as the machine (or operation) time per part. The tooling cost for this job order is estimated to be $500. Excluding tooling costs and machine overhead, other factory overhead costs (for indirect labour, utilities, indirect materials, etc.) are $9 per direct labour hour. If the average direct labour hour rate (including fringe benefits) is $10.75 per hour, determine (a) the total estimated costs for the job order of 5 000 units, and (b) the unit selling price if profit is to be 30% of the total cost. (9.5)

5. The welding department of a mining equipment manufacturing plant consists of four cost centres: manual arc welding (A), semiautomatic welding (B), furnace brazing and heat treating (C), and finishing (D). Some oxyacetylene cutting is also done in Centre B. Assume that it is possible to allocate departmental overhead expenses directly to each cost centre and that the following data for the welding department were compiled last year by the accounting system.

Cost Centre	Departmental Expenses	Direct Labour		Direct Material
		Hours	Cost	
A	$10 500	10 000	$52 000	$8 000
B	6 800	4 000	16 000	8 000
C	4 600	1 500	4 500	3 000
D	2 400	2 800	6 000	2 000

Compute the overhead rate (or rates) applicable by the following methods (9.5)

    (a) blanket (departmental) percentage of direct labour cost
    (b) blanket percentage of prime cost

(c) blanket hourly rate per direct labour hour
(d) percentage of direct labour cost for each cost centre
(e) percentage of prime cost for each cost centre
(f) rate per direct labour hour for each cost centre

6. Consider the welding department and four cost centres given in Problem 5. The direct labour hours and cost for each cost centre remain the same as in Problem 5, but new and additional data are given below (assume cost data are for the previous year).

Cost Centre	Square metres occupied	Cost of Machinery	Number of Direct Labour Employees
A	900	$ 5 000	5
B	400	6 000	2
C	600	9 000	1
D	500	3 000	1
Total	2 400	$23 000	9

Expenses other than for direct labour and materials chargeable to the welding department last year were:

Maintenance	$4 000
Gas and electricity	10 000
Supervision and other indirect labour	24 000
Miscellaneous supplies	5 000
Equipment depreciation	2 300
Building depreciation	4 000

Determine an overhead rate per direct labour hour for each cost centre if the welding department expenses above are first allocated to each cost centre as follows: (9.5)

(a) Maintenance expenses and equipment depreciation expenses are allocated according to the value of equipment (percent of total) in each cost centre.

(b) Building depreciation expenses are allocated according to the floor space occupied by each cost centre.

(c) Supervision and other indirect labour expenses are allocated according to the number of direct labour employees of each cost centre.

(d) Supplies and gas and electricity expenses are allocated according to the number of direct labour hours for each cost centre.

Chapter 10
Fundamental Economic Concepts

10.1 Introduction

The main goal of engineering economic analysis is to develop a logical methodology for choosing an engineering project from among several feasible, competing alternatives where the criterion for selection is a measure of economic effectiveness. Although the subject of engineering economic analysis can be studied on its own, some knowledge of basic economic principles and some experience in engineering problem solving are particularly helpful in understanding the objectives and the methods of the subject. In this chapter, two basic economic concepts closely related to the subject of engineering economic analysis are outlined; the supply and demand interaction and the theory of production. It is important that readers without any previous exposure to economic theory study the material presented in this chapter. However, readers who are already familiar with these fundamental economic concepts may omit it.

10.2 Supply and Demand

In a free-market economy, the laws of supply and demand determine the price of goods and the quantities of these goods that are sold. It is important for the practising engineer to understand the fundamental concepts of supply and demand, their interaction, the way this interaction determines prices and quantities of goods sold, and the method of predicting the change in price and quantity sold due to change in demand and supply.

10.2.1 Demand

The quantity of a commodity that would be purchased by all consumers in a market, termed the *quantity demanded*, is a specific quantity of goods desired per time period (e.g., 23 000 snowmobiles per year, 7 000 tractors per week, or 15 000 kg of chromium per month). Quantity demanded is also a function of a number of factors: the price of the commodity, the size of the population served by the market, the income of the population, prices of related commodities, tastes and preferences of the population, advertising, etc. The quantity demanded normally changes if any one of these factors changes. For example, the number of snowmobiles demanded in the Canadian market will increase if the price of the snowmobile drops, if the Canadian population increases, or if the average income of Canadians rises. Even an increase in air fares to Florida would tend to increase the number of snowmobiles demanded in the Canadian market.

Of the factors affecting quantity demanded, price is the most significant. Quantity demanded is the quantity desired at a given price, and it is not necessarily the same as the quantity actually purchased. For instance, let us assume that at a price of $3 000 per unit, 23 000 snowmobiles are demanded in a year, but only 18 000 units are available for sale at this price in the market. Thus, regardless of the fact that no more than 18 000 snowmobiles can be actually purchased in this particular year for the price of $3 000 per unit, the quantity demanded is still 23 000 snowmobiles per year.

The relationship between quantity demanded and price can be expressed in general as

$$Q_d = f(P, x_1, x_2, \ldots x_n)$$

where Q_d = the quantity demanded, P = the price, and x_1 to x_n = other factors.

The most important variable in this equation is the price of the commodity. If all factors except the price are held constant, the equation becomes a univariable relationship between quantity demanded and price

$$Q_d = f(P)$$

This relationship is called the demand function, or simply the *demand*. The terms *quantity demanded* and *demand* must be clearly distinguished. The quantity demanded refers to a specific quantity desired per time period at a given price, such as 23 000 snowmobiles per year at $3 000, whereas the demand is the functional relationship between price and quantity demanded. The demand function has the following important properties: as the price of the commodity falls, the quantity of the commodity demanded by consumers increases and as the price rises, the quantity demanded drops. The demand function of a com-

Table 10.1 Demand Schedule

Price ($ per unit)	Quantity Demanded (units per year)
1 000	55 000
2 000	35 000
3 000	23 000
4 000	15 000
5 000	10 000
6 000	9 000

modity is usually specified in a table called a *demand schedule,* or by a graph representing a *demand curve,* or occasionally by a mathematical function. A hypothetical demand schedule for snowmobiles in the Canadian market is given in Table 10.1, and the corresponding demand curve is shown in Figure 10.1.

The slope of this typical demand curve is negative at every point, reflecting the fundamental relationship that exists between price and quantity demanded: as the price drops, all other things being equal, the quantity demanded increases; and as the price rises, all other things being equal, the quantity demanded falls. Each point on the demand curve represents a specific price-quantity combination. At point A the price is $2 000 per unit, and the corresponding quantity demanded is 35 000 units per year. At point B, the price is $3000 per unit, and 23 000 units per year are demanded at this price. As the price rises from $2 000 to $3 000 per unit, the quantity demanded is reduced from 35 000 to 23 000 units per year. This change is represented by the arrow on the Figure pointing from A to B, and is called *movement along the demand curve.*

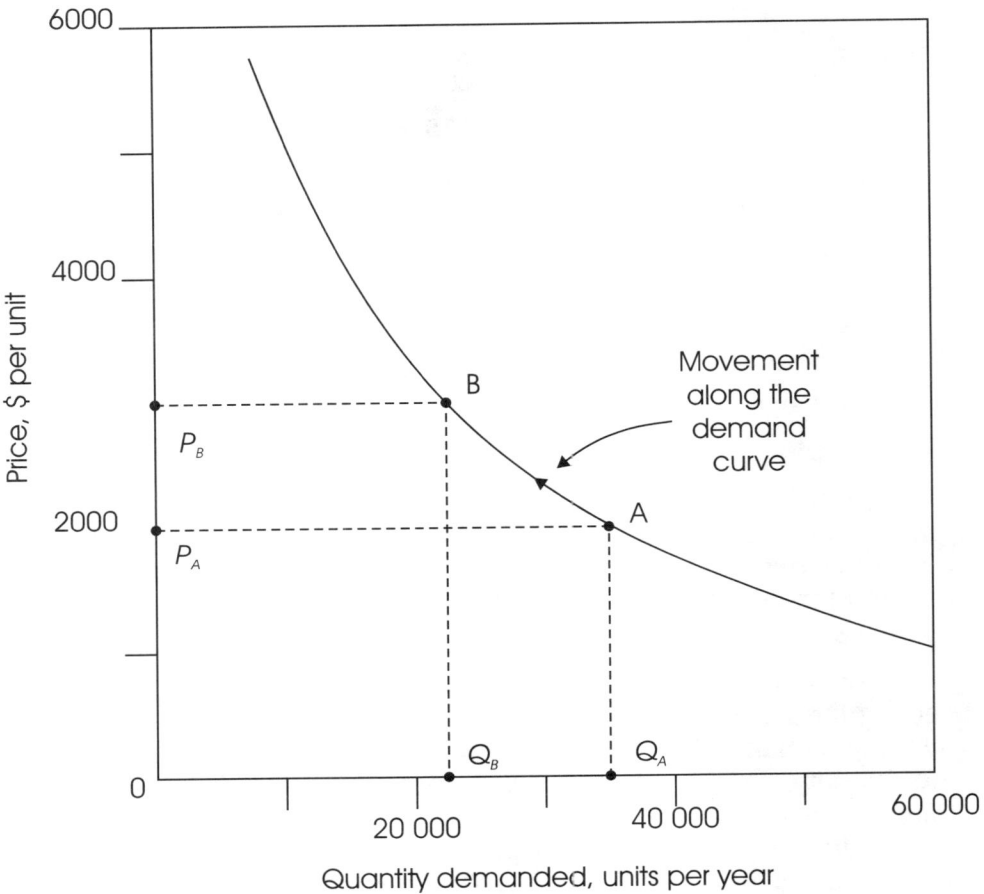

Fig. 10.1 Demand curve.

10.2.2 Shift in the Demand Curve

The demand curve is the graphical representation of the univariable relationship between price and quantity demanded of a commodity, as all other factors remain constant. If factors other than the price of the commodity change, the demand curve responds to that change. For example, if the Canadian population increases or if the average income of the population increases, the quantity of snowmobiles demanded at any given price will increase, and the demand curve of

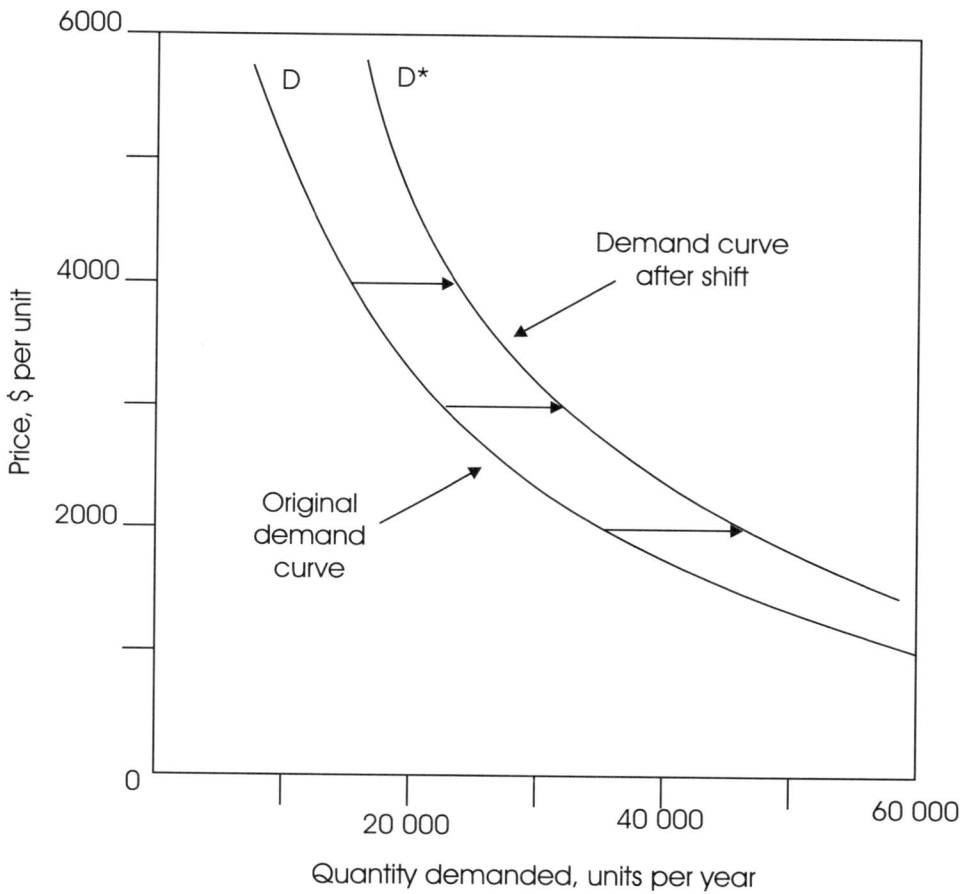

Fig. 10.2 Shift in the demand curve caused by changes in factors other than the price

snowmobiles will shift to the right as shown in Figure 10.2. Similarly, a change in factors resulting in a reduction in the quantity demanded at any given price shifts the demand curve to the left.

10.2.3 Supply

The quantity of a commodity supplied by all producers in a market, called the *quantity supplied*, is expressed as a rate or flow of goods using dimensions of "units per time," such as 10 000 tractors per week or 17 000 kg of chromium per month. Quantity supplied also depends on a number of factors: the price of the commodity, the cost of production, the state of technology, etc. The quantity supplied is a function of these factors and, in general, it is expressed as

$$Q_s = g(P, y_1, y_2, \ldots y_m)$$

where

$$Q_s = \text{the quantity supplied}$$
$$P = \text{the price}$$
$$y_1 \text{ to } y_m = \text{other factors}$$

The most important factor in this equation is the price of the commodity. If all other factors are constant, the quantity supplied becomes a univariable function of the price. This functional relationship, $Q_s = g(P)$, is called the *supply*. A clear distinction must be made between quantity supplied and supply. The *quantity supplied* is a specific quantity supplied per time period at a given price, such as 57 500 snowmobiles per year at $2 000, whereas the supply refers to the functional relationship between price and quantity supplied. The supply function of a commodity is normally specified in a table, called the supply schedule, or by a graph, called the supply curve. Table 10.2 shows a hypothetical supply schedule for snowmobiles in the Canadian market and the corresponding supply curve is shown in Figure 10.3.

Table 10.2 Supply Schedule

Price ($ per unit)	Quantity Supplied (units per year)
1 000	25 000
2 000	38 500
3 000	48 000
4 000	57 500
5 000	65 000
6 000	72 000

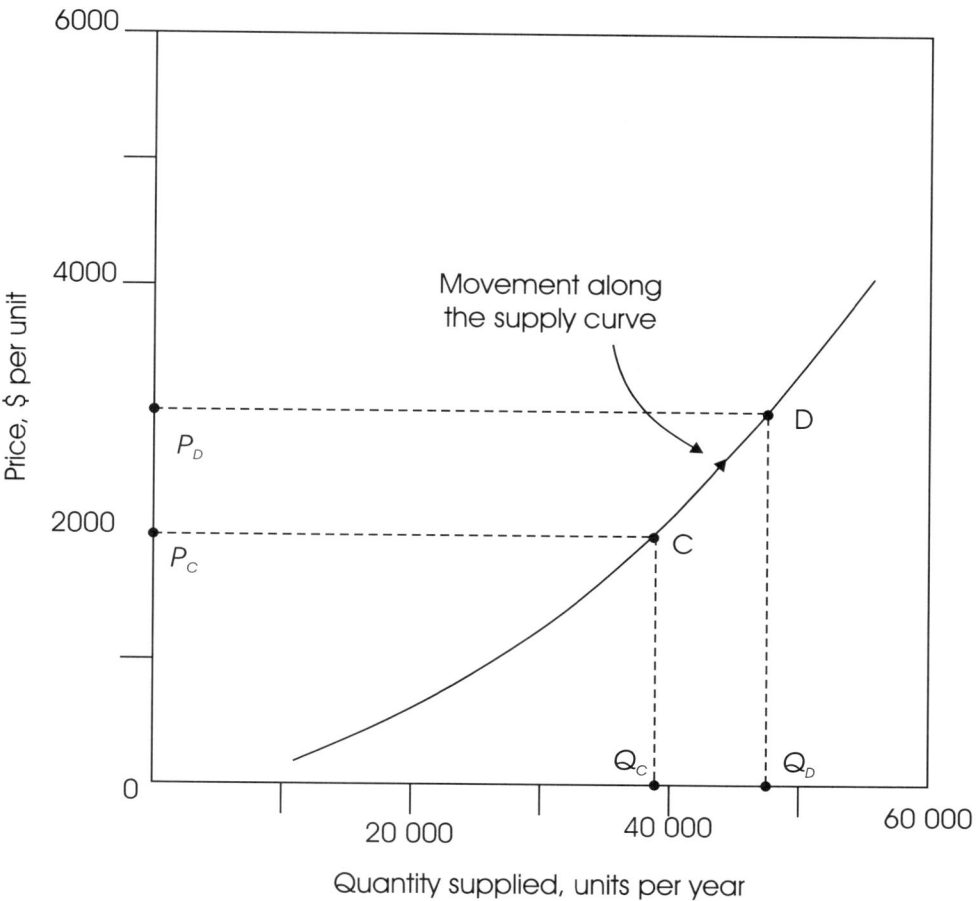

Fig. 10.3 Supply curve.

The slope of a typical supply curve is positive at each point of the curve, indicating that the amount of commodity available for sale (the quantity supplied) increases as the price increases if all other factors remain unchanged. On the other hand, the quantity supplied decreases if the price of the commodity falls, all other things being equal. A point on the curve represents a specific price-quantity combination. For example, at point C the price of a snowmobile is $2 000, and the corresponding quantity supplied is 38 500 units per year. As the price rises from $2 000 to $3 000 per unit the quantity supplied increases to 48 000 units per year. This change is *movement along the supply curve*, and it is indicated by the arrow pointing from point C to point D in Figure 10.3.

10.2.4 Shift in the Supply Curve

The supply curve is the graphical representation of the univariable relationship existing between the price of a commodity and quantity of this commodity supplied, all other factors remaining constant. If a factor other than the price changes, the quantity supplied responds to the change. For example, if the cost of production increases, the quantity supplied decreases, because a number of manufacturers or producers will no longer be able to manufacture or produce the commodity at a profit under these conditions. The reduced quantity supplied by the manufacturers and producers results in a shift in the supply curve to the left. However, other changes might tend to increase the quantity supplied at any given price, which would result in a shift of the supply curve to the right (see Figure 10.4). For example, an extended strike at Sudbury in the nickel mines would have a considerable effect on the nickel supply in Canada, resulting in a shift to the left in the supply curve of nickel. On the other hand, the rapid advances that have occurred in microprocessor technology have resulted in a shift to the right in the supply curve of computers.

10.2.5 Price

For the purpose of determining the price of the commodity at the marketplace, the demand and supply curves shown in Figures 10.1 and 10.3 are combined in Figure 10.5. It can be observed that at a price of, say $4 000 per unit, only 17 000 snowmobiles per year would be demanded, but 53 000 units would be supplied. This results in a large excess supply or *surplus* of 36 000 unsaleable snowmobiles. Sellers normally cut prices when there is a surplus, in order to be able to sell their products. The large surplus at the $4 000 price would result in a considerable downward pressure on the price, and the $4 000 price would fall rapidly. On the other hand, at a price of only $1 300 per unit, 52 000 units per year would be demanded, but since only 30 000 units per year would be supplied, there would be an excess demand or *shortage* of 22 000 units. Buyers normally bid prices up if there is a shortage of goods. The shortage at the $1 300 price would result in a

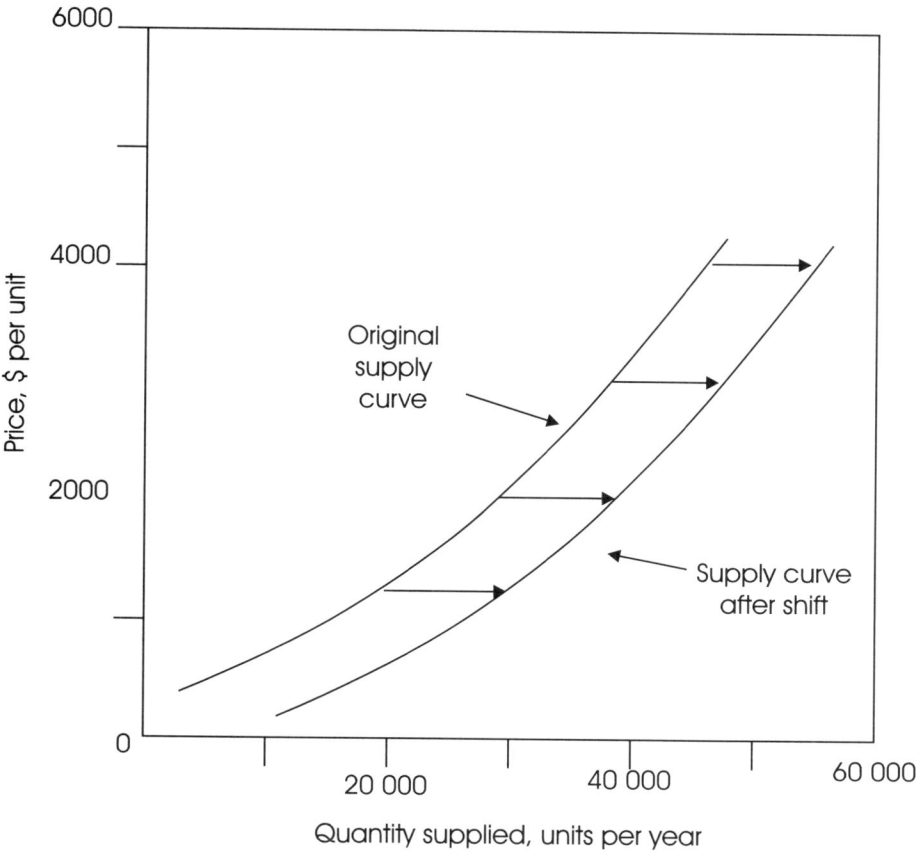

Fig. 10.4 Shift in the supply curve caused by changes in factors other than the price

considerable upward pressure on the price, and the $1 300 price would rise quickly. What is the stable (equilibrium) price of this commodity and how can this price be determined?

At the equilibrium price clearly neither shortage nor surplus can exist. Therefore, the quantity demanded at the equilibrium price must be exactly equal to the quantity supplied at that price. There is only one price that satisfies this condition, the price corresponding to the point of intersection of the supply and demand curves (point E in Figure 10.5).

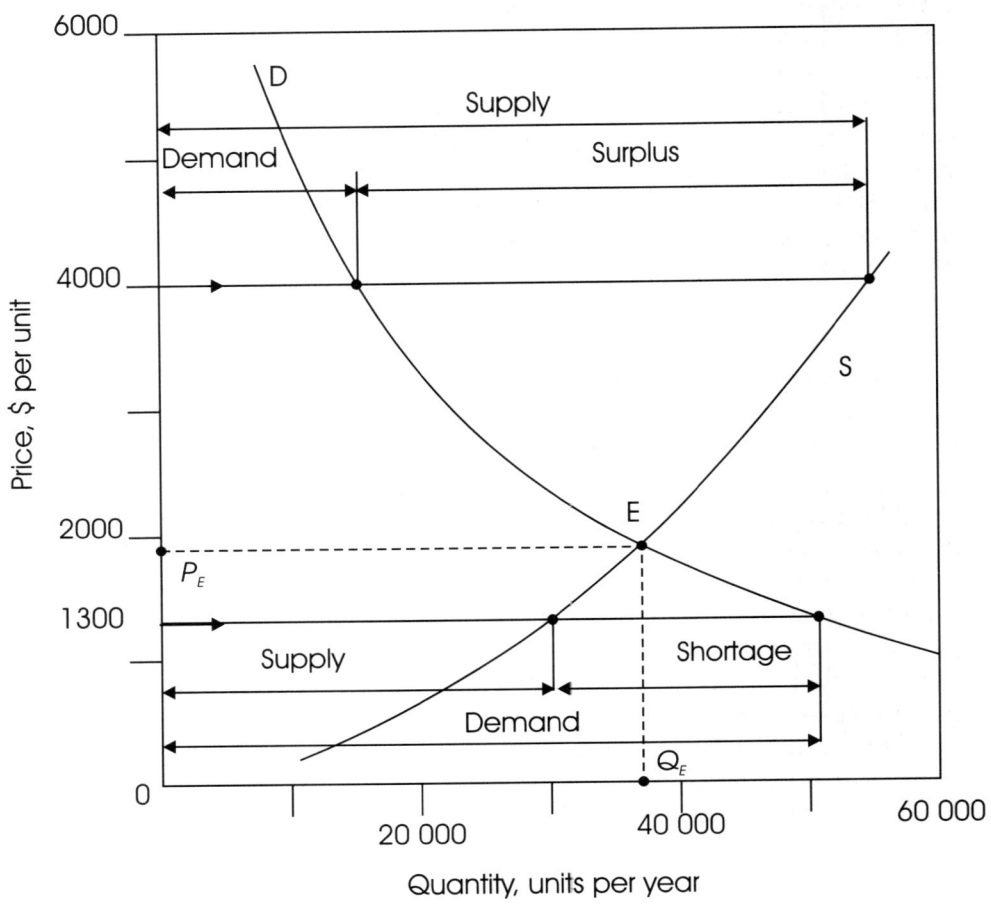

Fig. 10.5 The equilibrium point.

D = demand curve
S = supply curve
E = equilibrium point
P_E = equilibrium price
Q_E = equilibrium quantity

The point where the demand and supply curves intersect is called the equilibrium point, and the corresponding price and quantity are termed *equilibrium price* and *equilibrium quantity*. In Figure 10.5 the equilibrium price is $1 900 per unit and the equilibrium quantity is 37 000 units per year. At prices above the equilibrium price, there is a surplus and a downward pressure on the price. At prices below the equilibrium price, there is a shortage and an upward pressure on the price. The price moves toward the equilibrium price in a stable market free from governmental or monopolistic interference. Once the equilibrium price is reached, it remains stable unless disturbed by changes in market conditions (shifts in the demand or supply curves).

The equilibrium point moves if there is a shift in either demand or supply. If there is a shift to the right in the demand curve, a shortage develops at the original equilibrium price, creating an upward pressure on the price, and the equilibrium point moves upward along the supply curve (see Figure 10.6). At the new equilibrium point, E*, the equilibrium price is higher and the equilibrium quantity is larger than at the original equilibrium point, E. On the other hand, a shift to the left in the demand curve causes both equilibrium price and equilibrium quantity to fall. A shift to the right in the supply curve (see Figure 10.7) results in a surplus at the original equilibrium price and, therefore, the equilibrium point moves downward along the demand curve from point E to point E*. The new equilibrium price, P^*_E, is lower, and the equilibrium quantity, Q^*_E, is higher than the original values of P_E and Q_E. A shift to the left in the supply curve causes the equilibrium price to rise and the equilibrium quantity to fall.

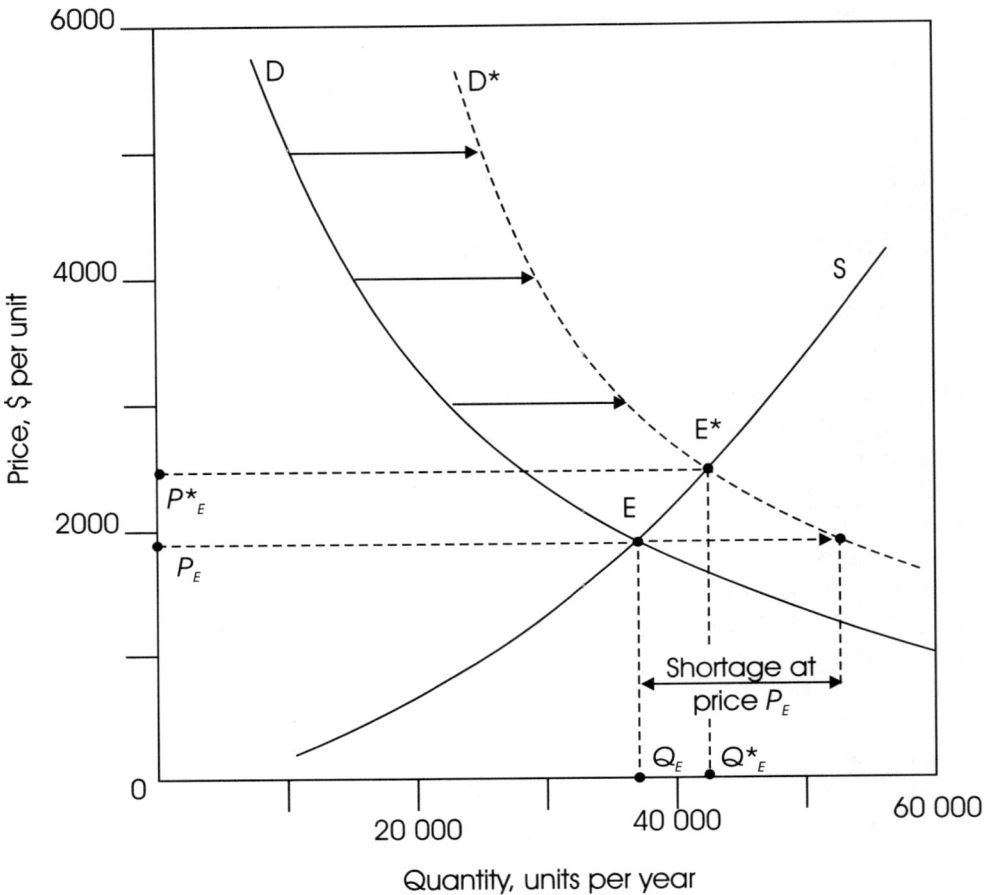

Fig. 10.6 The effect of rising demand on the equilibrium price.

As the demand rises (the demand curve shifts to the right from position D to position D*), a shortage develops at the original equilibrium price P_E, and the equilibrium point moves from point E to point E*, resulting in a higher equilibrium price, P^*_E, and a higher equilibrium quantity, Q^*_E.

Fundamental Economic Concepts

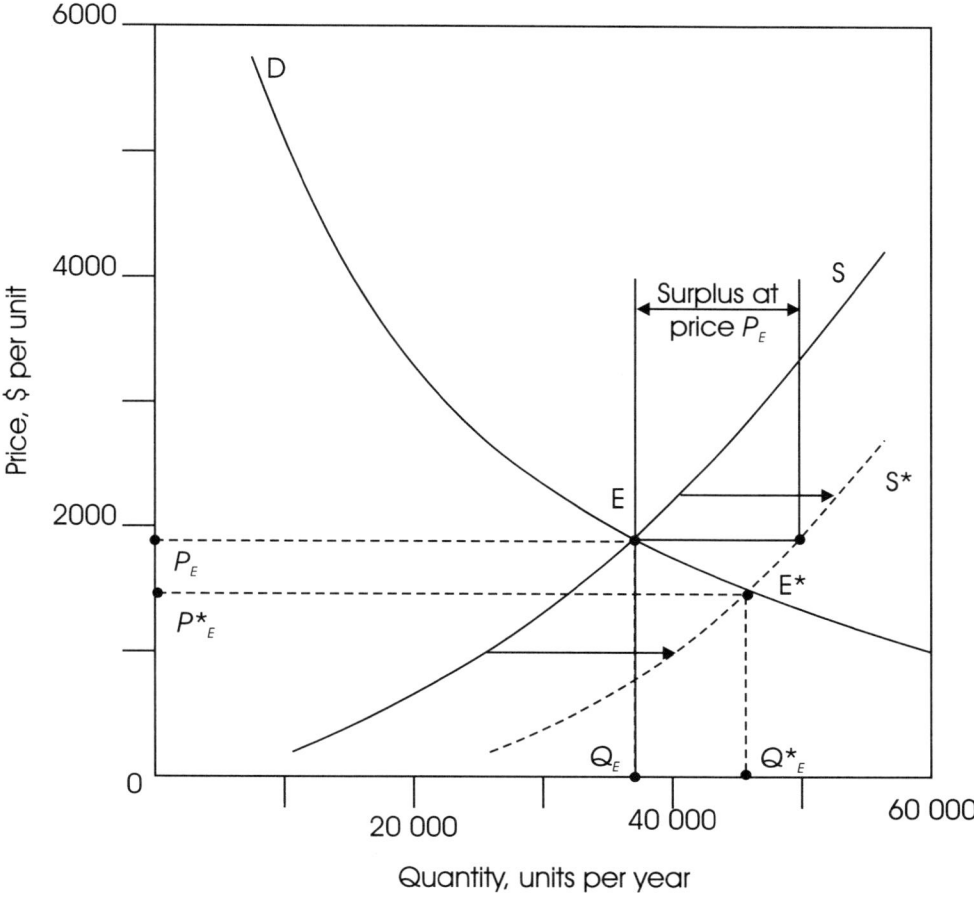

Fig. 10.7 The effect of rising supply on the equilibrium price.

As the supply rises (the supply curve shifts to the right from position S to position S*), a surplus develops at the original equilibrium price P_E, and the equilibrium point moves from point E to point E*, resulting in a lower equilibrium price, P^*_E, and a higher equilibrium quantity, Q^*_E.

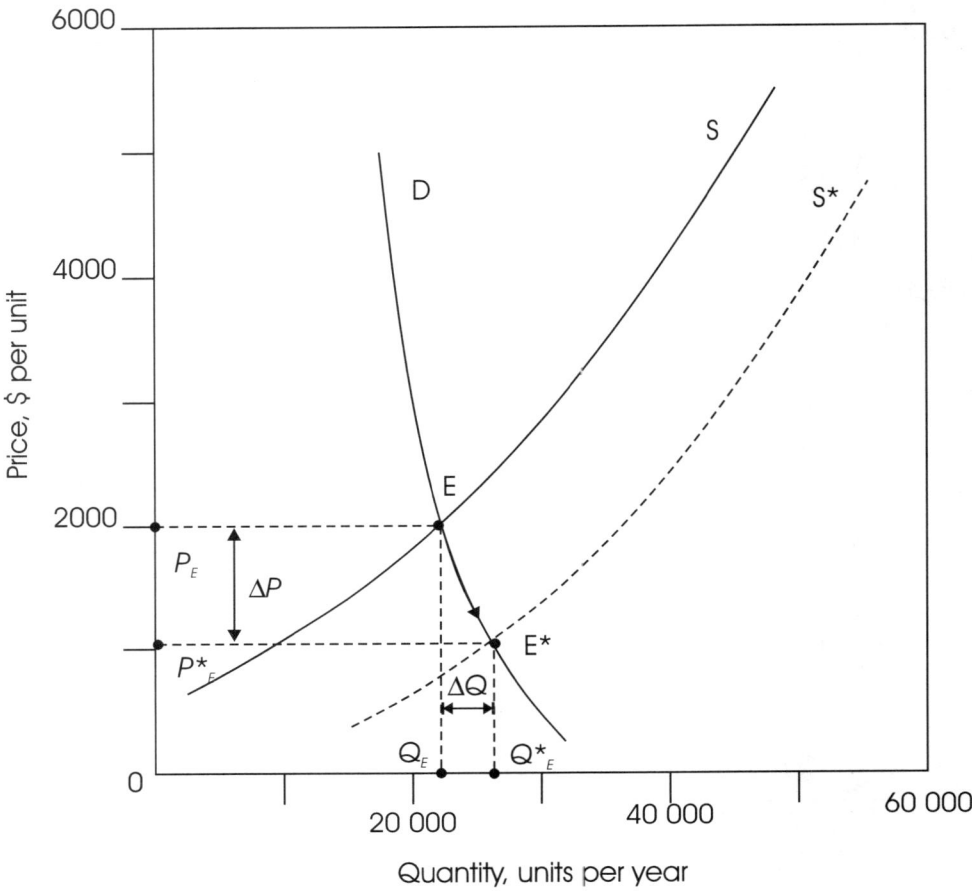

Fig. 10.8(a) The effect of a shift in supply: Steep demand curve

As the supply curve shifts from S to S*, the equilibrium point moves from E to E*.

10.2.6 Elasticity of Demand

As a result of a shift in supply, the point of equilibrium moves and the equilibrium price changes. For example, in Figure 10.8(a), a shift in supply causes the equilibrium price to drop sharply. However, in Figure 10.8(b), the same shift in supply results in a small drop in equilibrium price.

The change in equilibrium price and equilibrium quantity resulting from a shift in supply is clearly dependent on the magnitude of the shift in supply, and also on the shape of the demand curve. For exam-

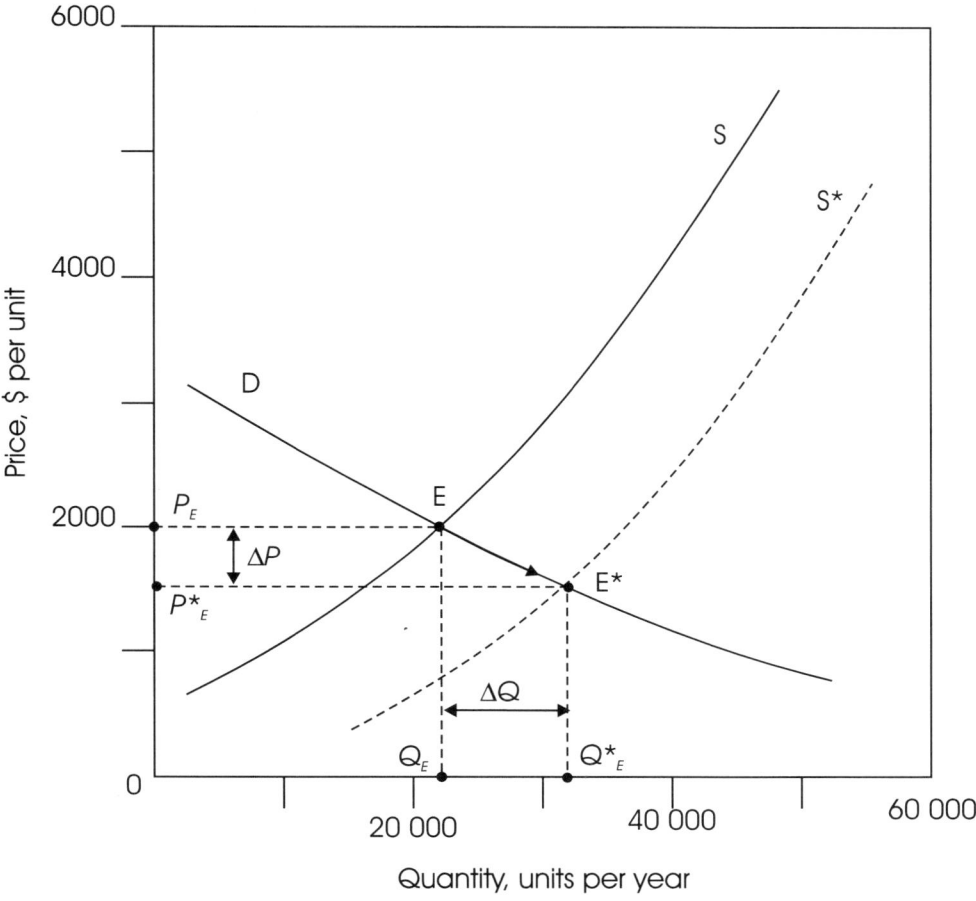

Fig. 10.8(b) The effect of a shift in supply: Flat demand curve

As the supply curve shifts from S to S*, the equilibrium point moves from E to E*.

ple, in Figures 10.8(a) and 10.8(b), the supply curves are shifted by identical amounts; however, the resulting changes in equilibrium prices and equilibrium quantities are different because of differences in the shape of the demand curves. Mathematicians and engineers generally characterize the shape of a curve by its slope and curvature. Economists characterize the shape of the demand curve by its elasticity.

The *elasticity of demand* (or *price elasticity*), which is dependent on the shape of the demand curve, measures the response of the quantity demanded to the change in price. It is defined as the ratio of the percentage change in quantity demanded to the corresponding percentage change in price.

$$E_d = -\frac{\Delta Q_d}{Q_d} \bigg/ \frac{\Delta P}{P} = -\frac{(\Delta Q_d)P}{(\Delta P)Q_d}$$

where

E_d = the elasticity of demand
P = the average price
Q_d = the average quantity demanded
ΔP = the change in price
ΔQ_d = the change in quantity demanded

The purpose of the negative sign in the equation is to make the demand elasticity positive (since the slope of the demand curve is negative).

The elasticity of demand is normally different at every point of the demand curve. Demand elasticity may vary in magnitude from zero to infinity. The demand is said to be

elastic if $E_d > 1$, and
inelastic if $E_d < 1$.

Example 10.1

Determine the elasticity of demand for snowmobiles at prices of $1 500 and $3 500. The demand schedule for snowmobiles is listed below:

Price $	Quantity Demanded (units per year)
1000	55 000
2000	35 000
3000	23 000
4000	15 000
5000	10 000

Fundamental Economic Concepts

At a price of $1 500

$$\Delta P = \$2\,000 - \$1\,000 = \$1\,000$$
$$Q_d = (55\,000 + 35\,000)/2 = 45\,000$$
$$\Delta Q_d = 35\,000 - 55\,000 = -20\,000$$

and therefore

$$E_d = -\frac{(-20\,000)1\,500}{1\,000(45\,000)} = 0.67$$

At a price of $3 500

$$\Delta P = \$4\,000 - \$3\,000 = \$1\,000$$
$$Q_d = (23\,000 + 15\,000)/2 = 19\,000$$
$$\Delta Q_d = 15\,000 - 23\,000 = -8\,000$$

and

$$E_d = -\frac{(-8\,000)3\,500}{1\,000(19\,000)} = 1.47$$

Therefore, the demand curve is inelastic in the vicinity of a price of $1 500, but elastic in the neighbourhood of a price of $3 500. Most demand curves are elastic at the high price end, inelastic at the low price end, and have an elasticity of one at a point between.

10.2.7 Total Expenditure

The total amount spent on a commodity by all consumers in the market is called the *total expenditure*, and it is expressed as

$$T = Q P$$

where

T = the total expenditure
Q = the quantity sold
and P = the price

For obvious reasons the total expenditure of consumers, which is the *total revenue* of producers, is quite important. The total expenditure generally changes as the equilibrium point moves because of a shift in either supply or demand.

Let us now examine the change in total expenditure resulting from a movement of the equilibrium point along the demand curve because of a shift in supply. As the price rises along the demand curve from P_A to P_B (see Figure 10.1), the change in total expenditure is

$$\Delta T = Q_B P_B - Q_A P_A$$
$$= (Q_B - P_A) P_B - Q_A (P_A - P_B)$$
$$= \Delta Q P_B + \Delta P Q_A$$

where $\Delta Q = Q_B - Q_A$ is the change in quantity demanded and $\Delta P = P_B - P_A$ is the change in price. In terms of average values

$$\Delta T = \Delta Q (P + \Delta P/2) + \Delta P (Q + \Delta Q/2)$$

where the average price is $P = (P_A + P_B)/2$, and the average quantity is $Q = (Q_A + Q_B)/2$, and since $\Delta P/2 < P$ and $\Delta Q/2 < Q$, we have the following approximate relation:

$$\Delta T = \Delta Q\, P + \Delta P\, Q$$
$$= \Delta P\, Q \left(1 + \frac{\Delta Q\, P}{\Delta P\, Q}\right)$$

or

$$\Delta T = \Delta P\, Q(1 - E_d) = P(\Delta Q)\left(1 - \frac{1}{E_d}\right)$$

During a movement along the demand curve from point A to point B (see Figure 10.1), the change in price (i.e., the quantity ΔP) is positive; therefore, the change in total expenditure is also positive if $E_d < 1$; zero if $E_d = 1$; and negative if $E_d > 1$. During a movement along the demand curve in the opposite direction, the price drops and ΔP is negative; therefore, the change in total expenditure is negative if $E_d < 1$; zero if $E_d = 1$; and positive if $E_d > 1$ (see Table 10.3).

Fundamental Economic Concepts

Table 10.3 The change in total expenditure during a movement along the demand curve

Price	Quantity	Elasticity of Demand	Total Expenditure
Rises	Decreases	Elastic, $E_d > 1$	Decreases
		Unitary, $E_d = 1$	Unchanged
		Inelastic, $E_d < 1$	Increases
Falls	Increases	Elastic, $E_d > 1$	Increases
		Unitary, $E_d = 1$	Unchanged
		Inelastic, $E_d < 1$	Decreases

The following conclusions can be reached by studying Table 10.3. If the demand is inelastic, a shift to the right in the supply curve, which causes the price to fall and the quantity demanded to increase, would result in a decrease in the total revenue. A shift to the left in the supply curve, causing the price to rise and the quantity to decrease, would also decrease the total expenditure provided that the demand was elastic.

One of the objectives of producers is to increase their total revenue. This interest is best served by a shift in the supply curve to the right if the demand is elastic, and to the left if the demand is inelastic.

Example 10.2

A company is engaged in manufacturing snowmobiles. The demand schedule for snowmobiles is specified in Example 10.1. The purchase of a number of automatic machines is being considered by this company. With the help of these machines production volume would be increased, although the unit cost of production would not change. In order to be able to sell the larger volume of snowmobiles produced, the company would reduce the price of snowmobiles by $100. What would the result of this action be on the total revenue received by the company if the current price of a snowmobile is

(a) $1 500
(b) $3 500?

The change in total revenue is $\Delta T = \Delta P \cdot Q(1 - E_d)$

(a) From Example 10.1 at a price of $1 500: $Q = 45\ 000$ and $E_d = 0.67$; thus

$$\Delta T_a = 100(45\,000)(1-0.67) = \$148\,500$$

(b) Also, from Example 10.1 at a price of $3 500: $Q = 19\,000$ and $E_d = 1.47$ and thus

$$\Delta T_b = 100(19\,000)(1-1.47) = -\$893\,000$$

Therefore, as a result of the proposed price reduction to the $3 400 price level, the total revenue would drop by $893 000; on the other hand, at the $1 500 price level, the total revenue would rise by $148 500. The former outcome would certainly be less desirable than the latter from the point of view of the company, and this result would probably have a considerable effect on the decision regarding the purchase of the automatic machines.

10.2.8 Elasticity of Supply

Elasticity of supply characterizes the supply curve and measures the responsiveness of the quantity supplied to the change in price. Elasticity of supply is the ratio of the percentage change in quantity supplied to the percentage change in price:

$$E_s = \frac{\Delta Q_s}{Q_s} : \frac{\Delta P}{P} = \frac{\Delta Q_s}{\Delta P} \frac{P}{Q_s}$$

where

E_s = the elasticity of supply
P = the average price
Q_s = the average quantity supplied
ΔP = the change in price
ΔQ_s = the change in quantity supplied

The elasticity of supply is generally different at each point of the supply curve. It may vary in magnitude from zero to infinity. The supply is said to be

elastic if $E_s > 1$, and
inelastic if $E_s < 1$.

10.2.9 Price Control

The government may interfer with the operation of the free market system by setting the minimum or maximum price of goods and services sold. The effect of this government action on production and consumption will now be examined by studying the supply-demand interaction.

Price Ceiling

The maximum price of goods or services may from time to time be set by the government. This action usually takes the form of rent control, price control, etc. The price ceiling (maximum price) has little effect on the market if it is set above the equilibrium price, since the commodity will continue to be traded at its equilibrium price (below the ceiling). If, on the other hand, the price ceiling is set below the equilibrium price (see Figure 10.9), the quantity supplied at the ceiling price (quantity Q_s) will be less than the quantity demanded at the price, quantity Q_d, and a shortage develops. In a free market without price control, the shortage quantity, $Q_d - Q_s$ (see Figure 10.9) would disappear after forcing the price to rise to its equilibrium level. However, as the price is not allowed to rise above the ceiling, the shortage will remain permanent. The quantity supplied at the maximum price would, in a free market, be normally sold at a price of P_B (see Figure 10.9) above the ceiling. A black market usually develops under these circumstances for the commodity to be sold above the legal maximum price. The black market price is often substantially higher than the equilibrium price that would exist in the absence of price controls.

Minimum Price

The minimum price of goods and services may also occasionally be set by the government. Government-controlled minimum wage falls into this category. The minimum price has little effect on the market if it is below the equilibrium price of the commodity, since the commodity will continue to be traded at its equilibrium price above the minimum price set. If, on the other hand, the minimum price is set above the equilibrium price (see Figure 10.10), a surplus develops. The quantity of the commodity supplied at the minimum price quantity Q_s is larger than the quantity Q_d, demanded at this price. In the absence of price control, the surplus, quantity $Q_s - Q_d$, on Figure 10.10, would disappear as the price fell to its equilibrium level. However, as the price is not allowed to fall below its controlled minimum value, the surplus

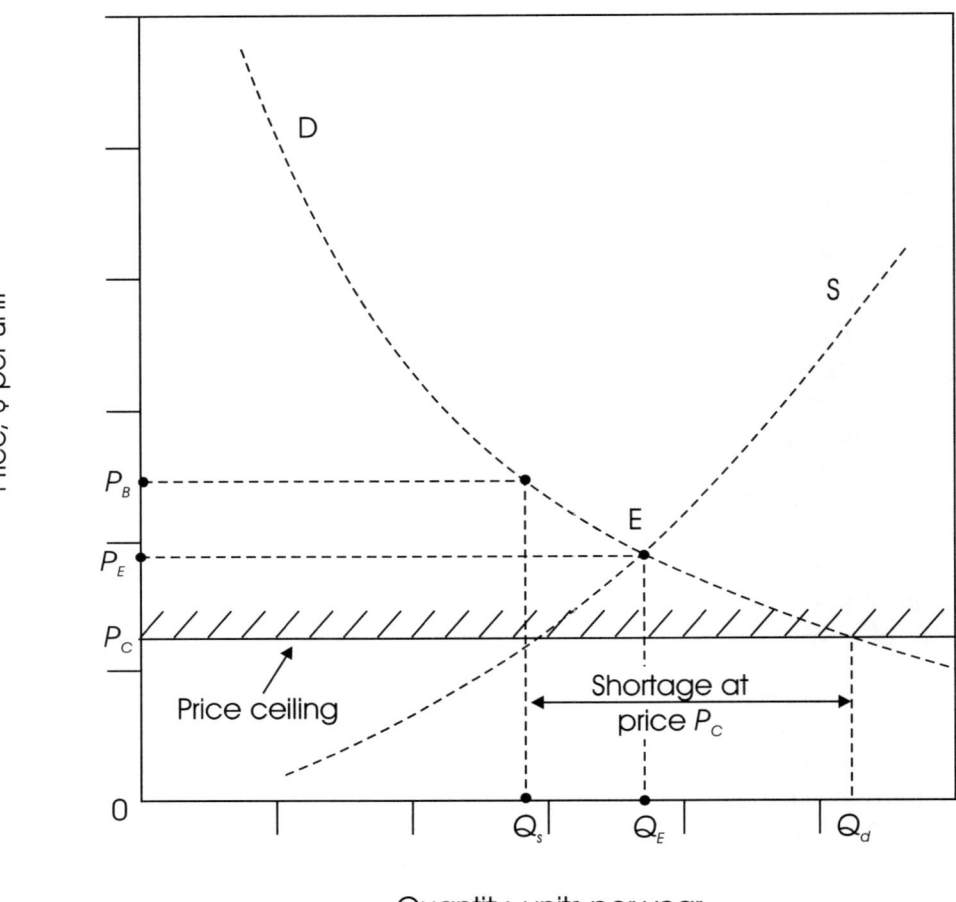

Fig. 10.9 Price ceiling.

P_E = equilibrium price
P_C = price ceiling (maximum price)
Q_E = equilibrium quantity
$Q_d - Q_s$ = shortage at the price ceiling

Fundamental Economic Concepts

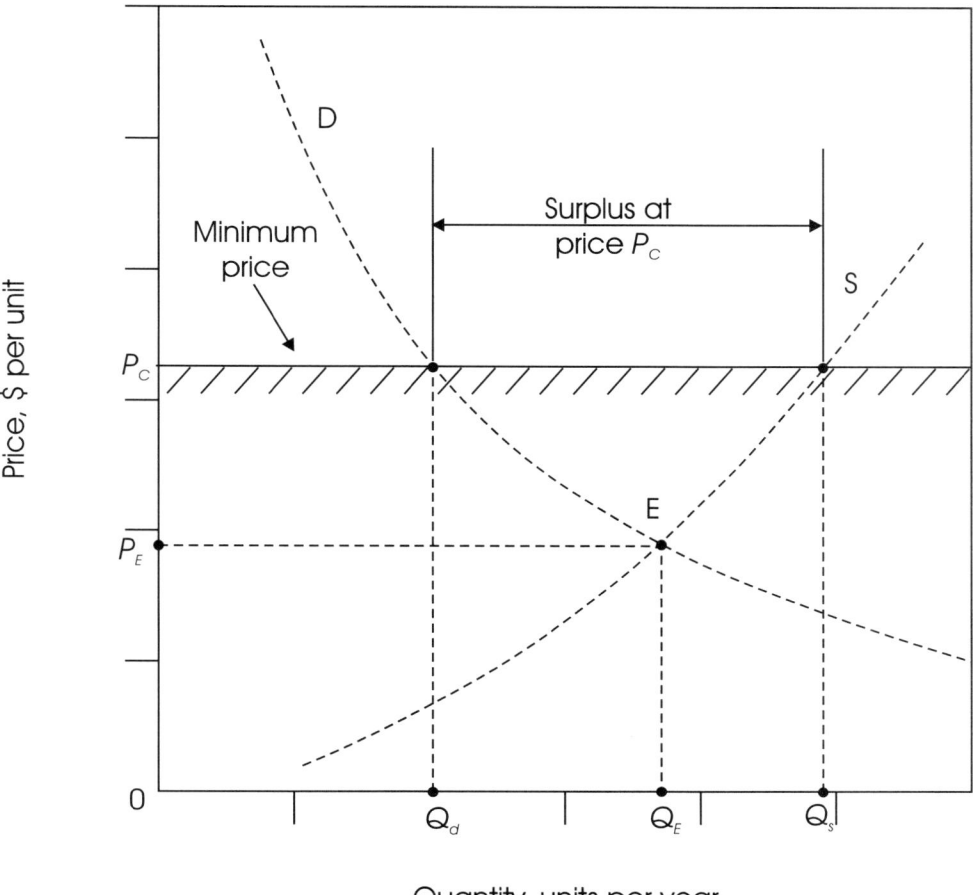

Fig. 10.10 Minimum price.

P_E = equilibrium price
P_C = controlled minimum price
Q_E = equilibrium quantity
$Q_s - Q_d$ = surplus at the minimum price

would be permanent. A minimum wage law, for example, tends to raise the wages of those who can keep their jobs, but it reduces the number of people employed, and creates a surplus of labour.

10.2.10 Price Support

The periodic (or seasonal) fluctuation in the supply of certain commodities, mainly agricultural commodities, results in widely fluctuating prices in the free market. Since wide fluctuation in the price of basic commodities is generally undesirable from the point of view of both producers and consumers, the government often intervenes in the operation of the free market in order to stabilize the price of basic commodities. Price fluctuation can be reduced (or eliminated) by introducing a price support scheme.

Suppose that the government is willing to support the price of a commodity at its equilibrium price level. The production of this commodity might fluctuate. In one particular period, production might rise above the equilibrium quantity to quantity Q_1 shown in Figure 10.11, and a surplus of quantity $Q_1 - Q_E$ is created at the equilibrium price level. This surplus would normally depress the price to P_1 below equilibrium price level. However, if the government intervened by purchasing the surplus from producers at the equilibrium price, the producers would be able to sell all of their product at the equilibrium price. The equilibrium quantity would be purchased by the consumers and the surplus quantity $(Q_1 - Q_E)$ by the government. In some other period, production might fall below the equilibrium quantity, to quantity Q_2. The price would then normally rise to P_2. However, if the government intervened by selling the quantity $Q_E - Q_2$ from its inventory at the equilibrium price, P_E, it would prevent the price of the commodity from rising. The equilibrium price, therefore, would always be maintained and the demand of consumers satisfied.

Price support schemes generally work well, provided that the price is supported at the equilibrium level. If, however, the price is supported above the equilibrium level the government subsidises producers, production increases, consumption drops, and the inventory held by the government grows. If, on the other hand, the price is supported below the equilibrium price level the government subsidises consumers, production drops, and consumption increases. For example, supporting oil prices in the home market below the international equilibrium price of the commodity for an extended period may cost billions of dollars annually to the government.

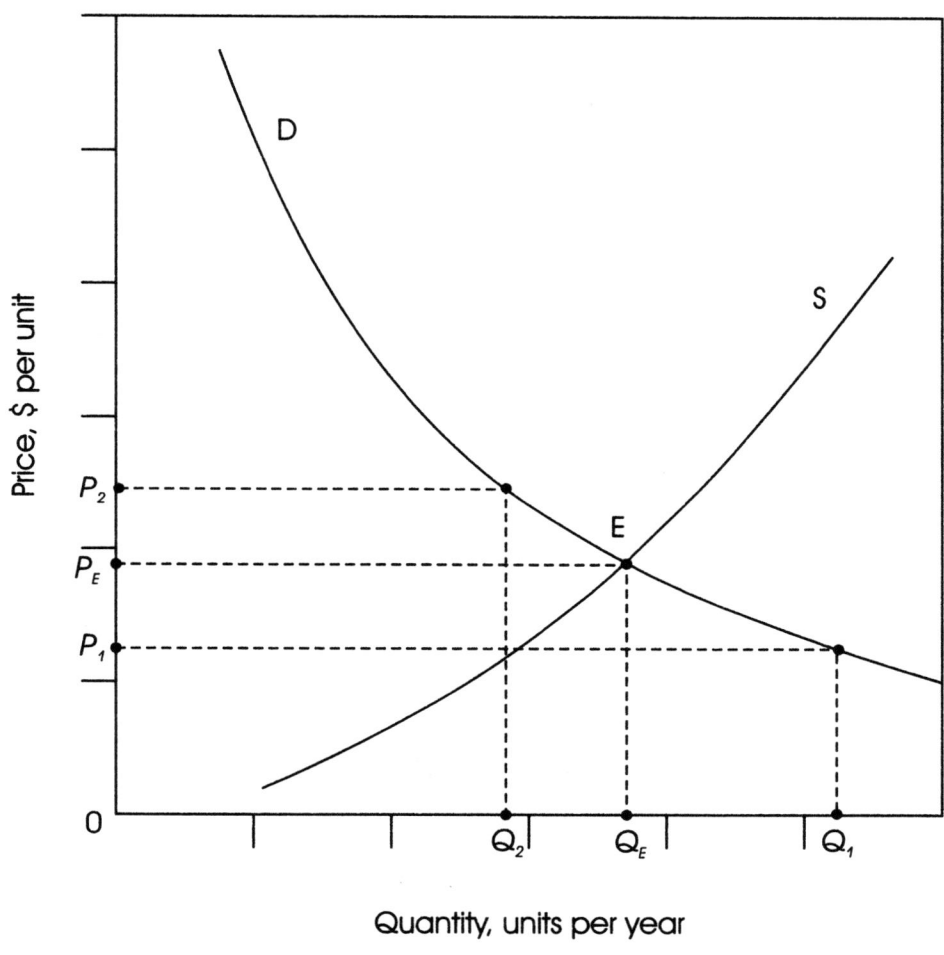

Fig. 10.11 Price support.

S = supply curve
D = demand curve

10.3 Production

The work of engineers is closely related to the production of a wide variety of goods and services. Engineers are frequently involved in important production decisions to determine the best way to use available plant and machinery, what new plant or facility to build and what machinery to acquire, what new technology to introduce, etc. To be effective in the area of production decision making, it is necessary to be familiar with the basic principles of the *theory of production.*

10.3.1 The Production Function

One of the primary objectives of an engineering company is to use the most economically efficient method of production. Using the *production function*, one can determine which factors need to be modified in order to achieve production at optimum level.

Firms that are engaged in production use manpower, machines, equipment, raw materials, etc., to produce goods and services. Manpower, machine, equipment, raw material, and other elements needed for production are called *inputs* into the production process. An input is either fixed or variable. A *variable input* is one that can be changed (increased or reduced) on short notice in response to a change in market conditions. *Fixed inputs*, on the other hand, cannot be changed easily. Factory buildings, power plants, oil refineries, and large computer installations are fixed inputs. Labour, machinery, tools, and raw materials are variable inputs. From the long-term point of view, all input is variable; everything can be changed given sufficient time. The size of the labour force or the quantity of raw material supply may be altered in a few weeks in response to changes in demand, but a change in the capacity of a nuclear power plant might take several years to implement. Consequently, some inputs are considered fixed for short-run decisions, even though all inputs are variable for long-run decisions.

The firm uses the inputs (fixed and variable) in its production process to produce goods and services called *outputs.* The relationship between inputs and outputs is the *production function*. The production function is an empirical relationship normally specified in the form of a table, a graph, or, in exceptional cases, a mathematical function. The production function is generally a multivariable function having at least two inputs (labour and capital). With the help of the production function, one is able to determine the best combination of inputs to produce a specified output.

10.3.2 The Univariable Production Function

A typical univariable production function is listed in the first two columns of Table 10.4. This function may represent the activity of an engineering company specializing in the production of automatic packaging machines. The variable input is labour, measured in thousands of labour hours per year, the fixed input is capital in the form of buildings, equipment, and machinery. The output, or *total product*, is the number of machines produced by the company yearly. The total product divided by the variable input (i.e., the number of machines produced per thousand labour hours in the present case) is the *average product*, or *productivity*. It can be expressed as

$$AP = \frac{TP}{L}$$

where

AP = the average product
TP = the total product (output)
L = the labour (input)

Table 10.4 The Production Function

(1) Input (Labour) 10^3 labour hours per year	(2) Output (Total Product) units per year	(3) Average Product units per 10^3 labour hours	(4) Marginal Product units per 10^3 labour hours
0	0	0.00	
10	5	0.50	} 0.50
20	12	0.60	} 0.70
30	23	0.77	} 1.10
40	45	1.12	} 2.30
50	60	1.20	} 1.50
60	68	1.13	} 0.80
70	69	0.99	} 0.10
80	67	0.84	} 0.20

Values of the average product were computed and listed in Column 3 of Table 10.4. The table shows that the average product increases initially, reaches a maximum value, and then begins to decrease. This is the general behaviour of a typical production function. Economists explain this behaviour using the *"law" of diminishing returns* which states that if increasing quantities of a variable factor are added to a given quantity of a fixed factor, a point will be reached after which each additional unit of variable factor will result in a smaller increase in output than did the previous unit (i.e., the average product begins to decrease).

Figure 10.12 shows a typical production function. Coordinates of a point on the curve are corresponding input and output values. The average product at a point on the curve is equal to the slope of the ray drawn from the origin to the point. For example, the average product at point C is $\frac{TP_C}{L_C}$, which is the slope of line OC. The average product increases until its maximum value is reached at point B. This point is called the *point of diminishing average productivity*. The average product drops if the input is increased beyond this point.

The *marginal product,* or *incremental product,* is the change in total product resulting from one unit increase in the variable input (i.e., the derivative of the total product), and may be expressed as

$$MP = \frac{\Delta TP}{\Delta L}$$

where

MP = the marginal product
ΔTP = the change in total product (output)
ΔL = the change in labour (variable input)

Values of the marginal product were computed and listed in Column 4 of Table 10.4 for the production function listed in the table. In Figure 10.12 the marginal product is equal to the slope of the tangent to the production function curve. It can be observed that the marginal product is positive and increasing (with increasing input) if $0 \leq L \leq L_C$. At point C the marginal product reaches its maximum value. In the range of $L_C \leq L \leq L_A$, the marginal product is positive but decreasing in value (with increasing input). At point A the total product is maximum and the marginal product is zero. It should also be observed that at point B the average product is maximum and the average and marginal prod-

Fundamental Economic Concepts

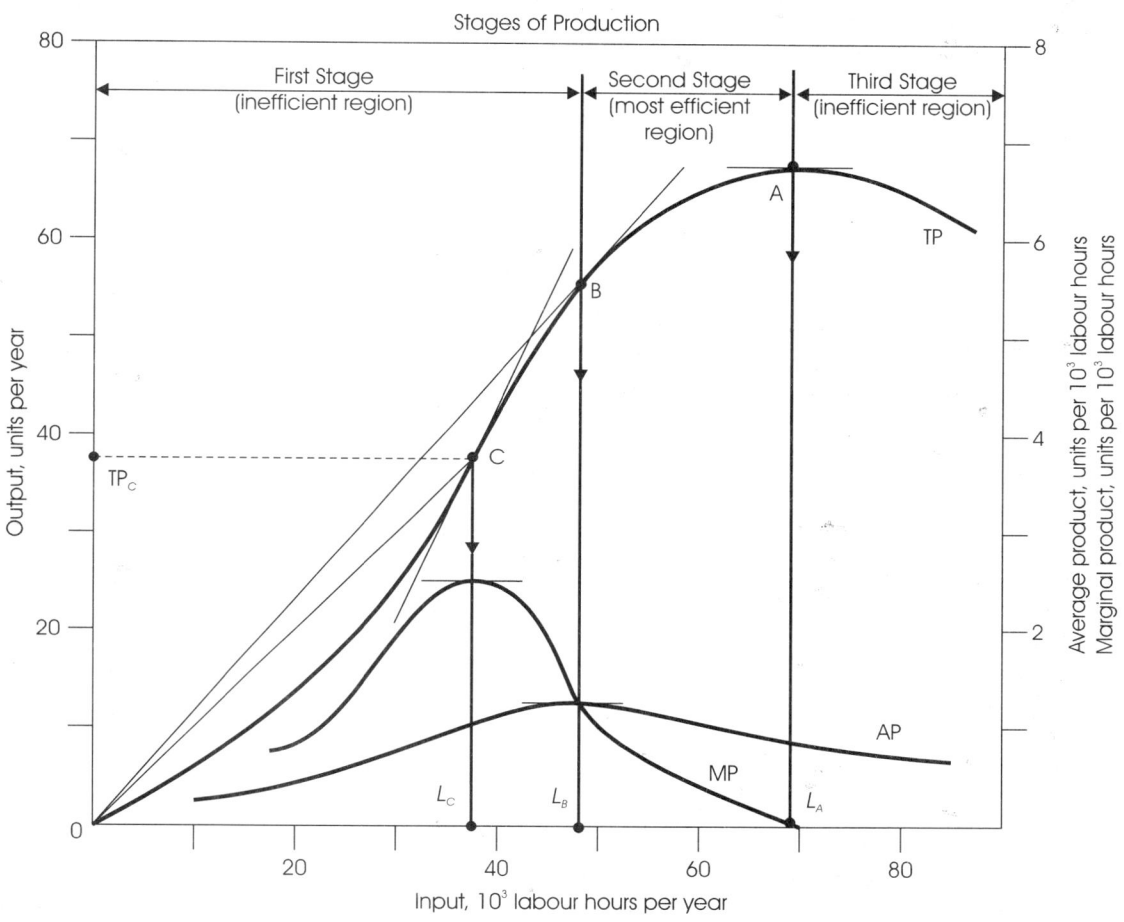

TP = total product curve
AP = average product curve
MP = marginal product curve

Fig. 10.12 The production function.

uct values are equal. Furthermore, in the region $L \geq L_A$, the marginal product is negative and the total product decreases with increasing input. Point C, where the marginal product is maximum, is the point of *diminishing marginal productivity*.

Three stages of production are normally distinguished on a typical production curve such as the one shown in Figure 10.12. In the first stage of production, productivity increases with increasing variable input, indicating that the fixed inputs are not fully utilized. It is advantageous to increase production in this stage at least until the limit of the region, point B, is reached. In stage three both the total product and the productivity decrease as the variable input is increased. This stage of production is rather inefficient. Production should not be sustained in this stage.

Production is normally performed in the second stage of the production function. If the desired output is not available in this stage, the fixed inputs should be changed to alter the shape of the production function, and to make the desired output available in the second stage of production. The characteristics of the three production stages are listed in Table 10.5.

	Table 10.5 The Stages of Production			
	Input	Total Product (Output)	Average Product (Productivity)	Marginal Product
First Stage	$0 \leq L < L_B$	Increases	Increases	Positive
Second Stage	$L_B \leq L < L_A$	Increases	Decreases	Positive
Third Stage	$L_A \leq L$	Decreases	Decreases	Negative

A typical property of the production function is that beyond a certain point both the marginal and the average products diminish as the variable inputs are increased, due to the *"law" of diminishing returns*.

One of the main objectives of the firm is to set up the best method of production. The best method of production is either the one that uses the smallest quantity of inputs, or the one that costs the least to obtain the required output. The method which uses the fewest inputs to produce a specified quantity of output is the *technologically most efficient method* of production. The method that costs the least to produce a specified output is the *economically most efficient method* of production.

10.3.3 Multivariable Production Function

Production generally requires a number of different inputs. For example, the firm engaged in producing automatic packaging machines uses raw materials (steel, plastic, rubber), finished goods (microprocessors, sensors, switches), machinery (lathes, milling machines, grinding machines), labour (engineers, technologists, machine operators), etc., in the production process. All of these items are inputs to the production. However, for the purpose of explanation, we will consider the simple case of only two inputs: labour and capital. The production function for two inputs is expressed as:

$$TP = f(L, K)$$

where

TP = the total product (output)
L = the labour (input)
K = the capital (input)

This function is represented by a curved surface, and the contour lines of this surface can be used for a two-dimensional graphical representation of the function. A contour line is defined by the specific relationship that exists between K and L for a given output quantity. This relationship, corresponding to a constant output and computed from the production function, can be represented by a two-dimensional curve (see Figure 10.13). This curve is called the *isoquant*, or *constant quantity curve*, and the output corresponding to any point on this curve is constant. An isoquant shows all alternative combinations of inputs for producing a given output. For example, at point A, the required labour input is L_A and the capital input is K_A. At point B, the labour input is L_B and the capital input K_B. As we move from point A to point B on the isoquant curve, capital is being substituted for labour with the output remaining unchanged.

In the first two stages of production, the marginal product is positive, indicating a reduction in output as the input is reduced. Therefore, if there are two inputs, as one input is reduced, the other must be increased to keep the output constant. Thus, the slope of the isoquant

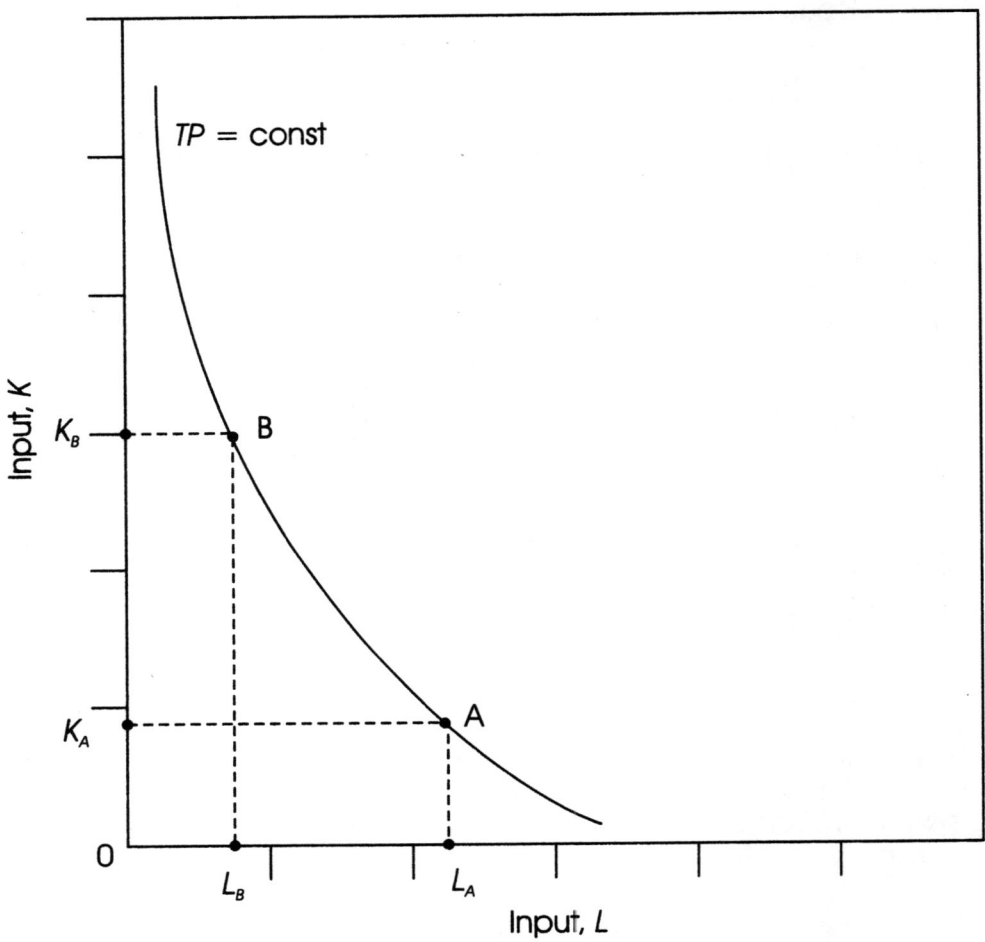

Fig. 10.13 The isoquant curve.

curve is negative. The slope of the isoquant curve is called the *marginal rate of substitution,* and it is expressed as

$$MRS = \frac{\Delta K}{\Delta L} = \frac{\Delta TP/\Delta L}{\Delta TP/\Delta K} = \frac{MP_L}{MP_K}$$

where

MRS = the marginal rate of substitution
ΔK = the change in input K
ΔL = the change in input L
ΔTP = the change in output (total product)
MP_L = the marginal product in terms of input L
MP_K = the marginal product in terms of input K

A group of isoquant curves obtained from a production function and representing different production outputs form an isoquant map, see Figure 10.14. The isoquant map is the contour map of the three-dimensional surface representing the production function. Isoquants corresponding to higher levels of output are further away from the origin on the isoquant map.

10.3.4 Cost of Production

The cost of production is the sum of the costs of all inputs. In the case of two inputs (labour and capital) the total cost can generally be expressed as

$$TC = p_L L + p_K K + FC$$

where

TC = the total cost
p_L = the unit cost of input L (labour)
p_K = the unit cost of input K (capital)
FC = the cost of all fixed inputs

This equation is valid only if the total cost is linearly dependent on the variable inputs, which is not always the case. However, for the sake of simplicity we will consider only linear cost functions in this

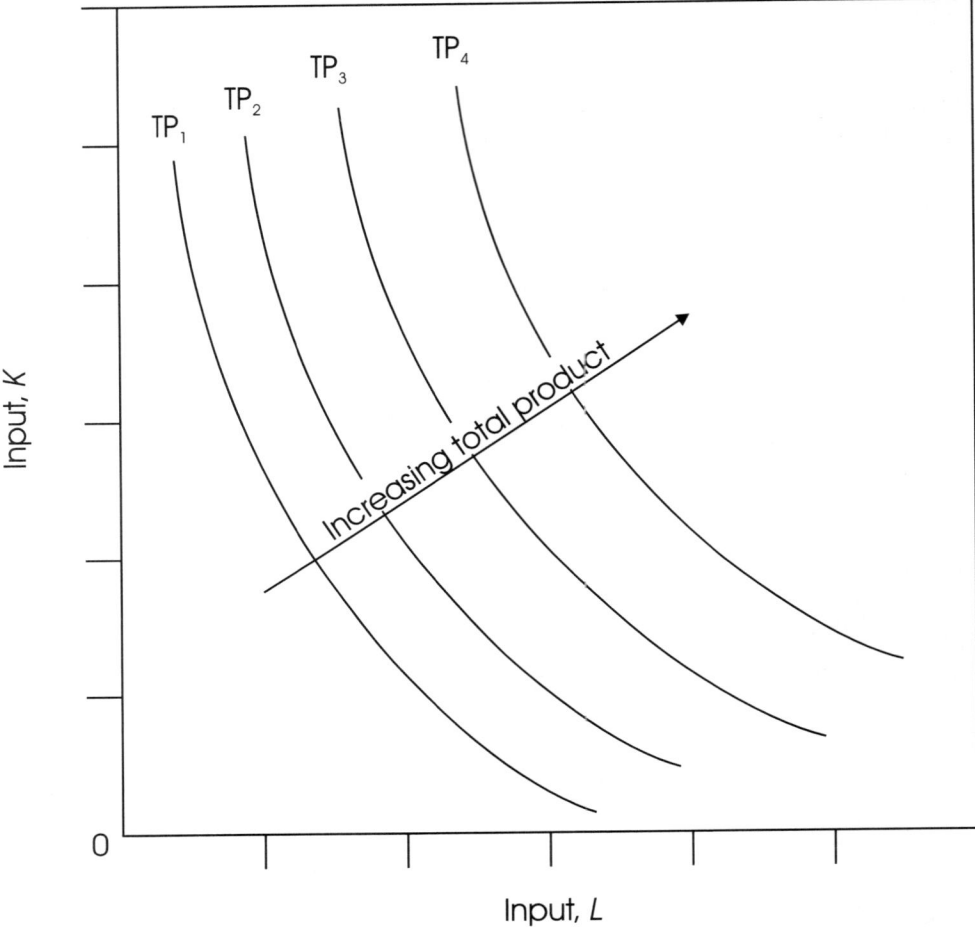

TP = total product
$TP_1 < TP_2 < TP_3 < TP_4$

Fig. 10.14 Isoquant map.

chapter. If the cost function is linear, the total cost can be represented by a plane surface. The two dimensional contour-map of this cost function is a family of parallel lines (see Figure 10.15). These lines are called isocost lines, and the total cost of production corresponding to any point on any one of these lines is constant. Each isocost line corresponds to a specific total cost value. Lines further away from the origin represent higher total cost values. The slope of the isocost lines is $-\frac{p_L}{p_K}$.

A family of isocost lines (corresponding to the same total cost function) forms an isocost map. The total production cost corresponding to any combination of inputs can be determined easily with the help of an isocost map. For example, the total cost of production corresponding to inputs L_A and K_A is TC_A in Figure 10.15.

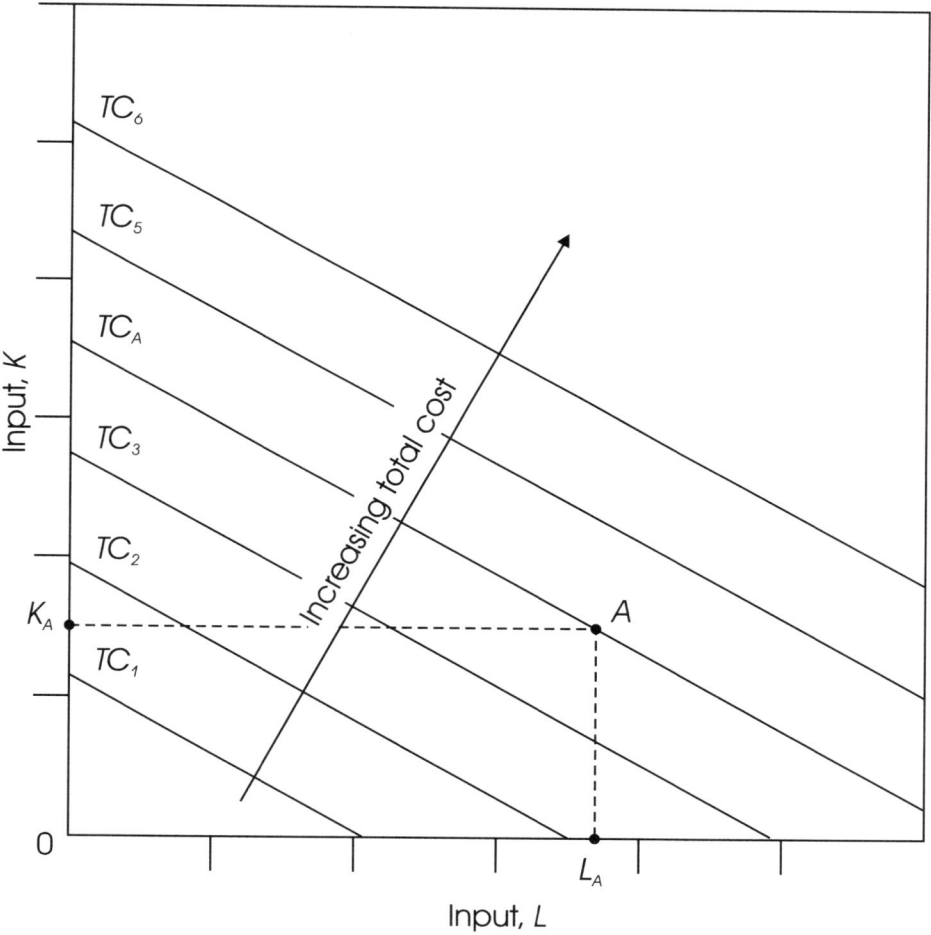

TC_1 = total cost values
$TC_1 < TC_2 < TC_3 < TC_A < TC_5$

Fig. 10.15 Isocost map.

In order to find the least costly combination of inputs to produce a specific output, it is necessary to combine the isoquant curve with the isocost map, as shown in Figure 10.16. Point A in Figure 10.16 represents the combination of inputs capable of producing a required output, TP_A, most economically. The cost of production corresponding to any other point on the same isoquant curve is clearly higher than the cost associated with point A. For example, the cost of production corresponding to points B or C is TC_3, higher than TC_A; the cost corresponding to point D is TC_4, also higher than TC_A. The point on the TP_A curve corresponding to the least production cost is the one where the slopes of the isocost and isoquant lines are equal. That is

$$-MRS = -\frac{p_L}{p_K}$$

or

$$\frac{MP_L}{MP_K} = \frac{p_L}{p_K}$$

where

MRS = the marginal rate of substitution
p_L = the unit cost of input L
p_K = the unit cost of input K

This result can easily be extended to a production function having more than two inputs. Let the linear cost function be

$$TC = FC + \sum_{i=1}^{N} p_i L_i$$

where

FC = the fixed cost
p_i = unit cost of input L_i
N = number of inputs

and the production function is

$$TP = TP(L_i)$$

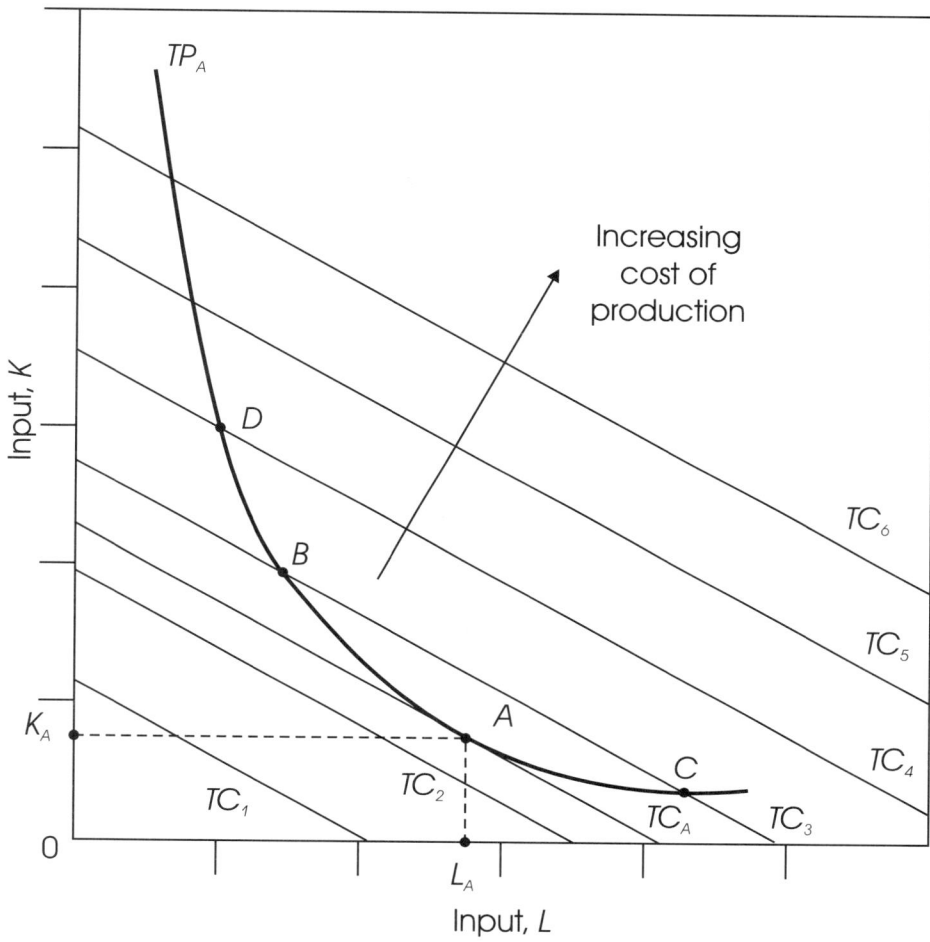

Fig. 10.16 The point of most economical production.

The point of most economical production is point A, where the slope of the isoquant curve is equal to the slope of the isocost lines. The corresponding optimum input quantities are L_A and K_A.

To determine the minimum cost of production at a specified output of TP_A, the following constrained optimization problem should be used:

$$TC = FC + \sum_{i=1}^{N} p_i L_i = \text{minimum}$$

and
$$TP_A = TP(L_i)$$

The equations can be solved using Lagrange's method

$$TC = FC + \sum_{i=1}^{N} p_i L_i + \lambda[TP_A - TP(L_i)] = \text{minimum}$$

where λ is a constant (Lagrange's multiplier), and the solution is

$$\frac{\partial TC}{\partial L_i} = p_i - \lambda \frac{\partial TP(L_i)}{\partial L_i} = 0; \quad i = 1, N$$

however,

$$\frac{\partial TP(L_i)}{\partial L_i} = MP_i$$

therefore,

$$\frac{1}{\lambda} = \frac{MP_i}{p_i}; \quad i = 1, N$$

or

$$\frac{MP_1}{p_1} = \frac{MP_2}{p_2} = \frac{MP_3}{p_3} = \frac{MP_i}{p_i}$$

where

MP_i = the marginal product in terms of input L_i
p_i = unit cost of input L_i

That is, the cost of production is minimum at the combination of inputs where the ratio of marginal product to unit input cost with respect to every input quantity is identical.

10.3.5 The Expansion Path

As the unit cost of inputs changes, the point of most economical production corresponding to a given output moves along the isoquant curve. For example, let us reexamine the production problem represented in Figure 10.16 in view of an increase in the unit cost of labour. As the unit cost of labour increases, the ratio of $\frac{p_L}{p_K}$ increases and, therefore, the slope of the isoquant curve at the point of most economical production will be steeper, changing from $MRS_1 = \frac{p_L}{p_K}$ to $MRS_2 = \frac{p_L{}^*}{p_K}$ (where $p_L{}^*$ is the increased unit labour cost). Thus the optimum point moves from A to B on the isoquant curve (see Figure 10.17). Therefore, the change in relative prices of inputs causes a partial replacement of the input that becomes relatively more expensive by the input that becomes relatively cheaper. In the present case, as labour cost increases, capital will replace labour.

For example, let us assume that the unit labour cost of our company manufacturing automatic packaging machines increases from $10 to $20, while the unit cost of capital remains unchanged. As a result of this change, the slope of the isocost curves will be steeper, and the optimum point of production moves from point A to point B on Figure 10.17. At point B the inputs required for production are L_B and K_B, $L_B < L_A$ and $K_B > K_A$. In order to reduce its total cost of production, the firm must undergo some adjustments. New, automated machinery will be purchased, and the labour force will be substantially reduced to approach the combination of inputs represented by Point B on the isoquant curve.

Let us now determine the point of most economical production using the method of Figure 10.16 for a number of different output levels. The results obtained are shown in Figure 10.18. The optimum points of the different isoquants, points A_1, A_2, A_3, etc., are joined by a continuous curve called the *expansion path*. The expansion path is the locus of optimum points and shows the best possible way to increase inputs in order to reach a certain level of output while at the same time minimizing the cost of production. For each point on the expansion path, the marginal rate of substitution is equal to the specified unit cost ratio. The curve in Figure 10.18 may be given two different applications.

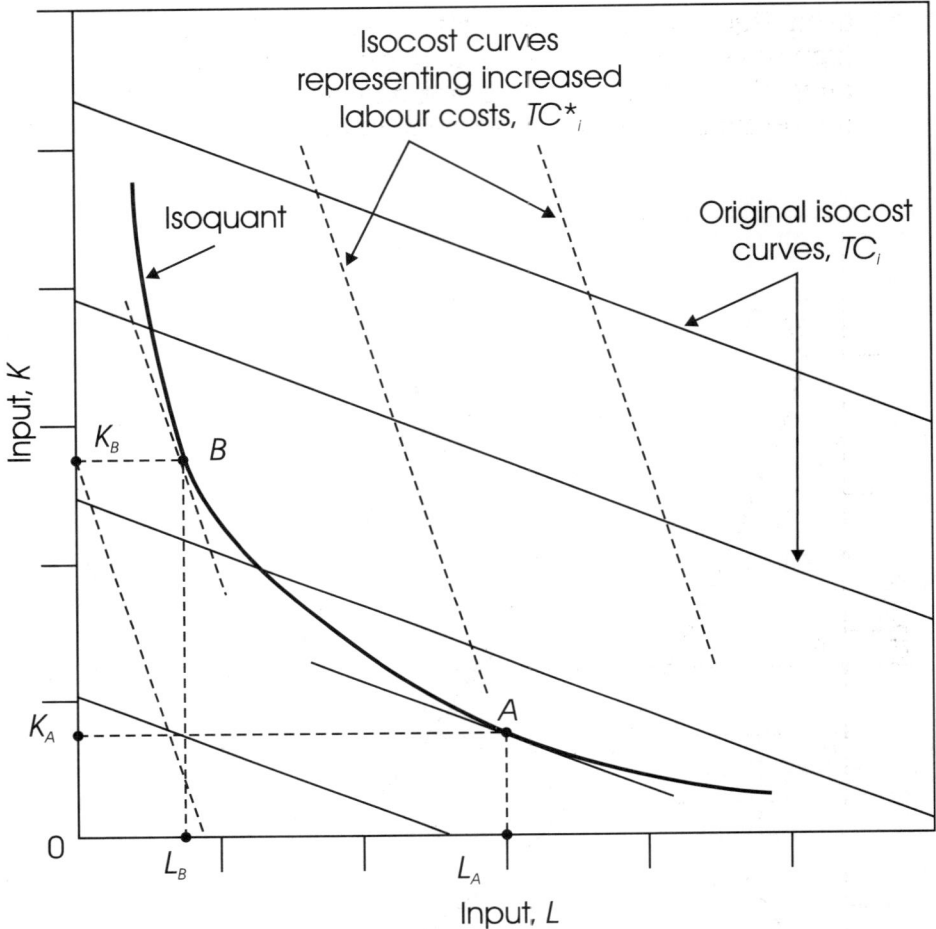

Fig. 10.17 The effect of changing unit input costs on the point of most economical production

1. It may be used to find the optimum combination of inputs to produce a specified output at minimum total cost. For example, if the production target is TP_2, the point of most economical production is A_2 (see Figure 10.18). The total cost of production corresponding to point A_2 is TC_2, and the inputs are L_2 units of labour and K_2 units of capital.

2. If a fixed budget for production is specified the highest level of output that can be obtained with this budget is determined

Fundamental Economic Concepts

using Figure 10.18. For example, if the budget limit is TC_3, the highest possible output level at this budget is TP_3, using L_3 units of labour and K_3 units of capital according to point A_3 on the expansion path (see Figure 10.18).

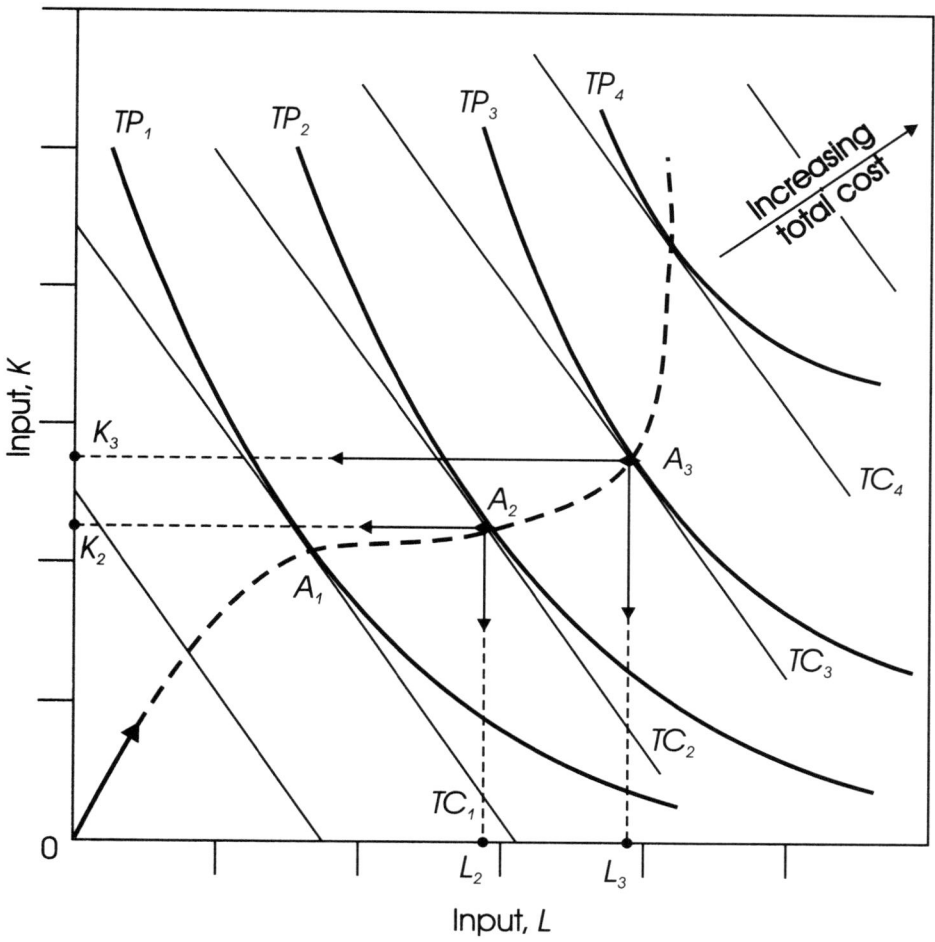

Fig. 10.18 The expansion path.

10.4 Summary

Two important basic concepts of economics closely related to the topic of engineering economic analysis were examined in detail in this chapter. The first is the *supply and demand* interaction, which is important to engineers for predicting the response of the market to changes in price or quantity supplied. The second is the *theory of production*, which is especially useful to engineers who play a role in a wide variety of production processes. The theory of production helps identify the combination of factors to be used for approaching the technologically or economically most efficient method of production. The reader will find both concepts useful in understanding some of the methods described and meeting some of the objectives outlined in this text.

Problems

1. The demand schedule for a commodity traded in the Canadian market is as follows:

Quantity demanded (10^6 kg per year)	2	4	6	8	10	12
Price ($ per kilogram)	95	62	40	24	13	6

   (a) Plot the demand curve.

   (b) Determine the elasticity of demand at a price of $50 per kilogram. (10.2.1, 10.2.6)

2. Determine the change in the total amount spent by all consumers on a commodity as its unit price rises from $20 to $30 during a movement along the demand curve. (See Problem 1 for the demand schedule.) (10.2.7)

3. The supply schedule for a commodity in the Canadian market is as follows:

Quantity supplied (10^6 kg per year)	2	4	6	8	10	12
Price ($ per kilogram)	20	40	60	80	100	120

(a) Plot the supply and demand curves (see Problem 1 for the demand schedule).

(b) Determine the equilibrium price.

(c) Determine the expected average monthly combined revenue derived from this commodity by all producers. (10.2.5, 10.2.7)

4. The maximum price of a commodity is set by the government at $35 per kilogram (see Problem 1 for the demand and Problem 3 for the supply schedule).

   (a) Determine the quantity supplied and the quantity demanded at the maximum price.

   (b) Determine the amount of shortage that will exist due to the price ceiling.

   (c) Determine the drop in production as a result of the introduction of the price ceiling.

   (d) Determine the loss of revenue the producers of this commodity will suffer as a result of the price ceiling.

   (e) What is the potential black market price of the commodity during the application of the price ceiling?

   (f) What would be the effect of raising the price ceiling to $55 per kilogram? (10.2.9)

5. The introduction of a new technology causes a shift in the supply curve of a commodity as follows:

	Price ($ per unit)	2	4	6	8	10
	Quantity demanded (10^4 units per year)	9.5	6.2	4.0	2.4	1.3
Quantity supplied (10^4 units per year)	With existing technology	2	4	6	8	10
	With new technology	3	5	7	9	11

Determine

(a) the change in price,

(b) the change in quantity sold, and

(c) the change in total revenue of all producers of the commodity resulting from the introduction of the new technology. (10.2.7)

6. The government supports the price of a commodity to eliminate price fluctuation at the level of $50 per kilogram. In 2001 the quantity of production of the commodity is 7×10^6 kg.

    (a) Determine the amount the government spends in 2001 supporting the price of this commodity.

    (b) How much would be the total revenue of all producers from the sale of this commodity in 2001 without price support?

    (c) How much would the consumers spend on this commodity in 2001 without price support? With price support? (See Problems 1 and 3 for the demand and supply schedules.) (10.2.10)

7. The production function of a machine shop is specified as:

Labour (10^4 working hours per year)	1	2	3	4	5	6	7
Output (units per year)	10	32	60	72	77	77	72

    (a) Plot the production curve.

    (b) Determine the points of maximum production, maximum productivity, and maximum marginal product.

    (c) What is the productivity at an input of 4×10^4 labour hours per year?

    (d) Specify the three different stages of production in terms of the input quantity. (10.3.2)

8. The labour input is increased from 45 000 labour hours to 50 000 labour hours per year in a machine shop. (See Problem 7 for the production function.) Determine the change in total product, average product, and marginal product. (10.3.2)

9. The isoquant curve corresponding to the desired output of a chemical plant is specified in terms of two inputs as:

Fundamental Economic Concepts

Labour (10^4 working hours per year)	1	2	3	4	5	6	7	8
Machinery (units)		85	60	44	33	24	16	10	6

The unit costs are $12 per working hour for labour and $18 000 per machine per year. Determine the minimum cost of production. (10.3.4)

10. In the chemical plant described in Problem 9, fifteen machines are available for production.

    (a) Determine the cost of production, and estimate the labour input requirements.

    (b) What would be the change in labour input if production were performed at optimum level? What would be the corresponding reduction in cost? (10.3.4)

11. It is necessary to increase the production of a copper mine by 30%. The isoquant curves corresponding to the existing and proposed production levels are specified in terms of the input quanties as:

Labour (10^4 working hours per year)		1	2	3	4	5	6	7	8
Machinery (units)	at existing production level	85	60	44	33	24	16	10	6
	at proposed production level	150	125	94	72	56	44	32	22

20 machines are used presently in the mine. Labour cost is $12 per hour. Machine cost is $18 000 per unit per year. Determine the required change in labour input to reach the proposed production level without any addition of machinery. What is the corresponding cost of production if the annual fixed cost is $800 000? (10.3.4)

12. The production in the mine, described in Problem 11, should be increased in the most economical way.

    (a) Determine the total cost of increased production, the required change in labour, and machinery inputs.

(b) Determine the total (minimum) cost of the increased production, and the required change in labour and machinery inputs if a 30% increase in labour cost is expected. (10.3.4)

13. The output of a production process is described by the following equation:

$$TP = 100LK$$

Where L is labour input in 1000 labour hours and K is the input of machinery. The unit costs of the inputs are: $10 per hour for labour, and $20 000 per machine per year. The fixed cost of production is $500 000 per year, including the costs of building and services. Determine the extension path for this production process. (10.3.5)

14. A fixed budget of $2 000 000 per year is allowed for production for the process described in Problem 13. Determine the maximum possible level of output and the corresponding input volumes. (10.3.5)

15. For the situation described in Problem 13, determine the total cost of production in terms of the output function. (10.3.5)

Appendix A
Discrete Compounding

Section I — Discrete Compound Interest Factors

Section II — Geometric Series Factors

TABLE A.1 Discrete Compounding: $i = 1\%$

	Single payment		Uniform series				Gradient series	
	Compound amount factor	Present worth factor	Compound amount factor	Sinking fund factor	Present worth factor	Capital recovery factor	Uniform series factor	Present worth factor
n	To find F given P $F\|P\ i,n$	To find P given F $P\|F\ i,n$	To find F given A $F\|A\ i,n$	To find A given F $A\|F\ i,n$	To find P given A $P\|A\ i,n$	To find A given P $A\|P\ i,n$	To find A given G $A\|G\ i,n$	To find P given G $P\|G\ i,n$
1	1.0100	0.9901	1.0000	1.0000	0.9901	1.0100	0.0000	0.0000
2	1.0201	0.9803	2.0100	0.4975	1.9704	0.5075	0.4975	0.9803
3	1.0303	0.9706	3.0301	0.3300	2.9410	0.3400	0.9934	2.9215
4	1.0406	0.9610	4.0604	0.2463	3.9020	0.2563	1.4876	5.8044
5	1.0510	0.9515	5.1010	0.1960	4.8534	0.2060	1.9801	9.6103
6	1.0615	0.9420	6.1520	0.1625	5.7955	0.1725	2.4710	14.3205
7	1.0721	0.9327	7.2135	0.1386	6.7282	0.1486	2.9602	19.9168
8	1.0829	0.9235	8.2857	0.1207	7.6517	0.1307	3.4478	26.3812
9	1.0937	0.9143	9.3685	0.1067	8.5660	0.1167	3.9337	33.6959
10	1.1046	0.9053	10.4622	0.0956	9.4713	0.1056	4.4179	41.8435
11	1.1157	0.8963	11.5668	0.0865	10.3676	0.0965	4.9005	50.8067
12	1.1268	0.8874	12.6825	0.0788	11.2551	0.0888	5.3815	60.5687
13	1.1381	0.8787	13.8093	0.0724	12.1337	0.0824	5.8607	71.1126
14	1.1495	0.8700	14.9474	0.0669	13.0037	0.0769	6.3384	82.4221
15	1.1610	0.8613	16.0969	0.0621	13.8651	0.0721	6.8143	94.4810
16	1.1726	0.8528	17.2579	0.0579	14.7179	0.0679	7.2886	107.2734
17	1.1843	0.8444	18.4304	0.0543	15.5623	0.0643	7.7613	120.7834
18	1.1961	0.8360	19.6147	0.0510	16.3983	0.0610	8.2323	134.9957
19	1.2081	0.8277	20.8109	0.0481	17.2260	0.0581	8.7017	149.8950
20	1.2202	0.8195	22.0190	0.0454	18.0456	0.0554	9.1694	165.4664
21	1.2324	0.8114	23.2392	0.0430	18.8570	0.0530	9.6354	181.6950
22	1.2447	0.8034	24.4716	0.0409	19.6604	0.0509	10.0998	198.5663
23	1.2572	0.7954	25.7163	0.0389	20.4558	0.0489	10.5626	216.0660
24	1.2697	0.7876	26.9735	0.0371	21.2434	0.0471	11.0237	234.1800
25	1.2824	0.7798	28.2432	0.0354	22.0232	0.0454	11.4831	252.8945
26	1.2953	0.7720	29.5256	0.0339	22.7952	0.0439	11.9409	272.1957
27	1.3082	0.7644	30.8209	0.0324	23.5596	0.0424	12.3971	292.0702
28	1.3213	0.7568	32.1291	0.0311	24.3164	0.0411	12.8516	312.5047
29	1.3345	0.7493	33.4504	0.0299	25.0658	0.0399	13.3044	333.4863
30	1.3478	0.7419	34.7849	0.0287	25.8077	0.0387	13.7557	355.0021
31	1.3613	0.7346	36.1327	0.0277	26.5423	0.0377	14.2052	377.0394
32	1.3749	0.7273	37.4941	0.0267	27.2696	0.0367	14.6532	399.5858
33	1.3887	0.7201	38.8690	0.0257	27.9897	0.0357	15.0995	422.6291
34	1.4026	0.7130	40.2577	0.0248	28.7027	0.0348	15.5441	446.1572
35	1.4166	0.7059	41.6603	0.0240	29.4086	0.0340	15.9871	470.1583
40	1.4889	0.6717	48.8864	0.0205	32.8347	0.0305	18.1776	596.8561
45	1.5648	0.6391	56.4811	0.0177	36.0945	0.0277	20.3273	733.7037
50	1.6446	0.6080	64.4632	0.0155	39.1961	0.0255	22.4363	879.4176
55	1.7285	0.5785	72.8525	0.0137	42.1472	0.0237	24.5049	1032.8148
60	1.8167	0.5504	81.6697	0.0122	44.9550	0.0222	26.5333	1192.8061
65	1.9094	0.5237	90.9366	0.0110	47.6266	0.0210	28.5217	1358.3903
70	2.0068	0.4983	100.6763	0.0099	50.1685	0.0199	30.4703	1528.6474
75	2.1091	0.4741	110.9128	0.0090	52.5871	0.0190	32.3793	1702.7340
80	2.2167	0.4511	121.6715	0.0082	54.8882	0.0182	34.2492	1879.8771
85	2.3298	0.4292	132.9790	0.0075	57.0777	0.0175	36.0801	2059.3701
90	2.4486	0.4084	144.8633	0.0069	59.1609	0.0169	37.8724	2240.5675
95	2.5735	0.3886	157.3538	0.0064	61.1430	0.0164	39.6265	2422.8811
100	2.7048	0.3697	170.4814	0.0059	63.0289	0.0159	41.3426	2605.7758

TABLE A.2 Discrete Compounding: $i = 2\%$

	Single payment		Uniform series				Gradient series	
	Compound amount factor	Present worth factor	Compound amount factor	Sinking fund factor	Present worth factor	Capital recovery factor	Uniform series factor	Present worth factor
n	To find F given P $F\|P\ i,n$	To find P given F $P\|F\ i,n$	To find F given A $F\|A\ i,n$	To find A given F $A\|F\ i,n$	To find P given A $P\|A\ i,n$	To find A given P $A\|P\ i,n$	To find A given G $A\|G\ i,n$	To find P given G $P\|G\ i,n$
1	1.0200	0.9804	1.0000	1.0000	0.9804	1.0200	0.0000	0.0000
2	1.0404	0.9612	2.0200	0.4950	1.9416	0.5150	0.4950	0.9612
3	1.0612	0.9423	3.0604	0.3268	2.8839	0.3468	0.9868	2.8458
4	1.0824	0.9238	4.1216	0.2426	3.8077	0.2626	1.4752	5.6173
5	1.1041	0.9057	5.2040	0.1922	4.7135	0.2122	1.9604	9.2403
6	1.1262	0.8880	6.3081	0.1585	5.6014	0.1785	2.4423	13.6801
7	1.1487	0.8706	7.4343	0.1345	6.4720	0.1545	2.9208	18.9035
8	1.1717	0.8535	8.5830	0.1165	7.3255	0.1365	3.3961	24.8779
9	1.1951	0.8368	9.7546	0.1025	8.1622	0.1225	3.8681	31.5720
10	1.2190	0.8203	10.9497	0.0913	8.9826	0.1113	4.3367	38.9551
11	1.2434	0.8043	12.1687	0.0822	9.7868	0.1022	4.8021	46.9977
12	1.2682	0.7885	13.4121	0.0746	10.5753	0.0946	5.2642	55.6712
13	1.2936	0.7730	14.6803	0.0681	11.3484	0.0881	5.7231	64.9475
14	1.3195	0.7579	15.9739	0.0626	12.1062	0.0826	6.1786	74.7999
15	1.3459	0.7430	17.2934	0.0578	12.8493	0.0778	6.6309	85.2021
16	1.3728	0.7284	18.6393	0.0537	13.5777	0.0737	7.0799	96.1288
17	1.4002	0.7142	20.0121	0.0500	14.2919	0.0700	7.5256	107.5554
18	1.4282	0.7002	21.4123	0.0467	14.9920	0.0667	7.9681	119.4581
19	1.4568	0.6864	22.8406	0.0438	15.6785	0.0638	8.4073	131.8139
20	1.4859	0.6730	24.2974	0.0412	16.3514	0.0612	8.8433	144.6003
21	1.5157	0.6598	25.7833	0.0388	17.0112	0.0588	9.2760	157.7959
22	1.5460	0.6468	27.2990	0.0366	17.6580	0.0566	9.7055	171.3795
23	1.5769	0.6342	28.8450	0.0347	18.2922	0.0547	10.1317	185.3309
24	1.6084	0.6217	30.4219	0.0329	18.9139	0.0529	10.5547	199.6305
25	1.6406	0.6095	32.0303	0.0312	19.5235	0.0512	10.9745	214.2592
26	1.6734	0.5976	33.6709	0.0297	20.1210	0.0497	11.3910	229.1987
27	1.7069	0.5859	35.3443	0.0283	20.7069	0.0483	11.8043	244.4311
28	1.7410	0.5744	37.0512	0.0270	21.2813	0.0470	12.2145	259.9392
29	1.7758	0.5631	38.7922	0.0258	21.8444	0.0458	12.6214	275.7064
30	1.8114	0.5521	40.5681	0.0246	22.3965	0.0446	13.0251	291.7164
31	1.8476	0.5412	42.3794	0.0236	22.9377	0.0436	13.4257	307.9538
32	1.8845	0.5306	44.2270	0.0226	23.4683	0.0426	13.8230	324.4035
33	1.9222	0.5202	46.1116	0.0217	23.9886	0.0417	14.2172	341.0508
34	1.9607	0.5100	48.0338	0.0208	24.4986	0.0408	14.6083	357.8817
35	1.9999	0.5000	49.9945	0.0200	24.9986	0.0400	14.9961	374.8826
40	2.2080	0.4529	60.4020	0.0166	27.3555	0.0366	16.8885	461.9931
45	2.4379	0.4102	71.8927	0.0139	29.4902	0.0339	18.7034	551.5652
50	2.6916	0.3715	84.5794	0.0118	31.4236	0.0318	20.4420	642.3606
55	2.9717	0.3365	98.5865	0.0101	33.1748	0.0301	22.1057	733.3527
60	3.2810	0.3048	114.0515	0.0088	34.7609	0.0288	23.6961	823.6975
65	3.6225	0.2761	131.1262	0.0076	36.1975	0.0276	25.2147	912.7085
70	3.9996	0.2500	149.9779	0.0067	37.4986	0.0267	26.6632	999.8343
75	4.4158	0.2265	170.7918	0.0059	38.6771	0.0259	28.0434	1084.6393
80	4.8754	0.2051	193.7720	0.0052	39.7445	0.0252	29.3572	1166.7868
85	5.3829	0.1858	219.1439	0.0046	40.7113	0.0246	30.6064	1246.0241
90	5.9431	0.1683	247.1567	0.0040	41.5869	0.0240	31.7929	1322.1701
95	6.5617	0.1524	278.0850	0.0036	42.3800	0.0236	32.9189	1395.1033
100	7.2446	0.1380	312.2323	0.0032	43.0984	0.0232	33.9863	1464.7527

TABLE A.3 Discrete Compounding: $i = 3\%$

	Single payment		Uniform series				Gradient series	
	Compound amount factor	Present worth factor	Compound amount factor	Sinking fund factor	Present worth factor	Capital recovery factor	Uniform series factor	Present worth factor
n	To find F given P $F\mid P\ i,n$	To find P given F $P\mid F\ i,n$	To find F given A $F\mid A\ i,n$	To find A given F $A\mid F\ i,n$	To find P given A $P\mid A\ i,n$	To find A given P $A\mid P\ i,n$	To find A given G $A\mid G\ i,n$	To find P given G $P\mid G\ i,n$
1	1.0300	0.9709	1.0000	1.0000	0.9709	1.0300	0.0000	0.000
2	1.0609	0.9426	2.0300	0.4926	1.9135	0.5226	0.4926	0.9426
3	1.0927	0.9151	3.0909	0.3235	2.8286	0.3535	0.9803	2.7729
4	1.1255	0.8885	4.1836	0.2390	3.7171	0.2690	1.4631	5.4383
5	1.1593	0.8626	5.3091	0.1884	4.5797	0.2184	1.9409	8.8888
6	1.1941	0.8375	6.4684	0.1546	5.4172	0.1846	2.4138	13.0762
7	1.2299	0.8131	7.6625	0.1305	6.2303	0.1605	2.8819	17.9547
8	1.2668	0.7894	8.8923	0.1125	7.0197	0.1425	3.3450	23.4806
9	1.3048	0.7664	10.1591	0.0984	7.7861	0.1284	3.8032	29.6119
10	1.3439	0.7441	11.4639	0.0872	8.5302	0.1172	4.2565	36.3088
11	1.3842	0.7224	12.8078	0.0781	9.2526	0.1081	4.7049	43.5330
12	1.4258	0.7014	14.1920	0.0705	9.9540	0.1005	5.1485	51.2482
13	1.4685	0.6810	15.6178	0.0640	10.6350	0.0940	5.5872	59.4196
14	1.5126	0.6611	17.0863	0.0585	11.2961	0.0885	6.0210	68.0141
15	1.5580	0.6419	18.5989	0.0538	11.9379	0.0838	6.4500	77.0002
16	1.6047	0.6232	20.1569	0.0496	12.5611	0.0796	6.8742	86.3477
17	1.6528	0.6050	21.7616	0.0460	13.1661	0.0760	7.2936	96.0280
18	1.7024	0.5874	23.4144	0.0427	13.7535	0.0727	7.7081	106.0137
19	1.7535	0.5703	25.1169	0.0398	14.3238	0.0698	8.1179	116.2788
20	1.8061	0.5537	26.8704	0.0372	14.8775	0.0672	8.5229	126.7987
21	1.8603	0.5375	28.6765	0.0349	15.4150	0.0649	8.9231	137.5496
22	1.9161	0.5219	30.5368	0.0327	15.9369	0.0627	9.3186	148.5094
23	1.9736	0.5067	32.4529	0.0308	16.4436	0.0608	9.7093	159.6566
24	2.0328	0.4919	34.4265	0.0290	16.9355	0.0590	10.0954	170.9711
25	2.0938	0.4776	36.4593	0.0274	17.4131	0.0574	10.4768	182.4336
26	2.1566	0.4637	38.5530	0.0259	17.8768	0.0559	10.8535	194.0260
27	2.2213	0.4502	40.7096	0.0246	18.3270	0.0546	11.2255	205.7309
28	2.2879	0.4371	42.9309	0.0233	18.7641	0.0533	11.5930	217.5320
29	2.3566	0.4243	45.2189	0.0221	19.1885	0.0521	11.9558	229.4137
30	2.4273	0.4120	47.5754	0.0210	19.6004	0.0510	12.3141	229.4137
31	2.5001	0.4000	50.0027	0.0200	20.0004	0.0500	12.6678	253.3609
32	2.5751	0.3883	52.5028	0.0190	20.3888	0.0490	13.0169	265.3993
33	2.6523	0.3770	55.0778	0.0182	20.7658	0.0482	13.3616	277.4642
34	2.7319	0.3660	57.7302	0.0173	21.1318	0.0473	13.7018	289.5437
35	2.8139	0.3554	60.4621	0.0165	21.4872	0.0465	14.0375	301.6267
40	3.2620	0.3066	75.4013	0.0133	23.1148	0.0433	15.6502	361.7499
45	3.7816	0.2644	92.7199	0.0108	24.5187	0.0408	17.1556	420.6325
50	4.3839	0.2281	112.7969	0.0089	25.7298	0.0389	18.5575	477.4803
55	5.0821	0.1968	136.0716	0.0073	26.7744	0.0373	19.8600	531.7411
60	5.8916	0.1697	163.0534	0.0061	27.6756	0.0361	21.0674	583.0526
65	6.8300	0.1464	194.3328	0.0051	28.4529	0.0351	22.1841	631.2010
70	7.9178	0.1263	230.5941	0.0043	29.1234	0.0343	23.2145	676.0869
75	9.1789	0.1089	272.6309	0.0037	29.7018	0.0337	24.1634	717.6978
80	10.6409	0.0940	321.3630	0.0031	30.2008	0.0331	25.0353	756.0865
85	12.3357	0.0811	377.8570	0.0026	30.6312	0.0326	25.8349	791.3529
90	14.3005	0.0699	443.3489	0.0023	31.0024	0.0323	26.5667	823.6302
95	16.5782	0.0603	519.2720	0.0019	31.3227	0.0319	27.2351	853.0742
100	19.2186	0.0520	607.2877	0.0016	31.5989	0.0316	27.8444	879.8540

TABLE A.4 Discrete Compounding: $i = 4\%$

	Single payment		Uniform series				Gradient series	
	Compound amount factor	Present worth factor	Compound amount factor	Sinking fund factor	Present worth factor	Capital recovery factor	Uniform series factor	Present worth factor
n	To find F given P $F\|P\,i,n$	To find P given F $P\|F\,i,n$	To find F given A $F\|A\,i,n$	To find A given F $A\|F\,i,n$	To find P given A $P\|A\,i,n$	To find A given P $A\|P\,i,n$	To find A given G $A\|G\,i,n$	To find P given G $P\|G\,i,n$
1	1.0400	0.9615	1.0000	1.0000	0.9615	1.0400	0.0000	0.0000
2	1.0816	0.9246	2.0400	0.4902	1.8861	0.5302	0.4902	0.9246
3	1.1249	0.8890	3.1216	0.3203	2.7751	0.3603	0.9739	2.7025
4	1.1699	0.8548	4.2465	0.2355	3.6299	0.2755	1.4510	5.2670
5	1.2167	0.8219	5.4163	0.1846	4.4518	0.2246	1.9216	8.5547
6	1.2653	0.7903	6.6330	0.1508	5.2421	0.1908	2.3857	12.5062
7	1.3159	0.7599	7.8983	0.1266	6.0021	0.1666	2.8433	17.0657
8	1.3686	0.7307	9.2142	0.1085	6.7327	0.1485	3.2944	22.1806
9	1.4233	0.7026	10.5828	0.0945	7.4353	0.1345	3.7391	27.8013
10	1.4802	0.6756	12.0061	0.0833	8.1109	0.1233	4.1773	33.8814
11	1.5395	0.6496	13.4864	0.0741	8.7605	0.1141	4.6090	40.3772
12	1.6010	0.6246	15.0258	0.0666	9.3851	0.1066	5.0343	47.2477
13	1.6651	0.6006	16.6268	0.0601	9.9856	0.1001	5.4533	54.4546
14	1.7317	0.5775	18.2919	0.0547	10.5631	0.0947	5.8659	61.9618
15	1.8009	0.5553	20.0236	0.0499	11.1184	0.0899	6.2721	69.7355
16	1.8730	0.5339	21.8245	0.0458	11.6523	0.0858	6.6720	77.7441
17	1.9479	0.5134	23.6975	0.0422	12.1657	0.0822	7.0656	85.9581
18	2.0258	0.4936	25.6454	0.0390	12.6593	0.0790	7.4530	94.3498
19	2.1068	0.4746	27.6712	0.0361	13.1339	0.0761	7.8342	102.8933
20	2.1911	0.4564	29.7781	0.0336	13.5903	0.0736	8.2091	111.5647
21	2.2788	0.4388	31.9692	0.0313	14.0292	0.0713	8.5779	120.3414
22	2.3699	0.4220	34.2480	0.0292	14.4511	0.0692	8.9407	129.2024
23	2.4647	0.4057	36.6179	0.0273	14.8568	0.0673	9.2973	138.1284
24	2.5633	0.3901	39.0826	0.0256	15.2470	0.0656	9.6479	147.1012
25	2.6658	0.3751	41.6459	0.0240	15.6221	0.0640	9.9925	156.1040
26	2.7725	0.3607	44.3117	0.0226	15.9828	0.0626	10.3312	165.1212
27	2.8834	0.3468	47.0842	0.0212	16.3296	0.0612	10.6640	174.1385
28	2.9987	0.3335	49.9676	0.0200	16.6631	0.0600	10.9909	183.1424
29	3.1187	0.3207	52.9663	0.0189	16.9837	0.0589	11.3120	192.1206
30	3.2434	0.3083	56.0849	0.0178	17.2920	0.0578	11.6274	201.0618
31	3.3731	0.2965	59.3283	0.0169	17.5885	0.0569	11.9371	209.9556
32	3.5081	0.2851	62.7015	0.0159	17.8736	0.0559	12.2411	218.7924
33	3.6484	0.2741	66.2095	0.0151	18.1476	0.0551	12.5396	227.5634
34	3.7943	0.2636	69.8579	0.0143	18.4112	0.0543	12.8324	236.2607
35	3.9461	0.2534	73.6522	0.0136	18.6646	0.0536	13.1198	244.8768
40	4.8010	0.2083	95.0255	0.0105	19.7928	0.0505	14.4765	286.5303
45	5.8412	0.1712	121.0294	0.0083	20.7200	0.0483	15.7047	325.4028
50	7.1067	0.1407	152.6671	0.0066	21.4822	0.0466	16.8122	361.1638
55	8.6464	0.1157	191.1592	0.0052	22.1086	0.0452	17.8070	393.6890
60	10.5196	0.0951	237.9907	0.0042	22.6235	0.0442	18.6972	422.9966
65	12.7987	0.0781	294.9684	0.0034	23.0467	0.0434	19.4909	449.2014
70	15.5716	0.0642	264.2905	0.0027	23.3945	0.0427	20.1961	472.4789
75	18.9453	0.0528	448.6314	0.0022	23.6804	0.0422	20.8206	493.0408
80	23.0498	0.0434	551.2450	0.0018	23.9154	0.0418	21.3718	511.1161
85	28.0436	0.0357	676.0901	0.0015	24.1085	0.0415	21.8569	526.9384
90	34.1193	0.0293	827.9833	0.0012	24.2673	0.0412	22.2826	540.7369
95	41.5114	0.0241	1012.7846	0.0010	24.3978	0.0410	22.6550	552.7307
100	50.5049	0.0198	1237.6237	0.0008	24.5050	0.0408	22.9800	563.1249

TABLE A.5 Discrete Compounding: $i = 5\%$

	Single payment		Uniform series				Gradient series	
	Compound amount factor	Present worth factor	Compound amount factor	Sinking fund factor	Present worth factor	Capital recovery factor	Uniform series factor	Present worth factor
n	To find F given P $F\mid P\ i,n$	To find P given F $P\mid F\ i,n$	To find F given A $F\mid A\ i,n$	To find A given F $A\mid F\ i,n$	To find P given A $P\mid A\ i,n$	To find A given P $A\mid P\ i,n$	To find A given G $A\mid G\ i,n$	To find P given G $P\mid G\ i,n$
1	1.0500	0.9524	1.0000	1.0000	0.9524	1.0500	0.0000	0.0000
2	1.1025	0.9070	2.0500	0.4878	1.8594	0.5378	0.4878	0.9070
3	1.1576	0.8638	3.1525	0.3172	2.7232	0.3672	0.9675	2.6347
4	1.2155	0.8227	4.3101	0.2320	3.5460	0.2820	1.4391	5.1028
5	1.2763	0.7835	5.5256	0.1810	4.3295	0.2310	1.9025	8.2369
6	1.3401	0.7462	6.8019	0.1470	5.0757	0.1970	2.3579	11.9680
7	1.4071	0.7107	8.1420	0.1228	5.7864	0.1728	2.8052	16.2321
8	1.4775	0.6768	9.5491	0.1047	6.4632	0.1547	3.2445	20.9700
9	1.5513	0.6446	11.0266	0.0907	7.1078	0.1407	3.6758	26.1268
10	1.6289	0.6139	12.5779	0.0795	7.7217	0.1295	4.0991	31.6520
11	1.7103	0.5847	14.2068	0.0704	8.3064	0.1204	4.5144	37.4988
12	1.7959	0.5568	15.9171	0.0628	8.8633	0.1128	4.9219	43.6241
13	1.8856	0.5303	17.7130	0.0565	9.3936	0.1065	5.3215	49.9879
14	1.9799	0.5051	19.5986	0.0510	9.8986	0.1010	5.7133	56.5538
15	2.0789	0.4810	21.5786	0.0463	10.3797	0.0963	6.0973	63.2880
16	2.1829	0.4581	23.6575	0.0423	10.8378	0.0923	6.4736	70.1597
17	2.2920	0.4363	25.8404	0.0387	11.2741	0.0887	6.8423	77.1405
18	2.4066	0.4155	28.1324	0.0355	11.6896	0.0855	7.2034	84.2043
19	2.5270	0.3957	30.5390	0.0327	12.0853	0.0827	7.5569	91.3275
20	2.6533	0.3769	33.0660	0.0302	12.4622	0.0802	7.9030	98.4884
21	2.7860	0.3589	35.7193	0.0280	12.8212	0.0780	8.2416	105.6673
22	2.9253	0.3418	38.5052	0.0260	13.1630	0.0760	8.5730	112.8461
23	3.0715	0.3256	41.4305	0.0241	13.4886	0.0741	8.8971	120.0087
24	3.2251	0.3101	44.5020	0.0225	13.7986	0.0725	9.2140	127.1402
25	3.3864	0.2953	47.7271	0.0210	14.0939	0.0710	9.5238	134.2275
26	3.5557	0.2812	51.1135	0.0196	14.3752	0.0696	9.8266	141.2585
27	3.7335	0.2678	54.6691	0.0183	14.6430	0.0683	10.1224	148.2226
28	3.9201	0.2551	58.4026	0.0171	14.8981	0.0671	10.4114	155.1101
29	4.1161	0.2429	62.3227	0.0160	15.1411	0.0660	10.6936	161.9126
30	4.3219	0.2314	66.4388	0.0151	15.3725	0.0651	10.9691	168.6226
31	4.5380	0.2204	70.7608	0.0141	15.5928	0.0641	11.2381	175.2333
32	4.7649	0.2099	75.2988	0.0133	15.8027	0.0633	11.5005	181.7392
33	5.0032	0.1999	80.0638	0.0125	16.0025	0.0625	11.7566	188.1351
34	5.2533	0.1904	85.0670	0.0118	16.1929	0.0618	12.0063	194.4168
35	5.5160	0.1813	90.3203	0.0111	16.3742	0.0611	12.2498	200.5807
40	7.0400	0.1420	120.7998	0.0083	17.1591	0.0583	13.3775	299.5452
45	8.8950	0.1113	159.7002	0.0063	17.7741	0.0563	14.3644	255.3145
50	11.4674	0.0872	209.3480	0.0048	18.2559	0.0548	15.2233	277.9148
55	14.6356	0.0683	272.7126	0.0037	18.6335	0.0537	15.9664	297.5104
60	18.6792	0.0535	353.5837	0.0028	18.9293	0.0528	16.6062	314.3432
65	23.8399	0.0419	456.7980	0.0022	19.1611	0.0522	17.1541	328.6910
70	30.4264	0.0329	588.5285	0.0017	19.3427	0.0517	17.6212	340.8409
75	38.8327	0.0258	756.6537	0.0013	19.4850	0.0513	18.0176	351.0721
80	49.5614	0.0202	971.2288	0.0010	19.5965	0.0510	18.3526	359.6460
85	63.2544	0.0158	1245.0871	0.0008	19.6838	0.0508	18.6346	366.8007
90	80.7304	0.0124	1594.6073	0.0006	19.7523	0.0506	18.8712	372.7488
95	103.0347	0.0097	2040.6935	0.0005	19.8059	0.0505	19.0689	377.6774
100	131.5013	0.0076	2610.0252	0.0004	19.8479	0.0504	19.2337	381.7492

TABLE A.6 Discrete Compounding: $i = 6\%$

	Single payment		Uniform series				Gradient series	
	Compound amount factor	Present worth factor	Compound amount factor	Sinking fund factor	Present worth factor	Capital recovery factor	Uniform series factor	Present worth factor
n	To find F given P $F\|P\,i,n$	To find P given F $P\|F\,i,n$	To find F given A $F\|A\,i,n$	To find A given F $A\|F\,i,n$	To find P given A $P\|A\,i,n$	To find A given P $A\|P\,i,n$	To find A given G $A\|G\,i,n$	To find P given G $P\|G\,i,n$
1	1.0600	0.9434	1.0000	1.0000	0.9434	1.0600	0.0000	0.0000
2	1.1236	0.8900	2.0600	0.4854	1.8334	0.5454	0.4854	0.8900
3	1.1910	0.8396	3.1836	0.3141	2.6730	0.3741	0.9612	2.5692
4	1.2625	0.7921	4.3746	0.2286	3.4651	0.2886	1.4272	4.9455
5	1.3382	0.7473	5.6371	0.1774	4.2124	0.2374	1.8836	7.9345
6	1.4185	0.7050	6.9753	0.1434	4.9173	0.2034	2.3304	11.4594
7	1.5036	0.6651	8.3938	0.1191	5.5824	0.1791	2.7676	15.4497
8	1.5938	0.6274	9.8975	0.1010	6.2098	0.1610	3.1952	19.8416
9	1.6895	0.5919	11.4913	0.0870	6.8017	0.1470	3.6133	24.5768
10	1.7908	0.5584	13.1808	0.0759	7.3601	0.1359	4.0220	29.6023
11	1.8983	0.5268	14.9716	0.0668	7.8869	0.1268	4.4213	34.8702
12	2.0122	0.4970	16.8699	0.0593	8.3838	0.1193	4.8113	40.3369
13	2.1329	0.4688	18.8821	0.0530	8.8527	0.1130	5.1920	45.9629
14	2.2609	0.4423	21.0151	0.0476	9.2950	0.1076	5.5635	51.7128
15	2.3966	0.4173	23.2760	0.0430	9.7122	0.1030	5.9260	57.5546
16	2.5404	0.3936	25.6725	0.0390	10.1059	0.0990	6.2794	63.4592
17	2.6928	0.3714	28.2129	0.0354	10.4773	0.0954	6.6240	69.4011
18	2.8543	0.3503	30.9057	0.0324	10.8276	0.0924	6.9597	75.3569
19	3.0256	0.3305	33.7600	0.0296	11.1581	0.0896	7.2867	81.3062
20	3.2071	0.3118	36.7856	0.0272	11.4699	0.0872	7.6051	87.2304
21	3.3996	0.2942	39.9927	0.0250	11.7641	0.0850	7.9151	93.1136
22	3.6035	0.2775	43.3923	0.0230	12.0416	0.0830	8.2166	98.9412
23	3.8197	0.2618	46.9958	0.0213	12.3034	0.0813	8.5099	104.7007
24	4.0489	0.2470	50.8156	0.0197	12.5504	0.0797	8.7951	110.3812
25	4.2919	0.2330	43.8645	0.0182	12.7834	0.0782	9.0722	115.9732
26	4.5494	0.2198	59.1564	0.0169	13.0032	0.0769	9.3414	121.4684
27	4.8223	0.2074	63.7058	0.0157	13.2105	0.0757	9.6029	126.8600
28	5.1117	0.1956	68.5281	0.0146	13.4062	0.0746	9.8568	132.1420
29	5.4184	0.1846	73.6398	0.0136	13.5907	0.0736	10.1032	137.3096
30	5.7435	0.1741	79.0582	0.0126	13.7648	0.0726	10.3422	142.3588
31	6.0881	0.1643	84.8017	0.0118	13.9291	0.0718	10.5740	147.2864
32	6.4534	0.1550	90.8898	0.0110	14.0840	0.0710	10.7988	152.0901
33	6.8406	0.1462	97.3432	0.0103	14.2302	0.0703	11.0166	156.7681
34	7.2510	0.1379	104.1838	0.0096	14.3681	0.0696	11.2276	161.3192
35	7.6861	0.1301	111.4348	0.0090	14.4982	0.0690	11.4319	165.7427
40	10.2857	0.0972	154.7620	0.0065	15.0463	0.0665	12.3590	185.9568
45	13.7646	0.0727	212.7435	0.0047	15.4558	0.0647	13.1413	203.1096
50	18.4202	0.0543	290.3359	0.0034	15.7619	0.0634	13.7964	217.4574
55	24.6503	0.0406	394.1720	0.0025	15.9905	0.0625	14.3411	229.3222
60	32.9877	0.0303	533.1282	0.0019	16.1614	0.0619	14.7909	239.0428
65	44.1450	0.0227	719.0829	0.0014	16.2891	0.0614	15.1601	246.9450
70	59.0759	0.0169	967.9322	0.0010	16.3845	0.0610	15.4613	253.3271
75	79.0569	0.0126	1300.9487	0.0008	16.4558	0.0608	15.7058	258.4527
80	105.7960	0.0095	1746.5999	0.0006	16.5091	0.0606	15.9033	262.5493
85	141.5789	0.0071	2342.9817	0.0004	16.5489	0.0604	16.0620	265.8096
90	189.4645	0.0053	3141.0752	0.0003	16.5787	0.0603	16.1891	268.3946
95	253.5463	0.0039	4209.1042	0.0002	16.6009	0.0602	16.2905	270.4375
100	339.3021	0.0029	5638.3681	0.0002	16.6175	0.0602	16.3711	272.0471

TABLE A.7 Discrete Compounding: $i = 7\%$

	Single payment		Uniform series				Gradient series	
	Compound amount factor	Present worth factor	Compound amount factor	Sinking fund factor	Present worth factor	Capital recovery factor	Uniform series factor	Present worth factor
n	To find F given P $F\|P\ i,n$	To find P given F $P\|F\ i,n$	To find F given A $F\|A\ i,n$	To find A given F $A\|F\ i,n$	To find P given A $P\|A\ i,n$	To find A given P $A\|P\ i,n$	To find A given G $A\|G\ i,n$	To find P given G $P\|G\ i,n$
1	1.0700	0.9346	1.0000	1.0000	0.9346	1.0700	0.0000	0.0000
2	1.1449	0.8734	2.0700	0.4831	1.8080	0.5531	0.4831	0.8734
3	1.2250	0.8163	3.2149	0.3111	2.6243	0.3811	0.9549	2.5060
4	1.3108	0.7629	4.4399	0.2252	3.3872	0.2952	1.4155	4.7947
5	1.4026	0.7130	5.7507	0.1739	4.1002	0.2439	1.8650	7.6467
6	1.5007	0.6663	7.1533	0.1398	4.7665	0.2098	2.3032	10.9784
7	1.6058	0.6227	8.6540	0.1156	5.3893	0.1856	2.7304	14.7149
8	1.7182	0.5820	10.2598	0.0975	5.9713	0.1675	3.1465	18.7889
9	1.8385	0.5439	11.9780	0.0835	6.5152	0.1535	3.5517	23.1404
10	1.9672	0.5083	13.8164	0.0724	7.0236	0.1424	3.9461	27.7156
11	2.1049	0.4751	15.7836	0.0634	7.4987	0.1334	4.3296	32.4665
12	2.2522	0.4440	17.8885	0.0559	7.9427	0.1259	4.7025	37.3506
13	2.4098	0.4150	20.1406	0.0497	8.3577	0.1197	5.0648	42.3302
14	2.5785	0.3878	22.5505	0.0443	8.7455	0.1143	5.4167	47.3718
15	2.7590	0.3624	25.1290	0.0398	9.1079	0.1098	5.7583	52.4461
16	2.9522	0.3387	27.8881	0.0359	9.4466	0.1059	6.0897	57.5271
17	3.1588	0.3166	30.8402	0.0324	9.7632	0.1024	6.4110	62.5923
18	3.3799	0.2959	33.9990	0.0294	10.0591	0.0994	6.7225	67.6219
19	3.6165	0.2765	37.3790	0.0268	10.3356	0.0968	7.0242	72.5991
20	3.8697	0.2584	40.9955	0.0244	10.5940	0.0944	7.3163	77.5091
21	4.1406	0.2415	44.8652	0.0223	10.8355	0.0923	7.5990	82.3393
22	4.4304	0.2257	49.0057	0.0204	11.0612	0.0904	7.8725	87.0793
23	4.7405	0.2109	53.4361	0.0187	11.2722	0.0887	8.1369	91.7201
24	5.0724	0.1971	58.1767	0.0172	11.4693	0.0872	8.3923	96.2545
25	5.4274	0.1842	63.2490	0.0158	11.6536	0.0858	8.6391	100.6765
26	5.8074	0.1722	68.6765	0.0146	11.8258	0.0846	8.8773	104.9814
27	6.2139	0.1609	74.4838	0.0134	11.9867	0.0834	9.1072	109.1656
28	6.6488	0.1504	80.6977	0.0124	12.1371	0.0824	9.3289	113.2264
29	7.1143	0.1406	87.3465	0.0114	12.2777	0.0814	9.5427	117.1622
30	7.6123	0.1314	94.4608	0.0106	12.4090	0.0806	9.7487	120.9718
31	8.1451	0.1228	102.0730	0.0098	12.5318	0.0798	9.9471	124.6550
32	8.7153	0.1147	110.2182	0.0091	12.6466	0.0791	10.1381	128.2120
33	9.3253	0.1072	118.9334	0.0084	12.7538	0.0784	10.3219	131.6435
34	9.9781	0.1002	128.2588	0.0078	12.8540	0.0778	10.4987	134.9507
35	10.6766	0.0937	138.2369	0.0072	12.9477	0.0772	10.6687	138.1353
40	14.9745	0.0668	199.6351	0.0050	13.3317	0.0750	11.4233	152.2928
45	21.0025	0.0476	285.7493	0.0035	13.6055	0.0735	12.0360	163.7559
50	29.4570	0.0339	406.5289	0.0025	13.8007	0.0725	12.5287	172.9051
55	41.3150	0.0242	575.9286	0.0017	13.9399	0.0717	12.9215	180.1243
60	57.9464	0.0173	813.5240	0.0012	14.0392	0.0712	13.2321	185.7677
65	81.2729	0.0123	1146.7552	0.0009	14.1099	0.0709	13.4760	190.1452
70	113.9894	0.0088	1614.1342	0.0006	14.1604	0.0706	13.6662	193.5185
75	159.8760	0.0063	2269.6574	0.0004	14.1964	0.0704	13.8136	196.1035
80	224.2344	0.0045	3189.0627	0.0003	14.2220	0.0703	13.9273	198.0748
85	314.5003	0.0032	4478.5761	0.0002	14.2403	0.0702	14.0146	199.5717
90	441.1030	0.0023	6287.1854	0.0002	14.2533	0.0702	14.0812	200.7042
95	618.6697	0.0016	8823.8535	0.0001	14.2626	0.0701	14.1319	201.5581
100	867.7163	0.0012	12381.6618	0.0001	14.2693	0.0701	14.1703	202.2001

TABLE A.8 Discrete Compounding: $i = 8\%$

	Single payment		Uniform series				Gradient series	
	Compound amount factor	Present worth factor	Compound amount factor	Sinking fund factor	Present worth factor	Capital recovery factor	Uniform series factor	Present worth factor
n	To find F given P $F\|P\,i,n$	To find P given F $P\|F\,i,n$	To find F given A $F\|A\,i,n$	To find A given F $A\|F\,i,n$	To find P given A $P\|A\,i,n$	To find A given P $A\|P\,i,n$	To find A given G $A\|G\,i,n$	To find P given G $P\|G\,i,n$
1	1.0800	0.9259	1.0000	1.0000	0.9259	1.0800	0.0000	0.0000
2	1.1664	0.8573	2.0800	0.4808	1.7833	0.5608	0.4808	0.8573
3	1.2597	0.7938	3.2464	0.3080	2.5771	0.3880	0.9487	2.4450
4	1.3605	0.7350	4.5061	0.2219	3.3121	0.3019	1.4040	4.6501
5	1.4693	0.6806	5.8666	0.1705	3.9927	0.2505	1.8465	7.3724
6	1.5869	0.6302	7.3359	0.1363	4.6229	0.2163	2.2763	10.5233
7	1.7138	0.5835	8.9228	0.1121	5.2064	0.1921	2.6937	14.0242
8	1.8509	0.5403	10.6366	0.0940	5.7466	0.1740	3.0985	17.8061
9	1.9990	0.5002	12.4876	0.0801	6.2469	0.1601	3.4910	21.8081
10	2.1589	0.4632	14.4866	0.0690	6.7101	0.1490	3.8713	25.9768
11	2.3316	0.4289	16.6455	0.0601	7.1390	0.1401	4.2395	30.2657
12	2.5182	0.3971	18.9771	0.0527	7.5361	0.1327	4.5957	34.6339
13	2.7196	0.3677	21.4953	0.0465	7.9038	0.1265	4.9402	39.0463
14	2.9372	0.3405	24.2149	0.0413	8.2442	0.1213	5.2731	43.4723
15	3.1722	0.3152	27.1521	0.0368	8.5595	0.1168	5.5945	47.8857
16	3.4259	0.2919	30.3243	0.0330	8.8514	0.1130	5.9046	52.2640
17	3.7000	0.2703	33.7502	0.0296	9.1216	0.1096	6.2037	56.5883
18	3.9960	0.2502	37.4502	0.0267	9.3719	0.1067	6.4920	60.8426
19	4.3157	0.2317	41.4463	0.0241	9.6036	0.1041	6.7697	65.0134
20	4.6610	0.2145	45.7620	0.0219	9.8181	0.1019	7.0369	69.0898
21	5.0338	0.1987	50.4229	0.0198	10.0168	0.0998	7.2940	73.0629
22	5.4365	0.1839	55.4568	0.0180	10.2007	0.0980	7.5412	76.9257
23	5.8715	0.1703	60.8933	0.0164	10.3711	0.0964	7.7786	80.6726
24	6.3412	0.1577	66.7648	0.0150	10.5288	0.0950	8.0066	84.2997
25	6.8485	0.1460	73.1059	0.0137	10.6748	0.0937	8.2254	87.8041
26	7.3964	0.1352	79.9544	0.0125	10.8100	0.0925	8.4352	91.1842
27	7.9881	0.1252	87.3508	0.0114	10.9353	0.0914	8.6363	94.4390
28	8.6271	0.1159	95.3388	0.0105	11.0511	0.0905	8.8289	97.5687
29	9.3173	0.1073	103.9659	0.0096	11.1584	0.0896	9.0133	100.5738
30	10.0627	0.0994	113.2832	0.0088	11.2578	0.0888	9.1897	103.4558
31	10.8677	0.0920	123.3459	0.0081	11.3498	0.0881	9.3584	106.2163
32	11.7371	0.0852	134.2135	0.0075	11.4350	0.0875	9.5197	108.8575
33	12.6760	0.0789	145.9506	0.0069	11.5139	0.0869	9.6737	111.3819
34	13.6901	0.0730	158.6267	0.0063	11.5869	0.0863	9.8208	113.7924
35	14.7853	0.0676	172.3168	0.0058	11.6546	0.0858	9.9611	116.0920
40	21.7245	0.0460	259.0565	0.0039	11.9246	0.0839	10.5699	126.0422
45	31.9204	0.0313	386.5056	0.0026	12.1084	0.0826	11.0447	133.7331
50	46.9016	0.0213	573.7702	0.0017	12.2335	0.0817	11.4107	139.5928
55	68.9139	0.0145	848.9232	0.0012	12.3186	0.0812	11.6902	144.0065
60	101.2571	0.0099	1253.2133	0.0008	12.3766	0.0808	11.9015	147.3000
65	148.7798	0.0067	1847.2481	0.0005	12.4160	0.0805	12.0602	149.7387
70	218.6064	0.0046	2720.0801	0.0004	12.4428	0.0804	12.1783	151.5326
75	321.2045	0.0031	4002.5566	0.0002	12.4611	0.0802	12.2658	152.8448
80	471.9548	0.0021	5886.9354	0.0002	12.4735	0.0802	12.3301	153.8001
85	693.4565	0.0014	8655.7061	0.0001	12.4820	0.0801	12.3772	154.4925
90	1018.9151	0.0010	12723.9386	0.0001	12.4877	0.0801	12.4116	154.9925
95	1497.1205	0.0007	18701.5069	0.0001	12.4971	0.0801	12.4365	155.3524
100	2199.7613	0.0005	27484.5157	0.0000	12.4943	0.0800	12.4545	155.6107

TABLE A.9 Discrete Compounding: $i = 9\%$

	Single payment		Uniform series				Gradient series	
	Compound amount factor	Present worth factor	Compound amount factor	Sinking fund factor	Present worth factor	Capital recovery factor	Uniform series factor	Present worth factor
n	To find F given P $F\|P\ i,n$	To find P given F $P\|F\ i,n$	To find F given A $F\|A\ i,n$	To find A given F $A\|F\ i,n$	To find P given A $P\|A\ i,n$	To find A given P $A\|P\ i,n$	To find A given G $A\|G\ i,n$	To find P given G $P\|G\ i,n$
1	1.0900	0.9174	1.0000	1.0000	0.9174	1.0900	0.0000	0.0000
2	1.1881	0.8417	2.0900	0.4785	1.7591	0.5685	0.4785	0.8417
3	1.2950	0.7722	3.2781	0.3051	2.5313	0.3951	0.9426	2.3860
4	1.4116	0.7084	4.5731	0.2187	3.2397	0.3087	1.3925	4.5113
5	1.5386	0.6499	5.9847	0.1671	3.8897	0.2571	1.8282	7.1110
6	1.6771	0.5963	7.5233	0.1329	4.4859	0.2229	2.2498	10.0924
7	1.8280	0.5470	9.2004	0.1087	5.0330	0.1987	2.6574	13.3746
8	1.9926	0.5019	11.0285	0.0907	5.5348	0.1807	3.0512	16.8877
9	2.1719	0.4604	13.0210	0.0768	5.9952	0.1668	3.4312	20.5711
10	2.3674	0.4224	15.1929	0.0658	6.4177	0.1558	3.7978	24.3728
11	2.5804	0.3875	17.5603	0.0569	6.8052	0.1469	4.1510	28.2481
12	2.8127	0.3555	20.1407	0.0497	7.1607	0.1397	4.4910	32.1590
13	3.0658	0.3262	22.9534	0.0436	7.4869	0.1336	4.8182	36.0731
14	3.3417	0.2992	26.0192	0.0384	7.7862	0.1284	5.1326	39.9633
15	3.6425	0.2745	29.3609	0.0341	8.0607	0.1241	5.4346	43.8069
16	3.9703	0.2519	33.0034	0.0303	8.3126	0.1203	5.7245	47.5849
17	4.3276	0.2311	36.9737	0.0270	8.5436	0.1170	6.0024	51.2821
18	4.7171	0.2120	41.3013	0.0242	8.7556	0.1142	6.2687	54.8860
19	5.1417	0.1945	46.0185	0.0217	8.9501	0.1117	6.5236	58.3868
20	5.6044	0.1784	51.1601	0.0195	9.1285	0.1095	6.7674	61.7770
21	6.1088	0.1637	56.7645	0.0176	9.2922	0.1076	7.0006	65.0509
22	6.6586	0.1502	62.8733	0.0159	9.4424	0.1059	7.2232	68.2048
23	7.2579	0.1378	69.5319	0.0144	9.5802	0.1044	7.4357	71.2359
24	7.9111	0.1264	76.7898	0.0130	9.7066	0.1030	7.6384	74.1433
25	8.6231	0.1160	84.7009	0.0118	9.8226	0.1018	7.8316	76.9265
26	9.3992	0.1064	93.3240	0.0107	9.9290	0.1007	8.0156	79.5863
27	10.2451	0.0976	102.7231	0.0097	10.0266	0.0997	8.1906	82.1241
28	11.1671	0.0895	112.9682	0.0089	10.1161	0.0989	8.3571	84.5419
29	12.1722	0.0822	124.1354	0.0081	10.1983	0.0981	8.5154	86.8422
30	13.2677	0.0754	136.3075	0.0073	10.2737	0.0973	8.6657	89.0280
31	14.4618	0.0691	149.5752	0.0067	10.3428	0.0967	8.8083	91.1024
32	15.7633	0.0634	164.0370	0.0061	10.4062	0.0961	8.9436	93.0690
33	17.1820	0.0582	179.8003	0.0056	10.4644	0.0956	9.0718	94.9314
34	18.7284	0.0534	196.9823	0.0051	10.5178	0.0951	9.1933	96.6935
35	20.4140	0.0490	215.7108	0.0046	10.5668	0.0946	9.3083	98.3590
40	31.4094	0.0318	337.8824	0.0030	10.7574	0.0930	9.7957	105.3762
45	48.3273	0.0207	525.8587	0.0019	10.8812	0.0919	10.1603	110.5561
50	74.3575	0.0134	815.0836	0.0012	10.9617	0.0912	10.4295	114.3251
55	114.4083	0.0087	1260.0918	0.0008	11.0140	0.0908	10.6261	117.0362
60	176.0313	0.0057	1944.7921	0.0005	11.0480	0.0905	10.7683	118.9683
65	270.8460	0.0037	2998.2885	0.0003	11.0701	0.0903	10.8702	120.3344
70	416.7301	0.0024	4619.2232	0.0002	11.0844	0.0902	10.9427	121.2942
75	641.1909	0.0016	7113.2321	0.0001	11.0938	0.0901	10.9940	121.9646
80	986.5517	0.0010	10950.5741	0.0001	11.0998	0.0901	11.0299	122.4306
85	1517.9320	0.0007	16854.8003	0.0001	11.1038	0.0901	11.0551	122.7533
90	2335.5266	0.0004	25939.1842	0.0000	11.1064	0.0900	11.0726	122.9758
95	3593.4971	0.0003	39916.6350	0.0000	11.1080	0.0900	11.0847	123.1287
100	5529.0408	0.0002	61422.6755	0.0000	11.1091	0.0900	11.0930	123.2335

TABLE A.10 Discrete Compounding: $i = 10\%$

	Single payment		Uniform series				Gradient series	
	Compound amount factor	Present worth factor	Compound amount factor	Sinking fund factor	Present worth factor	Capital recovery factor	Uniform series factor	Present worth factor
n	To find F given P $F\|P\,i,n$	To find P given F $P\|F\,i,n$	To find F given A $F\|A\,i,n$	To find A given F $A\|F\,i,n$	To find P given A $P\|A\,i,n$	To find A given P $A\|P\,i,n$	To find A given G $A\|G\,i,n$	To find P given G $P\|G\,i,n$
1	1.1000	0.9091	1.0000	1.0000	0.9091	1.1000	0.0000	0.0000
2	1.2100	0.8264	2.1000	0.4762	1.7355	0.5762	0.4762	0.8264
3	1.3310	0.7513	3.3100	0.3021	2.4369	0.4021	0.9366	2.3291
4	1.4641	0.6830	4.6410	0.2155	3.1699	0.3155	1.3812	4.3781
5	1.6105	0.6209	6.1051	0.1638	3.7908	0.2638	1.8101	6.8618
6	1.7716	0.5645	7.7156	0.1296	4.3553	0.2296	2.2236	9.6842
7	1.9487	0.5132	9.4872	0.1054	4.8684	0.2054	2.6216	12.7631
8	2.1436	0.4665	11.4359	0.0874	5.3349	0.1874	3.0045	16.0287
9	2.3579	0.4241	13.5795	0.0736	5.7590	0.1736	3.3724	19.4215
10	2.5937	0.3855	15.9347	0.0627	6.1446	0.1627	3.7255	22.8913
11	2.8531	0.3505	18.5312	0.0540	6.4951	0.1540	4.0641	26.3963
12	3.1384	0.3186	21.3843	0.0468	6.8137	0.1468	4.3884	29.9012
13	3.4523	0.2897	24.5227	0.0408	7.1034	0.1408	4.6988	33.3772
14	3.7975	0.2633	27.9750	0.0357	7.3667	0.1357	4.9955	36.8005
15	4.1772	0.2394	31.7725	0.0315	7.6061	0.1315	5.2789	40.1520
16	4.5950	0.2176	35.9497	0.0278	7.8237	0.1278	5.5493	43.4164
17	5.0545	0.1978	40.5447	0.0247	8.0216	0.1247	5.8071	46.5819
18	5.5599	0.1799	45.5992	0.0219	8.2014	0.1219	6.0526	49.6395
19	6.1159	0.1635	51.1591	0.0195	8.3649	0.1195	6.2861	52.5827
20	6.7275	0.1486	57.2750	0.0175	8.5136	0.1175	6.5081	55.4069
21	7.4002	0.1351	64.0025	0.0156	8.6487	0.1156	6.7189	58.1095
22	8.1403	0.1228	71.4027	0.0140	8.7715	0.1140	6.9189	60.6893
23	8.9543	0.1117	79.5430	0.0126	8.8832	0.1126	7.1085	63.1462
24	9.8497	0.1015	88.4973	0.0113	8.9847	0.1113	7.2881	65.4813
25	10.8347	0.0923	98.3471	0.0102	9.0770	0.1102	7.4580	67.6964
26	11.9182	0.0839	109.1818	0.0092	9.1609	0.1092	7.6186	69.7940
27	13.1100	0.0763	121.0999	0.0083	9.2372	0.1083	7.7704	71.7773
28	14.4210	0.0693	134.2099	0.0075	9.3066	0.1075	7.9137	73.6495
29	14.8631	0.0630	148.6309	0.0067	9.3696	0.1067	8.0489	75.4146
30	17.4494	0.0573	164.4940	0.0061	9.4269	0.1061	8.1762	77.0766
31	19.1943	0.0521	181.9434	0.0055	9.4790	0.1055	8.2962	78.6395
32	21.1138	0.0474	201.1378	0.0050	9.5264	0.1050	8.4091	80.1078
33	23.2252	0.0431	222.2515	0.0045	9.5694	0.1045	8.5152	81.4856
34	25.5477	0.0391	245.4767	0.0041	9.6086	0.1041	8.6149	82.7773
35	28.1024	0.0356	271.0244	0.0037	9.6442	0.1037	8.7086	83.9872
40	45.2593	0.0221	442.5926	0.0023	9.7791	0.1023	9.0962	88.9525
45	72.8905	0.0137	718.9048	0.0014	9.8628	0.1014	9.3740	92.4544
50	117.3909	0.0085	1163.9085	0.0009	9.9148	0.1009	9.5704	94.8889
55	189.0591	0.0053	1880.5914	0.0005	9.9471	0.1005	9.7075	96.5619
60	304.4816	0.0033	3034.8164	0.0003	9.9672	0.1003	9.8023	97.7010
65	490.3707	0.0020	4893.7073	0.0002	9.9796	0.1002	9.8672	98.4705
70	789.7470	0.0013	7887.4696	0.0001	9.9873	0.1001	9.9113	98.9870
75	1271.8954	0.0008	12708.9537	0.0001	9.9921	0.1001	9.9410	99.3317
80	2048.4002	0.0005	20474.0021	0.0000	9.9951	0.1000	9.9609	99.5606
85	3298.9690	0.0003	32979.6903	0.0000	9.9970	0.1000	9.9742	99.7120
90	5313.0226	0.0002	53120.2261	0.0000	9.9981	0.1000	9.9831	99.8118
95	8556.6760	0.0001	85556.7605	0.0000	9.9988	0.1000	9.9889	99.8773
100	13780.6123	0.0001	137796.1234	0.0000	9.9993	0.1000	9.9927	99.9202

TABLE A.11 Discrete Compounding: $i = 12\%$

	Single payment		Uniform series				Gradient series	
	Compound amount factor	Present worth factor	Compound amount factor	Sinking fund factor	Present worth factor	Capital recovery factor	Uniform series factor	Present worth factor
n	To find F given P $F\|P\,i,n$	To find P given F $P\|F\,i,n$	To find F given A $F\|A\,i,n$	To find A given F $A\|F\,i,n$	To find P given A $P\|A\,i,n$	To find A given P $A\|P\,i,n$	To find A given G $A\|G\,i,n$	To find P given G $P\|G\,i,n$
1	1.1200	0.8929	1.0000	1.0000	0.8929	1.1200	0.0000	0.0000
2	1.2544	0.7972	2.1200	0.4717	1.6901	0.5917	0.4717	0.7972
3	1.4049	0.7118	3.3744	0.2963	2.4018	0.4163	0.9246	2.2208
4	1.5735	0.6355	4.7793	0.2092	3.0373	0.3292	1.3589	4.1273
5	1.7623	0.5674	6.3528	0.1574	3.6048	0.2774	1.7746	6.3970
6	1.9738	0.5066	8.1152	0.1232	4.1114	0.2432	2.1720	8.9302
7	2.2107	0.4523	10.0890	0.0991	4.5638	0.2191	2.5515	11.6443
8	2.4760	0.4039	12.2997	0.0813	4.9676	0.2013	2.9131	14.4714
9	2.7731	0.3606	14.7757	0.0677	5.3282	0.1877	3.2574	17.3563
10	3.1058	0.3220	17.5487	0.0570	5.6502	0.1770	3.5847	20.2541
11	3.4785	0.2875	20.6546	0.0484	5.9377	0.1684	3.8953	23.1288
12	3.8960	0.2567	24.1331	0.0414	6.1944	0.1614	4.1897	25.9523
13	4.3635	0.2292	28.0291	0.0357	6.4235	0.1557	4.4683	28.7024
14	4.8871	0.2046	32.3926	0.0309	6.6282	0.1509	4.7317	31.3624
15	5.4736	0.1827	37.2797	0.0268	6.8109	0.1468	4.9803	33.9202
16	6.1304	0.1631	42.7533	0.0234	6.9740	0.1434	5.2147	36.3670
17	6.8660	0.1456	48.8837	0.0205	7.1196	0.1405	5.4353	38.6973
18	7.6900	0.1300	55.7497	0.0179	7.2497	0.1379	5.6247	40.9080
19	8.6128	0.1161	63.4397	0.0158	7.3658	0.1358	5.8375	42.9979
20	9.6463	0.1037	72.0524	0.0139	7.4694	0.1339	6.0202	44.9676
21	10.8038	0.0926	81.6987	0.0122	7.5620	0.1322	6.1913	46.8188
22	12.1003	0.0826	92.5026	0.0108	7.6446	0.1308	6.3514	48.5543
23	13.5523	0.0738	104.6029	0.0096	7.7184	0.1296	6.5010	50.1776
24	15.1786	0.0659	118.1552	0.0085	7.7843	0.1285	6.6406	51.6929
25	17.0001	0.0588	133.3339	0.0075	7.8431	0.1275	6.7708	53.1046
26	19.0401	0.0525	150.3339	0.0067	7.8957	0.1267	6.8921	54.4177
27	21.3249	0.0469	169.3740	0.0059	7.9426	0.1259	7.0049	55.6369
28	23.8839	0.0419	190.6989	0.0052	7.9844	0.1252	7.1098	56.7674
29	26.7499	0.0374	214.5828	0.0047	8.0218	0.1247	7.2071	57.8141
30	29.9599	0.0334	241.3327	0.0041	8.0552	0.1241	7.2974	58.7821
31	33.5551	0.0298	271.2926	0.0037	8.0850	0.1237	7.3811	59.6761
32	37.5817	0.0266	304.8477	0.0033	8.1116	0.1233	7.4586	60.5010
33	42.0915	0.0238	342.4294	0.0029	8.1354	0.1229	7.5302	61.2612
34	47.1425	0.0212	384.5210	0.0026	8.1566	0.1226	7.5965	61.9612
35	52.7996	0.0189	431.6635	0.0023	8.1755	0.1223	7.6577	62.6052
40	93.0510	0.0107	767.0914	0.0013	8.2438	0.1213	7.8988	65.1159
45	163.9876	0.0061	1358.2300	0.0007	8.2825	0.1207	8.0572	66.7342
50	289.0022	0.0035	2400.0182	0.0004	8.3045	0.1204	8.1597	67.7624

TABLE A.12 Discrete Compounding: $i = 15\%$

	Single payment		Uniform series				Gradient series	
	Compound amount factor	Present worth factor	Compound amount factor	Sinking fund factor	Present worth factor	Capital recovery factor	Uniform series factor	Present worth factor
n	To find F given P $F\|P\ i,n$	To find P given F $P\|F\ i,n$	To find F given A $F\|A\ i,n$	To find A given F $A\|F\ i,n$	To find P given A $P\|A\ i,n$	To find A given P $A\|P\ i,n$	To find A given G $A\|G\ i,n$	To find P given G $P\|G\ i,n$
1	1.1500	0.8696	1.0000	1.0000	0.8696	1.1500	0.0000	0.0000
2	1.3225	0.7561	2.1500	0.4651	1.6257	0.6151	0.4651	0.7561
3	1.5209	0.6575	3.4725	0.2880	2.2832	0.4380	0.9071	2.0712
4	1.7490	0.5718	4.9934	0.2003	2.8550	0.3503	1.3263	3.7864
5	2.0114	0.4972	6.7242	0.1483	3.3522	0.2983	1.7228	5.7751
6	2.3131	0.4323	8.7537	0.1142	3.7845	0.2642	2.0972	7.9368
7	2.6600	0.3759	11.0668	0.0904	4.1604	0.2404	2.4498	10.1924
8	3.0590	0.3269	13.7268	0.0729	4.4873	0.2229	2.7813	12.4807
9	3.5179	0.2843	16.7858	0.0596	4.7716	0.2096	3.0922	14.7548
10	4.0456	0.2472	20.3037	0.0493	5.0188	0.1993	3.3832	16.9795
11	4.6524	0.2149	24.3493	0.0411	5.2337	0.1911	3.6549	19.1289
12	5.3503	0.1869	29.0017	0.0345	5.4206	0.1845	3.9082	21.1849
13	6.1528	0.1625	34.3519	0.0291	5.5831	0.1791	4.1438	23.1352
14	7.0757	0.1413	40.5047	0.0247	5.7245	0.1747	4.3624	24.9725
15	8.1371	0.1229	47.5804	0.0210	5.8474	0.1710	4.5650	26.6930
16	9.3576	0.1069	55.7175	0.0179	5.9542	0.1679	4.7522	28.2960
17	10.7613	0.0929	65.0751	0.0154	6.0472	0.1654	4.9251	29.7828
18	12.3755	0.0808	75.8364	0.0132	6.1280	0.1632	5.0843	31.1565
19	14.2318	0.0703	88.2118	0.0113	6.1982	0.1613	5.2307	32.4213
20	16.3665	0.0611	102.4436	0.0098	6.2593	0.1598	5.3651	33.5822
21	18.8215	0.0531	118.8101	0.0084	6.3125	0.1584	5.4883	34.6448
22	21.6447	0.0462	137.6316	0.0073	6.3587	0.1573	5.6010	35.6150
23	24.8915	0.0402	159.2764	0.0063	6.3988	0.1563	5.7040	36.4988
24	28.6252	0.0349	184.1678	0.0054	6.4338	0.1554	5.7979	37.3023
25	32.9190	0.0304	212.7930	0.0047	6.4641	0.1547	5.8834	38.0314
26	37.8568	0.0264	245.7120	0.0041	6.4906	0.1541	5.9612	38.6918
27	43.5353	0.0230	283.5688	0.0035	6.5135	0.1535	6.0319	39.2890
28	50.0656	0.0200	327.1041	0.0031	6.5335	0.1531	6.0960	39.8283
29	57.5755	0.0174	377.1697	0.0027	6.5509	0.1527	6.1541	40.3146
30	66.2118	0.0151	434.7451	0.0023	6.5660	0.1523	6.2066	40.7526
31	76.1435	0.0131	500.9569	0.0020	6.5791	0.1520	6.2541	41.1466
32	87.5651	0.0114	577.1005	0.0017	6.5905	0.1517	6.2970	41.5006
33	100.6998	0.0099	664.6655	0.0015	6.6005	0.1515	6.3357	41.8184
34	115.8048	0.0086	765.3654	0.0013	6.6091	0.1513	6.3705	42.1033
35	133.1755	0.0075	881.1702	0.0011	6.6166	0.1511	6.4019	42.3586
40	267.8635	0.0037	1779.0903	0.0006	6.6418	0.1506	6.5168	43.2830
45	538.7693	0.0019	3585.1285	0.0003	6.6543	0.1503	6.5830	43.8051
50	1083.6574	0.0009	7217.7163	0.0001	6.6605	0.1501	6.6205	44.0958

TABLE A.13 Discrete Compounding: $i = 20\%$

	Single payment		Uniform series				Gradient series	
	Compound amount factor	Present worth factor	Compound amount factor	Sinking fund factor	Present worth factor	Capital recovery factor	Uniform series factor	Present worth factor
n	To find F given P $F\|P\ i,n$	To find P given F $P\|F\ i,n$	To find F given A $F\|A\ i,n$	To find A given F $A\|F\ i,n$	To find P given A $P\|A\ i,n$	To find A given P $A\|P\ i,n$	To find A given G $A\|G\ i,n$	To find P given G $P\|G\ i,n$
1	1.2000	0.8333	1.0000	1.0000	0.8333	1.2000	0.0000	0.0000
2	1.4400	0.6944	2.2000	0.4545	1.5278	0.6545	0.4545	0.6944
3	1.7280	0.5787	3.6400	0.2747	2.1065	0.4747	0.8791	1.8519
4	2.0736	0.4823	5.3680	0.1863	2.5887	0.3863	1.2742	3.2986
5	2.4883	0.4019	7.4416	0.1344	2.9906	0.3344	1.6405	4.9061
6	2.9860	0.3349	9.9299	0.1007	3.3255	0.3007	1.9788	6.5806
7	3.5832	0.2791	12.9159	0.0774	3.6046	0.2774	2.2902	8.2551
8	4.2998	0.2326	16.4991	0.0606	3.8372	0.2606	2.5756	9.8831
9	5.1598	0.1938	20.7989	0.0481	4.0310	0.2481	2.8364	11.4335
10	6.1917	0.1615	25.9587	0.0385	4.1925	0.2385	3.0739	12.8871
11	7.4301	0.1346	32.1504	0.0311	4.3271	0.2311	3.2893	14.2330
12	8.9161	0.1122	39.5805	0.0253	4.4392	0.2253	3.4841	15.4667
13	10.6993	0.0935	48.4966	0.0206	4.5327	0.2206	3.6597	16.5883
14	12.8392	0.0779	59.1959	0.0169	4.6106	0.2169	3.8175	17.6008
15	15.4070	0.0649	72.0351	0.0139	4.6755	0.2139	3.9588	18.5095
16	18.4884	0.0541	87.4421	0.0114	4.7296	0.2114	4.0851	19.3208
17	22.1861	0.0451	105.9306	0.0094	4.7746	0.2094	4.1976	20.0419
18	26.6233	0.0376	128.1167	0.0078	4.8122	0.2078	4.2975	20.6805
19	31.9480	0.0313	154.7400	0.0065	4.8435	0.2065	4.3861	21.2439
20	38.3376	0.0261	186.6880	0.0054	4.8696	0.2054	4.4643	21.7395
21	46.0051	0.0217	225.0256	0.0044	4.8913	0.2044	4.5334	22.1742
22	55.2061	0.0181	271.0307	0.0037	4.9094	0.2037	4.5941	22.5546
23	66.2474	0.0151	326.2369	0.0031	4.9245	0.2031	4.6475	22.8867
24	79.4968	0.0126	392.4842	0.0025	4.9371	0.2025	4.6943	23.1760
25	95.3962	0.0105	471.9811	0.0021	4.9476	0.2021	4.7352	23.4276
26	114.4755	0.0087	567.3773	0.0018	4.9563	0.2018	4.7709	23.6460
27	137.3706	0.0073	681.8528	0.0015	4.9636	0.2015	4.8020	23.8353
28	164.8447	0.0061	819.2233	0.0012	4.9697	0.2012	4.8291	23.9991
29	197.8136	0.0051	984.0680	0.0010	4.9747	0.2010	4.8527	24.1406
30	237.3763	0.0042	1181.8816	0.0008	4.9789	0.2008	4.8731	24.2628
31	284.8516	0.0035	1419.2579	0.0007	4.9824	0.2007	4.8908	24.3681
32	341.8219	0.0029	1704.1095	0.0006	4.9854	0.2006	2.0961	24.4588
33	410.1863	0.0024	2045.9314	0.0005	4.9878	0.2005	4.9194	24.5368
34	492.2235	0.0020	2456.1176	0.0004	4.9898	0.2004	4.9308	24.6038
35	590.6682	0.0017	2948.3411	0.0003	4.9915	0.2003	4.9406	24.6614
40	1469.7716	0.0007	7343.8578	0.0001	4.9966	0.2001	4.9728	24.8469
45	3657.2620	0.0003	18281.3099	0.0001	4.9986	0.2001	4.9877	24.9316
50	9100.4382	0.0001	45497.1908	0.0000	4.9995	0.2000	4.9945	24.9698

TABLE A.14 Discrete Compounding: $i = 25\%$

	Single payment		Uniform series				Gradient series	
	Compound amount factor	Present worth factor	Compound amount factor	Sinking fund factor	Present worth factor	Capital recovery factor	Uniform series factor	Present worth factor
n	To find F given P $F\|P\ i,n$	To find P given F $P\|F\ i,n$	To find F given A $F\|A\ i,n$	To find A given F $A\|F\ i,n$	To find P given A $P\|A\ i,n$	To find A given P $A\|P\ i,n$	To find A given G $A\|G\ i,n$	To find P given G $P\|G\ i,n$
1	1.2500	0.8000	1.0000	1.0000	0.8000	1.2500	0.0000	0.0000
2	1.5625	0.6400	2.2500	0.4444	1.4400	0.6944	0.4444	0.6400
3	1.9531	0.5120	3.8125	0.2623	1.9520	0.5123	0.8525	1.6640
4	2.4414	0.4096	5.7656	0.1734	2.3616	0.4234	1.2249	2.8928
5	3.0518	0.3277	8.2070	0.1218	2.6893	0.3718	1.5631	4.2035
6	3.8147	0.2621	11.2588	0.0888	2.9514	0.3388	1.8683	5.5142
7	4.7684	0.2097	15.0735	0.0663	3.1611	0.3163	2.1424	6.7725
8	5.9605	0.1678	19.8419	0.0504	3.3289	0.3004	2.3872	7.9469
9	7.4506	0.1342	25.8023	0.0388	3.4631	0.2888	2.6048	9.0207
10	9.3132	0.1074	33.2529	0.0301	3.5705	0.2801	2.7971	9.9870
11	11.6415	0.0859	42.5661	0.0235	3.6564	0.2735	2.9663	10.8460
12	14.5519	0.0687	54.2077	0.0184	3.7251	0.2684	3.1145	11.6020
13	18.1899	0.0550	68.7596	0.0145	3.7801	0.2645	3.2437	12.2617
14	22.7374	0.0440	86.9495	0.0115	3.8241	0.2615	3.3559	12.8334
15	28.4217	0.0352	109.6868	0.0091	3.8593	0.2591	3.4530	13.3260
16	35.5271	0.0281	138.1085	0.0072	3.8874	0.2572	3.5366	13.7482
17	44.4089	0.0225	173.6357	0.0058	3.9099	0.2558	3.6084	14.1085
18	55.5112	0.0180	218.0446	0.0046	3.9279	0.2546	3.6698	14.4147
19	69.3889	0.0144	273.5558	0.0037	3.9424	0.2537	3.7222	14.6741
20	86.7362	0.0115	342.9447	0.0029	3.9539	0.2529	3.7667	14.8932
21	108.4202	0.0092	429.6809	0.0023	3.9631	0.2523	3.8045	15.0777
22	135.5253	0.0074	538.1011	0.0019	3.9705	0.2519	3.8365	15.2326
23	169.4066	0.0059	673.6264	0.0015	3.9764	0.2515	3.8634	15.3625
24	211.7582	0.0047	843.0329	0.0012	3.9811	0.2512	3.8861	15.4711
25	264.6978	0.0038	1054.7912	0.0009	3.9849	0.2509	3.9052	15.5618
26	330.8722	0.0030	1319.4890	0.0008	3.9879	0.2508	3.9212	15.6373
27	413.5903	0.0024	1650.3612	0.0006	3.9903	0.2506	3.9346	15.7002
28	516.9879	0.0019	2063.9515	0.0005	3.9923	0.2505	3.9457	15.7524
29	646.2349	0.0015	2580.9394	0.0004	3.9938	0.2504	3.9551	15.7957
30	807.7936	0.0012	3227.1743	0.0003	3.9950	0.2503	3.9628	15.8316
31	1009.7420	0.0010	4034.9678	0.0002	3.9960	0.2502	3.9693	15.8614
32	1262.1774	0.0008	5044.7098	0.0002	3.9968	0.2502	3.9746	15.8859
33	1577.7218	0.0006	6306.8872	0.0002	3.9975	0.2502	3.9791	15.9062
34	1972.1523	0.0005	7884.6091	0.0001	3.9980	0.2501	3.9828	15.9229
35	2465.1903	0.0004	9856.7613	0.0001	3.9984	0.2501	3.9858	15.9367

TABLE A.15 Discrete Compounding: $i = 5\%$

	Geometric series present worth factor, $(P\mid A_1\ i,j,n)$				
n	$j=4\%$	$j=6\%$	$j=8\%$	$j=10\%$	$j=15\%$
1	0.9524	0.9524	0.9524	0.9524	0.9524
2	1.8957	1.9138	1.9320	1.9501	1.9955
3	2.8300	2.8844	2.9396	2.9954	3.1379
4	3.7554	3.8643	3.9759	4.0904	4.3891
5	4.6721	4.8535	5.0419	5.2375	5.7595
6	5.5799	5.8521	6.1383	6.4393	7.2604
7	6.4792	6.8602	7.2661	7.6983	8.9043
8	7.3699	7.8779	8.4261	9.0173	10.7047
9	8.2521	8.9053	9.6192	10.3991	12.6765
10	9.1258	9.9425	10.8464	11.8467	14.8362
11	9.9913	10.9896	12.1087	13.3632	17.2016
12	10.8485	12.0466	13.4070	14.9519	19.7922
13	11.6976	13.1137	14.7425	16.6163	22.6295
14	12.5386	14.1910	16.1161	18.3599	25.7371
15	13.3715	15.2785	17.5289	20.1866	29.1407
16	14.1966	16.3764	18.9821	22.1002	32.8683
17	15.0137	17.4848	20.4769	24.1050	36.9510
18	15.8231	18.6037	22.0143	26.2052	41.4226
19	16.6248	19.7332	23.5956	28.4055	46.3200
20	17.4189	20.8736	25.2222	30.7105	51.6838
21	18.2054	22.0247	26.8952	33.1253	57.5584
22	18.9844	23.1869	28.6160	35.6550	63.9925
23	19.7559	24.3601	30.3860	38.3053	71.0394
24	20.5202	25.5445	32.2066	41.0817	78.7575
25	21.2771	26.7401	34.0791	43.9904	87.2106
26	22.0269	27.9472	36.0052	47.0375	96.4687
27	22.7695	29.1657	37.9863	50.2298	106.6086
28	23.5050	30.3959	40.0240	53.5741	117.7142
29	24.2335	31.6377	42.1199	57.0776	129.8774
30	24.9551	32.8914	44.2757	60.7480	143.1991
31	25.6698	34.1571	46.4931	64.5931	157.7895
32	26.3777	35.4348	48.7739	68.6213	173.7695
33	27.0789	36.7246	51.1198	72.8414	191.2713
34	27.7734	38.0267	53.5328	77.2624	210.4400
35	28.4612	39.3413	56.0146	81.8940	231.4343
36	29.1426	40.6683	58.5674	86.7461	254.4280
37	29.8174	42.0080	61.1932	91.8292	279.6116
38	30.4858	43.3605	63.8939	97.1544	307.1937
39	31.1478	44.7258	66.6719	102.7332	337.4026
40	31.8036	46.1042	69.5291	108.5776	370.4886
41	32.4531	47.4957	72.4681	114.7004	406.7256
42	33.0964	48.9004	75.4910	121.1147	446.4138
43	33.7335	50.3185	78.6002	127.8344	489.8817
44	34.3647	51.7501	81.7983	134.8742	537.4895
45	34.9898	53.1953	85.0878	142.2491	589.6314
46	35.6089	54.6543	88.4713	149.9753	646.7391
47	36.2221	56.1272	91.9514	158.0693	709.2857
48	36.8296	57.6141	95.5310	166.5488	777.7891
49	37.4312	59.1152	99.2128	175.4321	852.8167
50	38.0271	60.6306	102.9998	184.7384	934.9897

TABLE A.16 Discrete Compounding: $i = 5\%$

Geometric series future worth factor, $(F\dagger A_1\ i, j, n)$

n	j = 4%	j = 6%	j = 8%	j = 10%	j = 15%
1	1.0000	1.0000	1.0000	1.0000	1.0000
2	2.0900	2.1100	2.1300	2.1500	2.2000
3	3.2761	3.3391	3.4029	2.4675	3.6325
4	4.5648	4.6971	4.8328	4.9719	5.3350
5	5.9629	6.1944	6.4349	6.6846	7.3508
6	7.4777	7.8423	8.2260	8.6293	9.7297
7	9.1169	9.6530	10.2241	10.8323	12.5292
8	10.8886	11.6393	12.4492	13.3227	15.8157
9	12.8016	13.8151	14.9225	16.1324	19.6655
10	14.8650	16.1953	17.6677	19.2970	24.1666
11	17.0885	18.7959	20.7100	22.8555	29.4205
12	19.4824	21.6340	24.0771	26.8514	35.5439
13	22.0576	24.7279	27.7992	31.3324	42.6714
14	24.8255	28.0972	31.9087	36.3513	50.9577
15	27.7985	31.7630	36.4414	41.9664	60.5813
16	30.9893	35.7477	41.4356	48.2420	71.7475
17	34.4118	40.0754	46.9333	55.2490	84.6925
18	38.0803	44.7720	52.9800	63.0660	99.6883
19	42.0101	49.8649	59.6250	71.7792	117.0482
20	46.2175	55.3838	66.9220	81.4840	137.1324
21	50.7195	61.3601	74.9290	92.2857	160.3556
22	55.5342	67.8277	83.7093	104.3003	187.1948
23	60.6808	74.8226	93.3313	117.6556	218.1993
24	66.1796	82.3835	103.3694	132.4927	254.0008
25	72.0519	90.5516	115.4040	148.9670	295.3260
26	78.3203	99.3710	128.0227	167.2501	343.0112
27	85.0088	108.8890	141.3202	187.5308	398.0186
28	92.1426	119.1558	156.8992	210.0173	461.4548
29	99.7484	130.2252	173.3713	234.9391	534.5932
30	107.8545	142.1549	191.3572	262.5492	618.8983
31	116.4906	155.0061	210.9877	293.1261	716.0550
32	125.6883	168.8445	232.4047	326.9767	828.0013
33	135.4807	183.7401	255.7620	364.4393	956.9664
34	145.9032	199.7677	281.2262	405.8864	1105.5146
35	156.9926	217.0071	308.9776	451.7284	1276.5951
36	168.7884	235.5436	339.2119	502.4173	1473.6004
37	181.3317	255.4680	372.1406	558.4508	1700.4322
38	194.6664	276.8775	407.9933	620.3773	1961.5785
39	208.8385	299.8756	447.0182	688.8005	2262.2007
40	223.8968	324.5729	489.4844	764.3853	2608.2356
41	239.8927	351.0873	535.6832	847.8639	3006.5109
42	256.8804	379.5445	585.9298	940.0422	3464.8795
43	274.9172	410.0788	640.5658	1041.8080	3992.3730
44	294.0635	442.8332	699.9607	1154.1385	4599.3787
45	314.3832	477.9603	764.5147	1278.1095	5297.8426
46	335.9435	515.6229	834.6609	1414.9055	6101.5040
47	358.8155	555.9946	910.8680	1565.8303	7026.1639
48	383.0741	599.2602	993.6434	1732.3193	8089.9944
49	408.7984	645.6171	1083.5362	1915.9525	9313.8949
50	436.0716	695.2754	1181.1404	2118.4691	10721.9004

TABLE A.17 Discrete Compounding: $i = 8\%$

	Geometric series present worth factor, $(P\|A_1\ i,j,n)$				
n	$j=4\%$	$j=6\%$	$j=8\%$	$j=10\%$	$j=15\%$
1	0.9259	0.9259	0.9259	0.9259	0.9259
2	1.8176	1.8347	1.8519	1.8690	1.9119
3	2.6762	2.7267	2.7778	2.8295	2.9617
4	3.5030	3.6021	3.7037	3.8079	4.0796
5	4.2992	4.4613	4.6296	4.8043	5.2699
6	5.0659	5.3046	5.5556	5.8192	6.5374
7	5.8042	6.1323	6.4815	6.8529	7.8871
8	6.5151	6.9447	7.4074	7.9057	9.3242
9	7.1997	7.7420	8.3333	8.9780	10.8545
10	7.8590	8.5246	9.2593	10.0702	12.4839
11	8.4939	9.2926	10.1852	11.1826	14.2190
12	9.1052	10.0465	11.1111	12.3157	16.0665
13	9.6939	10.7863	12.0370	13.4696	18.0338
14	10.2608	11.5125	12.9630	14.6450	20.1286
15	10.8067	12.2252	13.8889	15.8421	22.3592
16	11.3324	12.9248	14.8148	17.0614	24.7343
17	11.8386	13.6114	15.7407	18.3033	27.2634
18	12.3260	14.2852	16.6667	19.5682	29.9564
19	12.7954	14.9466	17.5926	20.8565	32.8239
20	13.2475	15.5957	18.5185	22.1687	35.8773
21	13.6827	16.2329	19.4444	23.5051	39.1286
22	14.1019	16.8582	20.3704	24.8663	42.5906
23	14.5055	17.4719	21.2963	26.2527	46.2771
24	14.8942	18.0743	22.2222	27.6648	50.2024
25	15.2685	18.6655	23.1481	29.1031	54.3822
26	15.6289	19.2458	24.0741	30.5679	58.8329
27	15.9760	19.8153	25.0000	32.0599	63.5721
28	16.3102	20.3743	25.9259	33.5796	68.6184
29	16.6321	20.9229	26.8519	35.1273	73.9919
30	16.9420	21.4614	27.7778	36.7038	79.7136
31	17.2404	21.9899	28.7037	38.3094	85.8061
32	17.5278	22.5086	29.6296	39.9477	92.2935
33	17.8046	23.0177	30.5556	41.6104	99.2015
34	18.0711	23.5173	31.4815	43.3069	106.5571
35	18.3277	24.0078	32.4074	45.0348	114.3895
36	18.5748	24.4891	33.3333	46.7947	122.7296
37	18.8128	24.9615	34.2593	48.5872	131.6102
38	19.0419	25.4252	35.1852	50.4129	141.0664
39	19.2626	25.8803	36.1111	52.2724	151.1355
40	19.4751	26.3269	37.0370	54.1663	161.8573
41	19.6797	26.7653	37.9630	56.0953	173.2739
42	19.8768	27.1956	38.8889	58.0600	185.4306
43	20.0665	27.6179	39.8148	60.0611	198.3752
44	20.2493	28.0324	40.7407	62.0993	212.1587
45	20.4252	28.4392	41.6667	64.1752	226.8357
46	20.5946	28.8385	42.5926	66.2896	242.4639
47	20.7578	29.2304	43.5185	68.4431	259.1051
48	20.9149	29.6150	44.4444	70.6365	276.8249
49	21.0662	29.9925	45.3704	72.8705	2.95.6932
50	21.2119	30.3630	46.2963	75.1459	315.7844

TABLE A.18 Discrete Compounding: $i = 8\%$

Geometric series future worth factor, $(F\dagger A_1\ i, j, n)$

n	$j = 4\%$	$j = 6\%$	$j = 8\%$	$j = 10\%$	$j = 15\%$
1	1.0000	1.0000	1.0000	1.0000	1.0000
2	2.1200	2.1400	2.1600	2.1800	2.2300
3	3.3712	3.4348	3.4992	3.5644	3.7309
4	4.7658	4.9006	5.0388	5.1806	5.5502
5	6.3169	6.5551	6.8024	7.0591	7.7433
6	8.0389	8.4178	8.8160	9.2343	10.3741
7	9.9473	10.5097	11.1081	11.7446	13.5171
8	12.0590	12.8541	13.7106	14.6329	17.2585
9	14.3923	15.4763	16.6584	17.9472	21.6982
10	16.9670	18.4039	19.9900	21.7409	26.9519
11	19.8046	21.6670	23.7482	26.0739	33.1536
12	22.9284	25.2987	27.9797	31.0129	40.4583
13	26.3638	29.3348	32.7362	36.6324	49.0452
14	30.1379	33.8145	38.0747	43.0152	59.1216
15	34.2806	38.7805	44.0579	50.2540	70.9270
16	38.8240	44.2795	50.7547	58.4515	84.7383
17	43.8029	50.3623	58.2410	67.7226	100.8749
18	49.2551	57.0840	66.6003	78.1949	119.7062
19	55.2213	64.5051	75.9244	90.0104	141.6582
20	61.7459	72.6911	86.3140	103.3271	167.2226
21	68.8766	81.7135	97.8801	118.3208	196.9669
22	76.6655	91.6501	110.7443	135.1867	231.5458
23	85.1687	102.5857	125.0404	154.1419	271.7142
24	94.4469	114.6123	140.9151	175.4276	318.3428
25	104.5660	127.8302	158.5295	199.3115	372.4354
26	115.5971	142.3485	178.0604	226.0912	435.1492
27	127.6173	158.2858	199.7015	256.0966	507.8179
28	140.7101	175.7710	223.6657	289.6944	591.9787
29	154.9656	194.9443	250.1861	327.2909	689.4026
30	170.4815	215.9583	279.5182	369.3373	802.1302
31	187.3634	238.9784	311.9424	416.3337	932.5124
32	205.7256	264.1848	347.7654	468.8347	1083.2569
33	225.6917	291.7730	387.3237	527.4552	1257.4826
34	247.3954	321.9554	430.9857	592.8768	1458.7810
35	270.9814	354.9629	479.1547	665.8546	1691.2883
36	296.6060	391.0460	532.2724	747.2254	1959.7669
37	324.4384	430.4769	590.3224	837.9162	2269.7001
38	354.6616	473.5512	655.3338	938.9534	2627.4007
39	387.4733	520.5895	726.3857	1051.4740	3040.1361
40	423.0875	571.9402	804.6119	1176.7367	3516.2718
41	461.7355	627.9811	890.7054	1316.1349	4065.4371
42	503.6674	689.1225	985.4243	1471.2109	4698.7151
43	549.1536	755.8093	1089.5977	1643.6714	5428.8619
44	598.4864	828.5245	1204.1322	1835.4052	6270.5578
45	651.9818	907.7919	1330.0187	2048.5017	7240.6974
46	709.9816	994.1799	1468.3407	2285.2723	8358.7225
47	772.8549	1088.3048	1620.2820	2548.2737	9647.0049
48	841.0011	1190.8351	1787.1366	2840.3330	11131.2877
49	914.8517	1302.4957	1970.3181	3164.5769	12841.1914
50	994.8732	1424.0729	2171.3709	3524.4620	14810.7976

TABLE A.19 Discrete Compounding: $i = 10\%$

	Geometric series present worth factor, $(P\mid A_1\ i, j, n)$				
n	$j = 4\%$	$j = 6\%$	$j = 8\%$	$j = 10\%$	$j = 15\%$
1	0.9091	0.9091	0.9091	0.9091	0.9091
2	1.7686	1.7851	1.8017	1.8182	1.8595
3	2.5812	2.6293	2.6780	2.7273	2.8531
4	3.3495	3.4428	3.5384	3.6364	3.8919
5	4.0759	4.2267	4.3831	4.5455	4.9779
6	4.7627	4.9821	5.2125	5.4545	6.1133
7	5.4120	5.7100	6.0269	6.3636	7.3002
8	6.0259	6.4115	6.8264	7.2727	8.5411
9	6.6063	7.0874	7.6113	8.1818	9.8385
10	7.1550	7.7388	8.3820	9.0909	11.1948
11	7.6738	8.3664	9.1387	10.0000	12.6127
12	8.1644	8.9713	9.8817	10.9091	14.0951
13	8.6281	9.5542	10.6111	11.8182	15.6449
14	9.0666	10.1158	11.3273	12.7273	17.2651
15	9.4811	10.6571	12.0304	13.6364	18.9590
16	9.8731	11.1786	12.7208	14.5455	20.7298
17	10.2436	11.6812	13.3986	15.4545	22.5812
18	10.5940	12.1656	14.0640	16.3636	24.5167
19	10.9252	12.6323	14.7174	17.2727	26.5402
20	11.2384	13.0820	15.3589	18.1818	28.6556
21	11.5345	13.5154	15.9888	19.0909	30.8672
22	11.8144	13.9330	16.6071	20.0000	33.1794
23	12.0791	14.3354	17.2143	20.9091	35.5966
24	12.3293	14.7232	17.8104	21.8182	38.1238
25	12.5659	15.0969	18.3957	22.7273	40.7658
26	12.7896	15.4570	18.9703	23.6364	43.5278
27	13.0011	15.8041	19.5345	24.5455	46.4155
28	13.2010	16.1385	20.0884	25.4545	49.4343
29	13.3900	16.4607	20.6322	26.3636	52.5905
30	13.5688	16.7712	21.1662	27.2727	55.8900
31	13.7377	17.0704	21.6904	28.1818	59.3396
32	13.8975	17.3588	22.2052	29.0909	62.9459
33	14.0485	17.6367	22.7105	30.0000	66.7162
34	14.1913	17.9044	23.2067	30.9091	70.6578
35	14.3264	18.1624	23.6938	31.8182	74.7786
36	14.4540	18.4111	24.1721	32.7273	79.0867
37	14.5747	18.6507	24.6417	33.6364	83.5907
38	14.6888	18.8816	25.1028	34.5455	88.2994
39	14.7967	19.1040	25.5555	35.4545	93.2221
40	14.8987	19.3184	25.9999	36.3636	98.3685
41	14.9951	19.5250	26.4363	37.2727	103.7489
42	15.0863	19.7241	26.8647	38.1818	109.3739
43	15.1725	19.9160	27.2854	39.0909	115.2545
44	15.2540	20.1009	27.6983	40.0000	121.4024
45	15.3311	20.2790	28.1038	40.9091	127.8298
46	15.4039	20.4507	28.5019	41.8182	134.5493
47	15.4728	20.6161	28.8928	42.7273	141.5743
48	15.5379	20.7755	29.2766	43.6364	148.9186
49	15.5995	20.9291	29.6534	44.5455	156.5967
50	15.6577	21.0772	30.0233	45.4545	164.6238

TABLE A.20 Discrete Compounding: $i = 10\%$

Geometric series future worth factor, $(F\dagger A_1\ i, j, n)$

n	$j=4\%$	$j=6\%$	$j=8\%$	$j=10\%$	$j=15\%$
1	1.0000	1.0000	1.0000	1.0000	1.0000
2	2.1400	2.1600	2.1800	2.2000	2.2500
3	3.4356	3.4996	3.5644	3.6300	3.7975
4	4.9040	5.0406	5.1806	5.3240	5.6981
5	6.5643	6.8071	7.0591	7.3205	8.0169
6	8.4374	8.8260	9.2343	9.6631	10.8300
7	10.5464	11.1272	11.7446	12.4009	14.2261
8	12.9170	13.7435	14.6329	15.5897	18.3087
9	15.5773	16.7117	17.9472	19.2923	23.1986
10	18.5583	20.0724	21.7409	23.5795	29.0363
11	21.8944	23.8705	26.0739	28.5312	35.9855
12	25.6233	28.1558	31.0129	34.2374	44.2364
13	29.7866	32.9836	36.6324	40.7996	54.0103
14	34.4304	38.4149	43.0152	48.3318	65.5641
15	39.6051	44.5172	50.2540	56.9625	79.1963
16	45.3665	51.3655	58.4515	66.8360	95.2530
17	51.7762	59.0424	67.7226	78.1145	114.1359
18	58.9017	67.6395	78.1949	90.9805	136.3107
19	66.8177	77.2577	90.0104	105.6384	162.3173
20	75.6063	88.0091	103.3271	122.3182	192.7807
21	85.3580	100.0172	118.3208	141.2775	228.4254
22	96.1726	113.4184	135.1857	162.8055	270.0894
23	108.1598	128.3638	154.1419	187.2263	318.7431
24	121.4405	145.0200	175.4276	214.9033	375.5089
25	136.1478	163.5709	199.3115	246.2433	441.6849
26	152.4284	184.2198	226.0912	281.7024	518.7724
27	170.4438	207.1912	256.0966	321.7908	608.5064
28	190.3715	232.7327	289.6944	367.0798	712.8924
29	212.4074	261.1176	327.2909	418.2088	834.2472
30	236.7667	292.6478	369.3373	475.8928	975.2474
31	263.6868	327.6560	416.3337	540.9315	1138.9839
32	293.4286	366.5098	468.8347	614.2190	1329.0258
33	326.2796	409.6141	527.4552	696.7546	1549.4935
34	362.5559	457.4161	592.8768	789.6553	1805.1427
35	402.6058	510.4088	665.8546	894.1684	2101.4617
36	446.8125	569.1357	747.2254	1011.6877	2444.7834
37	495.5976	634.1965	837.9162	1143.7692	2842.4136
38	549.4255	706.2523	938.9534	1292.1500	3302.7796
39	608.8069	786.0318	1051.4740	1458.7694	3835.6009
40	674.3039	874.3384	1176.7367	1645.7911	4452.0858
41	746.5353	972.0580	1316.1349	1855.6295	5165.1579
42	826.1819	1080.1667	1471.2109	2090.9776	5989.7168
43	913.9929	1199.7404	1643.6714	2354.8391	6942.9380
44	1010.7927	1331.9649	1835.4052	2650.5630	8044.6188
45	1117.4868	1478.1468	2048.5017	2981.8834	9317.5757
46	1235.0785	1639.7261	2285.2723	3352.9622	10788.1025
47	1364.6612	1818.2892	2548.2737	3768.4380	12486.4975
48	1507.4451	2015.5841	2840.3330	4233.4793	14447.6696
49	1664.7601	2233.5363	3164.5769	4753.8445	16711.8372
50	1838.0695	2474.2675	3524.4620	5335.9479	19325.3318

TABLE A.21 Discrete Compounding: $i = 15\%$

Geometric series present worth factor, $(P|A_1\ i,j,n)$

n	$j=4\%$	$j=6\%$	$j=8\%$	$j=10\%$	$j=15\%$
1	0.8696	0.8696	0.8696	0.8696	0.8696
2	1.6560	1.6711	1.6862	1.7013	1.7391
3	2.3671	2.4099	2.4531	2.4969	2.6087
4	3.0103	3.0908	3.1734	3.2579	3.4783
5	3.5919	3.7185	3.8498	3.9858	4.3478
6	4.1179	4.2971	4.4850	4.6821	5.2174
7	4.5936	4.8303	5.0816	5.3481	6.0870
8	5.0237	5.3219	5.6418	5.9851	6.9565
9	5.4128	5.7749	6.1680	6.5945	7.8261
10	5.7647	6.1926	6.6621	7.1773	8.6957
11	6.0828	6.5775	7.1261	7.7348	9.5652
12	6.3705	6.9323	7.5619	8.2681	10.4348
13	6.6307	7.2593	7.9712	8.7782	11.3043
14	6.8660	7.5608	8.3556	9.2661	12.1739
15	7.0789	7.8386	8.7165	9.7328	13.0435
16	7.2713	8.0947	9.0555	10.1792	13.9130
17	7.4454	8.3308	9.3739	10.6062	14.7826
18	7.6028	8.5484	9.6729	11.0146	15.6522
19	7.7451	8.7489	9.9537	11.4053	16.5217
20	7.8738	8.9338	10.2173	11.7790	17.3913
21	7.9903	9.1042	10.4650	12.1364	18.2609
22	8.0955	9.2613	10.6976	12.4783	19.1304
23	8.1907	9.4060	10.9160	12.8053	20.0000
24	8.2768	9.5395	11.1211	13.1181	20.8696
25	8.3547	9.6625	11.3137	13.4173	21.7391
26	8.4251	9.7759	11.4946	13.7035	22.6087
27	8.4888	9.8803	11.6645	13.9773	23.4783
28	8.5464	9.9767	11.8241	14.2392	24.3478
29	8.5985	10.0655	11.9739	14.4896	25.2174
30	8.6456	10.1473	12.1146	14.7292	26.0870
31	8.6882	10.2227	12.2468	14.9584	26.9565
32	8.7267	10.2922	12.3709	15.1776	27.8261
33	8.7615	10.3563	12.4874	15.3873	28.6957
34	8.7930	10.4154	12.5969	15.5878	29.5652
35	8.8215	10.4698	12.6997	15.7796	30.4348
36	8.8473	10.5200	12.7962	15.9631	31.3043
37	8.8706	10.5663	12.8869	16.1386	32.1739
38	8.8917	10.6089	12.9720	16.3065	33.0435
39	8.9107	10.6482	13.0520	16.4671	33.9130
40	8.9280	10.6845	13.1271	16.6207	34.7826
41	8.9436	10.7178	13.1976	16.7676	35.6522
42	8.9576	10.7486	13.2639	16.9082	36.5217
43	8.9704	10.7770	13.3261	17.0426	37.3913
44	8.9819	10.8031	13.3845	17.1712	38.2609
45	8.9923	10.8272	13.4393	17.2942	39.1304
46	9.0018	10.8495	13.4908	17.4118	40.0000
47	9.0103	10.8699	13.5392	17.5244	40.8696
48	9.0180	10.8888	13.5847	17.6320	41.7391
49	9.0250	10.9062	13.6273	17.7350	42.6087
50	9.0313	10.9222	13.6674	17.8334	43.4783

TABLE A.22 Discrete Compounding: $i = 15\%$

	Geometric series future worth factor, $(F{\dagger}A_1\ i,j,n)$				
n	$j = 4\%$	$j = 6\%$	$j = 8\%$	$j = 10\%$	$j = 15\%$
1	1.0000	1.0000	1.0000	1.0000	1.0000
2	2.1900	2.2100	2.2300	2.2500	2.3000
3	3.6001	3.6651	3.7309	3.7975	3.9675
4	5.2650	5.4059	5.5502	5.6981	6.0835
5	7.2246	7.4792	7.7433	8.0169	8.7450
6	9.5249	9.9394	10.3741	10.8300	12.0681
7	12.2190	12.8488	13.5171	14.2261	16.1914
8	15.3678	16.2797	17.2585	18.3087	21.2802
9	19.0415	20.3155	21.6982	23.1986	27.5312
10	23.3210	25.0523	26.9519	29.0363	35.1788
11	28.2994	30.6010	33.1536	35.9855	44.5011
12	34.0838	37.0895	40.4583	44.2364	55.8287
13	40.7974	44.6651	49.0452	54.0103	69.5533
14	48.5821	53.4978	59.1216	65.5641	86.1390
15	57.6011	63.7834	70.9270	79.1963	106.1356
16	68.0422	75.7474	84.7383	95.2530	130.1930
17	80.1215	89.6499	100.8749	114.1359	159.0796
18	94.0876	105.7902	119.7062	136.3107	193.7028
19	110.2266	124.5130	141.6582	162.3173	235.1336
20	128.8674	146.2156	167.2226	192.7807	284.6354
21	150.3886	171.3550	196.9669	228.4254	343.6973
22	175.2257	200.4579	231.5458	270.0894	414.0734
23	203.8795	234.1301	271.7142	318.7431	497.8292
24	236.9261	273.0694	318.3428	375.5089	597.3950
25	275.0283	318.0787	372.4354	441.6849	715.6294
26	318.9484	370.0824	435.1492	518.7724	855.8928
27	369.5631	430.1441	507.8179	608.5064	1022.1335
28	427.8810	499.4881	591.9787	712.8924	1218.9888
29	495.0618	579.5230	689.4026	834.2472	1451.9027
30	572.4398	671.8698	802.1302	975.2474	1727.2636
31	661.5491	778.3937	932.5124	1138.9839	2052.5649
32	764.1546	901.2409	1083.2569	1329.0258	2436.5932
33	882.2859	1042.8804	1257.4826	1549.4935	2889.6473
34	1018.2772	1206.1531	1458.7810	1805.1427	3423.7942
35	1174.8130	1394.3271	1691.2883	2101.4617	4053.1681
36	1354.9811	1611.1622	1959.7669	2444.7834	4794.3188
37	1562.3322	1860.9838	2269.7001	2842.4136	5666.6185
38	1800.9501	2148.7675	2627.4007	3302.7796	6692.7359
39	2075.5314	2480.2368	3040.1361	3835.6009	7899.1896
40	2391.4775	2861.9759	3516.2718	4452.0858	9316.9929
41	2755.0002	3301.5580	4065.4371	5165.1579	10982.4054
42	3173.2432	3807.6945	4698.7151	5989.7168	12937.8093
43	3654.4225	4390.4057	5428.8619	6942.9380	15232.7302
44	4207.9864	5061.2171	6270.5578	8044.6188	17925.0267
45	4844.8008	5833.3851	7240.6974	9317.5757	21082.2757
46	5577.3622	6722.1575	8358.7225	10788.1025	24783.3864
47	6420.0413	7745.0716	9647.0049	12486.4975	29120.4790
48	7389.3653	8922.2982	11131.2877	14447.6696	34201.0732
49	8504.3406	10277.0368	12841.1914	16711.8372	40150.6349
50	9786.8251	11835.9699	14810.7976	19325.3318	47115.5409

Appendix B
Continuous Compounding

Section I — Continuous Compounding Interest Factors

Section II — Continuous Compounding, Continuous Flow Interest Factors

Section III — Geometric Series Factors

TABLE B.1 Continuous Compounding: $r = 1\%$

	Single payment		Uniform series				Gradient series	
	Compound amount factor	Present worth factor	Compound amount factor	Sinking fund factor	Present worth factor	Capital recovery factor	Uniform series factor	Present worth factor
n	To find F given P $F\|P\ r,n$	To find P given F $P\|F\ r,n$	To find F given A $F\|A\ r,n$	To find A given F $A\|F\ r,n$	To find P given A $P\|A\ r,n$	To find A given P $A\|P\ r,n$	To find A given G $A\|G\ r,n$	To find P given G $P\|G\ r,n$
1	1.0101	0.9900	1.0000	1.0000	0.9900	1.0101	0.0000	0.0000
2	1.0202	0.9802	2.0101	0.4975	1.9702	0.5076	0.4975	0.9802
3	1.0305	0.9704	3.0303	0.3300	2.9407	0.3401	0.9933	2.9211
4	1.0408	0.9608	4.0607	0.2463	3.9015	0.2563	1.4875	5.8035
5	1.0513	0.9512	5.1015	0.1960	4.8527	0.2061	1.9800	9.6084
6	1.0618	0.9418	6.1528	0.1625	5.7945	0.1726	2.4708	14.3172
7	1.0725	0.9324	7.2146	0.1386	6.7269	0.1487	2.9600	19.9116
8	1.0833	0.9231	8.2871	0.1207	7.6500	0.1307	3.4475	26.3734
9	1.0942	0.9139	9.3704	0.1067	8.5639	0.1168	3.9333	33.6848
10	1.1052	0.9048	10.4646	0.0956	9.4688	0.1056	4.4175	41.8284
11	1.1163	0.8958	11.5698	0.0864	10.3646	0.0965	4.9000	50.7867
12	1.1275	0.8869	12.6860	0.0788	11.2515	0.0889	5.3809	60.5428
13	1.1388	0.8781	13.8135	0.0724	12.1296	0.0824	5.8600	71.0800
14	1.1503	0.8694	14.9524	0.0669	12.9990	0.0769	6.3376	82.3816
15	1.1618	0.8607	16.1026	0.0621	13.8597	0.0722	6.8134	94.4315
16	1.1735	0.8521	17.2645	0.0579	14.7118	0.0680	7.2876	107.2137
17	1.1853	0.8437	18.4380	0.0542	15.5555	0.0643	7.7601	120.7123
18	1.1972	0.8353	19.6233	0.0510	16.3908	0.0610	8.2310	134.9119
19	1.2092	0.8270	20.8205	0.0480	17.2177	0.0581	8.7002	149.7972
20	1.2214	0.8187	22.0298	0.0454	18.0364	0.0554	9.1677	165.3531
21	1.2337	0.8106	23.2512	0.0430	18.8470	0.0531	9.6336	181.5648
22	1.2461	0.8025	24.4848	0.0408	19.6495	0.0509	10.0978	198.4177
23	1.2586	0.7945	25.7309	0.0389	20.4441	0.0489	10.5604	215.8974
24	1.2712	0.7866	26.9895	0.0371	21.2307	0.0471	11.0213	233.9898
25	1.2840	0.7788	28.2608	0.0354	22.0095	0.0454	11.4805	252.6811
26	1.2969	0.7711	29.5448	0.0338	22.7806	0.0439	11.9381	271.9573
27	1.3100	0.7634	30.8417	0.0324	23.5439	0.0425	12.3941	291.8052
28	1.3231	0.7558	32.1517	0.0311	24.2997	0.0412	12.8484	312.2114
29	1.3364	0.7483	33.4748	0.0299	25.0480	0.0399	13.3010	333.1628
30	1.3499	0.7408	34.8112	0.0287	25.7888	0.0388	13.7520	354.6465
31	1.3634	0.7334	36.1611	0.0277	26.5222	0.0377	14.2013	376.6499
32	1.3771	0.7261	37.5245	0.0266	27.2484	0.0367	14.6490	399.1605
33	1.3910	0.7189	38.9017	0.0257	27.9673	0.0358	15.0950	422.1661
34	1.4049	0.7118	40.2926	0.0248	28.6791	0.0349	15.5394	445.6545
35	1.4191	0.7047	41.6976	0.0240	29.3838	0.0340	15.9821	469.6139
40	1.4918	0.6703	48.9370	0.0204	32.8034	0.0305	18.1710	596.0725
45	1.5683	0.6376	56.5475	0.0177	36.0563	0.0277	20.3190	732.6280
50	1.6487	0.6065	64.5483	0.0155	39.1505	0.0255	22.4261	877.9948
55	1.7333	0.5769	72.9593	0.0137	42.0938	0.0238	24.4926	1030.9885
60	1.8221	0.5488	81.8015	0.0122	44.8936	0.0223	26.5187	1190.5195
65	1.9155	0.5220	91.0971	0.0110	47.5568	0.0210	28.5045	1355.5862
70	2.0138	0.4966	100.8692	0.0099	50.0902	0.0200	30.4505	1525.2692
75	2.1170	0.4724	111.1424	0.0090	52.5000	0.0190	32.3567	1698.7256
80	2.2255	0.4493	121.9423	0.0082	54.7922	0.0183	34.2235	1875.1837
85	2.3396	0.4274	133.2960	0.0075	56.9727	0.0176	36.0513	2053.9382
90	2.4596	0.4066	145.2317	0.0069	59.0468	0.0169	37.8402	2234.3453
95	2.5857	0.3867	157.7794	0.0063	61.0198	0.0164	39.5907	2415.8187
100	2.7183	0.3679	170.9705	0.0058	62.8965	0.0159	41.3032	2597.8253

TABLE B.2 Continuous Compounding: $r = 2\%$

	Single payment		Uniform series				Gradient series	
	Compound amount factor	Present worth factor	Compound amount factor	Sinking fund factor	Present worth factor	Capital recovery factor	Uniform series factor	Present worth factor
n	To find F given P $F\mid P\ r,n$	To find P given F $P\mid F\ r,n$	To find F given A $F\mid A\ r,n$	To find A given F $A\mid F\ r,n$	To find P given A $P\mid A\ r,n$	To find A given P $A\mid P\ r,n$	To find A given G $A\mid G\ r,n$	To find P given G $P\mid G\ r,n$
1	1.0202	0.9802	1.0000	1.0000	0.9802	1.0202	0.0000	0.0000
2	1.0408	0.9608	2.0202	0.4950	1.9410	0.5152	0.4950	0.9608
3	1.0618	0.9418	3.0610	0.3267	2.8828	0.3469	0.9867	2.8443
4	1.0833	0.9231	4.1228	0.2426	3.8059	0.2628	1.4750	5.6137
5	1.1052	0.9048	5.2061	0.1921	4.7107	0.2123	1.9600	9.2330
6	1.1275	0.8869	6.3113	0.1584	5.5976	0.1786	2.4417	13.6676
7	1.1503	0.8694	7.4388	0.1344	6.4670	0.1546	2.9200	18.8838
8	1.1735	0.8521	8.5891	0.1164	7.3191	0.1366	3.3950	24.8488
9	1.1972	0.8353	9.7626	0.1024	8.1544	0.1226	3.8667	31.5309
10	1.2214	0.8187	10.9598	0.0912	8.9731	0.1114	4.3351	38.8995
11	1.2461	0.8025	12.1812	0.0821	9.7756	0.1023	4.8002	46.9247
12	1.2712	0.7866	13.4273	0.0745	10.5623	0.0947	5.2619	55.5776
13	1.2969	0.7711	14.6985	0.0680	11.3333	0.0882	5.7203	64.8302
14	1.3231	0.7558	15.9955	0.0625	12.0891	0.0827	6.1754	74.6554
15	1.3499	0.7408	17.3186	0.0577	12.8299	0.0779	6.6272	85.0269
16	1.3771	0.7261	18.6685	0.0536	13.5561	0.0738	7.0757	95.9191
17	1.4049	0.7118	20.0456	0.0499	14.2678	0.0701	7.5209	107.3074
18	1.4333	0.6977	21.4505	0.0466	14.9655	0.0668	7.9628	119.1679
19	1.4623	0.6839	22.8839	0.0437	15.6494	0.0639	8.4014	131.4774
20	1.4918	0.6703	24.3461	0.0411	16.3197	0.0613	8.8368	144.2135
21	1.5220	0.6570	25.8380	0.0387	16.9768	0.0589	9.2688	157.3545
22	1.5527	0.6440	27.3599	0.0365	17.6208	0.0568	9.6976	170.8792
23	1.5841	0.6313	28.9126	0.0346	18.2521	0.0548	10.1231	184.7675
24	1.6161	0.6188	30.4967	0.0328	18.8709	0.0530	10.5453	198.9995
25	1.6487	0.6065	32.1128	0.0311	19.4774	0.0513	10.9643	213.5562
26	1.6820	0.5945	33.7615	0.0296	20.0719	0.0498	11.3800	228.4192
27	1.7160	0.5827	35.4435	0.0282	20.6547	0.0484	11.7925	243.5707
28	1.7507	0.5712	37.1595	0.0269	21.2259	0.0471	12.2018	258.9933
29	1.7860	0.5599	38.9102	0.0257	21.7858	0.0459	12.6078	274.6705
30	1.8221	0.5488	40.6963	0.0246	22.3346	0.0448	13.0106	290.5860
31	1.8589	0.5379	42.5184	0.0235	22.8725	0.0437	13.4102	306.7243
32	1.8965	0.5273	44.3773	0.0225	23.3998	0.0427	13.8065	323.0704
33	1.9348	0.5169	46.2738	0.0216	23.9167	0.0418	14.1997	339.6097
34	1.9739	0.5066	48.2086	0.0207	24.4233	0.0409	14.5897	356.3280
35	2.0138	0.4966	50.1824	0.0199	24.9199	0.0401	14.9765	373.2119
40	2.2255	0.4493	60.6663	0.0165	27.2591	0.0367	16.8630	459.6713
45	2.4596	0.4066	72.2528	0.0138	29.3758	0.0340	18.6714	548.4862
50	2.7183	0.3679	85.0578	0.0118	31.2910	0.0320	20.4028	638.4254
55	3.0042	0.3329	99.2096	0.0101	33.0240	0.0303	22.0588	728.4707
60	3.3201	0.3012	114.8497	0.0087	34.5921	0.0289	23.6409	817.7873
65	3.6693	0.2725	132.1346	0.0076	36.0109	0.0278	25.1507	905.6984
70	4.0552	0.2466	151.2375	0.0066	37.2947	0.0268	26.5899	991.6629
75	4.4817	0.2231	172.3494	0.0058	38.4564	0.0260	27.9604	1075.2549
80	4.9530	0.2019	195.6817	0.0051	39.5075	0.0253	29.2640	1156.1476
85	5.4739	0.1827	221.4679	0.0045	40.4585	0.0247	30.5028	1234.0977
90	6.0496	0.1653	249.9660	0.0040	41.3191	0.0242	31.6786	1308.9328
95	6.6859	0.1496	281.4613	0.0036	42.0978	0.0238	32.7937	1380.5397
100	7.3891	0.1353	316.2689	0.0032	42.8023	0.0234	33.8499	1448.8552

TABLE B.3 Continuous Compounding: $r = 3\%$

	Single payment		Uniform series				Gradient series	
	Compound amount factor	Present worth factor	Compound amount factor	Sinking fund factor	Present worth factor	Capital recovery factor	Uniform series factor	Present worth factor
n	To find F given P $F/P\ r,n$	To find P given F $P/F\ r,n$	To find F given A $F/A\ r,n$	To find A given F $A/F\ r,n$	To find P given A $P/A\ r,n$	To find A given P $A/P\ r,n$	To find A given G $A/G\ r,n$	To find P given G $P/G\ r,n$
1	1.0305	0.9704	1.0000	1.0000	0.9704	1.0305	0.0000	0.0000
2	1.0618	0.9418	2.0305	0.4925	1.9122	0.5230	0.4925	0.9418
3	1.0942	0.9139	3.0923	0.3234	2.8261	0.3538	0.9800	2.7696
4	1.1275	0.8869	4.1865	0.2389	3.7131	0.2693	1.4625	5.4304
5	1.1618	0.8607	5.3140	0.1882	4.5738	0.2186	1.9400	8.8732
6	1.1972	0.8353	6.4758	0.1544	5.4090	0.1840	2.4125	13.0496
7	1.2337	0.8106	7.6730	0.1303	6.2196	0.1608	2.8801	17.9131
8	1.2712	0.7866	8.9067	0.1123	7.0063	0.1427	3.3427	23.4195
9	1.3100	0.7634	10.1779	0.0983	7.7696	0.1287	3.8002	29.5265
10	1.3499	0.7408	11.4879	0.0870	8.5104	0.1175	4.2529	36.1939
11	1.3910	0.7189	12.8378	0.0779	9.2294	0.1083	4.7005	43.3831
12	1.4333	0.6977	14.2287	0.0703	9.9270	0.1007	5.1433	51.0575
13	1.4770	0.6771	15.6621	0.0638	10.6041	0.0943	5.5811	59.1822
14	1.5220	0.6570	17.1390	0.0583	11.2612	0.0888	6.0139	67.7238
15	1.5683	0.6376	18.6610	0.0536	11.8988	0.0840	6.4419	76.6506
16	1.6161	0.6188	20.2293	0.0494	12.5176	0.0799	6.8649	85.9324
17	1.6653	0.6005	21.8454	0.0458	13.1181	0.0762	7.2831	95.5403
18	1.7160	0.5827	23.5107	0.0425	13.7008	0.0730	7.6964	105.4470
19	1.7683	0.5655	25.2267	0.0396	14.2663	0.0701	8.1048	115.6265
20	1.8221	0.5488	26.9950	0.0370	14.8151	0.0675	8.5084	126.0539
21	1.8776	0.5326	28.8171	0.0347	15.3477	0.0652	8.9072	136.7057
22	1.9348	0.5169	30.6947	0.0326	15.8646	0.0630	9.3012	147.5596
23	1.9937	0.5016	32.6295	0.0306	16.3662	0.0611	9.6904	158.5943
24	2.0544	0.4868	34.6232	0.0289	16.8529	0.0593	10.0748	169.7896
25	2.1170	0.4724	36.6776	0.0273	17.3253	0.0577	10.4545	181.1264
26	2.1815	0.4584	38.7946	0.0258	17.7837	0.0562	10.8294	192.5866
27	2.2479	0.4449	40.9761	0.0244	18.2285	0.0549	11.1996	204.1529
28	2.3164	0.4317	43.2240	0.0231	18.6603	0.0536	11.5652	215.8090
29	2.3869	0.4190	45.5404	0.0220	19.0792	0.0524	11.9261	227.5397
30	2.4596	0.4066	47.9273	0.0209	19.4858	0.0513	12.2823	239.3302
31	2.5345	0.3946	50.3869	0.0198	19.8803	0.0503	12.6339	251.1668
32	2.6117	0.3829	52.9214	0.0189	20.2632	0.0494	12.9810	263.0365
33	2.6912	0.3716	55.5331	0.0180	20.6348	0.0485	13.3235	274.9270
34	2.7732	0.3606	58.2243	0.0172	20.9954	0.0476	13.6614	286.8266
35	2.8577	0.3499	60.9975	0.0164	21.3453	0.0468	13.9948	298.7245
40	3.3201	0.3012	76.1830	0.0131	22.9459	0.0436	15.5953	357.8483
45	3.8574	0.2592	93.8259	0.0107	24.3235	0.0411	17.0874	415.6245
50	4.4817	0.2231	114.3242	0.0087	25.5092	0.0392	18.4750	471.2816
55	5.2070	0.1920	138.1397	0.0072	26.5297	0.0377	19.7623	524.2887
60	6.0496	0.1653	165.8094	0.0060	27.4081	0.0365	20.9538	574.3044
65	7.0287	0.1423	197.9570	0.0051	28.1641	0.0355	22.0541	621.1335
70	8.1662	0.1225	235.3072	0.0042	28.8149	0.0347	23.0677	664.6933
75	9.4877	0.1054	278.7019	0.0036	29.3750	0.0340	23.9996	704.9860
80	11.0232	0.0907	329.1193	0.0030	29.8570	0.0335	24.8543	742.0766
85	12.8071	0.0781	287.6961	0.0026	30.2720	0.0330	25.6368	776.0754
90	14.8797	0.0672	455.7526	0.0022	30.6291	0.0326	26.3516	807.1241
95	17.2878	0.0578	534.8229	0.0019	30.9365	0.0323	27.0032	835.3849
100	20.0855	0.0498	626.6895	0.0016	31.2010	0.0321	27.5963	861.0319

TABLE B.4 Continuous Compounding: $r = 4\%$

	Single payment		Uniform series				Gradient series	
	Compound amount factor	Present worth factor	Compound amount factor	Sinking fund factor	Present worth factor	Capital recovery factor	Uniform series factor	Present worth factor
n	To find F given P $F\|P\ r,n$	To find P given F $P\|F\ r,n$	To find F given A $F\|A\ r,n$	To find A given F $A\|F\ r,n$	To find P given A $P\|A\ r,n$	To find A given P $A\|P\ r,n$	To find A given G $A\|G\ r,n$	To find P given G $P\|G\ r,n$
1	1.0408	0.9608	1.0000	1.0000	0.9608	1.0408	0.0000	0.0000
2	1.0833	0.9231	2.0408	0.4900	1.8839	0.5308	0.4900	0.9231
3	1.1275	0.8869	3.1241	0.3201	2.7708	0.3609	0.9733	2.6970
4	1.1735	0.8521	4.2516	0.2352	3.6230	0.2760	1.4500	5.2534
5	1.2214	0.8187	5.4251	0.1843	4.4417	0.2251	1.9201	8.5238
6	1.2712	0.7866	6.6465	0.1505	5.2283	0.1913	2.3834	12.4615
7	1.3231	0.7558	7.9178	0.1263	5.9841	0.1671	2.8402	16.9962
8	1.3771	0.7261	9.2409	0.1082	6.7103	0.1490	3.2904	22.0792
9	1.4333	0.6977	10.6180	0.0942	7.4079	0.1350	3.7339	27.6606
10	1.4918	0.6703	12.0513	0.0830	8.0783	0.1238	4.1709	33.6935
11	1.5527	0.6440	13.5432	0.0738	8.7223	0.1146	4.6013	40.1339
12	1.6161	0.6188	15.0959	0.0662	9.3411	0.1071	5.0252	46.9405
13	1.6820	0.5945	16.7120	0.0598	9.9356	0.1006	5.4425	54.0747
14	1.7507	0.5712	18.3940	0.0544	10.5068	0.0952	5.8534	61.5004
15	1.8221	0.5488	20.1447	0.0496	11.0556	0.0905	6.2578	69.1838
16	1.8965	0.5273	21.9668	0.0455	11.5829	0.0863	6.6558	77.0932
17	1.9739	0.5066	23.8633	0.0419	12.0895	0.0827	7.0473	85.1991
18	2.0544	0.4868	25.8371	0.0387	12.5763	0.0795	7.4326	93.4738
19	2.1383	0.4677	27.8916	0.0359	13.0439	0.0767	7.8114	101.8918
20	2.2255	0.4493	30.0298	0.0333	13.4933	0.0741	8.1840	110.4291
21	2.3164	0.4317	32.2554	0.0310	13.9250	0.0718	8.5503	119.0633
22	2.4109	0.4148	34.5717	0.0289	14.3398	0.0697	8.9104	127.7737
23	2.5093	0.3985	36.9826	0.0270	14.7383	0.0679	9.2644	136.5412
24	2.6117	0.3829	39.4919	0.0253	15.1212	0.0661	9.6122	145.3477
25	2.7183	0.3679	42.1036	0.0238	15.4891	0.0646	9.9539	154.1768
26	2.8292	0.3535	44.8219	0.0223	15.8425	0.0631	10.2896	163.0132
27	2.9447	0.3396	47.6511	0.0210	16.1821	0.0618	10.6193	171.8427
28	3.0649	0.3263	50.5958	0.0198	16.5084	0.0606	10.9431	180.6522
29	3.1899	0.3135	53.6607	0.0186	16.8219	0.0594	11.2609	189.4298
30	3.3201	0.3012	56.8506	0.0176	17.1231	0.0584	11.5730	198.1645
31	3.4556	0.2894	60.1707	0.0166	17.4125	0.0574	11.8792	206.8460
32	3.5966	0.2780	63.6263	0.0157	17.6905	0.0565	12.1797	215.4651
33	3.7434	0.2671	67.2230	0.0149	17.9576	0.0557	12.4746	224.0135
34	3.8962	0.2567	70.9664	0.0141	18.2143	0.0549	12.7638	232.4833
35	4.0552	0.2466	74.8626	0.0134	18.4609	0.0542	13.0475	240.8676
40	4.9530	0.2019	96.8625	0.0103	19.5562	0.0511	14.3845	281.3065
45	6.0496	0.1653	123.7332	0.0081	20.4530	0.0489	15.5918	318.8989
50	7.3891	0.1353	156.5532	0.0064	21.1872	0.0472	16.6775	353.3480
55	9.0250	0.1108	196.6396	0.0051	21.7883	0.0459	17.6498	382.5581
60	11.0232	0.0907	245.6012	0.0041	22.2804	0.0449	18.5172	412.5715
65	13.4637	0.0743	305.4031	0.0033	22.6834	0.0441	19.2882	437.5217
70	16.4446	0.0608	378.4453	0.0026	23.0133	0.0435	19.9710	459.5987
75	20.0855	0.0498	467.6593	0.0021	23.2834	0.0429	20.5737	479.0243
80	24.5325	0.0408	576.6254	0.0017	23.5045	0.0425	21.1038	496.0344
85	29.9641	0.0334	709.7170	0.0014	23.6856	0.0422	21.5687	510.8663
90	36.5982	0.0273	872.2754	0.0011	23.8338	0.0420	21.9751	523.7508
95	44.7012	0.0224	1070.8247	0.0009	23.9552	0.0417	22.3295	534.9066
100	54.5982	0.0183	1313.3333	0.0008	24.0545	0.0416	22.6376	544.5370

TABLE B.5 Continuous Compounding: $r = 5\%$

	Single payment		Uniform series				Gradient series	
	Compound amount factor	Present worth factor	Compound amount factor	Sinking fund factor	Present worth factor	Capital recovery factor	Uniform series factor	Present worth factor
n	To find F given P $F/P\ r,n$	To find P given F $P/F\ r,n$	To find F given A $F/A\ r,n$	To find A given F $A/F\ r,n$	To find P given A $P/A\ r,n$	To find A given P $A/P\ r,n$	To find A given G $A/G\ r,n$	To find P given G $P/G\ r,n$
1	1.0513	0.9512	1.0000	1.0000	0.9512	1.0513	0.0000	0.0000
2	1.1052	0.9048	2.0513	0.4875	1.8561	0.5388	0.4875	0.9048
3	1.1618	0.8607	3.1564	0.3168	2.7168	0.3681	0.9667	2.6263
4	1.2214	0.8187	4.3183	0.2316	3.5355	0.2828	1.4375	5.0824
5	1.2840	0.7788	5.5397	0.1805	4.3143	0.2318	1.9001	8.1976
6	1.3499	0.7408	6.8237	0.1465	5.0551	0.1978	2.3544	11.9017
7	1.4191	0.7047	8.1736	0.1223	5.7598	0.1736	2.8004	16.1299
8	1.4918	0.6703	9.5926	0.1042	6.4301	0.1555	3.2382	20.8221
9	1.5683	0.6376	11.0845	0.0902	7.0678	0.1415	3.6678	25.9231
10	1.6487	0.6065	12.6528	0.0790	7.6743	0.1303	4.0892	31.3819
11	1.7333	0.5769	14.3015	0.0699	8.2512	0.1212	4.5025	37.1514
12	1.8221	0.5488	16.0347	0.0624	8.8001	0.1136	4.9077	43.1883
13	1.9155	0.5220	17.8569	0.0560	9.3221	0.1073	5.3049	49.4529
14	2.0138	0.4966	19.7724	0.0506	9.8187	0.1018	5.6941	55.9085
15	2.1170	0.4724	21.7862	0.0459	10.2911	0.0972	6.0753	62.5216
16	2.2255	0.4493	23.9032	0.0418	10.7404	0.0931	6.4487	69.2616
17	2.3396	0.4274	26.1287	0.0383	11.1678	0.0895	6.8143	76.1002
18	2.4596	0.4066	28.4683	0.0351	11.5744	0.0864	7.1720	83.0119
19	2.5857	0.3867	30.9279	0.0323	11.9611	0.0836	7.5221	89.9732
20	2.7183	0.3679	33.5137	0.0298	12.3290	0.0811	7.8646	96.9629
21	2.8577	0.3499	36.2319	0.0276	12.6789	0.0789	8.1996	103.9617
22	3.0042	0.3329	39.0896	0.0256	13.0118	0.0769	8.5270	110.9520
23	3.1582	0.3166	42.0938	0.0238	13.3284	0.0750	8.8471	117.9180
24	3.3201	0.3012	45.2519	0.0221	13.6296	0.0734	9.1599	124.8455
25	3.4903	0.2865	48.5721	0.0206	13.9161	0.0719	9.4654	131.7216
26	3.6693	0.2725	52.0624	0.0192	14.1887	0.0705	9.7638	138.5349
27	3.8574	0.2592	55.7317	0.0179	14.4479	0.0692	10.0551	145.2751
28	4.0552	0.2466	59.5891	0.0168	14.6945	0.0681	10.3395	151.9332
29	4.2631	0.2346	63.6443	0.0157	14.9291	0.0670	10.6170	158.5012
30	4.4817	0.2231	67.9074	0.0147	15.1522	0.0660	10.8877	164.9720
31	4.7115	0.2122	72.3891	0.0138	15.3644	0.0651	11.1517	171.3394
32	4.9530	0.2019	77.1006	0.0130	15.5663	0.0642	11.4091	177.5982
33	5.2070	0.1920	82.0536	0.0122	15.7584	0.0635	11.6601	183.7438
34	5.4739	0.1827	87.2606	0.0115	15.9411	0.0627	11.9046	189.7724
35	5.7546	0.1738	92.7346	0.0108	16.1149	0.0621	12.1429	195.6807
40	7.3891	0.1353	124.6132	0.0081	16.8646	0.0593	13.2435	223.3452
45	9.4877	0.1054	165.5462	0.0060	17.4484	0.0573	14.2024	247.8097
50	12.1825	0.0821	218.1052	0.0046	17.9032	0.0559	15.0329	269.1364
55	15.6426	0.0639	285.5923	0.0035	18.2573	0.0548	15.7480	287.5163
60	20.0855	0.0498	372.2475	0.0027	18.5331	0.0540	16.3604	303.2096
65	25.7903	0.0388	483.5149	0.0021	18.7479	0.0533	16.8822	316.5055
70	33.1155	0.0302	626.3851	0.0016	18.9152	0.0529	17.3245	327.6968
75	42.5211	0.0235	809.8341	0.0012	19.0455	0.0525	17.6979	337.0640
80	54.5982	0.0183	1045.3872	0.0010	19.1469	0.0522	18.0116	344.8665
85	70.1054	0.0143	1347.8435	0.0007	19.2260	0.0520	18.2742	351.3382
90	90.0171	0.0111	1736.2049	0.0006	19.2875	0.0518	18.4931	356.6861
95	115.5843	0.0087	2234.8710	0.0004	19.3354	0.0517	18.6751	361.0906

TABLE B.6 Continuous Compounding: $r = 6\%$

	Single payment		Uniform series				Gradient series	
	Compound amount factor	Present worth factor	Compound amount factor	Sinking fund factor	Present worth factor	Capital recovery factor	Uniform series factor	Present worth factor
n	To find F given P $F\|P\ r,n$	To find P given F $P\|F\ r,n$	To find F given A $F\|A\ r,n$	To find A given F $A\|F\ r,n$	To find P given A $P\|A\ r,n$	To find A given P $A\|P\ r,n$	To find A given G $A\|G\ r,n$	To find P given G $P\|G\ r,n$
1	1.0618	0.9418	1.0000	1.0000	0.9148	1.0618	0.0000	0.0000
2	1.1275	0.8869	2.0618	0.4850	1.8287	0.5468	0.4850	0.8869
3	1.1972	0.8353	3.1893	0.3135	2.6640	0.3754	0.9600	2.5575
4	1.2712	0.7866	4.3866	0.2280	3.4506	0.2898	1.4251	4.9173
5	1.3499	0.7408	5.6578	0.1767	4.1914	0.2386	1.8802	7.8806
6	1.4333	0.6977	7.0077	0.1427	4.8891	0.2045	2.3254	11.3690
7	1.5220	0.6570	8.4410	0.1185	5.5461	0.1803	2.7607	15.3113
8	1.6161	0.6188	9.9629	0.1004	6.1649	0.1622	3.1862	19.6428
9	1.7160	0.5827	11.5790	0.0864	6.7477	0.1482	3.6020	24.3047
10	1.8221	0.5488	13.2950	0.0752	7.2965	0.1371	4.0080	29.2441
11	1.9348	0.5169	15.1171	0.0662	7.8133	0.1280	4.4043	34.4126
12	2.0544	0.4868	17.0519	0.0586	8.3001	0.1205	4.7911	39.7668
13	2.1815	0.4584	19.1064	0.0523	8.7585	0.1142	5.1684	45.2677
14	2.3164	0.4317	21.2878	0.0470	9.1902	0.1088	5.5363	50.8800
15	2.4596	0.4066	23.6042	0.0424	9.5968	0.1042	5.8949	56.5719
16	2.6117	0.3829	26.0638	0.0384	9.9797	0.1002	6.2442	62.3153
17	2.7732	0.3606	28.6755	0.0349	10.3402	0.0967	6.5845	68.0848
18	2.9447	0.3396	31.4487	0.0318	10.6798	0.0936	6.9156	73.8580
19	3.1268	0.3198	34.3934	0.0291	10.9997	0.0909	7.2379	79.6147
20	3.3201	0.3012	37.5202	0.0267	11.3009	0.0885	7.5514	85.3374
21	3.5254	0.2837	40.8403	0.0245	11.5845	0.0863	7.8562	91.0105
22	3.7434	0.2671	44.3657	0.0225	11.8516	0.0844	8.1525	96.6203
23	3.9749	0.2516	48.1091	0.0208	12.1032	0.0826	8.4403	102.1550
24	4.2207	0.2369	52.0840	0.0192	12.3401	0.0810	8.7199	107.6044
25	4.4817	0.2231	56.3047	0.0178	12.5633	0.0796	8.9912	112.9595
26	4.7588	0.2101	60.7864	0.0165	12.7734	0.0783	9.2546	118.2129
27	5.0531	0.1979	65.5452	0.0153	12.9713	0.0771	9.5101	123.3583
28	5.3656	0.1864	70.5983	0.0142	13.1577	0.0760	9.7578	128.3904
29	5.6973	0.1755	75.9639	0.0132	13.3332	0.0750	9.9980	133.3049
30	6.0496	0.1653	81.6612	0.0122	13.4985	0.0741	10.2307	138.0986
31	6.4237	0.1557	87.7109	0.0114	13.6542	0.0732	10.4560	142.7688
32	6.8210	0.1466	94.1346	0.0106	13.8008	0.0725	10.6743	147.3136
33	7.2427	0.1381	100.9556	0.0099	13.9389	0.0717	10.8855	151.7318
34	7.6906	0.1300	108.1983	0.0092	14.0689	0.0711	11.0899	156.0228
35	8.1662	0.1225	115.8889	0.0086	14.1913	0.0705	11.2876	160.1863
40	11.0232	0.0907	162.0915	0.0062	14.7046	0.0680	12.1809	179.1156
45	14.8797	0.0672	224.4584	0.0045	15.0848	0.0663	12.9295	195.0399
50	20.0855	0.0498	308.6449	0.0032	15.3665	0.0651	13.5519	208.2453
55	27.1126	0.0369	422.2849	0.0024	15.5752	0.0642	14.0654	219.0716
60	36.5982	0.0273	575.6828	0.0017	15.7298	0.0636	14.4862	227.8648
65	49.4024	0.0202	782.7483	0.0013	15.8443	0.0631	14.8288	234.9516
70	66.6863	0.0150	1062.2574	0.0009	15.9292	0.0628	15.1060	240.6259
75	90.0171	0.0111	1439.5553	0.0007	15.9920	0.0625	15.3291	245.1437
80	121.5104	0.0082	1948.8543	0.0005	16.0386	0.0623	15.5078	248.7234
85	164.0219	0.0061	2636.3359	0.0004	16.0731	0.0622	15.6503	251.5478
90	221.4064	0.0045	3564.3390	0.0003	16.0986	0.0621	15.7633	253.7679
95	298.8674	0.0033	4817.0122	0.0002	16.1176	0.0620	15.8527	255.5073
100	403.4288	0.0025	6507.9442	0.0002	16.1316	0.0620	15.9232	256.8660

TABLE B.7 Continuous Compounding: $r = 7\%$

	Single payment		Uniform series				Gradient series	
	Compound amount factor	Present worth factor	Compound amount factor	Sinking fund factor	Present worth factor	Capital recovery factor	Uniform series factor	Present worth factor
n	To find F given P $F\|P\ r,n$	To find P given F $P\|F\ r,n$	To find F given A $F\|A\ r,n$	To find A given F $A\|F\ r,n$	To find P given A $P\|A\ r,n$	To find A given P $A\|P\ r,n$	To find A given G $A\|G\ r,n$	To find P given G $P\|G\ r,n$
1	1.0725	0.9324	1.0000	1.0000	0.9324	1.0725	0.0000	0.0000
2	1.1503	0.8694	2.0725	0.4825	1.8018	0.5550	0.4825	0.8694
3	1.2337	0.8106	3.2228	0.3103	2.6123	0.3828	0.9534	2.4905
4	1.3231	0.7558	4.4565	0.2244	3.3681	0.2969	1.4126	4.7579
5	1.4191	0.7047	5.7796	0.1730	4.0728	0.2455	1.8603	7.5766
6	1.5220	0.6570	7.1987	0.1389	4.7299	0.2114	2.2964	10.8619
7	1.6323	0.6126	8.7206	0.1147	5.3425	0.1872	2.7211	14.5376
8	1.7507	0.5712	10.3529	0.0966	5.9137	0.1691	3.1344	18.5361
9	1.8776	0.5326	12.1036	0.0826	6.4463	0.1551	3.5364	22.7968
10	2.0138	0.4966	13.9812	0.0715	6.9429	0.1440	3.9272	27.2661
11	2.1598	0.4630	15.9950	0.0625	7.4059	0.1350	4.3069	31.8962
12	2.3164	0.4317	18.1547	0.0551	7.8376	0.1276	4.6755	36.6450
13	2.4843	0.4025	20.4711	0.0488	8.2401	0.1214	5.0333	41.4753
14	2.6645	0.3753	22.9554	0.0436	8.6154	0.1161	5.3804	46.3544
15	2.8577	0.3499	26.6199	0.0390	8.9654	0.1115	5.7168	51.2535
16	3.0649	0.3263	28.4775	0.0351	9.2916	0.1076	6.0428	56.1477
17	3.2871	0.3042	31.5424	0.0317	9.5959	0.1042	6.3585	61.0152
18	3.5254	0.2837	34.8295	0.0287	9.8795	0.1012	6.6640	65.8374
19	3.7810	0.2645	38.3549	0.0261	10.1440	0.0986	6.9596	70.5979
20	4.0552	0.2466	42.1359	0.0237	10.3906	0.0962	7.2453	75.2833
21	4.3492	0.2299	46.1911	0.0216	10.6205	0.0942	7.5215	79.8818
22	4.6646	0.2144	50.5404	0.0198	10.8349	0.0923	7.7881	84.3838
23	5.0028	0.1999	55.2050	0.0181	11.0348	0.0906	8.0456	88.7813
24	5.3656	0.1864	60.2078	0.0166	11.2212	0.0891	8.2940	93.0679
25	5.7546	0.1738	65.5733	0.0153	11.3949	0.0878	8.5335	97.2385
26	6.1719	0.1620	71.3279	0.0140	11.5570	0.0865	8.7643	101.2891
27	6.6194	0.1511	77.4998	0.0129	11.7080	0.0854	8.9867	105.2170
28	7.0993	0.1409	84.1192	0.0119	11.8489	0.0844	9.2009	109.0202
29	7.6141	0.1313	91.2185	0.0110	11.9802	0.0835	9.4070	112.6976
30	8.1662	0.1225	98.8326	0.0101	12.1027	0.0826	9.6052	116.2488
31	8.7583	0.1142	106.9987	0.0093	12.2169	0.0819	9.7958	119.6742
32	9.3933	0.1065	115.7570	0.0086	12.3233	0.0811	9.9790	122.9744
33	10.0744	0.0993	125.1504	0.0080	12.4226	0.0805	10.1550	126.1507
34	10.8049	0.0926	135.2248	0.0074	12.5151	0.0799	10.3239	129.2049
35	11.5883	0.0863	146.0297	0.0068	12.6014	0.0794	10.4860	132.1389
40	16.4446	0.0608	213.0056	0.0047	12.9529	0.0772	11.2017	145.0937
45	23.3361	0.0429	308.0489	0.0032	13.2006	0.0758	11.7769	155.4611
50	33.1155	0.0302	442.9218	0.0023	13.3751	0.0748	12.2347	163.6396
55	46.9931	0.0213	634.3155	0.0016	13.4981	0.0741	12.5957	170.0178
60	66.6863	0.0150	905.9161	0.0011	13.5847	0.0736	12.8781	174.9458
65	94.6324	0.0106	1291.3358	0.0008	13.6458	0.0733	13.0973	178.7238
70	134.2898	0.0074	1838.2723	0.0005	13.6888	0.0731	13.2664	181.6014
75	190.5663	0.0052	2614.4121	0.0004	13.7192	0.0729	13.3959	183.7808
80	270.4264	0.0037	3715.8070	0.0003	13.7405	0.0728	13.4946	185.4235
85	383.7533	0.0026	5278.7607	0.0002	13.7556	0.0727	13.5695	186.6563
90	544.5719	0.0018	7496.6976	0.0001	13.7662	0.0726	13.6260	187.5782
95	772.7843	0.0013	10644.0999	0.0001	13.7737	0.0726	13.6685	188.2652
100	1096.6332	0.0009	15110.4764	0.0001	13.7790	0.0726	13.7003	188.7757

TABLE B.8 Continuous Compounding: $r = 8\%$

	Single payment		Uniform series				Gradient series	
	Compound amount factor	Present worth factor	Compound amount factor	Sinking fund factor	Present worth factor	Capital recovery factor	Uniform series factor	Present worth factor
n	To find F given P $F\|P\,r,n$	To find P given F $P\|F\,r,n$	To find F given A $F\|A\,r,n$	To find A given F $A\|F\,r,n$	To find P given A $P\|A\,r,n$	To find A given P $A\|P\,r,n$	To find A given G $A\|G\,r,n$	To find P given G $P\|G\,r,n$
1	1.0833	0.9231	1.0000	1.0000	0.9231	1.0833	0.0000	0.0000
2	1.1735	0.8521	2.0833	0.4800	1.7753	0.5633	0.4800	0.8521
3	1.2712	0.7866	3.2568	0.3071	2.5619	0.3903	0.9467	2.4254
4	1.3771	0.7261	4.5280	0.2208	3.2880	0.3041	1.4002	4.6038
5	1.4918	0.6703	5.9052	0.1693	3.9584	0.2526	1.8404	7.2851
6	1.6161	0.6188	7.3970	0.1352	4.5771	0.2185	2.2676	10.3790
7	1.7507	0.5712	9.0131	0.1109	5.1483	0.1942	2.6817	13.8063
8	1.8965	0.5273	10.7637	0.0929	5.6756	0.1762	3.0829	17.4973
9	2.0544	0.4868	12.6602	0.0790	6.1624	0.1623	3.4713	21.3914
10	2.2225	0.4493	14.7147	0.0680	6.6117	0.1512	3.8470	25.4353
11	2.4109	0.4148	16.9402	0.0590	7.0265	0.1423	4.2102	29.5832
12	2.6117	0.3829	19.3511	0.0517	7.4094	0.1350	4.5611	33.7950
13	2.8292	0.3535	21.9628	0.0455	7.7629	0.1288	4.8998	38.0364
14	3.0649	0.3263	24.7920	0.0403	8.0891	0.1236	5.2265	42.2781
15	3.3201	0.3012	27.8569	0.0359	8.3903	0.1192	5.5415	46.4948
16	3.5966	0.2780	31.1770	0.0321	8.6684	0.1154	5.8449	50.6653
17	3.8962	0.2567	34.7736	0.0288	8.9250	0.1120	6.1369	54.7719
18	4.2207	0.2369	38.6698	0.0259	9.1620	0.1091	6.4178	58.7997
19	4.5722	0.2187	42.8905	0.0233	9.3807	0.1066	6.6879	62.7365
20	4.9530	0.2019	47.4627	0.0211	9.5826	0.1044	6.9473	66.5725
21	5.3656	0.1864	52.4158	0.0191	9.7689	0.1024	7.1963	70.3000
22	5.8124	0.1720	57.7813	0.0173	9.9410	0.1006	7.4352	73.9130
23	6.2965	0.1588	63.5938	0.0157	10.0998	0.0990	7.6642	77.4069
24	6.8210	0.1466	69.8903	0.0143	10.2464	0.0976	7.8836	80.7789
25	7.3891	0.1353	76.7113	0.0130	10.3817	0.0963	8.0937	84.0270
26	8.0045	0.1249	84.1003	0.0119	10.5067	0.0952	8.2947	87.1502
27	8.6711	0.1153	92.1048	0.0109	10.6220	0.0941	8.4870	90.1487
28	9.3933	0.1065	100.7759	0.0099	10.7285	0.0932	8.6707	93.0230
29	10.1757	0.0983	110.1693	0.0091	10.8267	0.0924	8.8461	95.7747
30	11.0232	0.0907	120.3449	0.0083	10.9174	0.0916	9.0136	98.4055
31	11.9413	0.0837	131.3681	0.0076	11.0012	0.0909	9.1734	100.9178
32	12.9358	0.0773	143.3094	0.0070	11.0785	0.0903	9.3257	103.3143
33	14.0132	0.0714	156.2452	0.0064	11.1499	0.0897	9.4708	105.5978
34	15.1803	0.0659	170.3584	0.0059	11.2157	0.0892	9.6090	107.7717
35	16.4446	0.0608	185.4387	0.0054	11.2765	0.0887	9.7405	109.8392
40	24.5325	0.0408	282.5472	0.0035	11.5172	0.0868	10.3069	118.7070
45	36.5982	0.0273	427.4161	0.0023	11.6786	0.0856	10.7426	125.4580
50	54.5982	0.0183	643.5351	0.0016	11.7868	0.0848	11.0738	130.5242
55	81.4509	0.0123	965.9467	0.0010	11.8593	0.0843	11.3230	134.2826
60	121.5104	0.0082	1446.9283	0.0007	11.9079	0.0840	11.5088	137.0449
65	181.2722	0.0055	2164.4686	0.0005	11.9404	0.0837	11.6461	139.0594
70	270.4264	0.0037	3234.9129	0.0003	11.9623	0.0836	11.7469	140.5190
75	403.4288	0.0025	4831.8281	0.0002	11.9769	0.0835	11.8203	141.5706
80	601.8450	0.0017	7214.1457	0.0001	11.9867	0.0834	11.8735	142.3245
85	897.8473	0.0011	10768.1458	0.0001	11.9933	0.0834	11.9119	142.8628
90	1339.4308	0.0007	16070.0911	0.0001	11.9977	0.0833	11.9394	143.2456
95	1998.1959	0.0005	23979.6640	0.0000	12.0007	0.0833	11.9591	143.5171
100	2980.9580	0.0003	35779.3601	0.0000	12.0026	0.0833	11.9731	143.7089

TABLE B.9 Continuous Compounding: $r = 9\%$

	Single payment		Uniform series				Gradient series	
	Compound amount factor	Present worth factor	Compound amount factor	Sinking fund factor	Present worth factor	Capital recovery factor	Uniform series factor	Present worth factor
n	To find F given P $F\|P\ r,n$	To find P given F $P\|F\ r,n$	To find F given A $F\|A\ r,n$	To find A given F $A\|F\ r,n$	To find P given A $P\|A\ r,n$	To find A given P $A\|P\ r,n$	To find A given G $A\|G\ r,n$	To find P given G $P\|G\ r,n$
1	1.0942	0.9139	1.0000	1.0000	0.9139	1.0942	0.0000	0.0000
2	1.1972	0.8353	2.0942	0.4775	1.7492	0.5717	0.4775	0.8353
3	1.3100	0.7634	3.2914	0.3038	2.5126	0.3980	0.9401	2.3620
4	1.4333	0.6977	4.6014	0.2173	3.2103	0.3115	1.3878	4.4551
5	1.5683	0.6376	6.0347	0.1657	3.8479	0.2599	1.8206	7.0056
6	1.7160	0.5827	7.6030	0.1315	4.4306	0.2257	2.2388	9.9193
7	1.8776	0.5326	9.3190	0.1073	4.9632	0.2015	2.6424	13.1149
8	2.0544	0.4868	11.1966	0.0893	5.4500	0.1835	3.0316	16.5221
9	2.2479	0.4449	13.2510	0.0755	5.8948	0.1696	3.4065	20.0810
10	2.4596	0.4066	15.4490	0.0645	6.3014	0.1587	3.7674	23.7401
11	2.6912	0.3716	17.9586	0.0557	6.6730	0.1499	4.1145	27.4559
12	2.9447	0.3396	20.6498	0.0484	7.0126	0.1426	4.4479	31.1914
13	3.2220	0.3104	23.5945	0.0424	7.3229	0.1366	4.7680	34.9158
14	3.5254	0.2837	26.8165	0.0373	7.6066	0.1315	5.0750	38.6033
15	3.8574	0.2592	30.3419	0.0330	7.8658	0.1271	5.3691	42.2327
16	4.2207	0.2369	34.1993	0.0292	8.1028	0.1234	5.6507	45.7866
17	4.6182	0.2165	38.4200	0.0260	8.3193	0.1202	5.9201	49.2512
18	5.0531	0.1979	43.0382	0.0232	8.5172	0.1174	6.1776	52.6155
19	5.5290	0.1809	48.0913	0.0208	8.6981	0.1150	6.4234	55.8711
20	6.0496	0.1653	53.6202	0.0186	8.8634	0.1128	6.6579	59.0117
21	6.6194	0.1511	59.6699	0.0168	9.0144	0.1109	6.8815	62.0332
22	7.2427	0.1381	66.2893	0.0151	9.1525	0.1093	7.0945	64.9326
23	7.9248	0.1262	73.5320	0.0136	9.2787	0.1078	7.2971	67.7087
24	8.6711	0.1153	81.4568	0.0123	9.3940	0.1065	7.4900	70.3612
25	9.4877	0.1054	90.1280	0.0111	9.4994	0.1053	7.6732	72.8908
26	10.3812	0.0963	99.6157	0.0100	9.5957	0.1042	7.8471	75.2990
27	11.3589	0.0880	109.9969	0.0091	9.6838	0.1033	8.0122	77.5879
28	12.4286	0.0805	121.3558	0.0082	9.7642	0.1024	8.1686	79.7603
29	13.5991	0.0735	133.7844	0.0075	9.8378	0.1016	8.3168	81.8193
30	14.8797	0.0672	147.3835	0.0068	9.9050	0.1010	8.4572	83.7683
31	16.2810	0.0614	162.2632	0.0062	9.9664	0.1003	8.5899	85.6109
32	17.8143	0.0561	178.5442	0.0056	10.0225	0.0998	8.7155	87.3511
33	19.4919	0.0513	196.3585	0.0051	10.0738	0.0993	8.8340	88.9928
34	21.3276	0.0469	215.8504	0.0046	10.1207	0.0988	8.9460	90.5401
35	23.3361	0.0429	237.1780	0.0042	10.1636	0.0984	9.0516	91.9970
40	36.5982	0.0273	378.0038	0.0026	10.3285	0.0968	9.4950	98.0684
45	57.3975	0.0174	598.8626	0.0017	10.4336	0.0958	9.8207	102.4654
50	90.0171	0.0111	945.2382	0.0011	10.5006	0.0952	10.0569	105.6042
55	141.1750	0.0071	1488.4633	0.0007	10.5434	0.0948	10.2262	107.8193
60	221.4064	0.0045	2340.4098	0.0004	10.5707	0.0946	10.3464	109.3680
65	347.2344	0.0029	3676.5279	0.0003	10.5880	0.0944	10.4309	110.4424
70	544.5719	0.0018	5771.9782	0.0002	10.5991	0.0943	10.4898	111.1829
75	854.0588	0.0012	9058.2984	0.0001	10.6062	0.0943	10.5307	111.6904
80	1339.4308	0.0007	14212.2744	0.0001	10.6107	0.0942	10.5588	112.0365
85	2100.6456	0.0005	22295.3179	0.0000	10.6136	0.0942	10.5781	112.2715
90	3294.4681	0.0003	34972.0534	0.0000	10.6154	0.0942	10.5913	112.4306
95	5166.7544	0.0002	54853.1321	0.0000	10.6166	0.0942	10.6002	112.5378
100	8103.0839	0.0001	86032.8702	0.0000	10.6173	0.0942	10.6063	112.6099

TABLE B.10 Continuous Compounding: $r = 10\%$

	Single payment		Uniform series				Gradient series	
	Compound amount factor	Present worth factor	Compound amount factor	Sinking fund factor	Present worth factor	Capital recovery factor	Uniform series factor	Present worth factor
n	To find F given P $F\|P\ r,n$	To find P given F $P\|F\ r,n$	To find F given A $F\|A\ r,n$	To find A given F $A\|F\ r,n$	To find P given A $P\|A\ r,n$	To find A given P $A\|P\ r,n$	To find A given G $A\|G\ r,n$	To find P given G $P\|G\ r,n$
1	1.1052	0.9048	1.0000	1.0000	0.9048	1.1052	0.0000	0.0000
2	1.2214	0.8187	2.1052	0.4750	1.7236	0.5802	0.4750	0.8187
3	1.3499	0.7408	3.3266	0.3006	2.4644	0.4058	0.9334	2.3004
4	1.4918	0.6703	4.6764	0.2138	3.1347	0.3190	1.3754	4.3113
5	1.6487	0.6065	6.1683	0.1621	3.7412	0.2673	1.8009	6.7374
6	1.8221	0.5488	7.8170	0.1279	4.2900	0.2331	2.2101	9.4815
7	2.0138	0.4966	9.6391	0.1037	4.7866	0.2089	2.6033	12.4610
8	2.2255	0.4493	11.6528	0.0858	5.2360	0.1910	2.9806	15.6063
9	2.4596	0.4066	13.8784	0.0721	5.6425	0.1772	3.3423	18.8589
10	2.7183	0.3679	16.3380	0.0612	6.0104	0.1664	3.6886	22.1698
11	3.0042	0.3329	19.0563	0.0525	6.3433	0.1576	4.0198	25.4985
12	3.3201	0.3012	22.0604	0.0453	6.6445	0.1505	4.3362	28.8116
13	3.6693	0.2725	25.3806	0.0394	6.9170	0.1446	4.6381	32.0820
14	4.0552	0.2466	29.0499	0.0344	7.1636	0.1396	4.9260	35.2878
15	4.4817	0.2231	33.1051	0.0302	7.3867	0.1354	5.2001	38.4116
16	4.9530	0.2019	37.5867	0.0266	7.5886	0.1318	5.4608	41.4401
17	5.4739	0.1827	42.5398	0.0235	7.7713	0.1287	5.7086	44.3630
18	6.0496	0.1653	48.0137	0.0208	7.9366	0.1260	5.9437	47.1731
19	6.6859	0.1496	54.0634	0.0185	8.0862	0.1237	6.1667	49.8653
20	7.3891	0.1353	60.7483	0.0165	8.2215	0.1216	6.3780	52.4367
21	8.1662	0.1225	68.1383	0.0147	8.3440	0.1198	6.5779	54.8858
22	9.0250	0.1108	76.3045	0.0131	8.4548	0.1183	6.7669	57.2127
23	9.9742	0.1003	85.3295	0.0117	8.5550	0.1169	6.9454	59.4184
24	11.0232	0.0907	95.3037	0.0105	8.6458	0.1157	7.1139	61.5049
25	12.1825	0.0821	106.3269	0.0094	8.7278	0.1146	7.2727	63.4749
26	13.4637	0.0743	118.5094	0.0084	8.8021	0.1136	7.4223	65.3318
27	14.8797	0.0672	131.9731	0.0076	8.8693	0.1127	7.5630	67.0791
28	16.4446	0.0608	146.8528	0.0068	8.9301	0.1120	7.6954	68.7210
29	18.1741	0.0550	163.2975	0.0061	8.9852	0.1113	7.8197	70.2616
30	20.0855	0.0498	181.4716	0.0055	9.0349	0.1107	7.9365	71.7054
31	22.1980	0.0450	201.5572	0.0050	9.0800	0.1101	8.0459	73.0569
32	24.5325	0.0408	223.7551	0.0045	9.1208	0.1096	8.1485	74.3206
33	27.1126	0.0369	248.2876	0.0040	9.1576	0.1092	8.2446	75.5008
34	29.9641	0.0334	275.4003	0.0036	9.1910	0.1088	8.3345	76.6021
35	33.1155	0.0302	305.3644	0.0033	9.2212	0.1084	8.4185	77.6288
40	54.5982	0.0183	509.6290	0.0020	9.3342	0.1071	8.7620	81.7864
45	90.0171	0.0111	846.4044	0.0012	9.4027	0.1064	9.0028	84.6508
50	148.4132	0.0067	1401.6532	0.0007	9.4443	0.1059	9.1691	86.5959
55	244.6919	0.0041	2317.1038	0.0004	9.4695	0.1056	9.2826	87.9017
60	403.4288	0.0025	3826.4266	0.0003	9.4848	0.1054	9.3592	88.7701
65	665.1416	0.0015	6314.8791	0.0002	9.4940	0.1053	9.4105	89.3433
70	1096.6332	0.0009	10417.6438	0.0001	9.4997	0.1053	9.4444	89.7190
75	1808.0424	0.0006	17181.9591	0.0001	9.5031	0.1052	9.4668	89.9640
80	2980.9580	0.0003	28334.4297	0.0000	9.5051	0.1052	9.4815	90.1229
85	4914.7688	0.0002	46721.7452	0.0000	9.5064	0.1052	9.4910	90.2255
90	8103.0839	0.0001	77037.3034	0.0000	9.5072	0.1052	9.4972	90.2916
95	13359.7268	0.0001	127019.2091	0.0000	9.5076	0.1052	9.5012	90.3340
100	22026.4658	0.0000	209425.4400	0.0000	9.5079	0.1052	9.5038	90.3611

TABLE B.11 Continuous Compounding: $r = 12\%$

	Single payment		Uniform series				Gradient series	
	Compound amount factor	Present worth factor	Compound amount factor	Sinking fund factor	Present worth factor	Capital recovery factor	Uniform series factor	Present worth factor
n	To find F given P $F\|P\ r,n$	To find P given F $P\|F\ r,n$	To find F given A $F\|A\ r,n$	To find A given F $A\|F\ r,n$	To find P given A $P\|A\ r,n$	To find A given P $A\|P\ r,n$	To find A given G $A\|G\ r,n$	To find P given G $P\|G\ r,n$
1	1.1275	0.8869	1.0000	1.0000	0.8869	1.1275	0.0000	0.0000
2	1.2712	0.7866	2.1275	0.4700	1.6735	0.5975	0.4700	0.7866
3	1.4333	0.6977	3.3987	0.2942	2.3712	0.4217	0.9202	2.1820
4	1.6161	0.6188	4.8321	0.2070	2.9900	0.3344	1.3506	4.0383
5	1.8221	0.5488	6.4481	0.1551	3.5388	0.2826	1.7615	6.2336
6	2.0544	0.4868	8.2703	0.1209	4.0256	0.2484	2.1531	8.6673
7	2.3164	0.4317	10.3247	0.0969	4.4573	0.2244	2.5257	11.2576
8	2.6117	0.3829	12.6411	0.0791	4.8402	0.2066	2.8796	13.9379
9	2.9447	0.3396	15.2528	0.0656	5.1798	0.1931	3.2153	16.6546
10	3.3201	0.3012	18.1974	0.0550	5.4810	0.1824	3.5332	19.3654
11	3.7434	0.2671	21.5176	0.0465	5.7481	0.1740	3.8337	22.0367
12	4.2207	0.2369	25.2610	0.0396	5.9850	0.1671	4.1174	24.6429
13	4.7588	0.2101	29.4817	0.0339	6.1952	0.1614	4.3848	27.1646
14	5.3656	0.1864	34.2405	0.0292	6.3815	0.1567	4.6364	29.5874
15	6.0496	0.1653	39.6061	0.0252	6.5468	0.1527	4.8728	31.9016
16	6.8210	0.1466	45.6557	0.0219	6.6934	0.1494	5.0946	34.1007
17	7.6906	0.1300	52.4767	0.0191	6.8235	0.1466	5.3025	36.1812
18	8.6711	0.1153	60.1673	0.0166	6.9388	0.1441	5.4969	38.1417
19	9.7767	0.1023	68.8384	0.0145	7.0411	0.1420	5.6785	39.9828
20	11.0232	0.0907	78.6151	0.0127	7.1318	0.1402	5.8480	41.7064
21	12.4286	0.0805	89.6383	0.0112	7.2123	0.1387	6.0058	43.3156
22	14.0132	0.0714	102.0669	0.0098	7.2836	0.1373	6.1527	44.8142
23	15.7998	0.0633	116.0801	0.0086	7.3469	0.1361	6.2893	46.2066
24	17.8143	0.0561	131.8799	0.0076	7.4030	0.1351	6.4160	47.4977
25	20.0855	0.0498	149.6942	0.0067	7.4528	0.1342	6.5334	48.6926
26	22.6464	0.0442	169.7797	0.0059	7.4970	0.1334	6.6422	49.7966
27	25.5337	0.0392	192.4261	0.0052	7.5362	0.1327	6.7428	50.8148
28	28.7892	0.0347	217.9598	0.0046	7.5709	0.1321	6.8357	51.7527
29	32.4597	0.0308	246.7490	0.0041	7.6017	0.1315	6.9215	52.6153
30	36.5982	0.0273	279.2087	0.0036	7.6290	0.1311	7.0006	53.4077
31	41.2644	0.0242	315.8070	0.0032	7.6533	0.1307	7.0734	54.1347
32	46.5255	0.0215	357.0714	0.0028	7.6747	0.1303	7.1404	54.8010
33	52.4573	0.0191	403.5968	0.0025	7.6938	0.1300	7.2020	55.4110
34	59.1455	0.0169	456.0542	0.0022	7.7107	0.1297	7.2586	55.9690
35	66.6863	0.0150	515.1996	0.0019	7.7257	0.1294	7.3105	56.4788
40	121.5104	0.0082	945.2031	0.0011	7.7788	0.1286	7.5114	58.4296
45	221.4064	0.0045	1728.7205	0.0006	7.8079	0.1281	7.6392	59.6459
50	403.4288	0.0025	3156.3822	0.0003	7.8239	0.1278	7.7191	60.3933

TABLE B.12 Continuous Compounding: $r = 15\%$

	Single payment		Uniform series				Gradient series	
	Compound amount factor	Present worth factor	Compound amount factor	Sinking fund factor	Present worth factor	Capital recovery factor	Uniform series factor	Present worth factor
n	To find F given P $F\|P\ r,n$	To find P given F $P\|F\ r,n$	To find F given A $F\|A\ r,n$	To find A given F $A\|F\ r,n$	To find P given A $P\|A\ r,n$	To find A given P $A\|P\ r,n$	To find A given G $A\|G\ r,n$	To find P given G $P\|G\ r,n$
1	1.1618	0.8607	1.0000	1.0000	0.8607	1.1618	0.0000	0.0000
2	1.3499	0.7408	2.1618	0.4626	1.6015	0.6244	0.4626	0.7408
3	1.5683	0.6376	3.5117	0.2848	2.2392	0.4466	0.9004	2.0161
4	1.8221	0.5488	5.0800	0.1969	2.7880	0.3587	1.3137	3.6625
5	2.1170	0.4724	6.9021	0.1449	3.2603	0.3067	1.7029	5.5520
6	2.4596	0.4066	9.0191	0.1109	3.6669	0.2727	2.0685	7.5848
7	2.8577	0.3499	11.4787	0.0871	4.0168	0.2490	2.4110	9.6845
8	3.3201	0.3012	14.3364	0.0698	4.3180	0.2316	2.7311	11.7928
9	3.8574	0.2592	17.6565	0.0566	4.5773	0.2185	3.0295	13.8667
10	4.4817	0.2231	21.5139	0.0465	4.8004	0.2083	3.3070	15.8749
11	5.2070	0.1920	25.9956	0.0385	4.9925	0.2003	3.5645	17.7954
12	6.0496	0.1653	31.2026	0.0320	5.1578	0.1939	3.8028	19.6137
13	7.0287	0.1423	37.2522	0.0268	5.3000	0.1887	4.0228	21.3210
14	8.1662	0.1225	44.2809	0.0226	5.4225	0.1844	4.2255	22.9129
15	9.4877	0.1054	52.4471	0.0191	5.5279	0.1809	4.4119	24.3885
16	11.0232	0.0907	61.9348	0.0161	5.6186	0.1780	4.5829	25.7493
17	12.8071	0.0781	72.9580	0.0137	5.6967	0.1755	4.7394	26.9986
18	14.8797	0.0672	85.7651	0.0117	5.7639	0.1735	4.8823	28.1411
19	17.2878	0.0578	100.6448	0.0099	5.8217	0.1718	5.0126	29.1823
20	20.0855	0.0498	117.9326	0.0085	5.8715	0.1703	5.1312	30.1282
21	23.3361	0.0429	138.0182	0.0072	5.9144	0.1691	5.2390	30.9853
22	27.1126	0.0369	161.3542	0.0062	5.9513	0.1680	5.3367	31.7598
23	31.5004	0.0317	188.4669	0.0053	5.9830	0.1671	5.4251	32.4582
24	36.5982	0.0273	219.9673	0.0045	6.0103	0.1664	5.5050	33.0867
25	42.5211	0.0235	256.5655	0.0039	6.0338	0.1657	5.5771	33.6511
26	49.4024	0.0202	299.0866	0.0033	6.0541	0.1652	5.6420	34.1571
27	57.3975	0.0174	348.4890	0.0029	6.0715	0.1647	5.7004	34.6101
28	66.6863	0.0150	405.8865	0.0025	6.0865	0.1643	5.7529	35.0150
29	77.4785	0.0129	472.5728	0.0021	6.0994	0.1640	5.8000	35.3764
30	90.0171	0.0111	550.0513	0.0018	6.1105	0.1637	5.8421	35.6985
31	104.5850	0.0096	640.0684	0.0016	6.1201	0.1634	5.8799	35.9854
32	121.5104	0.0082	744.6534	0.0013	6.1283	0.1632	5.9136	36.2405
33	141.1750	0.0071	866.1638	0.0012	6.1354	0.1630	5.9437	36.4672
34	164.0219	0.0061	1007.3388	0.0010	6.1415	0.1628	5.9706	36.6684
35	190.5663	0.0052	1171.3607	0.0009	6.1467	0.1627	5.9945	36.8468
40	403.4288	0.0025	2486.6727	0.0004	6.1638	0.1622	6.0798	37.4747
45	854.0588	0.0012	5271.1883	0.0002	6.1719	0.1620	6.1264	37.8118
50	1808.0424	0.0006	11166.0078	0.0001	6.1757	0.1619	6.1515	37.9900

TABLE B.13 Continuous Compounding: $r = 20\%$

	Single payment		Uniform series				Gradient series	
	Compound amount factor	Present worth factor	Compound amount factor	Sinking fund factor	Present worth factor	Capital recovery factor	Uniform series factor	Present worth factor
n	To find F given P $F\|P\ r,n$	To find P given F $P\|F\ r,n$	To find F given A $F\|A\ r,n$	To find A given F $A\|F\ r,n$	To find P given A $P\|A\ r,n$	To find A given P $A\|P\ r,n$	To find A given G $A\|G\ r,n$	To find P given G $P\|G\ r,n$
1	1.2214	0.8187	1.0000	1.0000	0.8187	1.2214	0.0000	0.0000
2	1.4918	0.6703	2.2214	0.4502	1.4891	0.6716	0.4502	0.6703
3	1.8221	0.5488	3.7132	0.2693	2.0379	0.4907	0.8675	1.7679
4	2.2255	0.4493	5.5353	0.1807	2.4872	0.4021	1.2528	3.1159
5	2.7183	0.3679	7.7609	0.1289	2.8551	0.3503	1.6068	4.5874
6	3.3201	0.3012	10.4792	0.0954	3.1563	0.3168	1.9306	6.0934
7	4.0552	0.2466	13.7993	0.0725	3.4029	0.2939	2.2255	7.5730
8	4.9530	0.2019	17.8545	0.0560	3.6048	0.2774	2.4929	8.9863
9	6.0496	0.1653	22.8075	0.0438	3.7701	0.2652	2.7344	10.3087
10	7.3891	0.1353	28.8572	0.0347	3.9054	0.2561	2.9515	11.5267
11	9.0250	0.1108	36.2462	0.0276	4.0162	0.2490	3.1459	12.6347
12	11.0232	0.0907	45.2712	0.0221	4.1069	0.2435	3.3194	13.6326
13	13.4637	0.0743	56.2944	0.0178	4.1812	0.2392	3.4736	14.5239
14	16.4446	0.0608	69.7581	0.0143	4.2420	0.2357	3.6102	15.3144
15	20.0855	0.0498	86.2028	0.0116	4.2918	0.2330	3.7307	16.0114
16	24.5325	0.0408	106.2883	0.0094	4.3325	0.2308	3.8367	16.6229
17	29.9641	0.0334	130.8209	0.0076	4.3659	0.2290	3.9297	17.1569
18	36.5982	0.0273	160.7850	0.0062	4.3932	0.2276	4.0110	17.6214
19	44.7012	0.0224	197.3832	0.0051	4.4156	0.2265	4.0819	18.0240
20	54.5982	0.0183	242.0844	0.0041	4.4339	0.2255	4.1435	18.3720
21	66.6863	0.0150	296.6825	0.0034	4.4489	0.2248	4.1970	18.6719
22	81.4509	0.0123	363.3689	0.0028	4.4612	0.2242	4.2432	18.9298
23	99.4843	0.0101	444.8197	0.0022	4.4713	0.2237	4.2831	19.1509
24	121.5104	0.0082	544.3040	0.0018	4.4795	0.2232	4.3175	19.3402
25	148.4132	0.0067	665.8145	0.0015	4.4862	0.2229	4.3471	19.5019
26	181.2722	0.0055	814.2276	0.0012	4.4917	0.2226	4.3724	19.6398
27	221.4064	0.0045	995.4999	0.0010	4.4963	0.2224	4.3942	19.7572
28	270.4264	0.0037	1216.9063	0.0008	4.5000	0.2222	4.4127	19.8571
29	330.2996	0.0030	1487.3327	0.0007	4.5030	0.2221	4.4286	19.9419
30	403.4288	0.0025	1817.6323	0.0006	4.5055	0.2220	4.4421	20.0137
31	492.7490	0.0020	2221.0610	0.0005	4.5075	0.2219	4.4536	20.0746
32	601.8450	0.0017	2713.8101	0.0004	4.5092	0.2218	4.4634	20.1261
33	735.0952	0.0014	3315.6551	0.0003	4.5105	0.2217	4.4717	20.1697
34	897.8473	0.0011	4050.7503	0.0002	4.5116	0.2216	4.4787	20.2064
35	1096.6332	0.0009	4948.5976	0.0002	4.5125	0.2216	4.4847	20.2374
40	2980.9580	0.0003	13459.4438	0.0001	4.5151	0.2215	4.5032	20.3327
45	8103.0839	0.0001	36594.3225	0.0000	4.5161	0.2214	4.5111	20.3726
50	22026.4658	0.0000	99481.4427	0.0000	4.5165	0.2214	4.5144	20.3890

TABLE B.14 Continuous Compounding: $r = 25\%$

	Single payment		Uniform series				Gradient series	
	Compound amount factor	Present worth factor	Compound amount factor	Sinking fund factor	Present worth factor	Capital recovery factor	Uniform series factor	Present worth factor
n	To find F given P $F\|P\ r,n$	To find P given F $P\|F\ r,n$	To find F given A $F\|A\ r,n$	To find A given F $A\|F\ r,n$	To find P given A $P\|A\ r,n$	To find A given P $A\|P\ r,n$	To find A given G $A\|G\ r,n$	To find P given G $P\|G\ r,n$
1	1.2840	0.7788	1.0000	1.0000	0.7788	1.2840	0.0000	0.0000
2	1.6487	0.6065	2.2840	0.4378	1.3853	0.7218	0.4378	0.6065
3	2.1170	0.4724	3.9327	0.2543	1.8577	0.5383	0.8350	1.5513
4	2.7183	0.3679	6.0497	0.1653	2.2256	0.4493	1.1929	2.6549
5	3.4903	0.2865	8.7680	0.1141	2.5121	0.3981	1.5131	3.8009
6	4.4817	0.2231	12.2584	0.0816	2.7352	0.3656	1.7975	4.9166
7	5.7546	0.1738	16.7401	0.0597	2.9090	0.3438	2.0486	5.9592
8	7.3891	0.1353	22.4947	0.0445	3.0443	0.3285	2.2687	6.9066
9	9.4877	0.1054	29.8837	0.0335	3.1497	0.3175	2.4605	7.7498
10	12.1825	0.0821	39.3715	0.0254	3.2318	0.3094	2.6266	8.4885
11	15.6426	0.0639	51.5539	0.0194	3.2957	0.3034	2.7696	9.1278
12	20.0855	0.0498	67.1966	0.0149	3.3455	0.2989	2.8921	9.6755
13	25.7903	0.0388	87.2821	0.0115	3.3843	0.2955	2.9964	10.1407
14	33.1155	0.0302	113.0725	0.0088	3.4145	0.2929	3.0849	10.5333
15	42.5211	0.0235	146.1879	0.0068	3.4380	0.2909	3.1595	10.8626
16	54.5982	0.0183	188.7090	0.0053	3.4563	0.2893	3.2223	11.1373
17	70.1054	0.0143	243.3071	0.0041	3.4706	0.2881	3.2748	11.3655
18	90.0171	0.0111	313.4126	0.0032	3.4817	0.2872	3.3186	11.5544
19	115.5843	0.0087	403.4297	0.0025	3.4904	0.2865	3.3550	11.7101
20	148.4132	0.0067	519.0140	0.0019	3.4971	0.2860	3.3851	11.8381
21	190.5663	0.0052	667.4271	0.0015	3.5023	0.2855	3.4100	11.9431
22	244.6919	0.0041	857.9934	0.0012	3.5064	0.2852	3.4305	12.0289
23	314.1907	0.0032	1102.6853	0.0009	3.5096	0.2849	3.4474	12.0989
24	403.4288	0.0025	1416.8760	0.0007	3.5121	0.2847	3.4612	12.1559
25	518.0128	0.0019	1820.3048	0.0005	3.5140	0.2846	3.4725	12.2023
26	665.1416	0.0015	2338.3176	0.0004	3.5155	0.2845	3.4817	12.2399
27	854.0588	0.0012	3003.4592	0.0003	3.5167	0.2844	3.4892	12.2703
28	1096.6332	0.0009	3857.5180	0.0003	3.5176	0.2843	3.4953	12.2949
29	1408.1048	0.0007	4954.1512	0.0002	3.5183	0.2842	3.5002	12.3148
30	1808.0424	0.0006	6362.2560	0.0002	3.5189	0.2842	3.5042	12.3308
31	2321.5724	0.0004	8170.2984	0.0001	3.5193	0.2841	3.5075	12.3438
32	2980.9580	0.0003	10491.8708	0.0001	3.5196	0.2841	3.5101	12.3542
33	3827.6258	0.0003	13472.8288	0.0001	3.5199	0.2841	3.5122	12.3625
34	4914.7688	0.0002	17300.4546	0.0001	3.5201	0.2841	3.5139	12.3692
35	6310.6881	0.0002	22215.2235	0.0000	3.5203	0.2841	3.5153	12.3746

TABLE B.15 Continuous Compounding: $r = 4\%$

	Continuous flow, uniform series			
	Present worth factor	Capital recovery factor	Compound amount factor	Sinking fund factor
n	To find P given $\overline{A}$ $P\|\overline{A}\ r,n$	To find $\overline{A}$ given P $\overline{A}\|P\ r,n$	To find F given $\overline{A}$ $F\|\overline{A}\ r,n$	To find $\overline{A}$ given F $\overline{A}\|F\ r,n$
1	0.9803	1.0201	1.0203	0.9801
2	1.9221	0.5203	2.0822	0.4803
3	2.8270	0.3537	3.1874	0.3137
4	3.6964	0.2705	4.3378	0.2305
5	4.5317	0.2207	5.5351	0.1807
6	5.3343	0.1875	6.7812	0.1475
7	6.1054	0.1638	8.0782	0.1238
8	6.8463	0.1461	9.4282	0.1061
9	7.5581	0.1323	10.8332	0.0923
10	8.2420	0.1213	12.2956	0.0813
11	8.8991	0.1124	13.8177	0.0724
12	9.5304	0.1049	15.4019	0.0649
13	10.1370	0.0986	17.0507	0.0586
14	10.7198	0.0933	18.7668	0.0533
15	11.2797	0.0887	20.5530	0.0487
16	11.8177	0.0846	22.4120	0.0446
17	12.3346	0.0811	24.3469	0.0411
18	12.8312	0.0779	26.3608	0.0379
19	13.3083	0.0751	28.4569	0.0351
20	13.7668	0.0726	30.6385	0.0326
21	14.2072	0.0704	32.9092	0.0304
22	14.6304	0.0684	35.2725	0.0284
23	15.0370	0.0665	37.7323	0.0265
24	15.4277	0.0648	40.2924	0.0248
25	15.8030	0.0633	42.9570	0.0233
26	16.1636	0.0619	45.7304	0.0219
27	16.5101	0.0606	48.6170	0.0206
28	16.8430	0.0594	51.6214	0.0194
29	17.1628	0.0583	54.7483	0.0183
30	17.4701	0.0572	58.0029	0.0172
31	17.7654	0.0563	61.3903	0.0163
32	18.0491	0.0554	64.9160	0.0154
33	18.3216	0.0546	68.5855	0.0146
34	18.5835	0.0538	72.4048	0.0138
35	18.8351	0.0531	76.3800	0.0131
40	19.9526	0.0501	98.8258	0.0101
45	20.8675	0.0479	126.2412	0.0079
50	21.6166	0.0463	159.7264	0.0063
55	22.2299	0.0450	200.6253	0.0050
60	22.7321	0.0440	250.5794	0.0040
65	23.1432	0.0432	311.5935	0.0032
70	23.4797	0.0426	386.1162	0.0026
75	23.7553	0.0421	477.1384	0.0021
80	23.9809	0.0417	588.3133	0.0017
85	24.1657	0.0414	724.1025	0.0014
90	24.3169	0.0411	889.9559	0.0011
95	24.4407	0.0409	1092.5296	0.0009
100	24.5421	0.0407	1339.9538	0.0007

TABLE B.16 Continuous Compounding: $r = 5\%$

	Continuous flow, uniform series			
	Present worth factor	Capital recovery factor	Compound amount factor	Sinking fund factor
n	To find P given $\overline{A}$ $P\|\overline{A}\ r,n$	To find $\overline{A}$ given P $\overline{A}\|P\ r,n$	To find F given $\overline{A}$ $F\|\overline{A}\ r,n$	To find $\overline{A}$ given F $\overline{A}\|F\ r,n$
1	0.9754	1.0252	1.0254	0.9752
2	1.9033	0.5254	2.1034	0.4754
3	2.7858	0.3590	3.2367	0.3090
4	3.6254	0.2758	4.4281	0.2258
5	4.4240	0.2260	5.6805	0.1760
6	5.1836	0.1929	6.9972	0.1429
7	5.9062	0.1693	8.3814	0.1193
8	6.5936	0.1517	9.8365	0.1017
9	7.2474	0.1380	11.3662	0.0880
10	7.8694	0.1271	12.9744	0.0771
11	8.4610	0.1182	14.6651	0.0682
12	9.0238	0.1108	16.4424	0.0608
13	9.5591	0.1046	18.3108	0.0546
14	10.0683	0.0993	20.2751	0.0493
15	10.5527	0.0948	22.3400	0.0448
16	11.0134	0.0908	24.5108	0.0408
17	11.4517	0.0873	26.7929	0.0373
18	11.8686	0.0843	29.1921	0.0343
19	12.2652	0.0815	31.7142	0.0315
20	12.6424	0.0791	34.3656	0.0291
21	13.0012	0.0769	37.1530	0.0269
22	13.3426	0.0749	40.0833	0.0249
23	13.6673	0.0732	43.1639	0.0232
24	13.9761	0.0716	46.4023	0.0216
25	14.2699	0.0701	49.8069	0.0201
26	14.5494	0.0687	53.3859	0.0187
27	14.8152	0.0675	57.1485	0.0175
28	15.0681	0.0664	61.1040	0.0164
29	15.3086	0.0653	65.2623	0.0153
30	15.5374	0.0644	69.6338	0.0144
31	15.7550	0.0635	74.2294	0.0135
32	15.9621	0.0626	79.0606	0.0126
33	16.1590	0.0619	84.1396	0.0119
34	16.3463	0.0612	89.4789	0.0112
35	16.5245	0.0605	95.0921	0.0105
40	17.2933	0.0578	127.7811	0.0078
45	17.8920	0.0559	169.7547	0.0059
50	18.3583	0.0545	223.6499	0.0045
55	18.7214	0.0534	292.8526	0.0034
60	19.0043	0.0526	381.7107	0.0026
65	19.2245	0.0520	495.8068	0.0020
70	19.3961	0.0516	642.3090	0.0016
75	19.5296	0.0512	830.4216	0.0012
80	19.6337	0.0509	1071.9630	0.0009
85	19.7147	0.0507	1382.1082	0.0007
90	19.7778	0.0506	1780.3426	0.0006
95	19.8270	0.0504	2291.6857	0.0004
100	19.8652	0.0503	2948.2632	0.0003

TABLE B.17   Continuous Compounding: $r = 6\%$

	Continuous flow, uniform series			
	Present worth factor	Capital recovery factor	Compound amount factor	Sinking fund factor
n	To find P given $\overline{A}$ $P\|\overline{A}\ r,n$	To find $\overline{A}$ given P $\overline{A}\|P\ r,n$	To find F given $\overline{A}$ $F\|\overline{A}\ r,n$	To find $\overline{A}$ given F $\overline{A}\|F\ r,n$
1	0.9706	1.0303	1.0306	0.9703
2	1.8847	0.5306	2.1249	0.4706
3	2.7455	0.3642	3.2870	0.3042
4	3.5562	0.2812	4.5208	0.2212
5	4.3197	0.2315	5.8310	0.1715
6	5.0387	0.1985	7.2222	0.1385
7	5.7159	0.1750	8.6994	0.1150
8	6.3536	0.1574	10.2679	0.0974
9	6.9542	0.1438	11.9334	0.0838
10	7.5198	0.1330	13.7020	0.0730
11	8.0525	0.1242	15.5799	0.0642
12	8.5541	0.1169	17.5739	0.0569
13	9.0266	0.1108	19.6912	0.0508
14	9.4715	0.1056	21.9394	0.0456
15	9.8905	0.1011	24.3267	0.0411
16	10.2851	0.0972	26.8616	0.0372
17	10.6568	0.0938	29.5532	0.0338
18	11.0067	0.0909	32.4113	0.0309
19	11.3363	0.0882	35.4461	0.0282
20	11.6468	0.0859	38.6686	0.0259
21	11.9391	0.0838	42.0904	0.0238
22	12.2144	0.0819	45.7237	0.0219
23	12.4737	0.0802	49.5817	0.0202
24	12.7179	0.0786	53.6783	0.0186
25	12.9478	0.0772	58.0282	0.0172
26	13.1644	0.0760	62.6470	0.0160
27	13.3684	0.0748	67.5515	0.0148
28	13.5604	0.0737	72.7593	0.0137
29	13.7413	0.0728	78.2891	0.0128
30	13.9117	0.0719	84.1608	0.0119
31	14.0721	0.0711	90.3956	0.0111
32	14.2232	0.0703	97.0160	0.0103
33	14.3655	0.0696	104.0457	0.0096
34	14.4995	0.0690	111.5102	0.0090
35	14.6257	0.0684	119.4362	0.0084
40	15.1547	0.0660	167.0529	0.0060
45	15.5466	0.0643	231.3289	0.0043
50	15.8369	0.0631	318.0923	0.0031
55	16.0519	0.0623	435.2106	0.0023
60	16.2113	0.0617	593.3039	0.0017
65	16.3293	0.0612	806.7075	0.0012
70	16.4167	0.0609	1094.7722	0.0009
75	16.4815	0.0607	1483.6189	0.0007
80	16.5295	0.0605	2008.5070	0.0005
85	16.5651	0.0604	2717.0318	0.0004
90	16.5914	0.0603	3673.4403	0.0003
95	16.6109	0.0602	4964.4567	0.0002
100	16.6254	0.0601	6707.1466	0.0001

TABLE B.18 Continuous Compounding: $r = 8\%$

	Continuous flow, uniform series			
	Present worth factor	Capital recovery factor	Compound amount factor	Sinking fund factor
n	To find P given $\overline{A}$ $P\|\overline{A}\ r,n$	To find $\overline{A}$ given P $\overline{A}\|P\ r,n$	To find F given $\overline{A}$ $F\|\overline{A}\ r,n$	To find $\overline{A}$ given F $\overline{A}\|F\ r,n$
1	0.9610	1.0405	1.0411	0.9605
2	1.8482	0.5411	2.1689	0.4611
3	2.6672	0.3749	3.3906	0.2949
4	3.4231	0.2921	4.7141	0.2121
5	4.1210	0.2427	6.1478	0.1627
6	4.7652	0.2099	7.7009	0.1299
7	5.3599	0.1866	9.3834	0.1066
8	5.9088	0.1692	11.2060	0.0892
9	6.4156	0.1559	13.1804	0.0759
10	6.8834	0.1453	15.3193	0.0653
11	7.3152	0.1367	17.6362	0.0567
12	7.7138	0.1296	20.1462	0.0496
13	8.0818	0.1237	22.8652	0.0437
14	8.4215	0.1187	25.8107	0.0387
15	8.7351	0.1145	29.0015	0.0345
16	9.0245	0.1108	32.4580	0.0308
17	9.2917	0.1076	36.2024	0.0276
18	9.5384	0.1048	40.2587	0.0248
19	9.7661	0.1024	44.6528	0.0224
20	9.9763	0.1002	49.4129	0.0202
21	10.1703	0.0983	54.5694	0.0183
22	10.3494	0.0966	60.1555	0.0166
23	10.5148	0.0951	66.2067	0.0151
24	10.6674	0.0937	72.7620	0.0137
25	10.8083	0.0925	79.8632	0.0125
26	10.9384	0.0914	87.5559	0.0114
27	11.0584	0.0904	95.8892	0.0104
28	11.1693	0.0895	104.9166	0.0095
29	11.2716	0.0887	114.6959	0.0087
30	11.3660	0.0880	125.2897	0.0080
31	11.4532	0.0873	136.7658	0.0073
32	11.5337	0.0867	149.1977	0.0067
33	11.6080	0.0861	162.6650	0.0061
34	11.6766	0.0856	177.2540	0.0056
35	11.7399	0.0852	193.0581	0.0052
40	11.9905	0.0834	294.1566	0.0034
45	12.1585	0.0822	444.9779	0.0022
50	12.2711	0.0815	669.9769	0.0015
55	12.3465	0.0810	1005.6359	0.0010
60	12.3971	0.0807	1506.3802	0.0007
65	12.4310	0.0804	2253.4030	0.0004
70	12.4538	0.0803	3367.8301	0.0003
75	12.4690	0.0802	5030.3599	0.0002
80	12.4792	0.0801	7510.5630	0.0001
85	12.4861	0.0801	11210.5911	0.0001
90	12.4907	0.0801	16730.3846	0.0001
95	12.4937	0.0800	24964.9487	0.0000
100	12.4958	0.0800	37249.4748	0.0000

TABLE B.19 Continuous Compounding: $r = 10\%$

	Continuous flow, uniform series			
	Present worth factor	Capital recovery factor	Compound amount factor	Sinking fund factor
n	To find P given $\bar{A}$ $P\|\bar{A}\ r,n$	To find $\bar{A}$ given P $\bar{A}\|P\ r,n$	To find F given $\bar{A}$ $F\|\bar{A}\ r,n$	To find $\bar{A}$ given F $\bar{A}\|F\ r,n$
1	0.9516	1.0508	1.0517	0.9508
2	1.8127	0.5517	2.2140	0.4517
3	2.5918	0.3858	3.4986	0.2858
4	3.2968	0.3033	4.9182	0.2033
5	3.9347	0.2541	6.4872	0.1541
6	4.5119	0.2216	8.2212	0.1216
7	5.0341	0.1986	10.1375	0.0986
8	5.5067	0.1816	12.2554	0.0816
9	5.9343	0.1685	14.5960	0.0685
10	6.3212	0.1582	17.1828	0.0582
11	6.6713	0.1499	20.0417	0.0499
12	6.9881	0.1431	23.2021	0.0431
13	7.2747	0.1375	26.6930	0.0375
14	7.5340	0.1327	30.5520	0.0327
15	7.7687	0.1287	34.8169	0.0287
16	7.9810	0.1253	39.5303	0.0253
17	8.1732	0.1224	44.7395	0.0224
18	8.3470	0.1198	50.4965	0.0198
19	8.5043	0.1176	56.8589	0.0176
20	8.6466	0.1157	63.8906	0.0157
21	8.7754	0.1140	71.6617	0.0140
22	8.8920	0.1125	80.2501	0.0125
23	8.9974	0.1111	89.7418	0.0111
24	9.0928	0.1100	100.2318	0.0100
25	9.1792	0.1089	111.8249	0.0089
26	9.2573	0.1080	124.6374	0.0080
27	9.3279	0.1072	138.7973	0.0072
28	9.3919	0.1065	154.4465	0.0065
29	9.4498	0.1058	171.7415	0.0058
30	9.5021	0.1052	190.8554	0.0052
31	9.5495	0.1047	211.9795	0.0047
32	9.5924	0.1042	235.3253	0.0042
33	9.6312	0.1038	261.1264	0.0038
34	9.6663	0.1035	289.6410	0.0035
35	9.6980	0.1031	321.1545	0.0031
40	9.8168	0.1019	535.9815	0.0019
45	9.8889	0.1011	890.1713	0.0011
50	9.9326	0.1007	1474.1316	0.0007
55	9.9591	0.1004	2436.9193	0.0004
60	9.9752	0.1002	4024.2879	0.0002
65	9.9850	0.1002	6641.4163	0.0002
70	9.9909	0.1001	10956.3316	0.0001
75	9.9945	0.1001	18070.4241	0.0001
80	9.9966	0.1000	29799.5799	0.0000
85	9.9980	0.1000	49137.6884	0.0000
90	9.9988	0.1000	81020.8398	0.0000
95	9.9993	0.1000	133587.2683	0.0000
100	9.995	0.1000	220254.6579	0.0000

TABLE B.20 Continuous Compounding: $r = 15\%$

	Continuous flow, uniform series			
	Present worth factor	Capital recovery factor	Compound amount factor	Sinking fund factor
n	To find P given $\overline{A}$ $P\|\overline{A}\ r,n$	To find $\overline{A}$ given P $\overline{A}\|P\ r,n$	To find F given $\overline{A}$ $F\|\overline{A}\ r,n$	To find $\overline{A}$ given F $\overline{A}\|F\ r,n$
1	0.9286	1.0769	1.0789	0.9269
2	1.7279	0.5787	2.3324	0.4287
3	2.4158	0.4139	3.7887	0.2639
4	3.0079	0.3325	5.4808	0.1825
5	3.5176	0.2843	7.4467	0.1343
6	3.9562	0.2528	9.7307	0.1028
7	4.3337	0.2307	12.3843	0.0807
8	4.6587	0.2147	15.4674	0.0647
9	4.9384	0.2025	19.0495	0.0525
10	5.1791	0.1931	23.2113	0.0431
11	5.3863	0.1857	28.0465	0.0357
12	5.5647	0.1797	33.6643	0.0297
13	5.7182	0.1749	40.1913	0.0249
14	5.8503	0.1709	47.7745	0.0209
15	5.9640	0.1677	56.5849	0.0177
16	6.0619	0.1650	66.8212	0.0150
17	6.1461	0.1627	78.7140	0.0127
18	6.2186	0.1608	92.5315	0.0108
19	6.2810	0.1592	108.5852	0.0092
20	6.3348	0.1579	127.2369	0.0079
21	6.3810	0.1567	148.9071	0.0067
22	6.4208	0.1557	174.0843	0.0057
23	6.4550	0.1549	203.3359	0.0049
24	6.4845	0.1542	237.3216	0.0042
25	6.5099	0.1536	276.8072	0.0036
26	6.5317	0.1532	322.6830	0.0031
27	6.5505	0.1527	375.9830	0.0027
28	6.5667	0.1523	437.9089	0.0023
29	6.5806	0.1520	509.8564	0.0020
30	6.5926	0.1517	593.4475	0.0017

TABLE B.21 Continuous Compounding: $r = 20\%$

	Continuous flow, uniform series			
	Present worth factor	Capital recovery factor	Compound amount factor	Sinking fund factor
n	To find P given $\overline{A}$ $P\|\overline{A}\ r,n$	To find $\overline{A}$ given P $\overline{A}\|P\ r,n$	To find F given $\overline{A}$ $F\|\overline{A}\ r,n$	To find $\overline{A}$ given F $\overline{A}\|F\ r,n$
1	0.9063	1.1033	1.1070	0.9033
2	1.6484	0.6066	2.4591	0.4066
3	2.2559	0.4433	4.1106	0.2433
4	2.7534	0.3632	6.1277	0.1632
5	3.1606	0.3164	8.5914	0.1164
6	3.4940	0.2862	11.6006	0.0862
7	3.7670	0.2655	15.2760	0.0655
8	3.9905	0.2506	19.7652	0.0506
9	4.1735	0.2396	25.2482	0.0396
10	4.3233	0.2313	31.9453	0.0313
11	4.4460	0.2249	40.1251	0.0249
12	4.5464	0.2200	50.1159	0.0200
13	4.6286	0.2160	62.3187	0.0160
14	4.6959	0.2129	77.2332	0.0129
15	4.7511	0.2105	95.4277	0.0105
16	4.7962	0.2085	117.6627	0.0085
17	4.8331	0.2069	144.8205	0.0069
18	4.8634	0.2056	177.9912	0.0056
19	4.8881	0.2046	218.5059	0.0046
20	4.9084	0.2037	267.9908	0.0037
22	4.9386	0.2025	402.2543	0.0025
23	4.9497	0.2020	492.4216	0.0020
24	4.9589	0.2017	602.5521	0.0017
25	4.9663	0.2014	737.0658	0.0014
26	4.9724	0.2011	901.3612	0.0011
27	4.9774	0.2009	1102.0321	0.0009
28	4.9815	0.2007	1347.1320	0.0007
29	4.9849	0.2006	1646.4978	0.0006
30	4.9876	0.2005	2012.1440	0.0005

TABLE B.22 Continuous Compounding: $r = 25\%$

	Continuous flow, uniform series			
	Present worth factor	Capital recovery factor	Compound amount factor	Sinking fund factor
n	To find P given $\overline{A}$ $P\|\overline{A}\ r,n$	To find $\overline{A}$ given P $\overline{A}\|P\ r,n$	To find F given $\overline{A}$ $F\|\overline{A}\ r,n$	To find $\overline{A}$ given F $\overline{A}\|F\ r,n$
1	0.8848	1.1302	1.1361	0.8802
2	1.5739	0.6354	2.5949	0.3854
3	2.1105	0.4738	4.4680	0.2238
4	2.5285	0.3955	6.8731	0.1455
5	2.8540	0.3504	9.9614	0.1004
6	3.1075	0.3218	13.9268	0.0718
7	3.3049	0.3026	19.0184	0.0526
8	3.4587	0.2891	25.5562	0.0391
9	3.5784	0.2795	33.9509	0.0295
10	3.6717	0.2724	44.7300	0.0224
11	3.7443	0.2671	58.5705	0.0171
12	3.8009	0.2631	76.3421	0.0131
13	3.8449	0.2601	99.1614	0.0101
14	3.8792	0.2578	128.4618	0.0078
15	3.9059	0.2560	166.0843	0.0060
16	3.9267	0.2547	214.3926	0.0047
17	3.9429	0.2536	276.4216	0.0036
18	3.9556	0.2528	356.0685	0.0028
19	3.9654	0.2522	458.3371	0.0022
20	3.9730	0.2517	389.6526	0.0017
21	3.9790	0.2513	758.2651	0.0013
22	3.9837	0.2510	974.7677	0.0010
23	3.9873	0.2508	1252.7626	0.0008
24	3.9901	0.2506	1609.7152	0.0006
25	3.9923	0.2505	2068.0513	0.0005
26	3.9940	0.2504	2656.5665	0.0004
27	3.9953	0.2503	3412.2351	0.0003
28	3.9964	0.2502	4382.5326	0.0002
29	3.9972	0.2502	5628.4194	0.0002
30	3.9978	0.2501	7228.1697	0.0001

TABLE B.23 Continuous Compounding: $r = 5\%$

Geometric series present worth factor, $(P|A_1\ r, c, n)_\infty$

n	$c = 4\%$	$c = 6\%$	$c = 8\%$	$c = 10\%$	$c = 15\%$
1	0.9512	0.9512	0.9512	0.9512	0.9512
2	1.8930	1.9120	1.9314	1.9512	2.0025
3	2.8254	2.8825	2.9415	3.0025	3.1643
4	3.7485	3.8627	3.9823	4.1077	4.4484
5	4.6624	4.8527	5.0548	5.2695	5.8674
6	5.5673	5.8527	6.1600	6.4909	7.4357
7	6.4631	6.8628	7.2988	7.7749	9.1690
8	7.3500	7.8830	8.4723	9.1248	11.0845
9	8.2281	8.9134	9.6816	10.5439	13.2015
10	9.0975	9.9542	10.9276	12.0357	15.5412
11	9.9582	11.0055	12.2117	13.6040	18.1269
12	10.8103	12.0673	13.5348	15.2527	20.9845
13	11.6540	13.1398	14.8982	16.9860	24.1427
14	12.4893	14.2231	16.3032	18.8081	27.6331
15	13.3162	15.3173	17.7509	20.7236	31.4905
16	14.1350	16.4225	19.2427	22.7374	35.7536
17	14.9455	17.5388	20.7800	24.8544	40.4651
18	15.7481	18.6663	22.3641	27.0799	45.6721
19	16.5426	19.8051	23.9964	29.4196	51.4267
20	17.3292	20.9554	25.6784	31.8792	57.7865
21	18.1080	22.1172	27.4116	34.4649	64.8152
22	18.8791	23.2907	29.1977	37.1832	72.5831
23	19.6425	24.4760	31.0381	40.0408	81.1679
24	20.3982	25.6732	32.9346	43.0450	90.6557
25	21.1465	26.8825	34.8888	46.2032	101.1412
26	21.8873	28.1039	36.9026	49.5233	112.7296
27	22.6208	29.3376	38.9777	53.0136	125.5367
28	23.3469	30.5836	41.1159	56.6829	139.6907
29	24.0658	31.8422	43.3193	60.5404	155.3334
30	24.7776	33.1135	45.5898	64.5956	172.6211
31	25.4823	34.3975	47.9295	68.8587	191.7271
32	26.1800	35.6944	50.3404	73.3404	212.8424
33	26.8707	37.0044	52.8247	78.0518	236.1785
34	27.5546	38.3275	55.3847	83.0049	261.9688
35	28.2316	39.6640	58.0226	88.2118	290.4716
36	28.9019	41.0138	60.7409	93.6858	321.9720
37	29.5656	42.3772	63.5420	99.4404	356.7853
38	30.2226	43.7544	66.4284	105.4900	395.2600
39	30.8732	45.1453	69.4026	111.8499	437.7810
40	31.5172	46.5503	72.4675	118.5358	484.7741
41	32.1548	47.9694	75.6257	125.5644	536.7095
42	32.7861	49.4027	78.8801	132.9535	594.1069
43	33.4111	50.8504	82.2335	140.7214	657.5409
44	34.0299	52.3127	85.6892	148.8876	727.6463
45	34.6425	53.7897	89.2500	157.4724	805.1248
46	35.2490	55.2815	92.9193	166.4974	890.7517
47	35.8495	56.7883	96.7003	175.9852	985.3842
48	36.4441	58.3103	100.5965	185.9594	1089.9691
49	37.0327	59.8475	104.6114	196.4449	1205.5534
50	37.6154	61.4002	108.7485	207.4681	1333.2938

TABLE B.24 Continuous Compounding: $r = 5\%$

	Geometric series future worth factor, $(F\|A_1\ r, c, n)_\infty$				
n	$c = 4\%$	$c = 6\%$	$c = 8\%$	$c = 10\%$	$c = 15\%$
1	1.0000	1.0000	1.0000	1.0000	1.0000
2	2.0921	2.1131	2.1346	2.1564	2.2131
3	3.2826	3.3489	3.4175	3.4884	3.6764
4	4.5784	4.7179	4.8640	5.0171	5.4332
5	5.9867	6.2310	6.4905	6.7662	7.5339
6	7.5150	7.9003	8.3151	8.7618	10.0372
7	9.1716	9.7387	10.3575	11.0332	13.0114
8	10.9650	11.7600	12.6392	13.6126	16.5362
9	12.9043	13.9790	15.1837	16.5361	20.7041
10	14.9992	16.4118	18.0166	19.8435	25.6231
11	17.2601	19.0753	21.1659	23.5792	31.4185
12	19.6977	21.9881	24.6620	27.7923	38.2363
13	22.3237	25.1699	28.5381	32.5373	46.2464
14	25.1503	28.6419	32.8305	37.8748	55.6462
15	28.1905	32.4267	37.5786	43.8719	66.6654
16	31.4579	36.5489	42.8255	50.6030	79.5711
17	34.9673	41.0345	48.6178	58.1505	94.6740
18	38.7340	45.9116	55.0067	66.6059	112.3352
19	42.7743	51.2102	62.0476	76.0705	132.9744
20	47.1057	56.9626	69.8011	86.6566	157.0800
21	51.7464	63.2032	78.3329	98.4886	185.2191
22	56.7159	69.9691	87.7147	111.7044	218.0516
23	62.0347	77.2999	98.0244	126.4566	256.3440
24	67.7245	85.2381	109.3467	142.9144	300.9874
25	73.8085	93.8290	121.7740	161.2649	353.0176
26	80.3111	103.1215	135.4066	181.7157	413.6383
27	87.2579	113.1674	150.3535	204.4962	484.2484
28	94.6764	124.0227	166.7334	229.8606	566.4738
29	102.5954	135.7471	184.6753	258.0905	662.2039
30	111.0455	148.4043	204.3195	289.4972	773.6343
31	120.0591	162.0628	225.8184	324.4256	903.3165
32	129.6703	176.7957	249.3376	363.2572	1054.2155
33	139.9152	192.6812	275.0572	406.4143	1229.7767
34	150.8323	209.8029	303.1729	454.3643	1434.0037
35	162.4618	228.2503	333.8972	507.6241	1671.5485
36	174.8466	248.1191	367.4612	566.7660	1947.8169
37	188.0319	269.5116	404.1156	632.4230	2269.0901
38	202.0654	292.5371	444.1330	705.2953	2642.6664
39	216.9977	317.3125	487.8094	786.1577	3077.0262
40	232.8823	343.9627	535.4663	875.8673	3582.0230
41	249.7754	372.6212	587.4528	975.3722	4169.1061
42	267.7369	403.4307	644.1479	1085.7209	4851.5781
43	286.8296	436.5436	705.9633	1208.0733	5644.8957
44	307.1202	472.1228	773.3457	1343.7123	6567.0108
45	328.6790	510.3423	846.7805	1494.0568	7638.8114
46	351.5804	551.3878	926.7940	1660.6759	8884.5204
47	375.9028	595.4579	1013.9582	1845.3049	10332.3143
48	401.7293	642.7645	1108.8934	2049.8628	12014.9221
49	429.1474	693.5341	1212.2730	2276.4720	13970.3711
50	458.2495	748.0082	1324.8280	2527.4790	16242.8438

TABLE B.25 Continuous Compounding: $r = 8\%$

Geometric series present worth factor, $(P|A_1\ r, c, n)_\infty$

n	$c = 4\%$	$c = 6\%$	$c = 8\%$	$c = 10\%$	$c = 15\%$
1	0.9231	0.9231	0.9231	0.9231	0.9231
2	1.8100	1.8280	1.8462	1.8649	1.9132
3	2.6622	2.7149	2.7693	2.8257	2.9750
4	3.4809	3.5842	3.6925	3.8059	4.1138
5	4.2675	4.4364	4.6156	4.8059	5.3352
6	5.0233	5.2716	5.5387	5.8261	6.6452
7	5.7495	6.0904	6.4618	6.8669	8.0501
8	6.4471	6.8929	7.3849	7.9287	9.5570
9	7.1175	7.6795	8.3080	9.0120	11.1730
10	7.7615	8.4506	9.2312	10.1172	12.9063
11	8.3803	9.2064	10.1543	11.2447	14.7652
12	8.9748	9.9472	11.0774	12.3949	16.7589
13	9.5460	10.6733	12.0005	13.5685	18.8972
14	10.0948	11.3851	12.9236	14.7657	21.1905
15	10.6221	12.0828	13.8467	15.9871	23.6501
16	11.1287	12.7666	14.7699	17.2332	26.2881
17	11.6155	13.4370	15.6930	18.5044	29.1173
18	12.0832	14.0940	16.6161	19.8013	32.1517
19	12.5325	14.7380	17.5392	21.1245	35.4060
20	12.9642	15.3693	18.4623	22.4743	38.8964
21	13.3790	15.9881	19.3854	23.8514	42.6398
22	13.7775	16.5946	20.3086	25.2564	46.6546
23	14.1604	17.1892	21.2317	26.6897	50.9606
24	14.5283	17.7719	22.1548	28.1520	55.5788
25	14.8817	18.3431	23.0779	29.6438	60.5318
26	15.2213	18.9030	24.0010	32.1658	65.8440
27	15.5476	19.4518	24.9241	32.7185	71.5413
28	15.8611	19.9898	25.8473	34.3026	77.6518
29	16.1623	20.5171	26.7704	35.9187	84.2053
30	16.4517	21.0339	27.6935	37.5674	91.2340
31	16.7297	21.5404	28.6166	39.2494	98.7723
32	16.9968	22.0371	29.5397	40.9654	106.8572
33	17.2535	22.5239	30.4628	42.7161	115.5283
34	17.5001	23.0010	31.3860	44.5021	124.8282
35	17.7370	23.4686	32.3091	46.3242	134.8024
36	17.9647	23.9271	33.2322	48.1832	145.4998
37	18.1834	24.3764	34.1553	50.0796	156.9728
38	18.3935	24.8168	35.0784	52.0144	169.2778
39	18.5954	25.2485	36.0015	53.9883	182.4749
40	18.7894	25.6717	36.9247	56.0021	196.6289
41	18.9758	26.0865	37.8478	58.0565	211.8092
42	19.1548	26.4930	38.7709	60.1524	228.0903
43	19.3269	26.8916	39.6940	62.2907	245.5518
44	19.4922	27.2822	40.6171	64.4722	264.2794
45	19.6510	27.6651	41.5402	66.6977	284.3650
46	19.8036	28.0404	42.4634	68.9682	305.9069
47	19.9502	28.4083	43.3865	71.2846	329.0107
48	20.0910	28.7689	44.3096	73.6478	353.7898
49	20.2264	29.1223	45.2327	76.0587	380.3656
50	20.3564	29.4688	46.1558	78.5183	408.8683

TABLE B.26 Continuous Compounding: $r = 8\%$

Geometric series future worth factor, $(F \mid A, r, c, n)_\infty$

n	$c = 4\%$	$c = 6\%$	$c = 8\%$	$c = 10\%$	$c = 15\%$
1	1.0000	1.0000	1.0000	1.0000	1.0000
2	2.1241	2.145	2.1666	2.1885	2.2451
3	3.3843	3.4513	3.5205	3.5921	3.7820
4	4.7937	4.9359	5.0850	5.2412	5.6653
5	6.3664	6.6183	6.8856	7.1695	7.9592
6	8.1181	8.5194	8.9509	9.4154	10.7391
7	10.0654	10.6623	11.3125	12.0217	14.0932
8	12.2269	13.0722	14.0054	15.0367	18.1246
9	14.6224	15.7771	17.0683	18.5146	22.9543
10	17.2735	18.8071	20.5543	22.5162	28.7235
11	20.2040	22.1956	24.4810	27.1098	35.5975
12	23.4395	25.9790	28.9308	32.3718	43.7693
13	27.0078	30.1972	33.9521	38.3881	53.4643
14	30.9392	34.8937	39.6090	45.2546	64.9459
15	35.2667	40.1162	45.9728	53.0790	78.5212
16	40.0261	45.9170	53.1219	61.9814	94.5487
17	45.2562	52.3530	61.1429	72.0967	113.4466
18	50.9993	59.4865	70.1315	83.5754	135.7023
19	57.3014	67.3856	80.1932	96.5858	161.8843
20	64.2121	76.1247	91.4445	111.3160	192.6550
21	71.7857	85.7851	104.0137	127.9763	228.7862
22	80.0809	96.4553	118.0422	146.8012	271.1772
23	89.1614	108.2322	133.6861	168.0529	320.8754
24	99.0967	121.2214	151.1169	192.0237	379.1005
25	109.9619	135.5383	170.5240	219.0400	447.2729
26	121.8386	151.3086	192.1155	249.4657	527.0461
27	134.8154	168.6694	216.1207	283.7067	620.3446
28	148.9884	187.7705	242.7919	322.2155	729.4088
29	164.4621	208.7749	272.4066	365.4965	856.8454
30	181.3496	231.8605	305.2702	414.1118	1005.6880
31	199.7738	357.2211	341.7185	468.6875	1179.4660
32	219.8680	285.0681	382.1205	529.9210	1382.2852
33	241.7768	315.6315	426.8820	598.5891	1618.9221
34	265.6571	349.1623	476.4489	675.5565	1894.9324
35	291.6791	385.9336	531.3113	761.7857	2216.7776
36	320.0274	426.2430	592.0073	858.3481	2591.9728
37	350.9022	470.4147	659.1281	966.4356	3029.2571
38	384.5208	518.8015	733.3229	1087.3745	3538.7925
39	421.1186	571.7877	815.3045	1222.6399	4132.3956
40	460.9512	629.7914	905.8552	1373.8725	4823.8051
41	504.2955	693.2681	1005.8337	1542.8964	5628.9945
42	551.4520	762.7131	1116.1825	1731.7400	6566.5343
43	602.7463	838.6659	1237.9352	1942.6579	7658.0136
44	658.5318	921.7130	1372.2262	2178.1560	8928.5294
45	791.1914	1012.4930	1520.2993	2441.0191	10407.2556
46	785.1404	1111.7003	1683.5188	2734.3416	12128.1042
47	856.8290	1220.0904	1863.3805	3061.5612	14130.4931
48	934.7453	1338.4850	2061.5244	3426.4968	16460.2392
49	1019.4185	1467.7778	2279.7482	3833.3901	19170.5950
50	1111.4222	1608.9405	2520.0222	4286.9517	22323.4542

TABLE B.27 Continuous Compounding: $r = 10\%$

Geometric series present worth factor, $(P|A_1\ r, c, n)_\infty$

n	$c = 4\%$	$c = 6\%$	$c = 8\%$	$c = 10\%$	$c = 15\%$
1	0.9048	0.9048	0.9048	0.9048	0.9048
2	1.7570	1.7742	1.7918	1.8097	1.8561
3	2.5595	2.6095	2.6611	2.7145	2.8561
4	3.3153	3.4120	3.5133	3.6193	3.9073
5	4.0271	4.1830	4.3485	4.5242	5.0125
6	4.6974	4.9239	5.1673	5.4290	6.1743
7	5.3287	5.6356	5.9698	6.3339	7.3957
8	5.9232	6.3195	6.7564	7.2387	8.6798
9	6.4831	6.9765	7.5275	8.1435	10.0296
10	7.0104	7.6078	8.2832	9.0484	11.4487
11	7.5070	8.2143	9.0241	9.9532	12.9405
12	7.9746	8.7971	9.7502	10.8580	14.5088
13	8.4151	9.3570	10.4620	11.7629	16.1576
14	8.8298	9.8949	11.1597	12.6677	17.8908
15	9.2205	10.4118	11.8435	13.5726	19.7129
16	9.5883	10.9084	12.5138	14.4774	21.6285
17	9.9348	11.3855	13.1709	15.3822	23.6422
18	10.2611	11.8439	13.8149	16.2871	25.7592
19	10.5684	12.2843	14.4462	17.1919	27.9848
20	10.8577	12.7075	15.0650	18.0967	30.3244
21	11.1303	13.1141	15.6715	19.0016	32.7840
22	11.3869	13.5047	16.2660	19.9064	35.3697
23	11.6286	13.8800	16.8488	20.8113	38.0880
24	11.8563	14.2406	17.4200	21.7161	40.9457
25	12.0707	14.5870	17.9799	22.6209	43.9498
26	12.2726	14.9199	18.5287	23.5258	47.1080
27	12.4627	15.2397	19.0667	24.4306	50.4281
28	12.6418	15.5470	19.5939	25.3354	53.9185
29	12.8104	15.8422	20.1108	26.2403	57.5878
30	12.9692	16.1259	20.6174	27.1451	61.4452
31	13.1188	16.3984	21.1140	28.0500	65.5004
32	13.2597	16.6603	21.6007	28.9548	69.7635
33	13.3923	16.9119	22.0779	29.8596	74.2452
34	13.5172	17.1536	22.5455	30.7645	78.9567
35	13.6349	17.3858	23.0039	31.6693	83.9097
36	13.7457	17.6089	23.4533	32.5741	89.1167
37	13.8500	17.8233	23.8937	33.4790	94.5906
38	13.9483	18.0293	24.3254	34.3838	100.3452
39	14.0409	18.2272	24.7486	35.2887	106.3949
40	14.1280	18.4173	25.1634	36.1935	112.7547
41	14.2101	18.6000	25.5699	37.0983	119.4406
42	14.2874	18.7755	25.9684	38.0032	126.4693
43	14.3602	18.9442	26.3591	38.9080	133.8583
44	14.4288	19.1062	26.7420	39.8128	141.6262
45	14.4934	19.2619	27.1173	40.7177	149.7924
46	14.5542	19.4114	27.4852	41.6225	158.3773
47	14.6114	19.5551	27.8457	42.5274	167.4023
48	14.6654	19.6932	28.1992	43.4322	176.8900
49	14.7162	19.8259	28.5457	44.3370	186.8642
50	14.7640	19.9533	28.8853	45.2419	197.3498

TABLE B.28 Continuous Compounding: $r = 10\%$

Geometric series future worth factor, $(F|A_1\ r, c, n)_\infty$

n	$c = 4\%$	$c = 6\%$	$c = 8\%$	$c = 10\%$	$c = 15\%$
1	1.0000	1.0000	1.0000	1.0000	1.0000
2	2.1460	2.1670	2.1885	2.2103	2.2670
3	3.4550	3.5224	3.5921	3.6642	3.8553
4	4.9458	5.0901	5.2412	5.3994	5.8291
5	6.6395	6.8967	7.1695	7.4591	8.2642
6	8.5592	8.9718	9.4154	9.8923	11.2504
7	10.7306	11.3488	12.0217	12.7548	14.8932
8	13.1823	14.0643	15.0367	16.1100	19.3172
9	15.9458	17.1595	18.5146	20.0299	24.6689
10	19.0562	20.6802	22.5162	24.5960	31.1208
11	22.5521	24.6773	27.1098	29.9011	38.8755
12	26.4767	29.2074	32.3718	36.0500	48.1710
13	30.8773	34.3336	38.3881	43.1615	59.2869
14	35.8067	40.1260	45.2546	51.3702	72.5508
15	41.3232	46.6624	53.0790	60.8280	88.3472
16	47.4914	54.0295	61.9814	71.7070	107.1265
17	54.3826	62.3236	72.0967	84.2016	129.4163
18	62.0759	71.6514	83.5754	98.5311	155.8342
19	70.6589	82.1317	96.5858	114.9433	187.1032
20	80.2285	93.8963	111.3160	133.7179	224.0688
21	90.8917	107.0916	127.9763	155.1702	267.7198
22	102.7672	121.8800	146.8012	179.6557	319.2122
23	115.9863	138.4416	168.0529	207.5753	379.8967
24	130.6939	156.9766	192.0237	239.3804	451.3512
25	147.0508	177.7066	219.0400	275.5794	535.4184
26	165.2346	200.8779	249.4657	316.7448	634.2500
27	185.4417	226.7632	283.7067	363.5209	750.3571
28	207.8894	255.6652	322.2155	416.6325	886.6703
29	232.8182	287.9193	365.4965	476.8948	1046.6085
30	260.4938	323.8974	414.1118	545.2244	1234.1597
31	291.2103	364.0116	468.6875	622.6516	1453.9746
32	325.2928	408.7188	529.9210	710.3344	1711.4754
33	363.1008	458.5251	598.5891	809.5735	2012.9833
34	405.0318	513.9913	675.5565	921.8297	2365.8655
35	451.5256	575.7389	761.7857	1048.7435	2778.7077
36	503.0682	644.4560	858.3481	1192.1563	3261.5132
37	560.1970	720.9052	966.4356	1354.1347	3825.9360
38	623.5064	805.9308	1087.3845	1536.9976	4485.5507
39	693.6533	900.4679	1222.6399	1743.3462	5256.1676
40	771.3643	1005.5522	1373.8725	1976.0980	6156.1980
41	857.4424	1122.3303	1542.8964	2238.5242	7207.0797
42	952.7756	1252.0716	1731.7400	2534.2921	8433.7723
43	1058.3454	1396.1817	1942.6579	2867.5122	9865.3318
44	1175.2371	1556.2165	2178.1560	3242.7909	11535.5801
45	1304.6503	1733.8984	2441.0191	3665.2891	13483.8828
46	1447.9113	1931.1339	2734.3416	4140.7880	15756.0539
47	1606.4860	2150.0328	3061.5612	4675.7628	18405.4073
48	1781.9951	2392.9306	3426.4968	5277.4643	21493.9796
49	1976.2301	2662.4116	3833.3901	5954.0105	25093.9520
50	2191.1713	2961.3357	4286.9517	6714.4890	29289.3025

TABLE B.29 Continuous Compounding: $r = 15\%$

Geometric series present worth factor, $(P|A_1\ r, c, n)_\infty$

n	c = 4%	c = 6%	c = 8%	c = 10%	c = 15%
1	0.8607	0.8607	0.8607	0.8607	0.8607
2	1.6318	1.6472	1.6632	1.6794	1.7214
3	2.3225	2.3663	2.4115	2.4582	2.5821
4	2.9413	3.0233	3.1092	3.1991	3.4428
5	3.4956	3.6238	3.7597	3.9037	4.3035
6	3.9922	4.1726	4.3662	4.5741	5.1642
7	4.4370	4.6742	4.9317	5.2117	6.0250
8	4.8356	5.1326	5.4590	5.8182	6.8857
9	5.1926	5.5515	5.9507	6.3952	7.7464
10	5.5124	5.9344	6.4091	6.9440	8.6071
11	5.7989	6.2844	6.8365	7.4660	9.4678
12	6.0556	6.6042	7.2350	7.9626	10.3285
13	6.2855	6.8965	7.6066	8.4350	11.1892
14	6.4915	7.1636	7.9530	8.8843	12.0499
15	6.6760	7.4078	8.2761	9.3117	12.9106
16	6.8413	7.6309	8.5773	9.7183	13.7713
17	6.9894	7.8348	8.8581	10.1050	14.6320
18	7.1220	8.0212	9.1199	10.4729	15.4927
19	7.2409	8.1915	9.3641	10.8229	16.3535
20	7.3472	8.3472	9.5917	11.1557	17.2142
21	7.4427	8.4895	9.8040	11.4724	18.0749
22	7.5281	8.6195	10.0019	11.7736	18.9356
23	7.6046	8.7383	10.1864	12.0601	19.7963
24	7.6732	8.8470	10.3584	12.3326	20.6570
25	7.7346	8.9462	10.5189	12.5918	21.5177
26	7.7897	9.0369	10.6684	12.8384	22.3784
27	7.8389	9.1198	10.8079	13.0730	23.2391
28	7.8831	9.1956	10.9379	13.2961	24.0998
29	7.9227	9.2649	11.0591	13.5084	24.9605
30	7.9581	9.3282	11.1722	13.7103	25.8212
31	7.9898	9.3860	11.2776	13.9023	26.6819
32	8.0183	9.4389	11.3759	14.0850	27.5427
33	8.0438	9.4872	11.4675	14.2588	28.4034
34	8.0666	9.5313	11.5529	14.4241	29.2641
35	8.0870	9.5717	11.6326	14.5813	30.1248
36	8.1053	9.6086	11.7069	14.7309	30.9855
37	8.1218	9.6423	11.7761	14.8732	31.8462
38	8.1365	9.6731	11.8407	15.0085	32.7069
39	8.1496	9.7013	11.9009	15.1372	33.5676
40	8.1614	9.7270	11.9570	15.2597	34.4283
41	8.1720	9.7505	12.0094	15.3762	35.2890
42	8.1814	9.7720	12.0582	15.4870	36.1497
43	8.1899	9.7916	12.1037	15.5924	37.0104
44	8.1975	9.8096	12.1461	15.6926	37.8712
45	8.2043	9.8260	12.1856	15.7880	38.7319
46	8.2104	9.8410	12.2225	15.8787	39.5926
47	8.2159	9.8547	12.2569	15.9650	40.4533
48	8.2208	9.8672	12.2890	16.0471	41.3140
49	8.2252	9.8787	12.3189	16.1252	42.1747
50	8.2291	9.8891	12.3468	16.1995	43.0354

Appendix C
Answers to Even-Numbered Problems

Chapter 2

2. (a) $56.7/m^3
   (b) $73.5/m^3
   (c) $3.89/m^3 increase

4. (a) $130.22
   (b) $54.12
   (c) $173.63

6. $967 300

8. (a) $306 100
   (b) $316 190

10. (a) k = 68.5, s = − 0.1673
    (b) 46.60
    (c) 3 776.8

12. 46 898

Chapter 3

2. (a) $308, $333, $359
   (b) $80, $55, $29

4. 20%

6. $4 467.20

8. (a) $3 833.80
   (b) $4 063.83

10. (a) $11 559
    (b) $12 228.83

12. $1 296.10

14. (a) $1 759.98
    (b) $2 394.45

16. $10 000

18. $2 879.30

20. $8 618.46

22. $14 108.60

24. − $979.50

26. $669.10

28. $2 570.45

30. $1 021.84

32. $1 360.19

34. $53.20

36. $5 730.40

38. $3 090.60

40. 14 years

42. 42.576%

44. $107 654.07

46. $1 325.45

48. $6 025.70

50. $3 480.91

52. − $809.88

54. $1 000.22

56. $2 628.25

58. $1 788.70

60. 6.68%

62. $1 543.79

64. − $1 049.18

66. $9 558.69

68. $28 583.80

70. $6 929.18

72. (a) $6 665.07
    (b) 7.45%
    (c) 8.325%

74. $E_1 = \$887.05$ $I_1 = \$299.95$
    $E_2 = \$940.22$ $I_2 = \$246.78$
    $E_3 = \$996.61$ $I_3 = \$190.39$
    $E_4 = \$1\,056.43$ $I_4 = \$130.57$
    $E_5 = \$1\,119.82$ $I_5 = \$67.18$

76. $5 539.79

78. $1 408.10

80. $2 848.41

82. $1 858.36

84. (a) $4 303.45
    (b) 5.6%

86. $P = \$13\,889.75$ $A = \$3\,321.84$

88. $6 303.00

90. $5 373.84

92. $2 956.83

94. 14.36% annual compounding

96. $2 115.75

98. (a) $4 290.60
    (b) 12.61%

100. $37 921.78

102. $23 880.00

Chapter 4

2. A and D
   $PW = \$4\,252.50$

4. Do nothing.

6. $AW(X) = \$18\,031$
   $AW(W \text{ and } Z) = \$6\,364$
   $AW(Y \text{ and } Z) = \$8\,818$
   $AW(Z) = \$0$

8. 0%

10. $6 840.40

12. $6 068.20

14. (a) Do nothing; A; B; D; C and D.
    (c) $PW(\text{do nothing}) = \$0$
    $PW(A) = \$269\,470$
    $PW(B) = \$303\,365$
    $PW(D) = \$240\,945$
    $PW(C, D) = \$431\,890$

16. − $12 330

18. (a) $EUAC_A(0\%) = \$2\,108$
    $EUAC_B(0\%) = \$2\,032$
    $EUAC_C(0\%) = \$2\,525$
    (b) $EUAC_A(8\%) = \$2\,565$
    $EUAC_B(8\%) = \$2\,764$
    $EUAC_C(8\%) = \$3\,460$
    (c) $EUAC_A(10\%) = \$2\,693$
    $EUAC_B(10\%) = \$2\,962$
    $EUAC_C(10\%) = \$3\,713$

20. $EUAC_A(10\%) = \$2\,924.29$
    $EUAC_B(10\%) = \$4\,066.36$

22. $EUAC_A(10\%) = \$4\,888.27$
    $EUAC_B(10\%) = \$5\,140.28$

24. Holding pond yields an equivalent annual savings of $1 050 when compared to the storage tank alternative.

26. Alternative 2

28. 2-cm insulation

30. Compressor B is preferred.

Answers to Even-Numbered Problems

32. (a) (i-iv) contract
    (b) (i, ii) purchase equipment,
        (iii-iv) contract
    (c) (iii, iv) purchase equipment
    (d) (i-iv) purchase equipment

34. Compressor Y has the largest PW.
    $PW_X = -\$70\ 356$
    $PW_Y = -\$64\ 608$

36. $i^*_{B-A} < MARR$ Choose A.

38. Choose B.

40. $64 570 end-of-year lease payment or $58 700 beginning-of-year lease payment.

42. Select proposal C; $AW = \$12\ 300$.

44. Leasing is preferred by $4 500 per year.

46. $9 095 end-of-year lease payment or $8 268 beginning-of-year lease payment.

48. (a) Keep
    (b) Replace

50. Replace

52. Yes

54. Purchase X;
    annual savings = $12 695

56. Replace old mixer;
    annual savings = $352.80

58. $EUAC$(old lathe) = $8 182.80
    $EUAC$(new lathe) = $7 082.80
    $EUAC$(subcontract) = $6 681.00
    Subcontract!

60. PW(overhaul) = $-\$24\ 963.50$
    PW(replace) = $-\$20\ 540.60$

62. Subcontract;
    annual savings = $1 894.40

64. (a) Replace every 10 years with infinite planning period.
    (b) Replace after 9 years service; keep the replacement for 6 years.
    (c) Replace every 7 years with infinite planning period; replace after 7 or 8 years and keep the replacement for the balance of the planning period.
    (d) The greater the discount rate, the greater the economic life of an asset.

66. (a) Replace every 10 or 11 years.
    (b) The maximum life of 12 years is the economic life.

Chapter 5

2. (a) $20 000, $0 (no switch to straight line)
   (b) $24 116.89, $4 116.89
   (c) $20 000,00, $0.00

4. $10 000, $8 000, $6 400, $5 120

6. $1 306.30

8. $2 250, $1 200, $600, $450

10. (a) $9 000
    (b) $18 000
    (c) $38 000
    (d) $248 000

12. $2.2237/unit

14. $PW(12) = \$930.34$. Yes, invest.

16. $12 418.40

18. $PW(10) = \$288.01$. Yes, invest.

20. (a) A: −$11 448.20
        B: −$17 310.25
    (b) A: −$12 475.60
        B: −$15 286.00
    (c) A: −$11 448.20
        B: −$15 286.00

22. $191.76

24. Figures for year 2 only are shown.
    (a) $9 482, $3 138, $30 000,
        −$8 138, $4 069, $16 449
    (b) $10 000, $3 000, $30 000,
        −$8 000, −$4 069, $16 000

26. $5 559.76 per year.

28. $1 375, $1 350

30. −$100 000, $20 000, $20 000,
    $20 000, $20 000, $65 000

32. $33 261.58

34. $PW(10) = \$4\ 156.80$. Yes, purchase incinerator.

36. 10 years, $2 336.85 per year.

38. $42 765.09

40. Keep:
    $AW(25) = \$11\ 827.10$ per year.
    Subcontract:
    $AW(25) = \$15\ 690.94$ per year.
    Prefer to subcontract.

Chapter 6

2. $12.89 per hour.

4. 0.7412

6. Route B; Route C.

8. $B - C = \$30\ 351.40$ per year.
   Yes, build the new addition.

10. Projects A and B. Opportunity cost slightly less than $i = 20\%$.

12. $B - C = \$1\ 453\ 751.30$ for preferred project C at $i = 5\%$.
    $B - C = \$51\ 562.40$ for preferred project A at $i = 15\%$.
    Higher discounting emphasized higher yearly benefits of project A during early years.

14. $B/C = 1.6667$;
    $B - C = \$58\ 400$ per year.
    Yes, build.

16. $B - C = \$6\ 351.40$ per year.
    Yes, build.

18. The fallacy is that the 9 000 persons receiving benefits of $1.75 will not patronize the facility. Also, the entrance fee was not deducted from the benefits. True $B/C = 0.50$. True $B - C = -\$9\ 000$.

20. (a) 10; (b) 1.1286; (c) $45 000; (d) Prefer $B - C$ as it would remain invariant.

22. Design 1 best. Present worth cost = $31 530 320.

Chapter 7

2. 3 years.

4. (a) 37.3%, (b) 60.7%, (c) 88.0%, (d) 117.8%

6. $7 344.

8. If annual usage is less than 673.41 h, use Z. If annual usage is greater than 758.16 h, use X. Otherwise, use Y.

10. $132.10 per year.

12. If production volume is less than 7.5, use B. Otherwise, use A.

14. Pessimistic value = 156.1%.
    Optimistic value = 53.2%.

16. Sensitivity analysis yields the following breakeven equation:
    $Y = -0.0711 + 0.3664X$.
    Combinations of X and Y above the breakeven line result in $AW > 0$.

18. 0.25

20. 0.29

22.

PW	p(PW)
−2 000	0.0016
−1 000	0.0160
0	0.0696
1 000	0.1720
2 000	0.2641
3 000	0.2580
4 000	0.1566
5 000	0.0540
6 000	0.0081

24. (a) $E[PW] = 14\,000$
$V[PW] = 896$
$Pr(PW \geq 0) = 1.000$
$Pr(AW \geq 0) = Pr(PW \geq 0)$
(b) $E[PW] = -\$9762$
$V[PW] = 664.804 \times 10^4$
$Pr(PW \geq 0) = 0.0001$

26. Probability A is best $= 0.2269$.

28. Cumulative averge
$PW = 662.50$.

30.

Trial	ERR	Trial	ERR
1	16.17%	6	15.83%
2	6.83%	7	23.71%
3	14.02%	8	27.07%
4	17.88%	9	24.74%
5	21.87%	10	21.07%

32. $E[PW] = -\$18\,097.73$

34. (a) $3.88
 (b) − $100.0
 (c) − $0.000216
 (d) − $0.0098
 (e) $79.8

Chapter 8

2. If $0 \leq \alpha < 1/3$, choose A_1
$\alpha = 1/3$, choose either A_1 or A_3
$1/3 \leq \alpha < 1/2$, choose A_3
$\alpha = 1/2$, choose either A_3 or A_2
$1/2 \leq \alpha \leq 1.0$, choose A_2

4. Laplace principle: Choose A_1 with $E(A_1) = \$13\,000$.
Maximin principle: Choose either A_1 or A_5 with value $= \$8\,000$.
Maximax principle: Choose A_2 with value $= \$20\,000$.
Savage principle: Choose A_1 with minimax regret value $= \$4\,000$.
Hurwicz principle: Choose A_1 or A_5 if $0 \leq \alpha < 1/3$. Choose either A_1, A_2 or A_5 if $\alpha = 1/3$. Choose A_2 if $1/3 \leq \alpha \leq 1.0$.

6. Minimax principle: Choose A_2 with minimax value $= 60$.
Minimin principle: Choose A_1 with minimin value $= 10$.
Minimax regret: Choose either A_2 or A_3 with minimax regret value $= 50$.
Expected cost: Choose A_2 with $E(A_2) = 60$ and zero variance.
Hurwicz principle: Choose A_2 with Hurwicz value $= 60$.

8. Minimax principle: Choose A_2 with minimax value $= 200$.
Minimin principle: Choose A_1 with minimin value $= 100$.
Minimax regret: Choose A_3 with minimax regret value $= 75$.

10. Keep present equipment with $E(\text{Keep}) = \$84\,000$. The most probable future principle would select the purchase alternative, ignoring the $100 000 loss possibility.

12. (a) $p(S_1) = 0.216$; $p(S_2) = 0.432$;
    $p(S_3) = 0.288$; $p(S_4) = 0.064$

    (b)

Guess	0.216 S_1	0.432 S_2	0.288 S_3	0.064 S_4	$E(A_i)$
S_1	$3	−$1	−$1	−$1	−0.136
S_2	−$1	$2	−$1	−$1	0.296
S_3	−$1	−$1	$3	−$1	0.152
S_4	−$1	−$1	−$1	$4	−0.680

    Both the expectation principle and most probable future principle selects the "Guess S_2" Alternative.

14. A solution can be obtained using the conditional probability theorem directly. However, using Bayes's theorem, Supplier C is the most likely producer of Model R with $P(C|R) = 0.507$ and, similarly, $P(A|R) = 0.338$ and $P(B|R) = 0.155$.

16. (a) $P(S_1|O_1) = 0.8235$;
    $P(S_2|O_1) = 0.1765$;
    $P(S_1|O_2) = 0.2258$;
    $P(S_2|O_2) = 0.7742$;
    $P(S_1|O_3) = 1.0000$;
    $P(S_2|O_3) = 0.0000$
    (c) Choose A_1 with $E(A_1) = 1.2$ versus E(send recon. flight) = 2.232 and $E(A_2) = 8.4$.

18. (b) Keep present truck, with expected value = −$21 700.

20. Choose A_1 with
    $E(A_1) = 0.9078 > E(A_2) = 0.8799$.

Chapter 9

2. Net loss before taxes = $3 025

4. (a) Total cost = $24 752.80.
   (b) Unit selling price = $4.95.

6. Overhead rates are $2.44/h, $2.73/h, $4.91/h, and $2.36/h for cost centres A, B, C, and D, respectively.

Chapter 10

2. $40.5 $\times 10^6$ per year.

4. (a) $Q_d = 6.55 \times 10^6$ kg per year.
   $Q_s = 3.50 \times 10^6$ kg per year.
   (b) 3.05×10^6 kg per year.
   (c) 1.55×10^6 kg per year.
   (d) −$130 000 000 per year.
   (e) $69.5 per kg.
   (f) the commodity would trade at equilibrium price.

6. (a) $97 500 000
   (b) $217 000 000
   (c) $217 000 000
   (d) $252 500 000

8. 2 units, − 1.26 $\times 10^{-4}$ units per hour, − 3 $\times 10^{-4}$ units per hour,

10. (a) $1 014 000 per year, 62 000 h per year.
    (b) − 2000 h, $6 000 per year.

12. (a) $1 984 000 per year, 4 000 hours per year, 22 units of machinery.
    (b) $2 142 000 per year, − 10 000 hours per year, 44 units of machinery.

14. 281 200 units, 76 000 h per year, 37 units, or 281 200 units, 74 000 h per year, 38 units.

Appendix D
Glossary of Technical Terms[1]

Account payable
>An amount owing to creditors, representing a liability, for purchases of goods or services.
>
>*créditeur, compte-fournisseur*

Account receivable
>An amount claimed against a debtor, usually money rights arising from the sale of goods or services.
>
>*débiteur, compte-client, créance, compte à reçevoir*

Amortization
>1. (a) As applied to a capitalized asset, the distribution of the initial cost by periodic charges to operations as in depreciation. Most properly applies to assets with indefinite life. (b) The reduction of a debt by either periodic or irregular payments.
>
>2. A plan to pay off a financial obligation according to some pre-arranged program.
>
>*amortissement*

Annual equivalent
>1. In *time value of money* (q.v.), a uniform annual amount for a prescribed number of years that is equivalent in value to the present worth of any sequence of financial events for a given interest rate.
>
>2. One of a sequence of equal end-of-year payments which would have the same financial effect when interest is considered as another payment or sequence of payments which are not necessarily equal in amount or equally spaced in time.
>
>*équivalent annuel*

Annuity
>1. An amount of money payable to a beneficiary at regular intervals for a prescribed period of time out of a fund reserved for that purpose.
>
>2. A series of equal payments occurring at equal periods of time.
>
>*rente*

---

[1] Based on *American National Standard Publication ANSI Z94.5–1972* (American Society of Mechanical Engineers, 1972), *Terminology for Accountants* (The Canadian Institute of Chartered Accountants, 1976), *Dictionnaire de la comptabilité* (Fernand Sylvain, L'Institue Canadien des Comptables Agréés, 1977).

3. Amount paid annually, including reimbursement of borrowed capital and payment of interest.

annuité

Annuity factor
The function of interest rate and time that determines the amount of periodic annuity that may be paid out of a given fund.

facteur de l'annuité

Annuity fund
A fund that is reserved for payment of annuities. The present worth of funds required to support future annuity payments.

fonds de rentes

Annuity fund factor
The function of interest rate and time that determines the present worth of funds required to support a specified program of annuity payments.

facteur du fonds de rentes

Apportion to
In accounting or budgeting, to assign a cost responsibility to a specific individual, organization unit, product, project, or order.

répartir

Asset
An economic resource of entity (including money resources, physical resources, and intangible resources).

bien, valeur active, élément d'actif, avoir

Average-interest method
A method of computing required return on investment based on the average book value of the asset during its life or during a specified study period.

méthode de l'intérêt moyen

Balance sheet
A formal statement of financial position, showing the assets, liabilities, and owners' equity of the accounting entity at a particular moment of time.

bilan

Bond
1. A certificate of indebtedness, issued by a government or corporation, generally being one of a number of such certificates. The term usually implies that the assets have been pledged as security.

obligation

2. An obligation in writing, sometimes supported by collateral, given by an individual or company to another individual or company to pay damages or to indemnify against losses caused by a third party through nonperformance of a contract or other duties or by defalca-

tion (e.g., a builder's performance bond or an employee fidelity bond).

bon de garantie d'exécution

Book value

1. The recorded current value of an asset. First cost less accumulated depreciation, amortization, or depletion.

2. Original cost of an asset less the accumulated depreciation.

3. The worth of a property as shown on the accounting records of a company. It is ordinarily taken to mean the original cost of the property less the amounts that have been charged as depreciation expense.

valeur comptable

Break-even chart

A graphical representation of the relation between total income and total costs for various levels of production and sales indicating areas of profit and loss.

graphique de rentabilité, graphique du point mort

Break-even point(s)

1. (a) In business operations, the rate of operations, output, or sales at which income is sufficient to equal operating cost, or operating cost plus additional obligations that may be specified. (b) The operating condition, such as output, at which two alternatives are equal in economy.

2. The percentage of capacity operation of a manufacturing plant at which income will just cover expenses.

seuil de rentabilité, point mort, point d'équilibre

Burden

Usually a synonym for overhead.

frais généraux

Capacity factor

1. The ratio of average load to maximum capacity.

2. The ratio of average load to the total capacity of the apparatus, which is the optimum load.

3. The ratio of the average actual use to the available capacity.

facteur de capacité

Capital

1. The financial resources involved in establishing and sustaining an enterprise or project (v. *investment and working capital*).

2. A term describing wealth which may be utilized to economic advantage. The form that this wealth takes may be as cash, land, equipment, patents, raw materials, finished products, etc.

capital

Capital asset
　　1. An asset, whether tangible or intangible, intended for long-term use and held as such (v. *fixed asset*).
　　valeurs immobilisées

　　2. (government accounting) Any asset of the capital fund.
　　biens du fonds d'immobilisation, biens immobilisés

Capital cost allowance
　　A deduction, akin to depreciation, allowed in computing income for income tax purposes.
　　amortissement du coût en capital, amortissement fiscal

Capital recovery
　　1. Charging periodically to operations amounts that will ultimately equal the amount of capital expenditure (v. *amortization, depletion,* and *depreciation*).
　　2. The replacement of the original cost of an asset plus interest.
　　3. The process of regaining the net investment in a project by means of revenue in excess of the costs from the project. (Usually implies amortization of principal plus interest on the diminishing unrecovered balance.)
　　recouvrement du capital

Capital recovery factor
　　A factor used to calculate the sum of money required at the end of each of a series of periods to regain the net investment of a project plus the compounded interest on the unrecovered balance.
　　facteur de recouvrement du capital

Capital tax factor
　　A factor used to determine the present value of tax savings resulting from capital cost allowance.
　　facteur de coût en capital

Capitalized cost
　　1. The present worth of a uniform series of periodic costs that continue for an indefinitely long time (hypothetically infinite). Not to be confused with a capitalized expenditure, which is, in accounting, the charging of an expenditure to a capital asset account rather than to an expense account.

　　2. The value at the purchase date of the first life of the asset of all expenditures to be made in reference to this asset over an infinite period of time. This cost can also be regarded as the sum of capital which, if invested in a fund earning a stipulated interest rate, will be sufficient to provide for all payments required to maintain the asset in perpetual service.
　　coût capitalisé

Cash flow
1. The flowback of profit plus depreciation from a given project.
    marge d'autofinancement, fonds autogénérés

2. The real dollars passing into and out of the treasury of a financial venture.
    mouvements de la trésorerie

Common costs
Costs which cannot be identified with a given output of products, operations, or services.
    coûts communs

Compound amount
The future worth of a sum invested (or loaned) at compound interest.
    montant capitalisé

Compound amount factor
1. The function of interest rate and time that determines the compound amount from a stated initial sum.

2. A factor which, when multiplied by the single sum or uniform series of payments, will give the future worth at compound interest of such single sum or series.
    facteur de montant capitalisé

Compound interest
1. The type of interest that is periodically added to the amount of investment (or loan) so that subsequent interest is based on the cumulative amount.

2. The interest charges under the condition that interest is charged on any previous interest earned in any time period, as well as on the principal.
    intérêt composé

Compounding, continuous
1. A compound interest situation in which the compounding period is zero and the number of periods infinitely great. A mathematical concept that is practical for dealing with frequent compounding and small interest rates.

2. A mathematical procedure for evaluating compound interest factors based on a continuous interest function rather than discrete interest periods.
    capitalisation continué

Compounding period
The time interval between dates at which interest is paid and added to the amount of an investment or loan. Designates frequency of compounding.
    période de capitalisation

Cost of capital
: The cost, expressed as a yield rate, of investment funds obtained either through borrowing, equity investment, or retention of earnings.
    coût du capital

Cost, product
: All manufacturing costs associated with the product and thus included in determining costs of goods sold and inventory valuations. Includes directly and indirectly attributable expenses, general expenses, and depreciation that can be equitably attributed to a product.
    coût incorporable, frais imputés à la fabrication

Critical Path Method
: A method of network analysis in which normal duration time is estimated for each activity within a project. The critical path identifies the shortest completion period based on the most time-consuming sequence of activities from the beginning to the end of the network (v. PERT).
    méthode du chemin critique

Current asset
: An asset that in the normal course of operations is expected to be converted into cash or consumed in the production of revenue within one year (or within the normal operating cycle if that cycle is longer than one year).
    élément d'actif à court terme, bien à court terme

Current liability
: A liability whose regular and ordinary liquidation is expected to occur within one year or within the normal operating cycle if that cycle is longer than one year. (A liability otherwise classified as current, but for which provision has been made for payment from other than current resources should be excluded from this definition.)
    élément de passif à court terme, dette à court terme, dette exigible à moins d'un an

Debenture
: A certificate of indebtedness issued by a government or company, generally being one of a number of such certificates. The term usually implies an unsecured obligation (v. *bond* (1)).
    obligation non garantie, débenture

Decisions under certainty
: Simple decisions that assume complete information and no uncertainty connected with the analysis of the decisions.
    décision dans une situation certaine

Decisions under risk
: A decision problem in which the analyst elects to consider several possible futures, the probabilities of which can be estimated.
    décision dans une situation de risque

Decisions under uncertainty
A decision for which the analyst elects to consider several possible futures, the probabilities of which *cannot* be estimated.

décision dans une situation incertaine

Declining balance depreciation (also known as *percent on diminishing value*)
A method of computing depreciation in which the annual charge is a fixed percentage of the depreciated book value at the beginning of the year to which the depreciation applies.

méthode de l'amortissement dégressif (à taux constant)
méthode de l'amortissement décroissant (à taux constant)

Depletion
1. A form of capital recovery applicable to extractive property (e.g., mines). Can be on a unit-of-output basis the same as straight-line depreciation related to original or current appraisal of extent and value of deposit (known as *cost depletion*). Can also be a percentage of income received from extractions (known as *percentage depletion*).

2. A lessening of the value of an asset due to a decrease in the quantity available. It is similar to depreciation except that it refers to such natural resources as coal, oil, and timber in forests.

épuisement

Depreciated book value
The first cost of the capitalized asset minus the accumulation of annual depreciation cost charges.

valeur comptable amortie

Depreciation
1. (a) Decline in value of a capitalized asset.

dépréciation

(b) A form of capital recovery applicable to a property with two or more years' life span, in which an appropriate portion of the asset's value is periodically charged to current operations.

dotation aux amortissements

2. (a) The loss of value due to obsolescence or attrition.

dépréciation

(b) In accounting, depreciation is the allocation of this loss of value according to some plan.

dépréciation, dotation aux amortissements

Discounted cash flow
1. The present worth of a sequence in time of sums of money when the sequence is considered as a flow of cash into and/or out of an economic unit.

flux monétaire actualisé, méthode de l'actualisation du flux monétaire

2. An investment analysis which compares the present worth of projected receipts and disbursements occurring at designated future times in order to estimate the rate of return from the investment or project.
   mouvements de la trésorerie actualisés

Economic return
   The profit derived from a project or business enterprise without consideration of obligations to financial contributors and claims of others based on profit.
   rendement économique

Effective interest
   The true value of interest rate computed by equations for compound interest rate for a one-year period.
   intérêt réel

Endowment
   A fund established for the support of some project or succession of donations or financial obligations.
   dotation

Endowment method
   As applied to economic analysis, a comparison of alternatives based on the present worth of the anticipated financial events.
   méthode de dotation

Equipment cost, delivered
   Cost of equipment delivered to the construction site but not uncrated. (FOB cost plus transportation, taxes, and duties.)
   coût du matériel livré

Equipment cost, FOB
   Cost of equipment crated and on board the delivery vehicle at the equipment manufacturer's location. Does not include tax, import duties, freight, or shipping expense.
   coût du matériel franco à bord (F.A.B.)

Expected yield
   The ratio expected return/investment, usually expressed as a percentage on an annual basis.
   rendement prévu

Factory overhead
   All production costs other than direct material and direct labour costs. These terms include all costs necessary to the operation and maintenance of the plant, including wages of foremen, building and machinery upkeep, depreciation, light, heating, repairs, insurance, taxes, etc.
   frais généraux de fabrication

First cost
The initial cost of a capitalized property, including transportation, installation, preparation for service, and other related initial expenditures.

coût initial

Fixed asset
A tangible long-term asset, such as land, building, equipment, etc., held for use rather than for sale (v. *capital asset* (1)).

immobilisations (corporelles), actif immobilisé (corporel), valeurs immobilisées (corporelles)

Fixed capital
Investment in capital assets (v. *working capital*).

capital fixe

Fixed cost, fixed expense
An indirect cost that remains relatively unchanged in total regardless of the volume of production or activity within a fairly wide range of volume (v. *semivariable cost, variable cost*).

frais fixes, coût fixe, frais constants, charge fixe

Fringe benefits
Payroll costs other than wages not paid directly to the employee. These include holidays, pensions, insurance, savings plans, etc.

avantages sociaux

Future worth
1. The equivalent value at a designated future date based on *time value of money*.

2. The monetary sum, at a given future time, which is equivalent to one or more sums at given earlier times when interest is compounded at a given rate.

valeur capitalisée

Going-concern value
The difference between the value of a property as it stands possessed of its going elements and the value of the property alone as it would stand at completion of construction as a bare or inert assembly of physical parts.

valeur d'exploitation, valeur d'usage

Goodwill
An intangible asset of a business when the business has value in excess of the sum of its net identifiable assets. Goodwill has had a variety of definitions, some relating to the nature of the asset, others to its value. As to its nature, it has been said to fall into the three classes of commercial, industrial, and financial goodwill, which are the consequence of favourable attitudes on the part of customers, employees, and creditors, respectively. As to its value, the most common explanations emphasize the present value of expected future earnings in excess of the return required to induce investment.
achalandage, survaleur

Gross margin, gross profit
The difference between cost and selling price; excess of net sales over the cost of goods sold.
bénéfice brut, marge bénéficiare brute

Income statement
A financial statement summarizing the items of revenue, income, expenses, and losses for a stated period (v. *statement of revenue and expenditure, profit and loss statement, earnings statement*).
état des résultats, état des revenus

Increment cost
The additional cost that will be incurred as the result of increasing the output one more unit. Conversely, it can be defined as the cost that will not be incurred if the output is reduced one unit. More technically, it is the variation in output resulting from a unit change in input. It is also known as the marginal cost.
coût différentiel, coût marginal

Indirect overhead
Overhead costs which are not traceable to the specific part of the organization which is the focus of attention.
frais généraux indirects

Intangibles
1. In economic analysis, conditions or economic factors that cannot be readily evaluated in *quantitative* terms as in money.
2. In accounting, the assets that cannot be reliably evaluated (e.g., *goodwill*).
actif incorporel

Interest
1. (a) Financial share in a project or enterprise. (b) Periodic compensation for the lending of money. (c) In economic analysis, synonymous with *required return, expected profit,* or *charge* for the use of capital.
2. The cost for the use of capital. Sometimes referred to as the *time value of money* (q.v.).
intérêt

Interest rate
The ratio of the interest payment to the principal for a given unit of time, usually expressed as a percentage of the principal.
taux d'intérêt

Interest rate, effective
An interest rate for a stated period (per year unless otherwise specified) that is the equivalent of a smaller rate of interest that is more frequently compounded.
taux d'intérêt réel

Interest rate, nominal
The customary type of interest rate designation on an annual basis without consideration of compounding periods. The usual basis for computing periodic interest payments.
taux d'intérêt nominal

Internal rate of return
The discounting rate that equates the present value of expected cash outflows with the present value of expected cash inflows. Generally calculated after tax. May or may not include opportunity costs and book depreciation as expenses. (Interest rate of return, discounted cash flow, profitability index, investor's method.)
taux de rendement interne

Investment
1. As applied to an enterprise as a whole, the cost (or present value) of all the properties and funds necessary to establish and maintain the enterprise as a going concern. The *capital* tied up in the enterprise or project.

2. Any expenditure which has substantial and enduring value (at least two years' anticipated life) and which is therefore capitalized.
investissement

Irreducibles
1. A term that may be used for the class of intangible conditions or economy factors that can only be *qualitatively* appraised (e.g., ethical considerations).

2. Matters than cannot readily be reduced to estimated money receipts and disbursements.
facteurs impondérables

Investor's method (v. *discounted cash flow method*)
méthode de l'investisseur

Life
1. Economic: that period of time after which a machine or facility should be discarded or replaced because of its excessive costs or reduced profitability. The economic impairment may be absolute or relative.
durée économique

2. Physical: that period of time after which a machine or facility can no longer be repaired in order to perform its design function properly.

durée matériel

3. Service: the period of time that a machine or facility will satisfactorily perform its function without major overhaul.

durée de service

MAPI method

1. A procedure for replacement analysis sponsored by the Machinery and Allied Products Institute.

2. A method of capital investment analysis which has been formulated by the Machinery and Allied Products Institute. This method uses a fixed format and provides charts and graphs to facilitate calculations. A prominent feature of this method is that it explicitly includes obsolescence.

méthode MAPI

Marginal cost

1. The cost of one additional unit of production, activity, or service.

2. The rate of change of cost with production or output.

coût marginal, coût différentiel

Marginal analysis

An economic concept concerned with those elements of costs and revenues which are associated directly with a specific course of action, normally using available current costs and revenues as a base and usually independent of traditional accounting allocation procedures.

analyse marginale

Minimum acceptable rate of return (MARR)

The return on investment chosen as acceptable for discounting purposes. This includes an allowance for risk.

taux de rendement, minimum acceptable

Nominal interest

The number employed loosely to describe the annual interest rate.

taux nominal

Noncash expense

An expense for accounting purposes which does not require an outlay of funds of the same amount in the same year (e.g., depreciation).

défence hors caisse

Obsolescence

1. The condition of being out of date. A loss of value occasioned by new developments which place the older property at a competitive disadvantage. A factor in depreciation.

2. A decrease in the value of an asset brought about by the development of new and more economical methods, processes, and/or machinery.

3. The loss of usefulness or worth of a product or facility as a result of the appearance of better and/or more economical products, methods, or facilities.

obsolescence, désuétude

Opportunity cost

The value of benefits sacrificed in selecting a course of action among alternatives; the value of the next best opportunity foregone by deciding to do one thing rather than another.

valeur de renonciation, coût d'option, manque à gagner

Overhead

Expenses which are incurred to produce a commodity or render a service, but which cannot conveniently be attributed to individual units of production or service.

frais généraux

Payoff period

1. Regarding an investment, the number of years (or months) required for the related profit or savings in operating cost to equal the amount of said investment.

2. The period of time at which a machine, facility, or other investment has produced sufficient net revenue to recover its investment costs. Also known as payout time or payback period.

période de récupération, période de recouvrement, délai de récupération, délai de recouvrement

Perpetual endowment

An endowment with hypothetically infinite life (v. *capitalized cost* and *endowment*).

dotation permanente

PERT

(abbr.) Program evaluation and review technique. A method of network analysis in which three time estimates are made for each activity—the optimistic time, the most likely time, and the pessimistic time—and which gives an expected completion date for the project within a probability range (v. *Critical Path Method*).

méthode de programmation optimale, graphique PERT

Present worth

1. The equivalent value at the present, based on *time value of money*.

2. (a) The monetary sum which is equivalent to a future sum when interest is compounded at a given rate. (b) The discounted value of future sums.

valeur actualisé

Present worth factor
　　1. A mathematical expression also known as the present value of an annuity of one
　　2. One of a set of mathematical formulae used to facilitate calculation of present worth in economic analyses involving compound interest.
　　facteur de valeur actuelle

Price index
　　An indicator, derived by aggregating sets of representative unit prices with an appropriate weighting base, which measures changes in price levels over time. Price indexes are stated in terms of a base point of time where the price level is arbitrarily set at 100.
　　indice des prix

Profitability index
　　The rate of return in an economic analysis or investment decision when calculated by the *discounted cash flow method* or the *investor's method* (q.v.).
　　indice de rentabilité

Rate of return
　　1. The interest rate at which the present worth of the cash flows on a project is zero.
　　2. The interest rate earned by an investment.
　　taux de rendement

Replacement policy
　　A set of decision rules (usually optimal) for the replacement of facilities that wear out, deteriorate, or fall over a period of time. Replacement models are generally concerned with weighing the increasing operating costs (and possibly decreasing revenues) associated with aging equipment against the net proceeds from alternative equipment.
　　politique de remplacement

Required return
　　The *minimum* return or profit necessary to justify an investment. Often termed *interest, expected return or profit,* or *charge for the use of capital*. It is the minimum acceptable percentage, no more and no less.
　　rendement exigé

Retirement of debt
　　The termination of a debt obligation by appropriate settlement with lender—understood to be in full amount unless partial settlement is specified.
　　remboursement de la dette

Return on average investment
Ratio of average annual profits over the economic life to the average book value of the investment, including working capital.
rendement sur l'investissement

Salvage value
1. The cost recovered or which could be recovered from a used property when removed, sold, or scrapped. A factor in appraisal of property value and in computing depreciation.

2. The market value of a machine or facility at any point in time. Normally, an estimate of an asset's net market value at the end of its estimated life.
valeur de récupération

Semivariable cost, semivariable expense
An indirect cost which varies with production or activity but not in direct proportion to the volume (v. *fixed cost, variable cost*).
frais semivariables, coût semivariable, frais semiproportionnels

Sensitivity
The relative magnitude of the change in one or more elements of an engineering economics problem that will reverse a decision among alternatives.
sensibilité

Simple interest
1. Interest that is not compounded—is not added to the income-producing investment or loan.

2. The interest charges under the condition that interest in any time period is charged only on the principal.
intérêt simple

Sinking fund
1. A fund accumulated by periodic deposits and reserved exclusively for a specific purpose, such as retirement of a debt or replacement of a property.

2. A fund created by making periodic deposits (usually equal) at compound interest in order to accumulate a given sum at a given future time for some specific purpose.
fonds d'amortissement, fonds de rachat

Sinking fund deposit factor
The function of interest rate and time that determines the periodic deposit required to accumulate a specified future amount. Reciprocal of *sinking fund factor* (q.v.).
facteur de dépôt du fonds d'amortissement

Sinking fund depreciation
1. A method of computing depreciation in which the periodic amount is presumed to be deposited in a *sinking fund* that earns in-

terest at a specified rate. The sinking fund may be real but is usually hypothetical.

2. A method of depreciation where a fixed sum of money is regularly deposited at compound interest in a real or imaginary fund in order to accumulate an amount equal to the total depreciation of an asset at the end of the asset's estimated life. The sinking fund depreciation in any year equals the sinking fund deposit plus interest in the sinking fund balance.

méthode de l'amortissement à l'intérêt composé (dotation croissante)

Sinking fund factor

1. The function of interest rate and time that determines the cumulative amount of a sinking fund resulting from specified periodic deposits. Future worth per unit of uniform periodic amounts.

2. The mathematical formulae used to facilitate sinking fund calculations.

facteur de fonds d'amortissement

Straight-line depreciation

Method of depreciation whereby the amount to be recovered (written off) is spread uniformly over the estimated life of the asset in terms of time periods or units of output. May be designated as *percent of initial value*.

méthode de l'amortissement constant, méthode de l'amortissement linéaire

Study period

In economic analysis, the length of time that is presumed to be covered in the schedule of events and appraisal of results. Often the anticipated life of the project under consideration, but a shorter time may be more appropriate for decision making.

période examinée

Sum of the years' digits method (also known as the *sum of digits method*)

A method of computing depreciation in which the amount for any year is based on the ratio (years of remaining life/$(1 + 2 + 3 \ldots +n)$), n being the total anticipated life.

méthode de l'amortissement proportionnel à l'ordre numérique inversé des années

Sunk cost

1. The unrecovered balance of an investment. It is a cost, already paid, that is not relevant to the decision concerning the future that is being made. Capital already invested that for some reason cannot be retrieved.

2. A past cost which has no relevance with respect to future receipts and disbursements of a facility undergoing an engineering economic analysis. This concept implies that since a past outlay is the same regardless of the alternative selected, it should not influence the choice between alternatives.

coût irrécupérable

Tangibles
Things that can be *quantitatively* measured or valued, such as items of cost and physical assets.

actif corporel

Time value of money
1. The cumulative effect of elapsed time on the money value of an event, based on the earning power of equivalent invested funds (v. *future worth* and *present worth*).

2. The expected interest rate that capital should or will earn.

valeur de l'argent dans le temps

Valuation or appraisal
The art of estimating the fair-exchange value of specific properties.

évaluation, estimation, expertise

Variable cost, variable expense
A cost that varies directly with the volume of production or activity (v. *fixed cost, semivariable cost*).

frais variables, coût variable, frais proportionnels

Working capital
1. That portion of investment represented by *current assets* (assets that are not capitalized) less the *current liabilities*. The capital necessary to sustain operations.

2. Those funds that are required to make the enterprise or project a going concern.

fonds de roulement, actif net à court terme

Yield
The ratio of return or profit over the associated investment, expressed as a percentage or decimal, usually on an annual basis.

rendement, rapport, taux effectif, taux de rendement

Index

A

accelerated depreciation, 238, 253, 263
account
  payable, 589
  receivable, 589
acid-test ratio, 464
active business, 246
ad valorem taxes, 242
after-tax
  basis, 242
  cash flow, 248-250
amortization, 589
analyzing alternatives with no positive cash flows, 174, 179
annual equivalent, 589
annual worth method, 153, 166
annuity, 589
  factor, 590
  fund, 590
apportion to, 590
aspiration-level principle, 395
asset, 590
autocorrelated, 352
average cost, 28
average-interest method, 590

B

balance sheet, 459, 590
Bayes's theorem, 418, 421
before-tax cash flow, 249
beginning-of-period cash flows, 97
benefit, 286
benefit-cost analysis, 283, 285, 297, 305
beta distribution, 353
black box approach, 2
bond problems, 103-110
bonds, 24, 103
book value, 20
break-even
  analysis, 330
  chart, 591
  point, 330, 591
burden, 591

C

Canadian controlled private corporations, 246
capacity factor, 591
capital, 591
capital asset, 592
  tax treatment of, 260
capital cost
  allowance, 224, 230-232, 263, 592
  estimation, 35
  tax factor, 243
capital gains and losses, 259
capital property, 259
capital recovery, 592
  cost, 110
  factor, 61, 592
capital tax factor, 592
capitalized cost, 592
capitalized value, 99
carry-back and carry-forward rules, 258
cash flow, 5, 593
  profiles, 146-148
  rate of return, 157
CCA, 224, 230

CCTF, 243
changing interest rates, 94
classified accounts, 241
classifying objectives, 436
collection period, 467
common costs, 593
comparison of alternatives, 137, 196
composite accounts, 241
compound amount, 593
  factor, 579
compound interest, 593
  approach, 50
compounding
  period, 593
  see also continuous compounding
computer simulation, 363-370
computer-based planning models, 195
conditional probability theorem, 417
constant quantity curve, 513
continuous compounding, 76, 593
  geometric series future worth factor, 78
  geometric series present worth factor, 78
  single sum, future worth factor, 76
  single sum, present worth factor, 77
  uniform series present worth factor, 82
continuous flow, 80
corporate income tax, 242-248
cost
  accounting, 471
  index, 35
  of capital, 23, 594
  of goods sold, 24
  terminology, 18
  product, 594
cost-capacity relationship, 38
cost-effectiveness analysis, 283, 316 - 318, 320
critical path method, 594
current
  asset, 594
  liability, 594
  ratio, 464

D

debenture, 594
debt to equity ratio, 465
decisions
  under assumed certainty, 388
  under certainty, 594
  under risk, 390, 594
  under uncertainty, 395, 595
declining balance depreciation, 229, 595
demand, 484
  curve, 486
  function, 484
depletion, 269, 595
depreciated book value, 595
depreciation, 20, 223, 595
  charge, 20
  method, effect of, 251
detailed estimates, 32-33
diminishing marginal productivity, 512
direct
  costs, 24
  material and labour costs, 25
disbenefit, 286
discount rate, 148
discounted cash flow, 595
  method, 599
discrete flows, 77
disposal cost, 20
dominance, 391
double declining balance method of depreciation, 232

E

economic
  life of an item, 21
  return, 596
effect of interest on borrowed money, 254-258
effective interest, 73, 596

elasticity
  of demand, 496
  of supply, 502
end-of-period
  cash flows, 97
  compounding, 97
endowment, 596
  method, 596
equilibrium
  point, 493
  price, 493, 503
  quantity, 493
equipment cost
  delivered, 596
  FOB, 596
equity
  capital, 23
  funds, 24
  ratio, 464
equivalence, 83
equivalent uniform annual cost, 189
ERR, 161
estimation, 30
EUAC, 189
excise taxes, 242
expansion path, 521
expected
  profit given perfect information, 422
  value, 393
  value of sample information, 423
  yield, 596
external rate of return method, 161

F

factory overhead, 596
first cost, 597
fixed
  asset, 597
  capital, 597
  cost, 26, 597
fixed-percentage-on-the-declining-balance method of depreciation, 226
fringe benefits, 597

functional life of an item, 21
future
  costs, 22
  worth, 49, 597
  worth method, 154, 166
  worth of a uniform series, 62

G

geometric series
  cash flow series, 69
  future worth factor, 71
  present worth factor, 70
going-concern value, 597
goods and services taxes, 242
goodwill, 598
gradient series
  of cash flows, 64
  present worth factor, 66
  gradient-to-uniform series
    conversion factor, 66
gross
  margin, 598
  profit, 598
  rate of return on investment, 466
group accounts, 241

H

Hespos, Richard F., 404
Hillier, F. S., 359
Hurwicz principle, 398

I

income
  forecast method of depreciation, 241
  ratio, 465
  statement, 462, 598
  tax incentives, 263
Incremental
  cost, 598
  approaches, 167-172
  product, 510

indirect
  costs, 24-25
  overhead, 598
inflationary effects, 112, 115
intangibles, 12, 598
interest, 598
interest rate, 599
  effective, 599
  nominal, 599
  applicable to public projects, 301
  changing, 94
internal rate of return, 157, 599
inventory turnover ratio, 467
investment, 599
  tax credit, 263
investor's method, 599
IRR, 157
irreducibles, 599
isocost line, 516
isoquant curve, 513

K

Klausner, R. F., 369

L

Lagrange's method, 520
Laplace principle, 396
learning curve, 40-44
lease-buy considerations, 266-269
life of an asset, 599
life-cycle cost, 19
likelihood statements, 419
linear break-even analysis, 330
loan cash flow, 249
loan payments, 90

M

Magee, John F., 404
manufacturing and processing
  profits deduction, 246
MAPI method, 600

marginal
  analysis, 600
  cost, 29, 600
  product, 510
  rate of substitution, 515
MARR, 23, 149, 600
matrix decision model, 385
maximax principle, 397
maximin principle, 397
maximum price, 503
measure of merit, 151, 286
minimax principle, 397
minimax regret, 401
minimin principle, 398
minimum
  acceptable rate of return, 600
  attractive rate of return, 23, 149
  price, 503
most probable future principle, 393
multiple
  compounding periods in a year, 73
  objectives, 435-444
multiple-asset accounts, 241
multiple-use projects, 313
mutually exclusive alternative, 138

N

nominal interest, 73, 600
noncash expense, 600
nonlinear break-even analysis, 335
non-monetary considerations, 11
normal distribution, 353

O

obsolescence, 600
operating
  and maintenance costs, 20
  day method of depreciation, 240
  ratio, 465
opportunity cost, 22-23, 601
order-of-magnitude estimates, 32
overcounting, 308
overhead, 24, 472, 591, 601

P

past costs, 21
payback period method, 155, 166
payoff period, 601
percent on diminishing value, 601
perpetual endowment, 601
perpetuities, 99
PERT, 601
planning period, 141-146, 285
point of
  diminishing average productivity, 510
  indifference, 331
  view, 298
posterior probabilities, 420
preliminary estimates, 32-33
present
  value, 49
  worth, 601
  worth factor, 602
  worth method, 152, 165
price
  control, 503
  determining, 490
  index, 35, 602
  support, 506
prior expected profi, 421
problem-solving process, 2
production function, 508
profit and loss statement, 462
profitability index, 602
program evaluation and review technique (PERT), 601
project estimation, 32
  objectives in, 284
public projects, 283, 298-312

Q

quantity supplied, 488

R

ranking, 436
rate of return, 602
  method, 157
recaptured capital cost allowance, 261
regret values, 401
replacement
  analyses, 183, 195
  policy, 602
required return, 602
retirement of debt, 602
return on average investment, 603
risk
  aggregation, 355
  analysis, 351

S

salvage value, 21, 225, 603
sample information, 421
Savage principle, 401
Savage, L. J., 401
savings/investment ratio method, 163
scientific research expenditures, 263
scrap value, 21
semivariable cost, 603
sensitivity, 603
  analysis, 342
sequential decisions, 403
series of cash flows, 54
shortage, 490
simple interest, 603
  approach, 50
single sum
  future worth factor, 52
  present worth factor, 54
sinking fund, 603
  deposit factor, 603
  depreciation, 235, 603
  factor, 63, 604
small business tax credit, 246
sources of data, 34
standard costs, 44
statement of earnings, 462
straight-line depreciation, 226, 604

Strassman, Paul A., 404
study period, 604
sum of the years' digits method of depreciation, 228, 604
sunk costs, 21, 604
supplementary analyses, 172
supply, 488
  and demand, 483
  curve, 489
surplus, 490

T

tangibles, 605
taxable income, 247-248
time value of money, 7, 49, 148-150, 605
tolls, fees, and user charges, 311
total expenditure, 499
true rate of return, 157

U

unequal lives, 179-183, 309
uniform series
  of cash flows, 58
  future worth factor, 63
  present worth factor, 59

units of production method of depreciation, 240
univariable production function, 509
utility theory, 352

V

valuation or appraisal, 605
value
  of imperfect information, 422
  of perfect information, 421
variable costs, 26-27, 605
variance, 392

W

weighting objectives, 438
witholding tax, 244
working capital, 463, 605

Y

yield, 605